Advances in Intelligent Systems and Computing

Volume 1369

AF892872

The series "Advances in Intelligent Systems and Computing" contains publications on theory, applications, and design methods of Intelligent Systems and Intelligent Computing. Virtually all disciplines such as engineering, natural sciences, computer and information science, ICT, economics, business, e-commerce, environment, healthcare, life science are covered. The list of topics spans all the areas of modern intelligent systems and computing such as: computational intelligence, soft computing including neural networks, fuzzy systems, evolutionary computing and the fusion of these paradigms, social intelligence, ambient intelligence, computational neuroscience, artificial life, virtual worlds and society, cognitive science and systems, Perception and Vision, DNA and immune based systems, self-organizing and adaptive systems, e-Learning and teaching, human-centered and human-centric computing, recommender systems, intelligent control, robotics and mechatronics including human-machine teaming, knowledge-based paradigms, learning paradigms, machine ethics, intelligent data analysis, knowledge management, intelligent agents, intelligent decision making and support, intelligent network security, trust management, interactive entertainment, Web intelligence and multimedia.

The publications within "Advances in Intelligent Systems and Computing" are primarily proceedings of important conferences, symposia and congresses. They cover significant recent developments in the field, both of a foundational and applicable character. An important characteristic feature of the series is the short publication time and world-wide distribution. This permits a rapid and broad dissemination of research results.

Indexed by DBLP, INSPEC, WTI Frankfurt eG, zbMATH, Japanese Science and Technology Agency (JST).

All books published in the series are submitted for consideration in Web of Science.

More information about this series at http://www.springer.com/series/11156

Subhransu Sekhar Dash · Bijaya Ketan Panigrahi ·
Swagatam Das
Editors

Sixth International Conference on Intelligent Computing and Applications

Proceedings of ICICA 2020

Springer

Editors
Subhransu Sekhar Dash
Department of Electrical Engineering
Government College of Engineering
Keonjhar, Odisha, India

Bijaya Ketan Panigrahi
Department of Electrical Engineering
Indian Institute of Technology Delhi
New Delhi, India

Swagatam Das
Electronics and Communication
Sciences Unit
Indian Statistical Institute
Kolkata, India

ISSN 2194-5357 ISSN 2194-5365 (electronic)
Advances in Intelligent Systems and Computing
ISBN 978-981-16-1334-0 ISBN 978-981-16-1335-7 (eBook)
https://doi.org/10.1007/978-981-16-1335-7

This Springer imprint is published by the registered company Springer Nature Singapore Pte Ltd.
The registered company address is: 152 Beach Road, #21-01/04 Gateway East, Singapore 189721, Singapore

Preface

This AISC volume contains the papers presented at the Sixth International Conference on Intelligent Computing and Applications (ICICA 2020) held during December 22–23, 2010, at Government College of Engineering, Keonjhar, Odisha, India.

ICICA 2020 is the Sixth International Conference aiming at bringing together the researchers from academia and industry to report and review the latest progresses in the cutting-edge research on various research areas of electronic circuits, power systems, renewable energy applications, image processing, computer vision and pattern recognition, machine learning, data mining and computational life sciences, management of data including big data and analytics, distributed and mobile systems including grid and cloud infrastructure, information security and privacy, VLSI, antenna, computational fluid dynamics and heat transfer, intelligent manufacturing, signal processing, intelligent computing, soft computing, web security, privacy and e-commerce, e-governance, optimization, communications, smart wireless and sensor networks, networking and information security, mobile computing and applications, industrial automation and MES, cloud computing, green IT and finally to create awareness about these domains to a wider audience of practitioners.

ICICA 2020 received 170 paper submissions including two foreign countries across the globe. All the papers were peer-reviewed by the experts in the area in India and abroad, and comments have been sent to the authors of accepted papers. Finally, 58 papers were accepted for oral presentation in the conference. This corresponds to an acceptance rate of 34% and is intended to maintain the high standards of the conference proceedings. The papers included in this AISC volume cover a wide range of topics in intelligent computing and algorithms and their real-time applications in problems from diverse domains of science and engineering.

The conference was inaugurated by Prof. Ramazan Bayindir, Gazi University, Turkey, on December 22, 2020. The conference featured distinguished keynote speakers as follows: Dr. Nattachote Rugthaicharoencheep, Senior IEEE Member, Rajamangala University of Technology, Thailand; Dr. Roman Senkerik, Associate Professor, Tomas Bata University, Czech Republic; Dr. Bijaya Ketan Panigrahi, IIT Delhi, India; Dr. Swagatam Das, ISI, Kolkata, India; Dr. Ranjan Kumar Behera, Associate Professor, IIT, Patna; and Dr. Subhransu Sekhar Dash, Professor and Head, GCE, Keonjhar, Odisha, India.

We take this opportunity to thank the authors of the submitted papers for their hard work, adherence to the deadlines, and patience with the review process. The quality of a referred volume depends mainly on the expertise and dedication of the reviewers. We are indebted to the technical committee members, who produced excellent reviews in short time frames. First, we are indebted to Dr. Trilochan Sahu, Principal, GCE, Keonjhar. We specially thank Dr. Sangram Keshori Mohapatra, Associate Professor (EE) and General Chair, ICICA 2020, and Prof. N. C. Patel, Prof. Rakesh Ranjan Shukla, and Dr. Manoj Kumar Senapati, Program Chairs of ICICA 2020, for their excellent support and arrangements because without them, it is beyond imagination to conduct this conference. We thank the International Advisory Committee members for providing valuable guidelines and inspiration to overcome various difficulties in the process of organizing this conference. We would also like to thank the participants of this conference. The members of faculty and students of the Electrical Engineering Department, Government College of Engineering, Keonjhar, deserve special thanks because without their involvement, we would not have been able to face the challenges of our responsibilities. Finally, we thank all the volunteers who made great efforts in meeting the deadlines and arranging every detail to make sure that the conference could run smoothly. We hope the readers of these proceedings find the papers inspiring and enjoyable.

Odisha, India
Kolkata, India
New Delhi, India
December 2020

Subhransu Sekhar Dash
Swagatam Das
Bijaya Ketan Panigrahi

Contents

About the Editors

Dr. Subhransu Sekhar Dash is presently Professor and Head in the Department of Electrical Engineering, Government College of Engineering, Keonjhar, Odisha, India. He received his Ph.D. degree from the College of Engineering, Guindy, Anna University, Chennai, India. He has more than 22 years of research and teaching experience. His research areas are AI techniques application to power system, modeling of FACTS controller, power quality, and smart grid. He is Visiting Professor at Francois Rabelais University, POLYTECH, France. He has published more than 220 research articles in peer-reviewed international journals and conferences. Professor Dash has played a role as Convenor and Program or General Chair for many international conferences. He is also one of the editors of few of the books titled *Introduction to FACTS* and *Power Quality Management* of IRD Publication and Vijay Nicole Imprints Private Limited, respectively.

Bijaya Ketan Panigrahi is Professor in the Department of Electrical Engineering, Indian Institute of Technology Delhi. His research interests include the security of cyber-physical systems, digital signal processing, and soft computing applications to power systems. Panigrahi received a Ph.D. in Electrical Engineering. He is an associate editor for *IEEE Systems Journal* and a senior member of the IEEE.

Swagatam Das received the B.E. Tel.E., M.E. Tel.E. (Control Engineering specialization), and Ph.D. degrees, all from Jadavpur University, India, in 2003, 2005, and 2009, respectively. Swagatam Das is currently serving as Associate Professor at the Electronics and Communication Sciences Unit of the Indian Statistical Institute, Kolkata, India. His research interests include evolutionary computing, pattern recognition, multi-agent systems, and wireless communication. Dr. Das has published more than 300 research articles in peer-reviewed journals and international conferences. He is the founding co-editor-in-chief of Swarm and Evolutionary Computation, an international journal from Elsevier. He has also served as or is serving as the associate editors of the Pattern Recognition (Elsevier), Neurocomputing (Elsevier), Information Sciences (Elsevier), IEEE Transactions on Systems, Man, and Cybernetics: Systems, IEEE Computational Intelligence Magazine, IEEE Access, and so on. He is an editorial board member of Progress in Artificial Intelligence (Springer),

Applied Soft Computing (Elsevier), Engineering Applications of Artificial Intelligence (Elsevier), and Artificial Intelligence Review (Springer). Dr. Das has 20,500+ Google Scholar citations and an H-index of 67 till date. He has been associated with the international program committees and organizing committees of several regular international conferences including IEEE CEC, IEEE SSCI, SEAL, GECCO, and SEMCCO. He has acted as the guest editors for special issues in journals like IEEE Transactions on Evolutionary Computation and IEEE Transactions on SMC, Part C. He is the recipient of the 2012 Young Engineer Award from the Indian National Academy of Engineering (INAE). He is also the recipient of the 2015 Thomson Reuters Research Excellence India Citation Award as the highest cited researcher from India in Engineering and Computer Science category between 2010 and 2014.

Closed-Loop Vision-Based Ball Balancer

M. Sakthi Ganesh, J. Anitha Roseline, G. Sowmya, and V. Vishwa Raj

Abstract This paper presents the mathematical modeling, design, and implementation of a ball balancer. The aim is to balance the ball at a desired position on the base plate. Ball balancer with two degrees of freedom uses machine vision as feedback and DC servomotors as actuators. Mathematical modeling of ball balancer system is done based on Lagrange–Euler equation, processing of the real-time image, and finding the co-ordinates of the ball is done using Open Source Computer Vision Library software. The co-ordinates of the ball are sent to Arduino through universal serial bus (USB) to transistor—transistor logic (TTL) converter. Perfect balancing of the ball is achieved using a proportional—derivative (PD) controller and pole placement technique. Stability of the designed ball balancer system is verified by its time response analysis using MATLAB/Simulink.

Keywords Arduino controller · Ball balancer system · Image processing · Machine vision · MATLAB/Simulink · Open CV software

1 Introduction

Stabilizing a system is one of the challenging problems in control field which includes systems such as cart-pole, ball-beam, and doubly inverted pendulums. The ball balancer is an advanced version of ball on beam system. The ball balancer model is otherwise called a ball-plate system which is nonlinear and unstable. It has four degrees of freedom, two in ball's position on the balancer plate and two in balancers inclination angle. The objective is to stabilize the ball in a predefined co-ordinate on the plate. Ball's position and its stability are controlled employing a PD controller and pole placement technique.

M. Sakthi Ganesh (✉) · J. Anitha Roseline · G. Sowmya · V. Vishwa Raj
Sri Sivasubramaniya Nadar College of Engineering, Kalavakkam, Chennai 603110, India

J. Anitha Roseline
e-mail: anithar@ssn.edu.in

S. S. Dash et al. (eds.), *Sixth International Conference on Intelligent Computing and Applications*, Advances in Intelligent Systems and Computing 1369,
https://doi.org/10.1007/978-981-16-1335-7_1

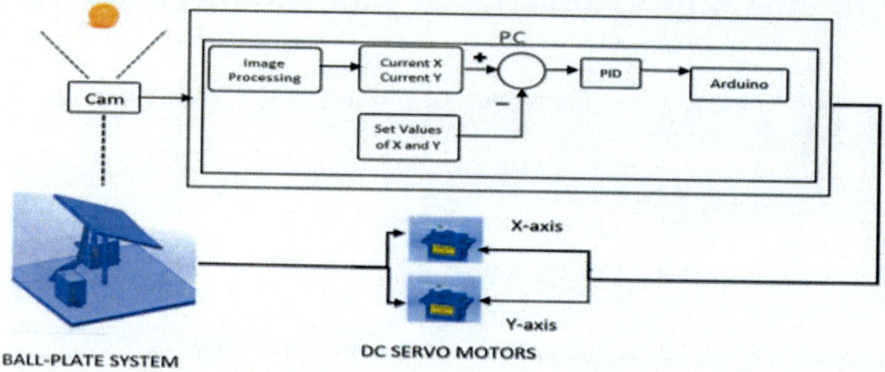

Fig. 1 Block diagram of vision-based ball balancer system

Figure 1 shows the block diagram of the proposed vision-based ball balancer system.

Ball balancing system has been developed using various control strategies in literature, basically a horizontal plate is tilted along each of the two axes in order to stabilize the position of the ball. Tilting of the base plate is achieved using motors controlled by position feedback system and the ball's position is sensed through a resistive touch screen [1]. Proportional derivative integral (PID) controller with filter is used to control the position of ball's axis using servomotor as its actuator and webcam as its position sensor [2]. It has been observed that a PID controller with filter gives better results than PID controller despite its slow response. LEGO MINDSTORMS is also used for model predictive control where a linear Kalman filter is used to estimate the position and velocity of the ball on the plate. The position is measured using mind sensor's NXTcam, and a linear controller is implemented in NXC making the whole system independent. The plate is actuated by the NXT servomotors [3]. LEGO's fire wire webcams were used to estimate the position the ball and balance using PID controller [4]. A combination of fuzzy logic controller and switching servocontroller is also used for controlling the position of the ball [5]. Symbol-based control of a ball on plate mechanical system is discussed in [6]. Application of several optimization technique for ball and plate balancing system is proposed in [7–10, 12]. Adaptive embedded control and modeling based on first principle of optimal control and AU-based control is discussed in [11, 13, 14]. Detailed mathematical modeling of the proposed electromechanical vision-based ball balancer system and its linearization technique is discussed in Sect. 2. Section 3 describes the controller design and the pole placement technique which makes the system stable. Section 4 describes the real-time implementation and the time response analysis of the vision-based ball balancer system. Section 5 concludes the paper highlighting its future scope.

2 Mathematical Modeling of Ball Balancer System

System dynamics is concerned with the study of forces and torques and their effect on motion. Mathematical modeling of the ball balancer system is done based on the following assumptions: The ball is always in contact with the plate and it does not slip. The ball is homogenous and symmetric. Joints of the motor to tilt the axis are perfectly rigid. The friction between the ball and the balancer is negligible.

The Euler–Lagrange equation for the ball balancer system is as following

$$\frac{\mathrm{d}}{\mathrm{d}t}\frac{\partial T}{\partial \dot{q}_i} - \frac{\partial T}{\partial q_i} + \frac{\partial V}{\partial q_i} = Q_i \tag{1}$$

where

q_i—Direction co-ordinate
T—Kinetic energy of the ball balancer system
V—Potential energy of the ball balancer system
Q_i—Composite force.

Assuming the XY co-ordinates, ball's position on the plate will be at the center. The T of the ball comprises of translational energy and rotational energy with respect to its center given by

$$T_b = \frac{1}{2} m_b\left(\dot{x}^2 + \dot{y}^2\right) + \frac{1}{2} I_b\left({}_x^2 + {}_y^2\right) \tag{2}$$

where $\dot{x}$ is translational velocity along x-axis, $\dot{y}$ is translational velocity along y-axis, ω_x is ball's rotational velocity along x-axis, ω_y is ball's rotational velocity along y-axis, is radius of the ball. The translational and rotational velocities of the ball are given by

$$\dot{x} = r_{bx}, \quad \dot{y} = r_{by} \tag{3}$$

The T of the plate consists of its rotational energy with respect to its center of mass. The equation is as follows,

$$T_p = \frac{1}{2}\left(I_p + I_b\right)\left(\dot{\alpha}^2 + \dot{\beta}^2\right) + \frac{1}{2} m_b\left(x\dot{\alpha} + y\dot{\beta}\right)^2 \tag{4}$$

where α and β are plate's angle of inclination along X-axis and Y-axis, respectively, as shown in Fig. 2. Therefore, $\dot{\alpha}$ and $\dot{\beta}$ are the plate's rotational velocity. Hence, the T of system is given by

$$T = T_b + T_p \tag{5}$$

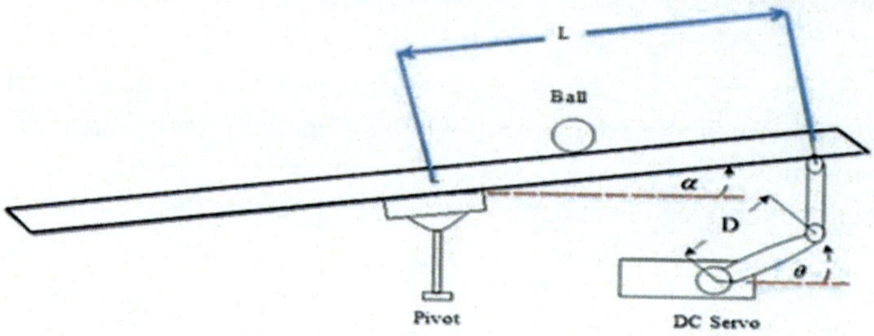

Fig. 2 Ball balancer model

The V of the ball relative to horizontal plane at the center of the inclined balancer is given by

$$V_b = m_b gh = m_b g(x \sin\alpha + y \sin\beta) \tag{6}$$

Substituting Eqs. (5) and (6) in (1), the nonlinear differential equations for the ball balancer are given by (7) and (8)

$$\left(m_b + \frac{I_b}{r_b^2}\right)\ddot{x} - m_b\left(x\dot{\alpha}^2 + y\dot{\alpha}\dot{\beta}\right) + m_b g \sin\alpha = 0 \tag{7}$$

$$\left(m_b + \frac{I_b}{r_b^2}\right)\ddot{y} - m_b\left(y\dot{\beta}^2 + x\dot{\alpha}\dot{\beta}\right) + m_b g \sin\beta = 0 \tag{8}$$

Now, the nonlinear terms in the system equation are eliminated by linearization technique. We can linearize the system equation by assuming small angle of inclination for the plate I.e. α and β are less than 1, sin α and sin β is approximately equal to α and β respectively. Slow rate of change of the plate, i.e., $\dot{\alpha} \ll 0, \dot{\beta} \ll 0$ and $\dot{\alpha}\dot{\beta} \cong 0$, $\dot{\alpha}^{2} \cong 0$ and $\dot{\beta}^2 \cong 0$. Moment of inertia of a hollow ball is:

$$I_b = \frac{2}{3} m_b r_b^2 \tag{9}$$

Substitute these values in Eqs. (7) and (8) and rearranging the equations. Let θ_x and θ_y be the input angle provided by the X- and Y-axis servomotor. Hence, the relationship between θ, α and β is

$$\alpha = \frac{D}{L}\theta_x \quad \text{and} \quad \beta = \frac{D}{L}\theta_y$$

By substituting and rearranging the equation, we get,

$$\ddot{x} = -\frac{3}{5}\frac{D}{L} g\theta_x - \frac{3}{5}\frac{B}{m_b}\dot{x} \tag{10}$$

Table 1 System parameters

Symbol	Description	Value
G	Gravitational constant	9.81 m/s^2
m_b	Ball's mass	0.0027 kg
m_p	Plate's mass	0.5 kg
r_b	Ball's radius	0.02 m
I_b	Ball's inertia	7.2 × 10^{-7} kg.m^2
h	Height at which the camera is fixed	0.9 m
B	Damping co-efficient of the system	0.05
L	Length of the plate from center	0.15 m
D	Length of the lever	0.03 m
A	Dimensions of the plate	0.3 × 0.3 m^2

$$\ddot{y} = -\frac{3}{5}\frac{D}{L}g\theta_y - \frac{3}{5}\frac{B}{m_b}\dot{y} \tag{11}$$

Substituting the system parameter values from Table 1, in (10) and (11), we obtain two separate differential equations.

$$\ddot{x} = -1.1771\theta_x - 11.11\dot{x} \tag{12}$$

$$\ddot{y} = -1.1771\theta_y - 11.11\dot{y} \tag{13}$$

The general state variable format is given by:

$$\dot{X} = AX + B\theta \tag{14}$$

$$Y = CX + D\theta \tag{15}$$

In our system: $x_1 = x$ and $x_2 = \dot{x}_1 = \dot{x}$

$$\begin{bmatrix} \dot{x}_1 \\ \dot{x}_2 \end{bmatrix} = \begin{bmatrix} 0 & 1 \\ 0 & -11.11 \end{bmatrix}\begin{bmatrix} x \\ \dot{x} \end{bmatrix} + \begin{bmatrix} 0 \\ -1.1772 \end{bmatrix}\theta_x \tag{16}$$

$$y_x = \begin{bmatrix} 1 & 0 \end{bmatrix}\begin{bmatrix} x_1 \\ x_2 \end{bmatrix} + [0][\theta_x] \tag{17}$$

We obtain similar equations for y_y.
Characteristic equation is:

$$|SI - A| = 0 \tag{18}$$

$$s^2 + 11.11s = 0 \tag{19}$$

3 Controller Design

The control law states that:

$$\theta = -Kx \tag{20}$$

where K_x is the feedback matrix and

$$K_x = \begin{bmatrix} K_{x1} & K_{x2} \end{bmatrix} \tag{21}$$

Then, the general state equation becomes,

$$\dot{x} = (A - BK)x \tag{22}$$

$$|SI - (A - BK)| = 0 \tag{23}$$

Then, the closed-loop characteristic equation of the system is

$$S^2 + S(11.11 - 1.1772K_{x2}) + 1.1772K_{x1} = 0 \tag{24}$$

Using pole placement technique, the poles are placed at -1 and -10, thus obtaining the desired characteristic equation as,

$$(s + 1)(s + 10) = 0 \tag{25}$$

The control law gives value between 0° and 360°.

$$\theta = -K_P\mathrm{e} - K_D\dot{\mathrm{e}} \tag{26}$$

But values between 0° and 90° are required; hence to map the obtained values to the required values, the equation is divided by a correction factor of 40. On comparing Eqs. (22) and (23) Kp_x, $K\mathrm{d}_x$, Kp_x and $K\mathrm{d}_y$, values are obtained as follows.

$$K_{px} = \frac{K_{x1}}{40} = 0.2123 \quad K_{\mathrm{dx}} = \frac{K_{x2}}{40} = 0.002336 \tag{27}$$

$$K_{py} = \frac{K_{y1}}{40} = 0.2123 \quad K_{\mathrm{dy}} = \frac{K_{y2}}{40} = 0.002336 \tag{28}$$

Table 2 PD controller values

Proportional gain (K_p)	Derivative gain (K_d)	% Error
0.2123	0.002336	32
0.23	0.002336	38
0.25	0.001336	15
0.25	0.00010	7

We have obtained a K_p and K_d values for the controller, and they are 0.2123 and 0.002336, respectively. They are slightly tuned to reach the desired proportional gain (Table 2).

4 Hardware Implementation

The prototype of the ball balancer is shown in Fig. 3. An acrylic base balancer plate of dimension 30 × 30 × 0.5 cm is attached to a universal joint which is mounted on a central beam. The central beam is fixed to a wooden ground plate of 2 cm thickness. The servomotors are also placed on the wooden ground plate as shown in Fig. 3. Each servomotor arm is connected to the acrylic base through a ball-sphere joint which is attached to the base.

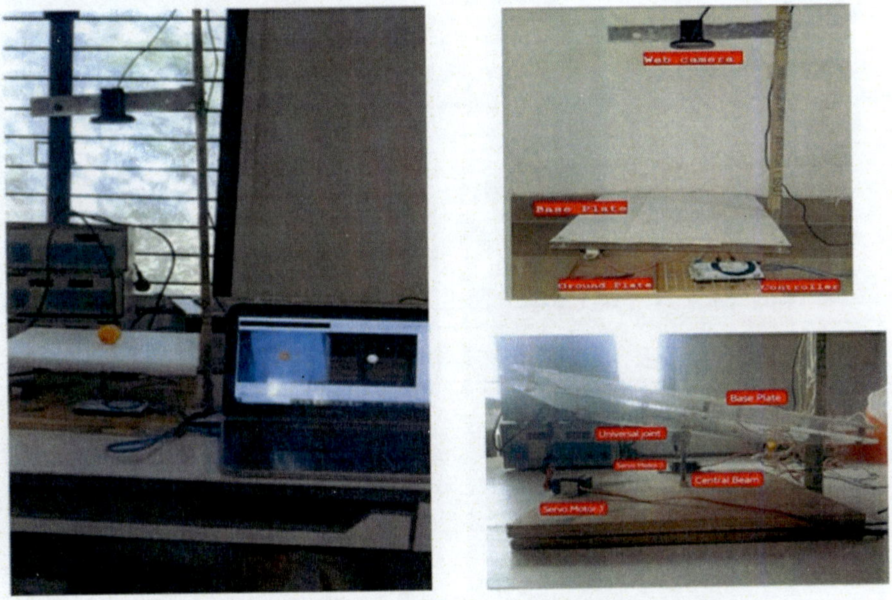

Fig. 3 Front view, side view, and main setup of the ball balancer system

The servomotors tilt the base in X and Y directions providing 2 degrees of freedom to the balancer. The ball's position is detected by capturing frames with a single web camera that is placed on top of the base and has an adjustable arm attached to it as shown in Fig. 3. A PCB was built to act as Arduino mega shield to control the two servomotors and USB to TTL. The wiring diagram and PCB of the ball balancer system are shown in Fig. 4. Once the image frames from the camera is obtained, it is converted from RGB to Hue saturation value as shown in Fig. 5. Later, mask is applied to filter out the unwanted colors and to detect only the color of the ball that is used. This is done by using already determined HSV values of the ball. Now to find the centroid of the masked image already obtained, HSV image is converted into binary image with the help of in-built function in Open CV. Having found the centroid of the ball, it is fed to the port at which Arduino is connected using USB to TTL converter. The algorithm for detecting the ball's center from the image is given in Fig. 6.

Fig. 4 Wiring diagram of the PCB and PCB for the ball balancer system

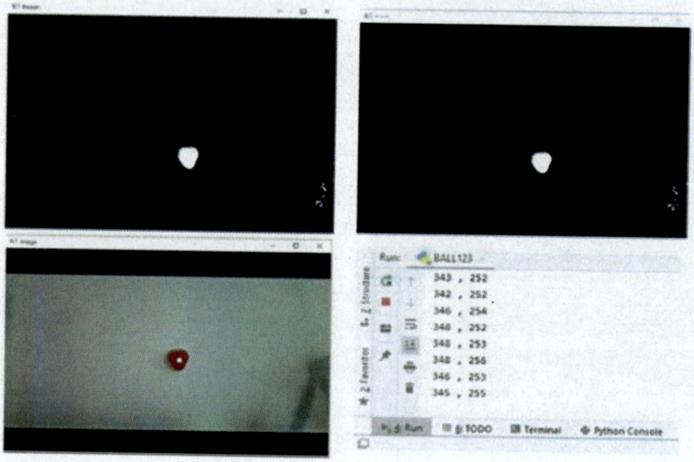

Fig. 5 RGB to HSV conversion

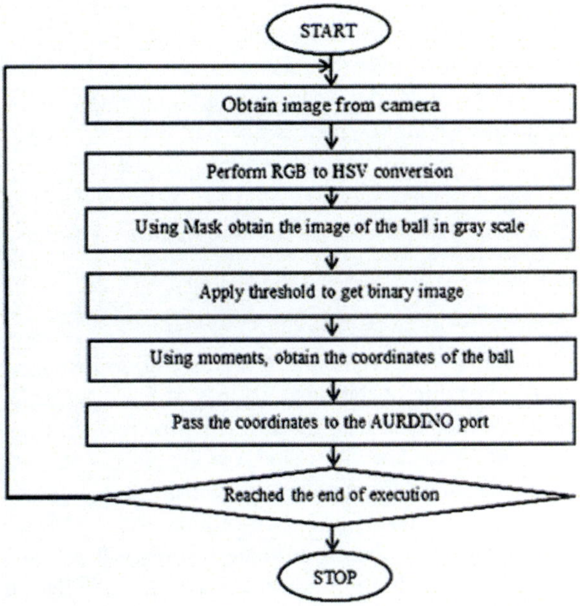

Fig. 6 Algorithm for detecting the ball from the image

Time response analysis of the ball balancer system is done using MATLAB. Step response was used to estimate both the closed and open-loop response of the balancer. Figure 7a shows the open-loop step response plot of the ball balancer system. It can be observed that the system is unstable. Therefore, to make it stable, the design parameter considered for this ball balancer system are settling time should be less

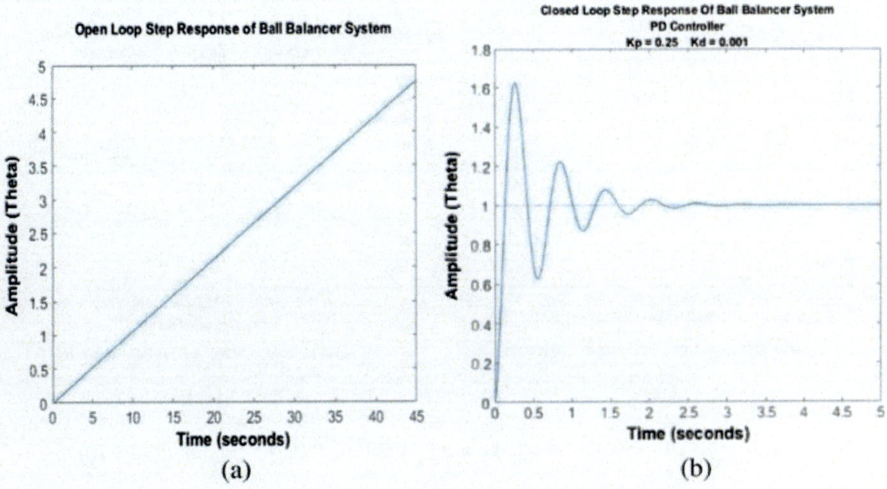

Fig. 7 **a** Open-loop **b** Closed-loop step response of ball balancer system

than 3 s and overshoot less than should be 5%. The damping ratio (ζ) and natural frequency (ωn) were found to be 0.7 and 1.9, respectively, using Eqs. (26) and (27) which relate them to maximum percent overshoot (*MP*) and settling time (*TS*) requirements.

$$M_P = \mathrm{e}^{-\zeta\pi/\sqrt{1-\zeta^2}} \tag{26}$$

$$T_S = \frac{4}{\zeta\omega_n} \tag{27}$$

It can be observed from the closed-loop response that the ball balancer system becomes stable at the desired time of 3 s. The time response of the ball balancer system is also analyzed by collecting the real-time data in terms of XY co-ordinates of the ball's center with respect to the plate (450 × 450) using open CV in python. Figure 8a shows the state trajectory followed by the ball to reach the desired location on the plate in terms of *XY* co-ordinate (330,270) when released from the position corresponding to the co-ordinate (20,100). It can be observed that the ball reaches the set point with minimum overshoot and oscillation as designed. The co-ordinate (330,270) is much closer to the center of the balancer plate.

In order to validate its performance, co-ordinate (50,350) which is far away from the center of the balancer plate or co-ordinate (50,350) which corresponds to the corner of the balancer plate is chosen. Figure 8b shows the state trajectory followed by the ball to reach the desired location on the plate in terms of XY co-ordinate when released from the co-ordinate (225,225) which corresponds to the center of the balancer plate. It can be observed that ball balances perfectly with minimum overshoot and oscillation.

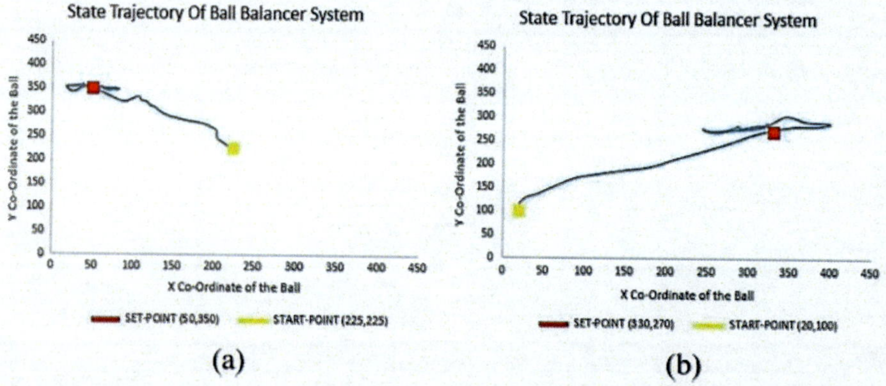

Fig. 8 State trajectory of the ball to reach the **a** set point (50, 350), **b** set point (330, 270)

5 Conclusion

Mathematical model of vision-based ball balancer was obtained based on the objective which is to drop the ball on the balancer and wait for the system to respond either to move/stabilize the ball at a desired location on the base. A PD controller was designed to control the balancer using servomotors which used image processing to track the position of the ball. Balancing at different location on the balancer was tested and found that the controller was robust and able to position the ball at any desired location on the balancer. Primary practical application of the ball balancer system is in solar panels fitted with solar trackers. The balancer once built and tested can be further used as a testbed for analyzing other control schemes for other balancing systems like inverted pendulums and self-balancing bot, rocket gimbals.

References

1. Andrews G, Colasuonno C, Herrmann A (2004) Ball on plate balancing system. Final year project, Rensselaer Polytechnic Institute, New York
2. Adiprasetya A, Surya Wibowo A (2016) Implementation of PID controller and pre-filter to control non-linear ball and plate system. In: International conference on control, electronics, renewable energy and communications, pp 174–178
3. https://robotics.benedettelli.com/ball-on-plate-1-nxt/,last accessed 2020
4. Wang H, Tian Y, Sui Z, Zhang X, Ding C (2007) Tracking control of ball and plate system with a double feedback loop structure. In: International conference on mechatronics and automation, pp 1114–1119
5. Ali E, Aphiratsakun N (2016) AU ball on plate balancing robot. In: IEEE International conference on robotics and biomimetics. pp 2031–2034
6. Yip P (2004) Symbol-based control of a ball-on-plate mechanical system. M.S. thesis, Department of Mechanical Engineering, University of Maryland, USA
7. Debono D, Bugeja M (2018) Application of sliding mode control to the ball and plate problem. Thesis, Department of Systems and Control Engineering, University of Malta
8. Cheng C, Tsai C (2016) Visual servo control for balancing a ball plate system. Int J Mech Eng Robot Res 5(1):28–32
9. Jezierski E, Łuczak P, Smyczyński P, Zarychta D (2019) Human–robot cooperation in sorting of randomly distributed objects. Arch Control Sci 29(4):603–615
10. Andrzejewski K, Czyżniewski M, Zielonka M, Langowski R, Zubowicz T (2019) A comprehensive approach to double inverted pendulum. Archives of Control Sci 29(3):459–483
11. Xiao J, Buttazzo G (2016) Adaptive embedded control for a ball and plate system. In: The eighth international conference on adaptive and self-adaptive systems and applications, pp 40–45
12. Frank J, Gomez J, Vikram K (2015) Using tablets in the vision-based control of ball and beam test-bed. In: Proceedings of 12th international conference of informatics in control, automation and robotics. France, pp 92–102
13. Dusek F, Honc D, Sharma R (2017) Modeling of ball and plate system based on first principle model and optimal control. In: 21st International conference on process control, pp 216–221
14. Ali E, Aphiratsakun N, AU ball on plate balancing robot. In: IEEE International conference on robotics and biomimetics, pp 2031–203

A Novel Network Learning for Image Compressive Sensing

V. Shyamala Susan and A. P. Bambang Sugiyono

Abstract Most traditional algorithms suffer from intensive computation for compressive sensing image reconstruction. In recent times, algorithms of deep learning-based reconstruction have been proposed to minimize the complexities of time rather than algorithms of iterative reconstruction. However, the deep learning networks' training phase has to reduce the complexity of the time. An innovative selective deep residual reconstruction network (SDRRN) learning algorithm is proposed to recreate the image from its measurement of compressively sensed (CS). The SDRRN contains two components: linear mapping of the network and residual networking. Simple fully interconnected layers are incorporated in the linear mapping network. A novel learning strategy is proposed which only propagates the selected training components in the residual learning process. Extensive experimentation shows that the SDRRN outperforms conventional iterative methods and also low-time complexity when compared with other deep learning-based methods.

Keywords Compressive sensing · Deep learning · Reconstruction · Convolution neural network · Residual learning · Network mapping

1 Introduction

The popular methods in multimedia content compression are transformation and prediction among the different coding systems. JPEG [1] is the most general standard for image compression that includes fundamental components of transformation/prediction. The input image in JPEG is segmented across non-overlapping blocks, but every non-overlapping block is further translated into a frequency domain. For each translated unit, the coefficients of the discrete cosine transform (DCT) are then condensed into a discrete stream using linearization and entropy coding. Video

V. S. Susan (✉)
A.P.C. Mahalaxmi College for Women, Tuticorin, India

A. P. B. Sugiyono
State Polytechnics of Malang, Malang, Indonesia

S. S. Dash et al. (eds.), *Sixth International Conference on Intelligent Computing and Applications*, Advances in Intelligent Systems and Computing 1369,
https://doi.org/10.1007/978-981-16-1335-7_2

coding principles are primarily implemented by the transform-prediction-based hybrid video compression frameworks.

Some of the coding schemes use much more intra-projection modes from adjacent reconstructed blocks in spatial space rather than discrete cosine representations. Other than intra-projection, inter-prediction is extremely successful for additional coding gains in video control, to locate the most comparative squares as expectations for the square to be coded where it uses motion approximation. For each unique coding process, the individual optimization strategy limits the increase in compression efficiency compared to end-to-end compression encoding. Essentially, there is yet another avenue of technological evolution based on neural network image and video compression techniques. With the revival of neural networks, the marriage of digital compression image/video and convolution neural network (CNN) is further advancing its development. With the advantage of the reconstruction-based algorithms, the complexity can be reduced drastically in deep leaning. So, it is feasible to evolve a methodology to reduce the learning networks' training phase complexity.

2 Literature Review

Several works recently incorporate the deep learning into the reconstruction of image/video. The techniques of machine learning for image compression, particularly from the perspective of the neural network, emerge from the late 1980s [2]. Rendering to the past evolution of neural network strategies, the technological roadmap for image compression applications is the multilayer perceptron (MLP), random neural network (RNN).

Kulkarni et al. [3] proposes a CNN-based ReconNet for compressive image recovery, which achieves regeneration of non-iterative compressive sensing data. Stacked denoising auto-encoders (SDAs) are utilized to develop a plottingamong CS measurements and image blocks [4]. Iliadis et al. for video recovery compression. The encoder assumes a binary sensing mask in the deep fully-connected network [5, 6], and the video reconstruction is evaluated by the decoder. To avoid inefficient computation in traditional approaches, there is an effectively achieved promising efficiency in image/video reconstruction. To limit the amount of coding bits regardless of the frequency estimate calculation in CNN, Toderici et al. first suggested an RNN-based image compression scheme [7] with the use of a scaled-additive coding framework [8]. More precisely, the proposed methodology in [7] is an architecture of multi-iteration compression that supports progressive-style variational compression of bitrates. Presently,

Minnen et al. given a spatially adaptive image compression system [9] that also improves image compression on the basis of RNN. A technique of optimizing complexity-distortion within bandwidth limitations for the challenge of video coding is further proposed in [10], which can be generalized to the optimization method of CNN model compression in combination with computational complexity and video processing performance. Compressive sensing is a popular technique for digital data collection and retrieval, e.g., videos and images. This collects the data as measurements of compressively sensed (CS) and then from those measurements of CS used to recreate the original data since the number of measurements needed is smaller in reliability than the theory of Nyquist. In many applications, for example, in medical scanners, cameras, etc., CS is very attractive. By achieving enormous output from CS measurements in image reconstruction, several algorithms have been proposed in recent years [11–15]. Several of these recent studies have used iterative optimization techniques to model and solve an optimization problem with the formal maximum likelihood assumption of the image signal.

Neural network-based compression algorithms have demonstrated potential outcome in image and video compression errors based on the conclusions. Even though memory consumption and computational complexity still have many issues, their concise representations and high efficiency in predicting image and video signals have allowed neural networks achieve significant coding in addition to state-of-the-art coding terminology.

This article suggests a learning of selective deep residual reconstruction network (SDRRN), which therefore improves the efficiency of reconstruction from quick speed measures of compressively sensed (CS). The summarization of the understandings as follows:

- This paper suggests, a novel SDRRN, a compressed sensing learning technique that surpasses related methods with a low complexity in time.
- Linear mapping gives an effectively relatively good rebuilding image with faster speed and lower computation complexity. Additionally, the residual learning increases the efficiency of reconstruction with a selective learning approach.

3 Methodology

According to the deep residual reconstruction network strategy, it takes the input CS measurement and feed forward them via linear mapping network $N^L(.)$ as well as residual network $N^R(.)$. The residual network $N^R(.)$ is independent network which takes the input as initial reconstructed image from linear mapping network $N^L(.)$. Figure 1 shows the architecture for the proposed SDRRN.

3.1 Linear Mapping and Residual Network

Compressive sensing is signal reconstruction process which recovers the signal from its CS measurement [16]. Consider the signal S and the compressive sensed measurement C, then the linear mapping will be termed as,

$$S = \phi C \tag{1}$$

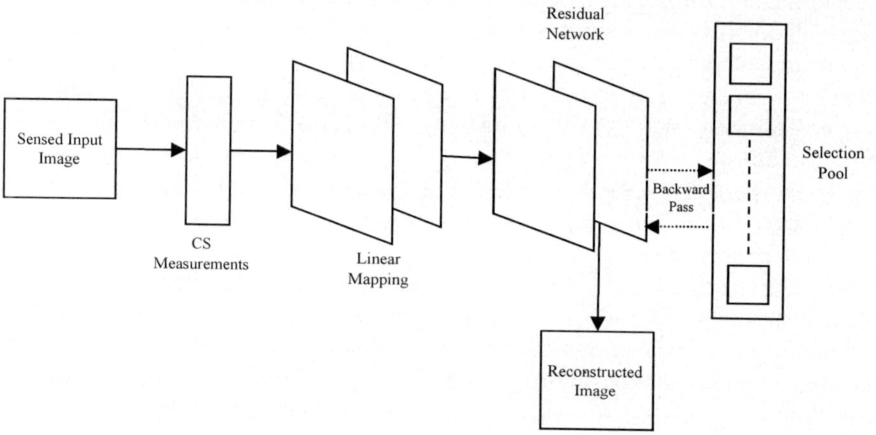

Fig. 1 Proposed SDRRN architecture

The mapping matrix ϕ^L for the network $N^L(.)$ can be estimated as minimization problem as bellow, the learned linear mapping network $N^L(.)$ produce an initial rebuilt image $\hat{I}$ could be gained for any assumed CS measurement C. The residual network estimates the other residual $\hat{r}$ based on $\hat{I}$.

$$\phi^L = \arg \quad \min \quad \|S - C\| \tag{2}$$

3.2 Selective Residual Learning

The input image I must be learned by the residual network $N^R(.)$ by minimizing the convolution layers' activation error. This process of minimization is performed many numbers of back propagations sequentially in each iteration the proposed selective residual learning is drastically reduces the back propagation attempts by sensitively selecting the principal patches from the selection pool generated from the training ground truth. This novel process of residual learning can be termed as follows,

$$N^R(I) \subseteq \Im(I) - I \tag{3}$$

where $\Im(I)$ denotes the output of feed forward operation. Hence, the residual learning parameters are derived as subset using the selection pool. Based on the application domain, the section pool will be determined and section rule also prescribed based on the sensing scenario.

3.3 SDRRN Architecture

SDRRN processes the $m*m$ sized image block which was computed from CS measurement. The CS calculation $y \in R^{m*1}$ is taken as an input by the linear mapping network $N^L(.)$, where m is a range of measuring rates (MRs). One totally linked layer comprising m*m neurons is the linear mapping network. With the size of m * m, the preliminary reconstructed image will be produced.

The residual network $N^R(.)$ feeds with the initial outcome as input, and outputs an image with size $m*m$. $N^R(.)$ contain four residual learning segments where each segment consists of three convolutional layers. In each segment, the primary convolution layer generates a feature maps. The second layer generates further enhanced features maps. The third layer produces a feature map with different kernel. A feature map with a different kernel is generated by the third layer. To generate the feature map with size $m*m$, the padding layers have been added in each layer, and batch normalization has also been added in the first two convolution layers of each block. With each convolutional layer, excluding the next convolutional layer, a ReLU layer is adopted.

3.4 Training Strategy

SDRRNtraining procedure involves training the two sub-networks $N^L(.)$ and $N^R(.)$. A usual step strategy with large learning rate is followed to train the network $N^L(.)$. Learning parameters have assigned as maximum iterations = 1,000,000, learning rate = 0.00, step size = 200,000, and gamma = 0.5. With this trained weights W^f, the residual network $N^R(.)$ will be trained smaller learning rate. Here, the proposed selected residual learning strategy is adopted to ease the training complexity.

3.5 Reconstruction Procedure

First the input image will be divided as $m*m$ tiles with no overlap. The CS measurement has been taken for tile. Then, SDRRN input with the CS measurements which has already been taken. The output of the final layer results the reconstructed image. To overcome the artifacts, effect the final reconstructed image involved with the simple BM3D [17] processing which produce the final result.

Table 1 Experimental results in terms of accuracy and speed (Convergence Time in Minutes) among the proposed SDRRN, DR^2N [16], and FCMN [18] methods

Test set	Proposed SDRRN		DR^2N [16]		FCMN [18]	
	Accuracy	Speed	Accuracy	Speed	Accuracy	Speed
1	0.9687	1.676	0.9461	1.945	0.9134	2.786
2	0.9232	1.255	0.9118	1.546	0.8956	1.959
3	0.8933	1.973	0.8593	2.431	0.8579	2.933
4	0.8767	2.814	0.8263	3.718	0.8214	3.895
5	0.9123	1.536	0.8919	1.623	0.8759	1.995

4 Experimental Results

To evaluate the SDRRN in terms of speed and robustness, the experimental analysis has been carried out for with few state-of-the-art methods such as DR^2N [16], FCMN [18]. For the fair comparison, the tests are directed on an Intel Core i3-7100 CPU at 3.90 GHz with 8 GB of memory. The test data sets are gathered from different sources with variable number of images. Totally five groups of test sets have been constituted by manual annotation process.

The high-performance reconstruction is achieved through high matching of the reconstructed image with ground truth or sensed input image. The speed of the learning process is referred as the ability to minimize the error in relatively minimum number of iterations. These measures are tabulated by experiment the reconstruction and training phases with respective samples. The comparative analysis shows the insight of the construction and training ability with different performance aspects. Table 1 shows the experimental evaluation of few state-of-the-art methods such as DR^2N [16], FCMN [18] in terms of accuracy and speed. Based on the results, the ability of the proposed SDRRN architecture is reliably better than the existing techniques [16, 18]. And Fig. 2 shows the graphical representation of the experimental results from the result one could understood the performance ability of the proposed selection strategy.

5 Conclusion

This research presents a novel selective deep residual reconstruction network for reconstruction of high-quality compressive sensing images. The system utilizes a fully convolutionary architecture and reconstructs the picture in two stages such as linear mapping and mapping of residuals. An innovative section strategy is chosen in the phase of residual learning for fast and enhanced learning accuracy. The network provides the highest results in the reconstruction function compared to other approaches with this upgrade. And they log and evaluate the tests. The generic

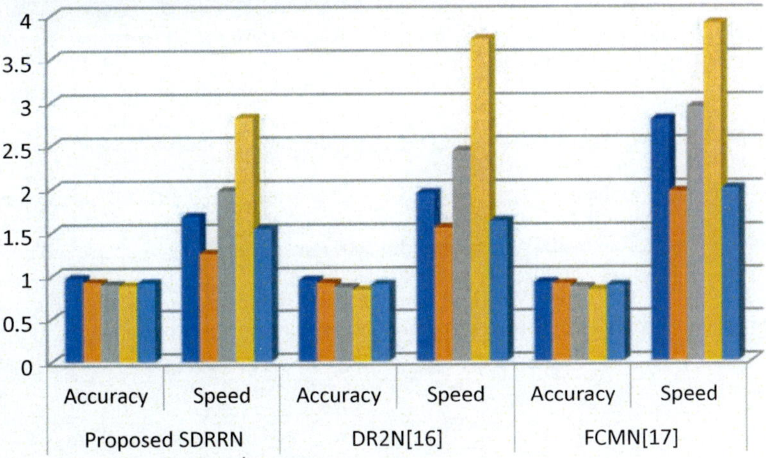

Fig. 2 The graphical representation of the experimental results

classification rues can be used in future research for all kinds of restoration and compressive sensing issues.

References

1. Wallace GK (1991) The JPEG still picture compression standard. Commun ACM 34(4):30–44
2. Chua L, Lin T (1988) A neural network approach to transform image coding. Int J Circuit Theory Appl 16(3):317–324
3. Kulkarni K, Lohit S, Turaga P, Kerviche R, Ashok A (2016) Reconnet: non-iterative reconstruction of images from compressively sensed random measurements. CVPR
4. Mousavi A, Patel AB, Baraniuk RG (2015) A deep learning approach to structured signal recovery. In: 2015 53rd Annual allerton conference on communication, control, and computing (Allerton). IEEE, pp 1336–1343
5. Iliadis M, Spinoulas L, Katsaggelos AK (2016) Deepbinarymask: learning a binary mask for video compressive sensing. arXiv preprint arXiv:1607.03343
6. Iliadis M, Spinoulas L, Katsaggelos AK (2018) Deep fully-connected networks for video compressive sensing. Digital Signal Process 72:9–18
7. Toderici G, Vincent D, Johnston N, Hwang SJ, Minnen D, Shor J, Covell M (2017) Full resolution image compression with recurrent neural networks. In: CVPR, pp 5435–5443
8. Balle J, Laparra V, Simoncelli EP (2016) End-to-end optimized image compression. arXiv preprint arXiv:1611.01704
9. Minnen D, Toderici G, Covell M, Chinen T, Johnston N, Shor J, Hwang SJ, Vincent D, Singh S (2018) Spatially adaptive image compression using a tiled deep network. arXiv preprint arXiv:1802.02629
10. Su L, Lu Y, Wu F, Li S, Gao W (2009) Complexity-constrained H.264 video encoding. IEEE Trans Circuits Syst Video Technol 19(4):477–490
11. Daubechies I, Defrise M, De Mol C (2004) An iterative thresholding algorithm for linear inverse problems with a sparsity constraint. Commun Pure Appl Math 57(11):1413–1457
12. Li C, Yin W, Jiang H, Zhang Y (2013) An efficient augmented lagrangian method with applications to total variation minimization. Comput Optim Appl 56(3):507–530

13. Mairal J, Bach F, Ponce J, Sapiro G, Zisserman A (2009) Non-local sparse models for image restoration. In: ICCV. IEEE, pp 2272–2279
14. Mallat SG, Zhang Z (1993) Matching pursuits with time-frequency dictionaries. TSP 41(12):3397–3415
15. Tropp JA, Gilbert AC (2007) Signal recovery from random measurements via orthogonal matching pursuit. TIT 53(12):4655–4666
16. Yao H, Feng D, Shiliang Z, Yongdong Z, Qi T, Changsheng X (2019) Dr2-net: deep residual reconstruction network for image compressive sensing. Neurocomputing 359:483–493
17. Dabov K, Foi A, Katkovnik V, Egiazarian K (2007) Image denoising by sparse 3-d transform-domain collaborative filtering. TIP 16(8):2080–2095
18. Du J, Xuemei X, Chenye W, Guangming S, Xun X, Yuxiang W (2019) Fully convolutional measurement network for compressive sensing image reconstruction. Neurocomputing 328:105–112

Performance of Photovoltaic-Based ZETA Converter Water Pumping Application

Ananthi Christy, D. Christy Priscilla, E. Maheswari, and S. S. Dash

Abstract Solar energy is eco-friendly in nature, and non-conventional energy provides best solution for energy demand. The energy source of PV panel is a nonlinear system and produces less amount of energy. Hence, a high setup ZETA converter along with solar PV used as a source to obtain high voltage and power with great efficiency. Maximum power point tracker (MPPT) employed in this paper to achieve high output. PV-based ZETA converter needs less compensation with more efficiency. Irradiance data helps to compare results of other converters using MATLAB simulations.

Keywords PV panel · ZETA converter · MPPT · Efficiency. DC-DC ZETA converter

1 Introduction

The best solution based on energy demand is non-conventional energy. Since, many energies available today but solar energy plays a vital role. Solar energy assumed to be eco-friendly in nature. PV panel energy source is a nonlinear device, but produces less power and efficiency. Some limitations of PV panel result in high cost, more maintenance, and less energy efficiency. Boost converter preferred along with PV panel to extract maximum amount of power. The boost converter is placed between

A. Christy (✉) · D. C. Priscilla
Department of EEE, Saveetha School of Engineering, SIMATS, Chennai, India

D. C. Priscilla
e-mail: dcpriscilla@outlook.com

E. Maheswari
Department of EEE, Sri Sairam Institute of Technology, Chennai, India
e-mail: maheswari.eee@sairamit.edu.in

S. S. Dash
Department of EEE GCE, Keonjhar, Odisha, India
e-mail: Subhransudash_fee@gcekjr.ac.in

S. S. Dash et al. (eds.), *Sixth International Conference on Intelligent Computing and Applications*, Advances in Intelligent Systems and Computing 1369,
https://doi.org/10.1007/978-981-16-1335-7_3

the load and PV panel to extract most power. The energy of the system is low by adjusting the duty ratio. DC-DC ZETA converter [1] employed between the load and power generation level to increase voltage. Semiconductor devices utilized as switch in converters, which in turn results in less efficiency, more loss, and voltage stress. Hence, soft switching method helps to rectify the switching losses by using additional inductor. Interleaved PWM pulses reduce the amount of ripples in output. The project uses MPPT technique using ZETA converter to obtain maximum power.

DC-DC converter accepts DC input voltage and acquires DC output voltage. ZETA converter comprises two inductors, diode and series of capacitor. The role of capacitor is used to reduce the supply of power, and the converter results step-up or step-down function at output. Assumptions in ZETA converter reveal that line frequency in DC voltage neglected. The switching device of semiconductors tends to be ideal in nature and operates in inductor current mode. Solar cells are arranged in parallel and series to build solar PV module. PV system is static in nature and possesses low-conversion efficiency. Load voltage, temperature, and solar insulation affect the output of PV models.

The entire PV system produces maximum output power and high efficiency on V-P or V-I curve and refers to maximum power point (MPP). Operating point of PV is maintained by using maximum power point tracking (MPPT). MPPT is an electronic device, which functions on PV modules. The method of MPPT helps to extract maximum amount of energy from photovoltaic (PV) system. Some types of MPPT method are short circuit current, open circuit voltage, *P* and *O*, and IC. The design of DC–DC converter is verified using PSpice simulations.

1.1 Related Works

According to paper [2], for discrete and continuous operation mode, the steady-state results were analyzed for pulse width modulated DC-DC power converters. The forward converter is replaced by coupling capacitor. The converter is dual to SEPIC converter. Considering the measures of ON switch duty cycle, the circuit performs as step-up or step-down converter. The transfer function of positive DC-DC voltage depends on transformer less version. The circuit is well suits for distributed power devices. The derivations of every circuit elements are designed well. Measurements of result at 100 kHz are compared with theoretical statement. In paper [3], ZETA converter comprises of two capacitors and two inductors. The fourth order of DC-DC converter refers to ZETA converter and operates in step-down or step-up mode. Control and active model of ZETA converter is compared with another converter. A small-signal dynamic device of converter is analyzed using a technique called state-space averaging (SSA). The output voltage regulates based on control to output transfer function. On step load and start-up change, the simulation proves that converter presents best performance level.

In paper [4], ZETA power converter helps to carry out switched power supply mode with four outputs. Pulse width modulation (PWM) controls the converter. A laboratory model containing 200 W results validated using converters. The instrument helps to obtain low volume, low cost, and great efficiency.

According to paper [5], intermediate ZETA converter corrects the power factor. In order to check load voltage and load disturbances, a technique called average current control used. The ZETA converter manages a power of 250 W and results simulated in PSIM. A bidirectional charging system called ZETA-SEPIC converter works between low voltage battery and high DC link voltage device [6]. Some advantages of the converter is to protect coupling capacitor and performs transformer less operation.

1.2 Contribution of This Work

In this work, we use a converter to obtain maximum output power along with PV panel. The setup of ZETA converter helps to obtain the results with high voltage and power. Maximum power point tracker (MPPT) tracks output with more efficiency and low cost [7]. The simulations of the results verified using software called MATLAB SIM. The main role of the software is to integrate DLL blocks into circuit containing codes. DLL reduces the simulation time and increases the quality.

2 Methodology

This section describes the components and working of corresponding device. Solar PV panels majorly helps in obtaining the essential results. Due to some disadvantages, ZETA converter helps to obtain higher-efficiency output with low cost. The maximum power point tracker results in high power output. The block diagram of the corresponding device comprises of solar panel, ZETA converter, capacitor filter, DC motor, driver circuit, and PIC 16F877A (Fig. 1).

Solar panel receives 12 V of solar energy, and the received solar energy converts into electrical energy [8]. The change in output is due to changes in temperature cell and irradiance. The solar panel utilizes an algorithm called maximum power point tracker (MPPT), which helps to obtain maximum amount of power from source. The maximum power is then transferred to ZETA converter. The main role of ZETA converter refers to DC-DC boost converter and contributes the voltage up to 76 V. The solar panel sense current and voltage. The sensed parameters then given to a PIC microcontroller. With the help of step-down transformer, +5 V supply given to PIC microcontroller. The vital role of PIC microcontroller is to increase the voltage level up to 12 V and to produce PWM pulses.

PWM pulses help to trigger the switch of MOSFET present in ZETA converter. Boost operation takes place when the switch triggers. Some trigger seems to present in the output of ZETA converter. Capacitive filter decreases the amount of ripples from ZETA converter. Later, a DC voltage from output of the ZETA converter is

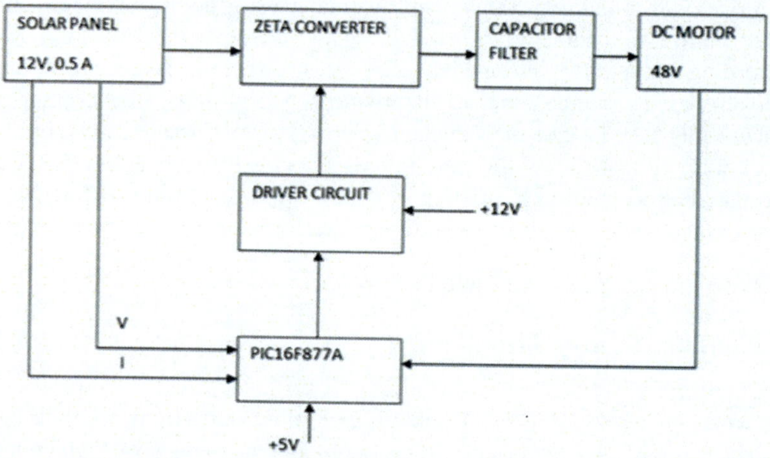

Fig. 1 Block diagram of PV water pumping system

supplied to DC motor. The DC motor helps to sense the current and voltage due to the closing.

3 Results and Discussions

The simulations of ZETA converter perform by MATLAB SIM. Simulation software designed in terms of power electronics and used to alter the parameter values. MATLAB SIM digital control module helps to design and simulate the digital power supplies. The custom C code or block diagram is carried out using digital control. The main characteristics of the software integrate DLL blocks into circuit by utilizing codes (Fig. 2).

The above graph represents the simulations between voltage and time. The given input voltage is 10 V. The sensed current and voltage then given to PIC microcontroller in order to adjust the pulses produced by PWM (Fig. 3).

The graph represents the output voltage plotted between voltage and time. Using ZETA converter, maximum amount of voltage is obtained (Table 1).

The comparison is made among three converters, namely SEPIC converter, forward converter, and ZETA converter. The single-ended primary inductor converter (SEPIC) is a kind of DC-DC converter, which allows voltage at output. The output voltage is lesser than, greater than, or equal to its input. There are two modes of SEPIC converter, namely continuous and discontinuous mode. During continuous mode, the inductor value never falls to zero. At discontinuous mode, the inductor value falls to zero. Some demerits are negative output polarity, and the converter is difficult to handle. In forward converter, the positive voltage is given to the core and flux increases. The output voltage obtained is greater than SEPIC converter.

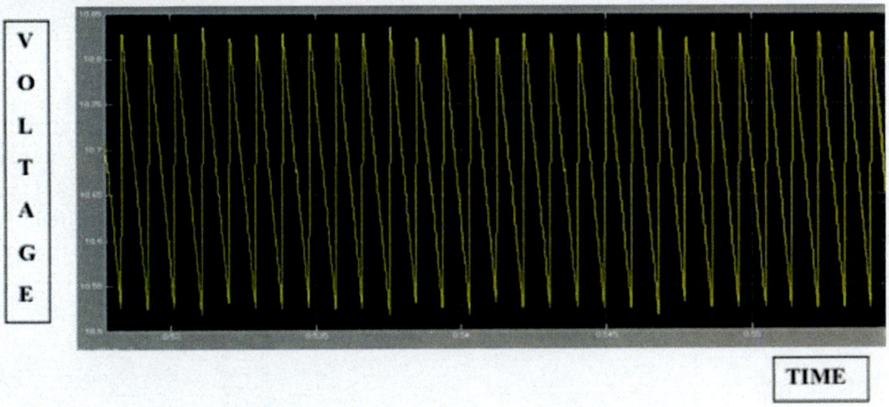

Fig. 2 ZETA converter simulation input

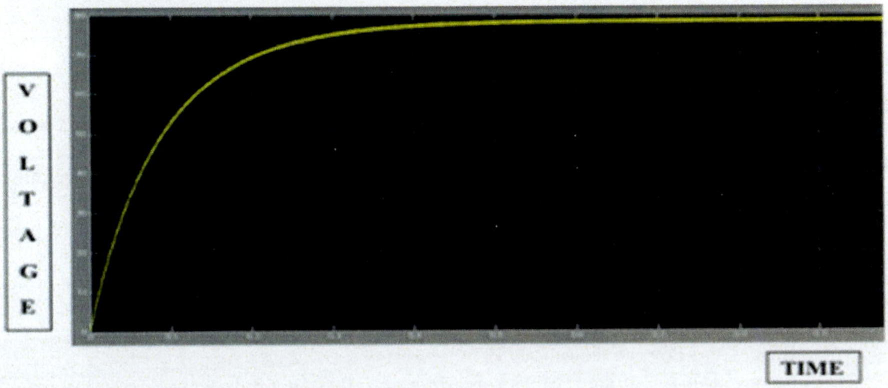

Fig. 3 ZETA converter simulation output

Table 1 Comparison table of converters

Type of converter	Input voltage (V)	Output voltage (V)
ZETA converter	10	78
Forward converter	10	56
SEPIC converter	10	37

The output of ZETA converter is greater than SEPIC and forward converter. MPPT algorithm is utilized in order to obtain maximum power. ZETA converter along with solar PV helps to achieve huge voltage conversion ratio. In pumping applications, coupled inductor plays a vital role to obtain more energy.

4 Conclusion

The best converter in the field of solar applications refers to ZETA converter. The concept of ZETA converter utilized in water pumping is a best system. The output produces more power and high accuracy. Maximum amount of power and voltage obtained using an algorithm is called maximum power point tracker. The results obtained from the converter are verified using MATLAB simulations. In future, interleaved boost converter is used instead of ZETA converter.

References

1. Pijit K, Sarawut S (2010) Isolated zeta converter: principle of operation and design in continuous conduction mode.Wseas Trans Circuits Syst 9(7); Bhim S, Ganesh DC (2008) analysis design, modeling, simulation and development of single-switch AC-DC converters for power factor and efficiency ımprovement. J Power Electron 8(1)
2. Johni BS, Sunil B et al (2012) Low crest power factor corrected electronicballast for fluorescent lamp. VSRD Int J Electri Electron Commun Eng 2
3. Keshri RK, Kadwane SG, Karan BM (2008) Power factor correction with neural network based zeta converter. IEEE Region 10 colloquium and the third ICIIS, Kharagpur, India
4. Denizar CM, Fernando de Souza C, Ivo B (1996) Zeta converter with high power factor operating in continuous conduction mode. In: Proceedings IEEE IECON. pp 1802–1807
5. Huai W, Issa B (1998) Comparison of basic converter topologies for power factor correction. IEEE
6. Bhim S, Mahima A, Sanjeet D (2008) Analysis design and implementation of a single-phase power-factor corrected AC-DC zeta converter with high frequency isolation
7. Oscar G, José AC, Roberto P, Pedro A, Javier U (2003) single phase power factor correction: a survey. IEEE Trans Power Electron 18(3)
8. Dan L, Viorel P, Dan N, Adrian P, Mihaela L, Mircea B (2005) In: Modelling analysis simulation and experimental

Performance Analysis of Radial Distribution System by Optimal Deployment of DG and DSTATCOM Considering Network Reconfiguration Using a SAR Algorithm

Srinivas Nagaballi, Vijay S. Kale, P. Sai Nandini, and Deepika Duppala

Abstract This paper demonstrates the search and rescue (SAR) optimization algorithm for enhancing the performance of the radial distribution system (RDS). Efficient performance can be obtained by optimal network reconfiguration, optimal deployment of distributed generation (DG) and distribution static compensator (DSTATCOM) units into the RDS. These are the effective methods to reduce the power loss in the system. Reduction in real power loss is considered as the fitness function in this paper. This study addresses the various power loss diminution approaches in the network. SAR is a metaheuristic algorithm developed based on human behaviour during search and rescue operations. To demonstrate the potency of the recent SAR algorithm, IEEE 33-bus RDS is utilized for study purposes. The obtained simulation results are compared with the results of other optimization methods. The results obtained are encouraging, with reduced power loss and improved voltage profile in the system.

Keywords Distributed generation · DSTATCOM · Radial distribution system · Search and rescue optimization · Electrical power losses

1 Introduction

The depletion of fossil fuels, increasing load demand and expansion of distribution systems lead to increased power losses in the distribution system. The majority of power system losses occur in the distribution systems. It became a significant concern for distribution utilities. Among the various power loss reducing methods in RDS feeder reconfiguration, optimal installation of DG and DSTATCOM is the significant reduction methods. The primary objective function considered in this paper is to mitigate the active power loss in the system. Reconfiguration is the process of

S. Nagaballi (✉) · V. S. Kale · P. S. Nandini · D. Duppala
Electrical Engineering Department, Visvesvaraya National Institute of Technology, Nagpur, Maharashtra 440010, India

S. S. Dash et al. (eds.), *Sixth International Conference on Intelligent Computing and Applications*, Advances in Intelligent Systems and Computing 1369,
https://doi.org/10.1007/978-981-16-1335-7_4

changing the status of its switches to alter the distribution network topology. The distribution system consists of two types of switches—the normally open switch (called tie switch), normally close switch (called sectionalizing switch). In reconfiguration, the switch changes status optimally to obtain an optimal configuration with reduced electric network loss. The feeder reconfiguration problem is a complex combinatorial optimization problem.

Network reconfiguration, placement of DGs and DSTATCOMs in RDS with different objective functions had been studied individually in various research papers. Civanlar et al. [1] proposed the branch exchange heuristic method, where a change in the power loss is calculated by changing the load between the feeders. In [2], the branch exchange method was used for the network reconfiguration. They are many optimization techniques applied to obtain an optimal configuration to reduce system power losses. Methods include binary particle swarm optimization (BPSO), genetic algorithm (GA), ant colony optimization (ACO), artificial bee colony (ABC), harmony search algorithm (HSA), etc. DG integration into the distribution system plays a vital role in mitigating electrical power losses. The DG is a power generating unit usually installed in the utility grid. Photovoltaic (PV) cells, wind turbines, fuel cells, gas turbines are a few different types of DG units. Similarly, DSTATCOM is a power electronic device employed in the distribution to control reactive power. Its quick response and contribution of fewer harmonics made the DSTATCOM device be preferred over conventional shunt capacitors. For improving the RDS performance, optimal size and location of DG and DSTATCOM units play an important role. Improper placement of DG and DSTATCOM can harm the system. Several metaheuristic optimization methods are employed for optimizing DG and for DSTATCOM into the RDS. In [3], network reconfiguration with capacitors and DGs was introduced to reduce the system's high loss. Simultaneous allocation of DG and feeder reconfiguration was implemented using HSA in [4]. A similar problem can be seen in [5]. This paper aims to mitigate power loss with optimal network reconfiguration with combined DG and DSTATCOM integration into RDS. Here, in this paper, DG is considered to be operating at unity power, and DSTATCOM serves as a reactive power supply device. The study is performed with different scenarios using the recent SAR optimization algorithm on the standard IEEE 33-bus test system.

2 Problem Formulation

The main goal of this paper is to mitigate the real power loss in the RDS. The objective function (F) is expressed as given below.

$$\text{Minimize } F = \sum_{n=1}^{n} I_n^2 \cdot R_n \tag{1}$$

where R_n & I_n represent the resistance value and current flow of the nth branch, respectively.

Constraints:

i. Node voltage constraints: The bus voltage must be within the permissible range for maintaining power quality.

$$V_{\text{mininum}} \leq V_i \leq V_{\text{maximum}}$$

where V_i represents the voltage at the ith bus of the system.

ii. Node compensation limit: The active power and reactive power injected by DG and DSTATCOM should be in the following permissible range.

$$P_{\min}^{DG} \leq P_n^{DG} \leq P_{\max}^{DG}, \quad \forall n \in N$$
$$Q_{\min}^{DS} \leq Q_n^{DS} \leq Q_{\max}^{DS}, \quad \forall n \in N$$

where $P_{\min}^{DG}$ & $P_{\max}^{DG}$ are minimum and maximum values of the DG real power, respectively. Similarly, $Q_{\min}^{DS}$ &$Q_{\max}^{DS}$ are minimum and maximum values of the reactive power provided by the DSTATCOM, respectively.

iii. Radial topology constraint: To retain the radial nature of the distribution system, whenever a tie switch is closed, among all the sectionalizing switches in the loop, only one switch should be opened. The radiality network should meet the following criteria.

$$N_{\text{loops}} = (N_{\text{br}} - N_{\text{bus}}) + 1$$

where N_{loops}, N_{br} and N_{bus} represent the total number of loops, branches and buses in the system, respectively.

3 Overview of Search and Rescue Optimization Algorithm

SAR is a metaheuristic optimization algorithm that mimics human beings while searching as groups for missed people, detecting food sources, or hunting. SAR is one such group of operations [6]. The solutions to the optimization problem are equal to the human's position. These solutions are the clues of the humans' solutions.

i. *Clues:* In the search operation, the clue data is collected by the group members. When better clues are found, old clues are stacked away in the matrix M (memory matrix) to enhance search operations, while the human positions with better clues are stored in the matrix X (human positions matrix). The dimensions of both the matrices are equal, i.e. $N \times D$. N represents the number of humans' positions, and D is the dimension of the problem. Matrix C is engaged in incorporating the found clues positions. The obtained matrix C is shown in the below equation.

$$C = \begin{bmatrix} X \\ M \end{bmatrix} = \begin{bmatrix} X_{11} \cdots X_{1D} \\ \cdots \\ X_{N1} \cdots X_{ND} \\ M_{11} \cdots M_{1D} \\ \cdots \\ M_{N1} \cdots M_{ND} \end{bmatrix} \tag{2}$$

where X_{1D} is the position of the first human at the Dth dimension. Similarly, M_{ND} is the position of the Nth memory at the Dth dimension. Every solution undergoes the social phase and individual phase to create new solutions.

ii. *Social phase:* New solution is created using the following equation.

$$X_{\text{new } p,q} \begin{cases} = C_{r,q} + r1 \times \left(X_{p,q} - C_{r,q}\right), \\ \text{if } f(C_r) < f\left(X_p\right), \text{ otherwise} \\ = X_{r,q} + r2 \times \left(X_{p,q} - C_{r,q}\right), \\ \quad \text{if } r2 < SE \text{ or } q = q_{\text{rand}}, \\ \quad \text{otherwise} \\ \quad = X_{p,q} \end{cases} \tag{3}$$

where $X_{\text{new } p,q}$ is the pth human new position for the qth dimension. $C_{r,q}$ is the position of the clue found at the qth dimension. Here, $f(C_r)$ & $f\left(X_p\right)$ are the fitness function values for the solutions C_r & X_p, respectively. $r1$ and $r2$ are the random numbers within the range [−1 1] and [0 1], respectively. q_{rand} is an arbitrary integer number ranging in [1 D]. Social effect (SE) is an algorithm parameter in the range [0 1].

iii. *Individual phase:* In this phase, the human search for the new position considering his/her current position and knowledge gained from the clues found in the social phase. Thus, the new position of the pth human is evaluated by the below equation.

$$X_{\text{new}\,p} = X_p + r3 \times (C_r - C_m),\;\; p \neq r \neq m \tag{4}$$

where both the subscripts r and m are random integers ranging within [1 2N], r3 is a random number within the range [0 1].

iv. *Boundary control:* While creating a new solution, values in the dimensions might exceed the specified range. To avoid this, the below equation is utilized to modify the human position.

$$X_{\text{new}\,p,q} = \begin{cases} \frac{X_{p,q}+X_q^{\max}}{2} & \text{if } X_{\text{new}\,p,q} > X_q^{\max} \\ \frac{X_{p,q}+X_q^{\min}}{2} & \text{if } X_{\text{new}\,p,q} < X_q^{\min} \end{cases} \tag{5}$$

where $X_q^{\min}\&X_q^{\max}$ respectively, represent the minimum and maximum threshold values for the qth dimensions.

v. *Updating data and positions:* If the fitness function value in position $X\text{new}_p(f(X_{\text{new}p}))$ is better than the preceding position $f(X_p)$, it will be stored in the memory matrix (M) at a random position using Eq. (6). Further, they are updated as the human new position as in Eq. (7). Otherwise, search continuous with the next position, and no updating is required.

$$M_n = \begin{cases} X_p, & \text{if } f\left(X_{\text{new}p}\right) < f\left(X_p\right) \\ M_n, & \text{otherwise} \end{cases}, \tag{6}$$

$$X_p = \begin{cases} X_{\text{new}\,p}, & \text{if } f\left(X_{\text{new}\,p}\right) < f\left(X_p\right), \\ X_p, & \text{otherwise} \end{cases} \tag{7}$$

where M_n is the memory position of the nth clue in the matrix M and subscript 'n' is a random integer ranging within [1 N]. This changing memory method enhances the algorithm diversity and its ability to find the global optimum solution.

vi. *Abandoning clues:* If a human cannot find better clues in his/her search operations, the human gets out of the current position and finds a new position. This behaviour is modelled using a parameter named unsuccessful search number (USN). Initially, the USN number for each human is set to 0. The USN is evaluated by using the following equation.

$$USN_p = \begin{cases} USN_p + 1, & \text{if } f\left(X_{\text{new}\,p}\right) < f\left(X_p\right), \\ 0 & \text{otherwise} \end{cases} \tag{8}$$

$$X_{p,q} = X_q^{\min} + r4 \times \left(X_q^{\max} - X_q^{\min}\right), \quad j = 1, \ldots, D \tag{9}$$

where $r4$ is a random number ranging within [0 1].

vii. *Control parameters:* SAR algorithm has two control parameters; they are SE and MU. According to the research carried out in [10], SE = 0.05, MU = 70 × D are advisable values for best results.

4 Implementation of SAR Optimization Algorithm for the Power Loss Minimization

The computation procedure of the SAR method for the power loss minimization for reconfiguring the network with DG and DSTATCOM units' presence is as follows. This approach is a complex combinational optimization problem.

Step 1:	Input the system data (i.e. line data, bus data and limitations) and SAR algorithm parameters
Step 2:	Compute the base case real power loss of the system using the backward–forward sweep load flow method
Step 3:	Initially, generate a random population of human positions (i.e. DG size, DSTATCOM size and locations) at different dimensions (number of devices) in the search space. Set the iteration count = 0
Step 4:	Evaluate the fitness function (i.e. power loss) for each human position
Step 5:	A group of humans with the best fitness function are stored in matrix M, and the remaining are store in matrix *X*
Step 6:	In search space, update the humans' position in the matrix *X* using Eq. (3). Compute the USN value for each human solution
Step 7:	In the individual phase, again update the humans' position in the matrix *X* using Eq. (4)
Step 8:	Check for the boundary limit (for DG, DSTATCOM size limits and bus positions with min. and max. range) for human positions in both social and individual phases. If it exceeds the limit, update the new positions using Eq. (5)
Step 9:	In the current iteration, '*i*' suppose the human position's fitness value is better than the previous fitness value. The new matrices *M* and *X* are updated using Eqs. 6 and 7, respectively
Step 10:	Compute the USN value for the updated human position in both the phases using Eq. (8)
Step 11:	Check if the USN is greater than the MU value, then modify the human's position using Eq. (9)
Step 12:	Check for the maximum iteration limit. If yes, go to the next step. Otherwise, increase the iteration count by one and go to step 4

(continued)

(continued)

Step 13:	Print out the best human position results such as DG and DSTATCOM units' best capacity and corresponding bus positions
Step 14:	Place the DG and DSTATCOM units obtained from step 13 into RDS, then reconfigure the network topology by changing the switches' position to minimize the power loss without violating the radiality constraint
Step 15:	Display the switches opened for optimal network topology with minimum real power loss in the network

5 Numerical Test Results and Discussions

The test system considered for the study purposes in this paper is the IEEE 33-bus system. It is a standard network voltage of 12.66 kV, consisting of 33 nodes, 32 sectionalizing switches and five tie-switches. The system total active and reactive power loads are 3715 KW and 2300 KVAr, respectively. The standard IEEE 33-bus RDS before reconfiguration is shown in Fig. 1.

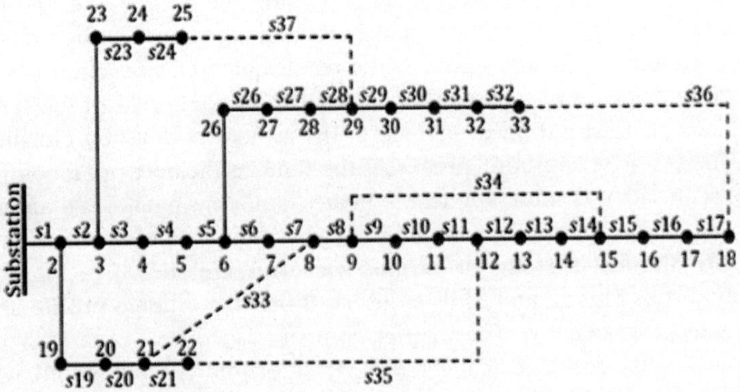

Fig. 1 Original network topology of IEEE 33-bus RDS

The system performance with simultaneous integration of DG and DSTATCOM units considering reconfiguration using the SAR algorithm is analysed on the IEEE 33 bus test system to show its efficacy. The simulation is performed in six different scenarios, which are as follows.

Scenario I:	The test system without any changes (base case)
Scenario II:	Only reconfiguring the test system by opening/closing the sectionalized and tie-switches positions
Scenario III:	A system with integration of DG units only
Scenario IV:	Integrating only DSTATCOM units into the system
Scenario V:	Simultaneously integration of DG and DSTATCOM units in the system
Scenario VI:	System reconfiguration with combined DG and DSTATCOM units installation

All the scenarios are programmed using MATLAB software, and Table 1 shows the simulation results obtained.

From Table 1, it is observed that the real power loss is reduced to 33.85, 65.50, 33.42, 92.94 and 92.54% for scenarios II, III, IV, V and VI, respectively concerning the base case. Similarly, the reactive power loss is reduced to 28.44, 64.60, 33.03, 91.44 and 90.75% for scenarios II, III, IV, V, VI, respectively. This study shows that high power loss reduction has been obtained using scenario V (i.e. the case with simultaneous DG and DSTATCOM placement). In scenario VI, the percentage of power loss reduction is nearly equal to the result obtained in scenario V. In RDS, under the circumstances of fault contingencies, the whole or part of the system may run out of power. Restoration of service to the system is done by closing the tie-switches. The service restoration process is the same as the network reconfiguration. Therefore, scenario V is more applicable in any fault contingencies than utilizing it for power loss minimization in the system. The graph of voltage profile and power losses plot of the test system for various scenarios are shown in Figs. 2 and 3; respectively. From Figs. 2 and 3, it is observed that the voltage profile and power losses obtained for scenario V are better compared to other scenarios' results. In Table 2, the results obtained using the SAR algorithm are compared with other optimization algorithms for various scenarios. Algorithms include genetic algorithm (GA), refined genetic algorithm (RGA), Krill Herd algorithm (KHA), bacterial foraging optimization algorithm (BFOA), cuckoo search algorithm (CSA), intersect mutation differential evolution(IMDE), gravitational search algorithm (GSA) and fuzzy-BFOA.

6 Conclusion

This paper address the various scenarios of the power loss reduction approach in RDS. IEEE 33-bus test system is utilized for study purposes. The objective of the

Table 1 Simulation results of IEEE 33-bus test system using SAR algorithm

Scenario	P_{LOSS} (KW)	Q_{LOSS} (KVAr)	V_{min} (p.u)	Switches opened	Size of DG in KW (Bus number)	Size of DSTATCOM in KVAr (Bus number)
Scenario I (base case)	211	143.12	0.9038	s33, s34, s35, s36, s37	….	….
Scenario II (Reconfiguration only)	111.18	106.07	0.9105	s4, s14, s15, s22, s33	….	….
Scenario III (DG only)	72.79	50.66	0.9684	s33, s34, s35, s36, s37	798 (13)	….
					1071.9 (24)	
					1049.6 (30)	
Scenario VI (DSTATCOM only)	140.48	95.84	0.9337	s33, s34, s35, s36, s37	….	341.1 (13)
						675.91 (26)
						727.87 (30)
Scenario V (Both DG and DSTATCOM)	14.89	12.25	0.9911	s33, s34, s35, s36, s37	773.6 (14)	555.15 (6)
					1130.1 (24)	244.31 (17)
					1074.6 (30)	715.75 (30)
Scenario VI (Reconfiguration with combined DG and DSTATCOM)	15.73	13.23	0.9876	s7, s34, s35, s36, s37	773.6 (14)	555.15 (6)
					1130.1 (24)	244.31 (17)
					1074.6 (30)	715.75 (30)

optimization problem in this paper was to reduce real power loss in the system. A recent SAR optimization is utilized in finding the optimal solution in this paper. The simulation results obtained for simultaneous placement of DG and DSTATCOM units into the system are much encouraging compared to other scenarios. Reconfiguration is a nonlinear constrained optimization problem. After implementing scenario V and reconfiguring the system, the simulation results obtained are nearly equal to the results obtained as in scenario V. Optimal deployment of DG and DSTATCOM units placement in RDS without violating the limits will lead to positive benefits to the system. Significant technical benefits include power loss reduction and enhance voltage profile. In case of fault contingencies in the system, reconfiguration is more helpful in maintaining a reliable power supply with reducing losses. The

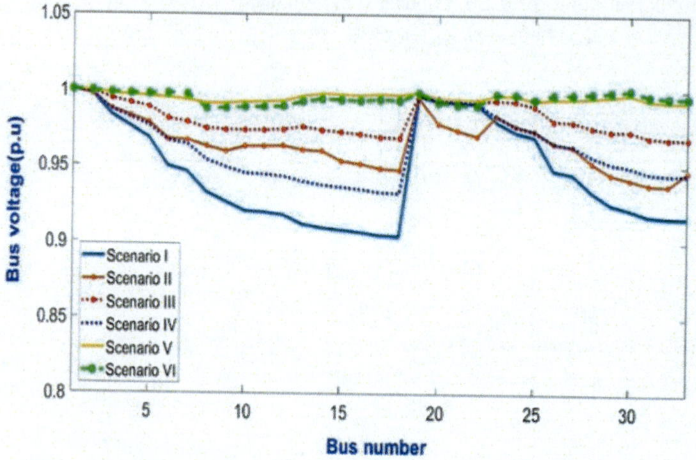

Fig. 2 Voltage profile of the test system for various scenarios

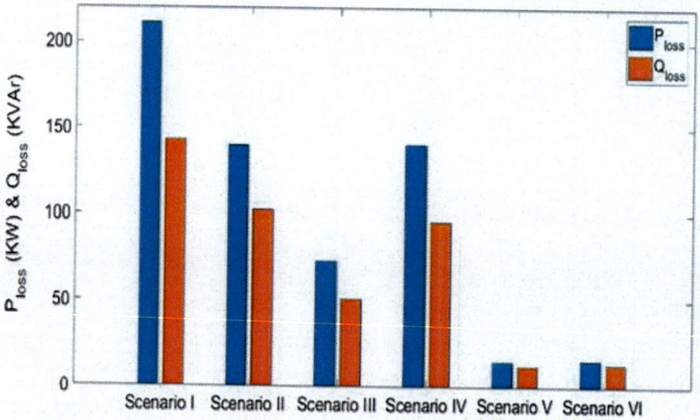

Fig. 3 Plot of real and reactive power losses for various scenarios

test system results obtained are compared with the other optimization methods for various scenarios to show the SAR algorithm efficacy.

Table 2 Comparative simulation results of IEEE 33-bus test system

Scenario	Method	P_{loss} (KW)	V_{min} (p.u)	Switches opened	Size of DG in KW (Bus number)	Size of DSTATCOM in KVAr (Bus number)
Scenario II (Reconfiguration only)	SAR	111.18	0.9378	s7, s9, s14, s32, s37	….	….
	GA [7]	141.6	0.931	s9, s28, s33, s34, s36		
	RGA [7]	139.46	0.9351	s7, s9, s14, s32, s37		
Scenario III (DG only)	SAR	72.79	0.9684	s33, s34, s35, s36, s37	798 (13), 1071.9 (24), 1049.6 (30)	….
	CSA [8]	75.16	0.9712	s33, s34, s35, s36, s37	776.1 (13), 1201 (24), 1302.6 (30)	
	KHA [9]	75.49	0.961	s33, s34, s35, s36, s37	810.7 (13), 836.8 (25), 841 (30)	
	BFOA [10]	98.3	….	s33, s34, s35, s36, s37	630 (17), 1110 (24), 1040 (30)	
	IMDE [11]	84.28	0.971	s33, s34, s35, s36, s37	840 (14), 1130 (30)	
Scenario VI (DSTATCOM only)	SAR	140.48	0.9337	s33, s34, s35, s36, s37	….	341 (13), 675.9 (26), 727.87 (30)
	BFOA [10]	144.04	….	s33, s34, s35, s36, s37		350 (18), 820 (30), 280 (33)

(continued)

Table 2 (continued)

Scenario	Method	P_{loss} (KW)	V_{min} (p.u)	Switches opened	Size of DG in KW (Bus number)	Size of DSTATCOM in KVAr (Bus number)
Scenario V (Both DG and DSTATCOM)	SAR	14.89	0.9911	s33, s34, s35, s36, s37	773.6 (14), 1130 (24), 1074 (30)	555.1 (6), 244.3 (17), 715.7 (30)
	BFOA [10]	41.41	….	s33, s34, s35, s36, s37	540 (17), 160 (18), 900 (33)	160 (18), 540 (30), 340 (33)
	GABC [12]	93.72	….	s33, s34, s35, s36, s37	1100 (28), 130 (29), 610 (30)	300 (16), 150 (17), 150 (18)
Scenario VI (Reconfiguration with combined DG and DSTATCOM)	SAR	15.73	0.9876	s7, s34, s35, s36, s37	773.6 (14), 1130 (24), 1074 (30)	555.1 (6), 244.3 (17), 715.7 (30)
	Fuzzy-BFO [13]	45.65	….	s9, s14, s27, s33, s36	758 (14), 1045 (24), 987 (30)	150 (8), 150 (18), 300 (30)

References

1. Civanlar S, Grainger J, Yin H, Lee S (1989) Distribution feeder reconfiguration for loss reduction. IEEE Trans Power Del 4(2):1492–1498
2. Baran ME, Wu FF (1989) Network Reconfiguration in distribution systems for loss reduction and load balancing. IEEE Trans Power Deliv 4(2):1401–1407
3. Hung DQ, Mithulananthana N, Bansal RC (2013) A combined practical approach for distribution system loss reduction. Int J Ambient Energy 1–9
4. Srinivasa RR, Kumudhini R, Satish K, Narasimham SVL (2013) Power loss minimization in distribution system using network reconfiguration in the presence of distributed generation. IEEE Trans Power Syst 28:317–325
5. Davodi A, Esapour K, Zare A, Rostami M-A (2015) A modified KH algorithm to solve the optimal reconfiguration problem in the presence of distributed generations. J Intell Fuzzy Syst 28:383–391
6. Amir S, Behrouz A, Saeed AG, Miguel AS, Adriana G (2019) A new optimization algorithm based on search and rescue operations. Math Problems Eng 1–23
7. Rao RS, Ravindra K, Satish K, Narasimham SVL (2013) Power loss minimization distribution system using network reconfiguration in the presence of distributed generation. IEEE Trans Power Syst 28(1):317–325

8. Moumita G, Kumar S, Mandal S, Mandal KK (2017) Optimal sizing and placement of DG units in radial distribution system using cuckoo search algorithm. Int J Appl Eng Res 12:362–369
9. Sultana S, Roy PK (2015) Krill herd algorithm for optimal location of distributed generator in radial distribution system. Appl Soft Comput J
10. Imran AM, Kowsalya M (2014) Optimal distributed generation and capacitor placement in power distribution networks for power loss minimization. In: International conference on advances in electrical engineering (ICAEE), Vellore, India, pp 1-6
11. Khodabakhshian A, Andishgar MH (2016) Simultaneous placement and sizing of DGs and shunt capacitors in distribution systems by using IMDE algorithm. Int J. Elect Power Energy Syst 82:599–607
12. Dixit M, Kundu P, Jariwala HR (2017) Incorporation of distributed generation and shunt capacitor in radial distribution system for techno-economic benefits. Eng Sci Technol Int J 20(2):482-493
13. Mohammadi M, Mohammadi Rozbahani A, Bahmanyar S (2017): Power loss reduction of distribution systems using BFO based optimal reconfiguration along with DG and shunt capacitor placement simultaneously in fuzzy framework. J Cent South Univ 24:90–103

Slope Instability Detection Using IoT with the Help of Ecological Indicators

Priyanka, Sunil Semwal, Piyush Dhuliya, Vikalp Joshi, and Diwaker Pant

Abstract Land sliding is a common problem of hilly areas; the problem is very critical when land sliding occurs at road side which leads serious damages. The slope instability detection is a sort of prediction which tells us the conditions which can make the slope fail resulting in land slide. It involves movement of the entire debris from the slope. Internet of things (IoT) gaining popularity nowadays and detection of slope instability using Internet of things (IoT) have the capability to quickly capture, process and transmit data for effective damage control. This in turn can save precious lives and huge economic losses. In this paper, the detection of slope instability is materialized with the help of Internet of things (IoT) using a combination of pore pressure and moisture sensor. Detection system involves a sensor column consisting of pore pressure & moisture sensor and data logger with an antenna. A data logger (Mote) collects the field data by the sensors. Real-time software (visual basic) represents the quantity of the soil parameters of the slope in the form of an online graph.

Keywords IoT · Sensor column · Slope instability · Landslide · Data logger

1 Introduction

As per Geological Survey of India, 12.6% land area of India, i.e. about 0.42 million km^2, is susceptible to landslide hazards [1]. Following Map Fig. 1 show cases the landslide prone states of India.

The mass movement of debris results in slope instability such as land sinking, landslides and landslips creep movement. This is mainly due to three factors or a

Priyanka (✉)
Graphic Era Hill University, Dehradun, India

S. Semwal · P. Dhuliya · D. Pant
Tula's Institute, Dehradun, India

V. Joshi
Sara Industries Pvt Ltd, Dehradun, India

S. S. Dash et al. (eds.), *Sixth International Conference on Intelligent Computing and Applications*, Advances in Intelligent Systems and Computing 1369,
https://doi.org/10.1007/978-981-16-1335-7_5

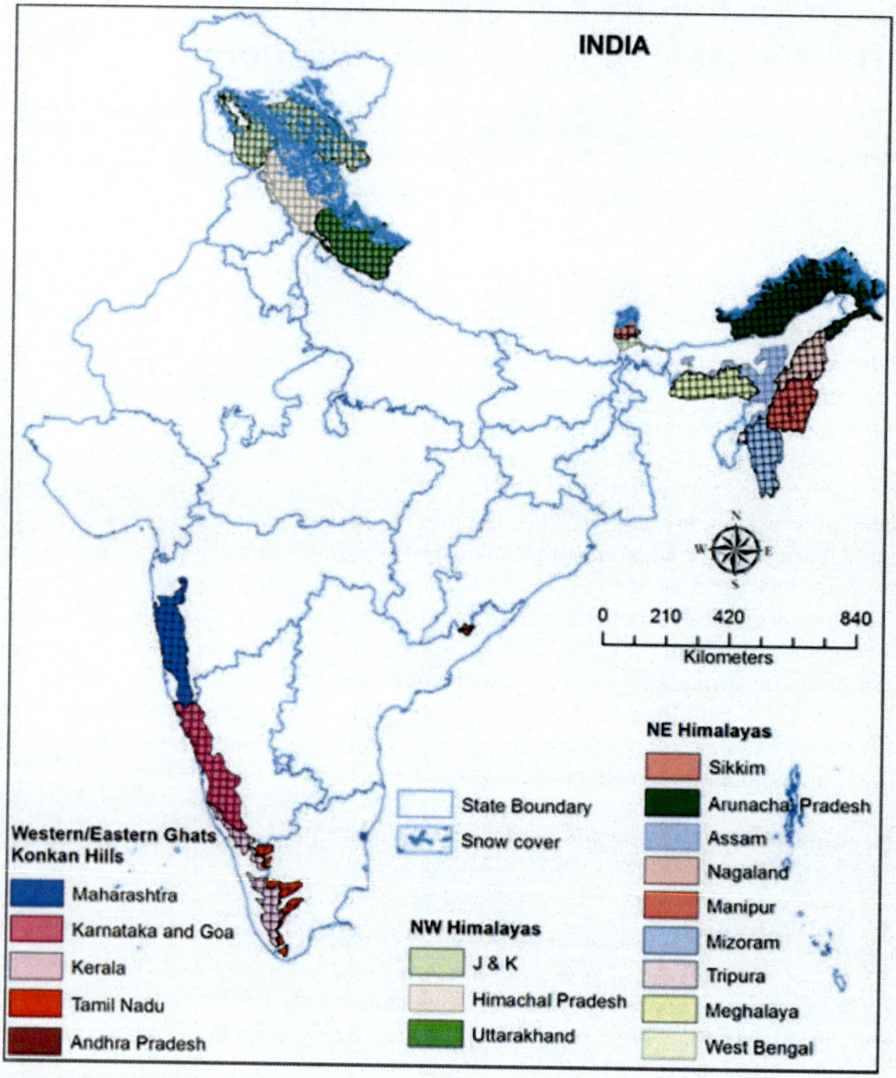

Fig. 1 Major landslide prone areas of India (0.42 Million km^2) [1]

combination of them, i.e. water, seismic activity or volcanic activity [2]. In India, this is mainly due to extreme rainfall and earthquake. There are various motivators for the rainfall and earthquakes which lead to slope instability they are numerated below.

a. ***Factor of safety of slope***
The factor of safety (FS) is defined as the ratio of shear strength to shear stress. When this ratio is greater than 1.25, the slope is considered safe. When this

ratio is close to 1, the slope is close to failure, if FS is less than 1, the slope stability is failed and unsafe. It means the greater the ratio the safer it is [3].

b. ***Soil shear strength***
The resistance to internal sliding in soil is measured by soil shear strength. For slope, the shear strength is pressure or load per unit area that it can withhold before shearing failure occurs [1, 4].

c. ***Soil Shear stress***
The shear stress is resistance of slope which opposes the magnitude of shear strength [2].

d. ***Gravitational force***
It is a natural phenomenon in which physical bodies with relevant mass appear to attract each other with a force proportional to the product of their masses [2].

e. ***Pore water pressure***
It is the pressure of the water in the voids of soil [2, 5].

f. ***Soil Moisture***
This is amount of water absorbed by soil [5].

g. ***Force due to seepage of water***
As water moves through the soil skeleton, there occurs a frictional drag which results in seepage force [5].

h. ***Snow melt***
It is a key mechanism for the initiation of land slide in cold mountainous regions. This can occur when there is sudden increase in temperature which leads to rapid melting of the snow. The melted water infiltrate into the ground, having impermeable layers below the surface due to still-frozen soil or rock, it leads to increases in pore water pressure, ultimately resulting in landslide activity [1].

i. ***Dams***
Due to increasing geological pressure from rising water large dams increase the possibility for earthquakes [2, 6].

j. ***Seismicity***
Seismic shaking and pore water pressure generation are two separated but interconnected process during earthquakes due to which landslides occur. The gravitational load on the slope changes due to seismic shaking in the passage of the earthquake waves through the rock and soil producing a complex set of accelerations resulting in land slide [2].

k. ***Shocks and Vibrations***
Earthquakes of 4.0 magnitudes and above weaken the slopes due to stress creation which in turn produces the largest and most destructive landslides. Vibration from blasting, pile driving and traffic are some of the activities that have been the cause for a number of slides [2].

l. ***Toe removals***
Water-level changes along coastlines, earth dams and the banks of lakes, reservoirs, canals and rivers result in slope saturation. This is due to intense rainfall, snowmelt and changes in groundwater levels resulting in landslides [2].

m. ***Excavations or cut increase***
The slope angle fill operations carried out without specialist advice [2].

In previous research, the authors used several parameters to find out the slope instability. In this paper, we have considered soil moisture and pore water pressure to detect slope instability as the potential parameter to identify the slope instability. A data logger system has been designed to collect the above parameter and identify the critical land sliding area in hilly region [7]. This system is IoT enabled so the communication module can transmit the collected data on web/mobile. Soil moisture and pore water pressure sensors are associated with a sensor column and mote. A mote is device which have an antenna and capable to transmit the field data to data management Centre (DMC). A mote has microcontroller, power unit and signal processing device.

2 Methodology

Laboratory setup has prepared according to Fig. 2, and the methodology of this system is divided into hardware and software module. The sensors' column is placed inside the slope in horizontal position (Fig. 3), and the sensor system is capable to collect the slope. The sensor column associated with various sensors like moisture sensor, pour water pressure sensor and small mote. The mote has microcontroller, a power-generated solar battery and communication unit with small antenna. A mote

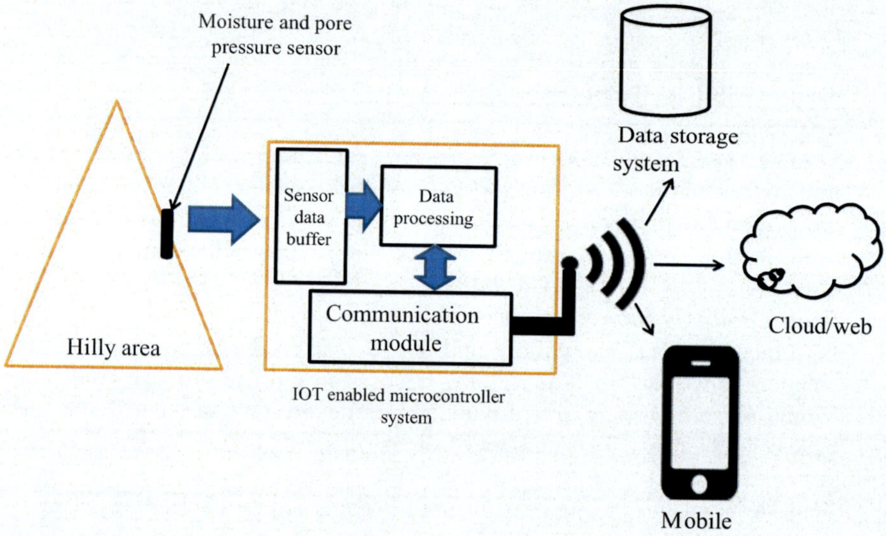

Fig. 2 Framework of proposed system

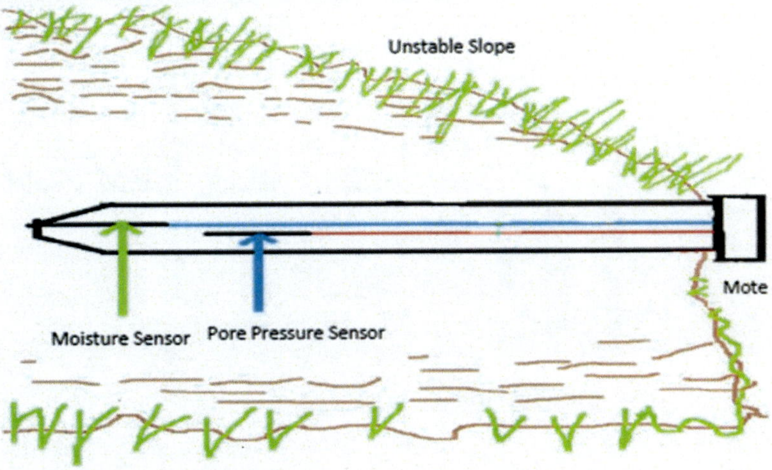

Fig. 3 Sensor column placed in the slope

transmits analog data from sensors to data collection system or data management centre (Fig. 3).

Soil Moisture sensor

It measures the content of water in soil. The dielectric constant of the soil is measured by the sensor to find its volumetric water content (VWC). VWC is measured by utilizing capacitance/frequency domain technology. Taking into account that the dielectric constant of air or soil minerals is much lower than that of water, and the VWC can be measured effectively using the dielectric constant of the soil [8].

Working of Soil Moisture Sensor Systems

Capacitance is used by soil moisture sensor to measure dielectric permittivity of the surrounding medium, i.e. soil, which in turn is a function of VWC. A voltage proportional to the dielectric permittivity is created by the sensor, enabling the measurement of water content of the soil [8]. The sensor takes into account the entire length of the sensor thus averaging the water content across the entire length of the sensor (Fig. 4).

Small–scale slope model

To study the effects of water or rain on the slope, a prototype is created in the laboratory which is small-scale model test box having dimension of $1 \times 1 \times 1$ft and 6 kg of soil. Test box consists sand and soil which is separated (Fig. 5).

For the small-scale slope instability test used a small-scale test box, an artificial slope is created on it. First the sensor columns were installed inside the box. The soil moisture sensor placed in the soil slope. Sensor transmits field data to mote & mote transmits analog data to PC via RF communication.

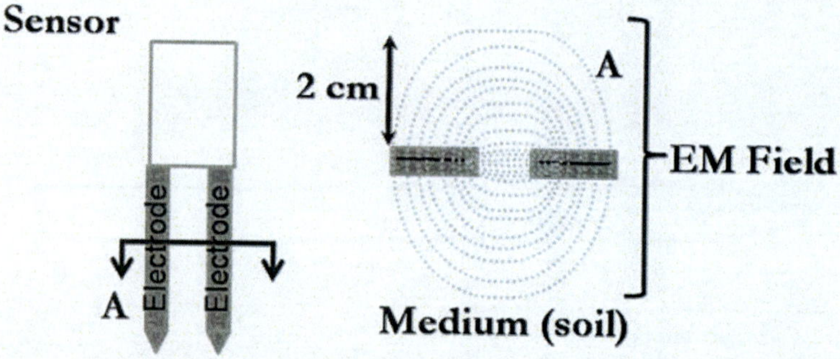

Fig. 4 Electromagnetic field lines interpreted in the sensor

Fig. 5 Small-scale experimental setup

3 Experimentation and Results

The calibration of moisture sensor and pore water pressure is done. For calibration, we take a cylindrical box of diameter 6 cm and height 5 cm (Table 1).

Theoretical results of calibration:

$$\begin{aligned}\text{Volume of soil container}(V_{sc}) &= \pi r^2 h \\ &= 3.14 \times 3^2 \times 5\,\text{cm}^3 \\ &= 141.3\,\text{cm}^3\end{aligned}$$

Table 1 Quantity of sample in a soil container for calibration

S.No.	Sample	Weight (gm)	Density (gm/cc)	Volume (cc)
1	Wet soil	302	2.33	129.61
2	Initial weight of water jar	84	1.00	84.00
3	Final weight of water jar	56	1.00	56.00
4	Weight of water	28	1.00	28.00

$$\begin{aligned}\text{Density of wet soil}(\text{d}_{ws}) &= (\text{Mass of soil}(M_s) \\ &\quad + \text{Mass of water to be added in soil}(M_w) \\ &\quad /(\text{Volume of soil container}(V_{sc})) \\ &= (302 + 28)/141.3\ \text{gm/cm}^3 \\ &= 2.33\ \text{gm/cm}\end{aligned}$$

$$\begin{aligned}\text{Volume of wet soil}(V_{ws}) &= (M_s)/(\text{d}_{ws}) \\ &= 302/2.33 \\ &= 129.61\ \text{cm}^3\end{aligned}$$

$$\begin{aligned}&\%\ \text{Moisture in Wet soil}(m_{ws}) \\ &= \text{volume of water}(V_w)/\text{volume of wet soil}(V_{ws}) \times 100 \\ &= 28/129.61 \times 100 \\ &= 21.60\%\end{aligned}$$

So, the theoretical calibration points the % moisture in soil in a given container to be 21.60.

The theoretical value is very near to the practical value which we have observed in laboratory (Fig. 6), and it shows that is sensor system which is well-calibrated.

A small-scale slope model (SSSM) was constructed to measure pore pressure and soil moisture using an IoT. Experiments were performed to check the operational ability of the sensors. First the sensor columns were installed inside the box for the SSSM test. The moisture sensor placed at the middle of SSSM while pore water pressure column was placed at the rightmost side. The SSSM contained sand and soil which is separated by an insulator sheet. The online graph also represents initial conditions of experiment at time 10:10:02 am. This time moisture around 1–2% and pore water pressure between 0.2–0.6 kPa, and the water contains very less around 0.028 L.

Figure 7 shows experiment starts 10.10.02 am with soil sample. Amount of soil is 2.25 kg. After some periods, the percentage of moisture increases as water is added. Graph shows that slope instability starts at 25% soil moisture. After sometime, slowly fixed amount of water is added to SSSM test box which increments soil moisture

Fig. 6: 21% soil moisture shows on LCD screen of the mote

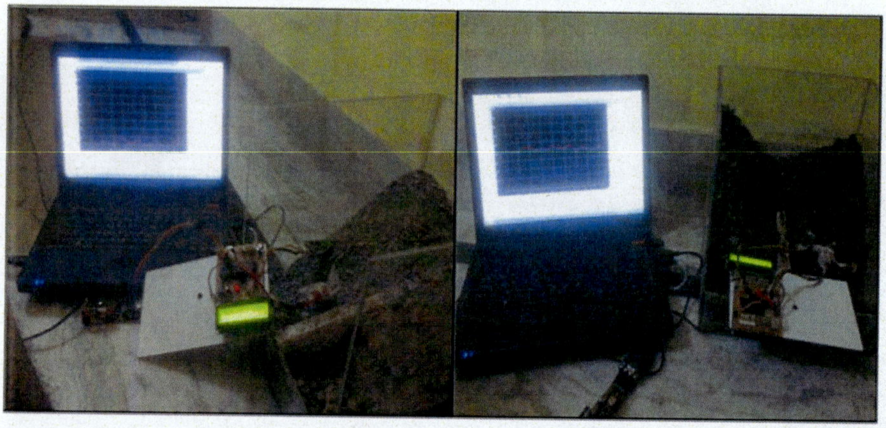

Fig. 7 Online graph between soil moisture and pore water pressure at different time

and pore water pressures. The time duration of water added is approximate 2.00 min. Another picture shows the soil moisture and pore water pressure, respectively, around 048-055 and 4.762–5.00 at time 10:12:20am. The same process again repeated. The water is continuously added for around 4 min to SSSM.

Total water added 0.8 liter in SSSM. Soil absorbs more amount of water, so at the end, the moisture level is high. The cracks appear in the slope at 55% soil moisture.

Instability of slope ends at 70% soil moisture. Figure 8 is the snapshot of the software interface of the proposed system which shows graph representation of the pore water pressure and soil moisture. This interface is developed in visual basics, which acquire the real-time data.

Table 2 represents the change in soil moisture and pore pressure and the effects on the slope during rain fall. It has been observed that when moisture reached 68%

Fig. 8 Graph between soil moisture and pore water pressure

Table 2 Changes in soil parameters

Time	Status of slope	Change in soil moisture %	Change in pore water pressure (KPa)	Water (ml)
10:10:02	Start of experiment	1.0–3.0	0.2–0.6	0.028
10:10:30	Start increasing the water	3.0–15	0.6–0.9	0.068
10:10:58	Water seepage reaches middle of slope	15.0–28.0	0.9–1.1	0.178
10:11:00	Saturation of slope's toe	28.0–35.0	1.2–1.9	0.328
10:11:02	Water reaches at other end of the box	35.0–43.0	1.9–2.7	0.628
10:11:50	Water level increases again	43.0–58.0	2.7–3.3	0.678
10:16:00	Start of landslide	58.0–68.0	3.3–4.9	0.728
10:16:12	End of landslide	68.0–70.0	4.9–5.5	0.828

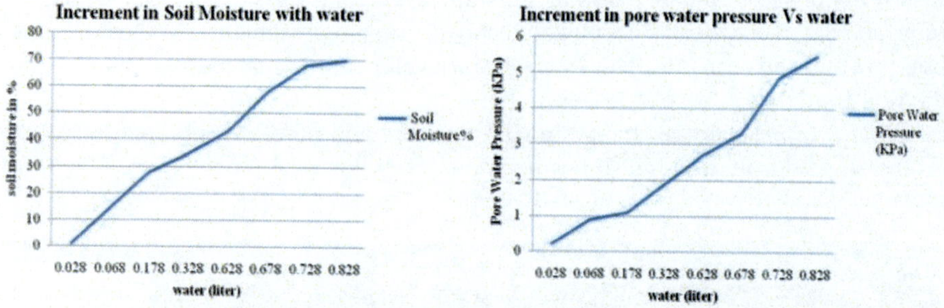

Fig. 9 Increment of soil moisture and pore water pressure with water contain

and pore pressure is 4.9 KPa, then the land start sliding. Figure 9 is the graphical representation of Table 2 which shows how the amounts of moisture and pore water pressure changes with time.

4 Conclusion

It has been observed that by using IoT and sensor column, we can get the information for the land slide in different condition. The system is useful for hilly region where the land sliding occurs frequently. Early warning of the land sliding is very useful and can save lots of lives and also economic losses. To make this more effective, the sensitivity of the sensors is very important. Increasing more sensor columns at the slopes can lead more accuracy. As there will be more sensor columns, the predictions will also enhance at various points of slope.

References

1. National Landslide Risk Management Strategy, National Disaster Management Authority Ministry of Home Affairs Government of India, September (2019)
2. Lynn M. Highland, Peter B (2008) In: The landslide handbook—a guide to understanding landslides, pubs.er.usgs.gov
3. Seed RB, James KM, Bolton Seed H (1990) Kettleman hills waste landfill slope failure. II: stability analyses. J Geotech Eng 116.4:669–690
4. Eid HT (2011) Shear strength of geosynthetic composite systems for design of landfill liner and cover slopes. Geotext Geomembr 29(3):335–344
5. Hammer TA, Fosså KT (2006) Influence of entrained air voids on pore water pressure and volume change of concrete before and during setting. Mater Struct 39(9):801–808
6. Zoback MD, Gorelick SM (2012) Earthquake triggering and large-scale geologic storage of carbon dioxide. Proc Natl Acad Sci 109(26):10164–10168

7. Pant D, Sandeep V, Piyush D (2017) A study on disaster detection and management using WSN in Himalayan region of Uttarakhand. In: 2017 3rd International conference on advances in computing, communication and automation (ICACCA)(Fall). IEEE
8. Yoshikawa K, Pier PO (2005) Comparing unfrozen water content measurements of frozen soil using recently developed commercial sensors. Cold Reg Sci Technol 42.3:250–256

Predicting Student Grade Using Machine Learning Techniques

S. Kaliraj, V. Sivakumar, D. Vivek, B. Hariharan, and R. Siva

Abstract In higher educational organizations, early evaluation prediction is a significant place of enthusiasm as it permits educators to improve performance in their courses by giving uncommon consideration at the beginning stages. AI procedures can be used for evaluations in various courses. In any case, the demonstration of these methods is exceptionally needful on the nature of information that made the determination of model a difficult activity. In this way, we assess distinctive AI procedures for students grade evaluation expectation. The anticipated evaluations by these strategies envision vulnerability on understudy learning and can be utilized for certainty gains, grade qualification arranging, customized prompting and to empower teachers to distinguish potential evaluation who may require help with relevant courses. Students can access their chances to enter certain universities, get scholarship if needed, and students can take timely action to get better grade on final exam. Parents can understand where their children need extra support and coaching and provide it in a timely manner to ensure good results for the final exam. Schools can identify problematic students early and take action to ensure that students pass exams with good grades to improve preparation.

Keywords College education · Machine learning · Results prediction · Multi linear algorithm

1 Introduction

The educational institute's main goal is to provide good quality of education to the students. As the prediction of student grade is more important for educational institutions because it is the ability to meet student needs [1], education is predominant for student life because these are the major factors that affect the future of students.

S. Kaliraj (✉) · D. Vivek · B. Hariharan · R. Siva
SRM Institute of Science and Technology, Kattankulathur, Tamilnadu, India

V. Sivakumar
Vel Tech Rangarajan Dr. Sagunthala R&D Institute of Science and Technology, Avadi, India

S. S. Dash et al. (eds.), *Sixth International Conference on Intelligent Computing and Applications*, Advances in Intelligent Systems and Computing 1369,
https://doi.org/10.1007/978-981-16-1335-7_6

Machine learning is an ability of instinct learning system from past experiences and improves the performance of students [2]. Currently, education plays a major role in modern society because education brings awareness to people [3]. To improve the student's performance using machine learning gained major attention. If the performance of the student is predicted well in advance, then we can improve student performance in an educational institution [4]. So, we are using different regression algorithm to evaluate the student performance such as multilinear regression, lasso regression, ridge regression, effective net regression, random forest, and K-nearest neighbor. All these regression algorithms were under the subset of supervised machine learning. Regression algorithm predicts the product values based on inserted features from the data feed in the system. These features have been collected from school to predict the final exam grade. The best feature is taken into account and put into a different machine learning algorithm.

The objective of the project is to find out a suitable machine learning algorithm for the student grade and predict the grade of a student in advance. The grade of a student identified ahead of time, and then, the educational institution can support students in a weak zone at the right time by using the pedagogies. Parents can also support their children to ensure good results for the final exam.

2 Motivation

All this work was motivated by students to perform well in their academics. All institutions consider that student's performance is one of the most crucial and important parts because most of the institutions have the best record of academic performances and mainly focus on generating graduates with good academic performances as well as extracurricular activities. This research aims to predict student performance to engage in improving ranking.

The main objectives of this study are as follows:

- Failure and success rate of students are predicted.
- Regression algorithm is used for student data.
- Final grade of student is calculated.

3 Problem Description

In education system, nowadays all are supported to study to improve their studies and performance in the examinations. If the performance of students was predicted in an advance than required, suggestion and support can be given to students by teachers and parents. Teachers can know the student performance in advance, and then, they can advise students personally and interactively. To predict the actual grade, we are using machine learning. Among the machine learning algorithms, selecting the accurate algorithm gives the best result.

4 Literature Review

Predicts the students' academic performance on building course: Comparing the foreseeing grade with various numerical models. The work introduced predicts the presentation at degree level and course level. Strikingly, it recommends a sort of upper destined for the exactness that can be accomplished while anticipating execution toward the finish of a degree. The four numerical models are utilized, namely different direct relapse, multilayer discernment systems, spiral premise capacities and backing vector machines to foresee students academic presentation in a building course [5]. It has dealt with the information of 323 college understudies who took the course of the elements at Utah State University in four semesters. The four models have accomplished a normal forecast exactness of 81–91%. This work shows that past imprints can anticipate the evaluation of a course with high exactness.

Mohan [6] fundamentally utilizes two sorts of methods for the general expectation of the understudies' presentation over an immense volume of significant worth and information. Those methods are learning analytics and predictive analytics. Learning analytics for the most part manages the information assortment and information preprocessing stage where the necessary information is based on the forecast model gathered from the CBSE schools, and they have utilized MySQL worker for putting away the information. At that point, the information pre-preparing step like information cleaning, information change and so forth is finished by utilizing the apache HIVE system. From the pre-handled information, required data is found out utilizing the MapReduce calculation in the Hadoop system. At that point comes the predictive investigation part where the genuine expectations are made utilizing the numerous direct relapse model.

Vyas [7] utilizes a choice tree model for foreseeing the academic performance of understudies. The information required for building the model is gathered, and preprocessing is performed where the nonstop qualities are changed over and the invalid qualities are dispensed with. At that point by utilizing the CART calculation, the choice tree forecast model is being manufactured. Furthermore, the understudies with horrible showing are anticipated out. Alana M is fundamentally focusing on two methodologies, namely clustering and relapse. Bunching is utilized for pre-preparing, and relapse is utilized for making forecasts. In bunching, information is assembled, and the information order is being finished. The relapse technique is for doing the forecasts.

Huda Al-Shehri [8], The work introduced in two forecast models for the assessment of students performance in the last assessment. The work was done on the dataset gave by the University of Minho in Portugal, and it comprises 395 information tests. The performance of understudies can be helpful for early insurances, moment activities or choosing an understudy that is good for a specific errand. The greater part of the work on the dataset utilized the K-nearest neighbor calculation and accomplished low outcomes, and the support vector machine calculation was once in a while utilized, which happens to be an exceptionally mainstream and incredible expectation method. To guarantee better examination, we applied both the support vector machine calculation and K-nearest neighbor calculation on the dataset to foresee the students evaluation and afterward thought about their exactness. Observational investigations result demonstrated that support vector machine accomplished somewhat better outcomes with a connection coefficient of 0.96, while the K-nearest neighbor accomplished a relationship coefficient of 0.95.

5 Proposed Framework

Figure 1 is the overall system architecture of the system. Our proposed framework will be applied here to predict the student performance.

Dataset. The dataset which is taken for predicting student grade is taken to the database. It is the place where the data stores.

Data preprocessing. The dataset consist of null values and more features. By preprocessing the data, we are analyzing the dataset completely using Python inbuilt libraries of data visualization. Data visualization helps to understand the dataset by graphical representation. By using null functions of average value of columns, the null values are replaced. The columns consisting of more null values can be removed (Fig. 2).

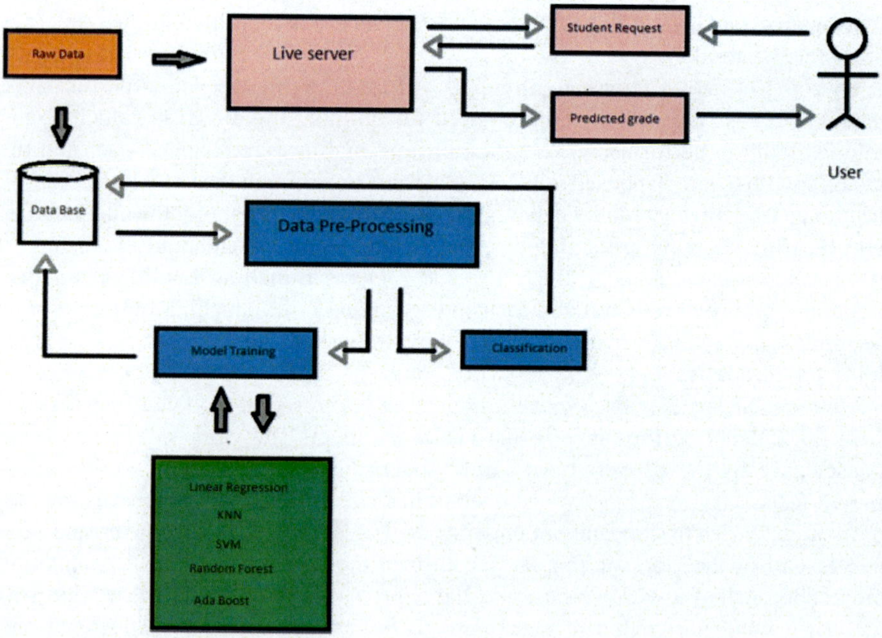

Fig. 1 System architecture

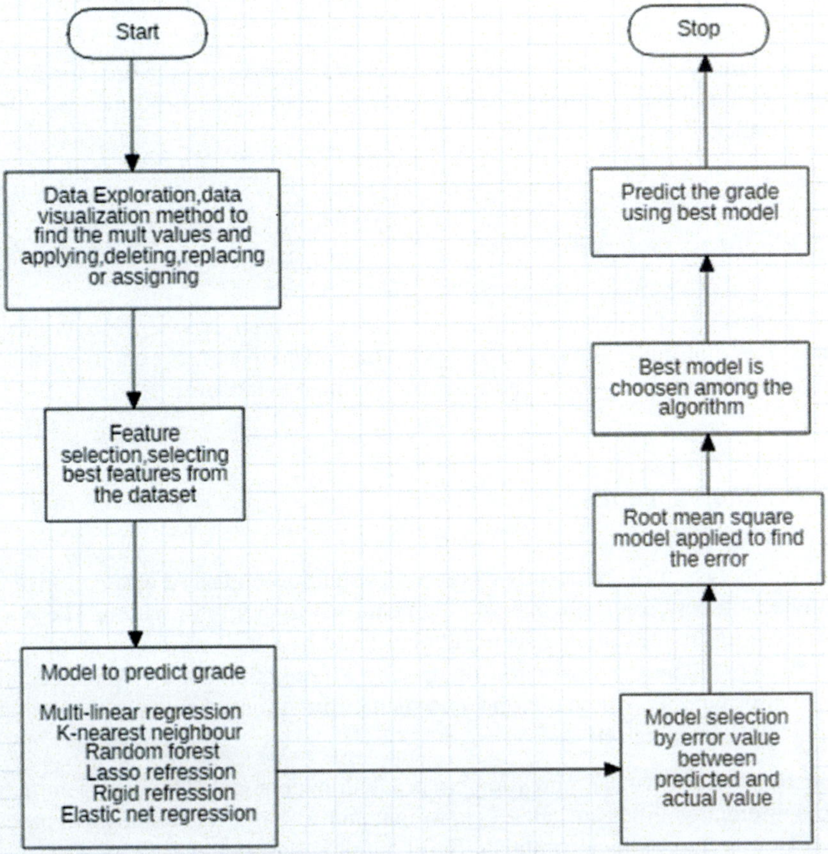

Fig. 2 Proposed framework for predicting the student grade

Feature selection. Dataset consists of many features; by using feature selection, we are choosing only few features which contribute more to predict the student grade correlation matrix and univariate analysis [9] which helps to select the primer features.

Model Training. The regression algorithms are multilinear regression, ridge regression, K-nearest neighbor, lasso regression, random forest and elastic net. All these algorithms were taking the primer features and train individually to predict the student grade [10].

Model Selection. Using root mean square error, we are predicting the error rate between the predicted value and actual value [11], so we know how accurate it predicts the value of student grade. So, less error rate value is chosen and used for prediction.

Grade prediction. The model is chosen and used for predicting the final grade of a student. As a user requests the software by the required input, then model which is trained predicts the student value.

5.1 Machine Learning

It is the capacity of a system to study for a fact in which experience is typically given in a type of information sources [12]. The system discovers conditions in the information that is excessively confusing for a human to perform.

5.2 Predictive Analytics

Predictive analysis utilizes recently gathered information, and an AI calculation finds the connection between properties of significant worth and information. The point is to foresee understudy grade dependent on test not by taking a glimpse at different highlights [13]. The demonstration of making a prediction model from recently realized information is called training, and such information is known as the training information or training set.

Linear Regression. Linear regression is a prediction method that predicts a numeric worth. Different qualities are sales, age, weight and so forth. It is a statistical procedure [14]. This means to accomplish useful autonomous factors that permit action preventive desires for a reliant variable for expectation.

$$Y = a + bX \tag{1}$$

Multilinear Regression. It is a statistical method that utilizes regression factors to foresee the result of the variables [15]. The different variables are the connection between the informative factors and reaction factors.

$$\hat{Y} = b_0 + b_1 X_1 + b_2 X_2 + \ldots + b_p X_p \tag{2}$$

Lasso Regression. Lasso regression is a sort that utilizes shrinkage which implies that information values are contracted toward main issues like a mean. Lasso regression performs L1 regularization, and it includes a factor of total estimation of coefficients in the streamlining objective [16]. Lasso regression advances the accompanying

$$\min(||Y - X\theta|||_2^2 + \lambda||\theta||_1 \tag{3}$$

Elastic Net Regression. Elastic net regression originally emerged because of evaluation on the lasso, whose factors determination can be subjected to information and make them precarious. The arrangement is to join the penalities of edge regression and lasso to get the best.

$$\min(\left|\left|Y - X\theta\right|\right|_2^2 + \lambda_1\left|\left|\theta\right|\right|_1 + \lambda_2||\theta||_2^2 \tag{4}$$

Rigid Regression. In rigid regression, the loss function is represented in such a way that we not only minimize the squared values but also penalize the size of the parameter to shrink them toward zero.

$$\min(|\left|Y - X\theta\right|||_2^2 + \left|\lambda_2||\theta||_2^2\right. \tag{5}$$

KNN. The K-nearest neighbors algorithm is a simple, supervised machine learning algorithm that can be used to solve both classification and regression problems [17]. The algorithm relies on input data to learn a function that produces an appropriate output when an unlabeled data is given.

$$\text{Euclidean Distance } (x, x_i) = \text{sqrt}(\text{sum}\left(\left(x_j - x_{ij}\right)^2\right)) \tag{6}$$

Random Forest. Random forest [18] is a multiple decision trees which merges them to get a more accurate and stable prediction.

$$F(x) = \frac{1}{J}\sum_{j=1}^{J} C_{j\text{full}} + \sum_{k=1}^{K}\left(\frac{1}{J}\sum_{j=1}^{J}\text{contribution}_j(x, k)\right) \tag{7}$$

5.3 Feature Selection

Student attributes and their description:

The variables of the dataset have two types: Numerical and categorical data type. To predict the Grade 3, find the highly correlated features which accurately give the G3 as a result.

The attributes are G1, G2, G3, age, mother education, father education, study time, Internet and travel time. To select the best features from the dataset is feature selection [19]. Using exploratory data analysis, visualization method which is shown in Fig. 3 helps to identify the highly correlated features.

Fig. 3 Future selection steps

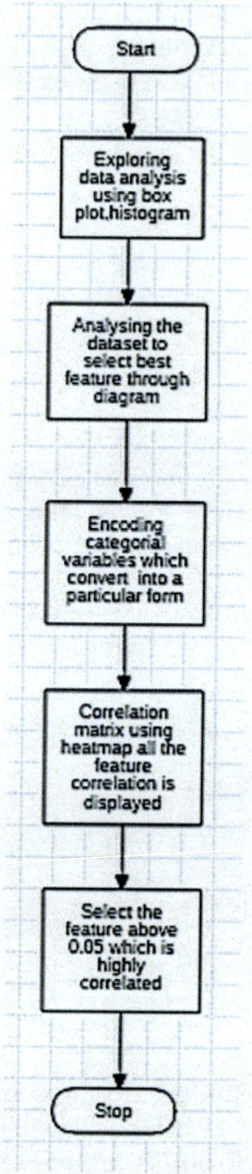

5.4 Correlation Matrix

The linear relationship between two variables value is shown between −1 and + 1. All correlated feature values are displayed using this algorithm [20]. Select the features above 0.05 which is highly correlated.

$$\mathrm{Corr}(x, y) = \frac{\sum_{i=1}^{n}(x_i - x')(y_i - y')}{\sqrt{\sum_{i=1}^{n}(x_i - x')^2 \sum_{i=1}^{n}(y_i - y')^2}} \quad (8)$$

5.5 *Heatmap*

A heatmap is a two-dimensional graphical representation of data where the individual value is contained in a matrix and represented as colors. The Seaborn Python package allows the creation of annotated heatmaps which can be used to Matplotlib tools as per the creator requirement [21].

5.6 *Grade Prediction*

The regression is utilized to discover student grade expectation. The methods like multilinear regression, SVM regression, random forest and decision tree are utilized to approve the models with one another [22]. The standard model predicts the characteristics of the training set of a given occasion of properties. Notwithstanding foreseeing student's presentation, the multi-regression model can be utilized to dissect how the various highlights add to the anticipated evaluation and hence increase some data about the students' performance [23]. For a legitimate examination, the assessed model is more helpful than the characteristics that have non-negative qualities which make all the model's segments add to the predictive evaluations.

6 Experiment and Result Analysis

Data Source: (Fig. 4)

When facing various challenges in machine learning through datasets, finding a suitable datasets relevant to use case is an essential. Dataset is taken from the educational institutions as shown in Fig 4. Size and flexibility characterize the dataset. Flexibility supports the number of tasks. The datasets which are problem-specific, to be preprocessed and cleaned are clear to understand and perform the operations.

Multilinear regression model. This is a prediction technique that predicts a numeric. Various attributes are like G1, G2, mother/father education, failures, study time, age, going out (time spend outside), etc. It is a statistical methodology [24]. The task aims to attain a functional independent variables that allow computing conditional expectations of a dependent variable for prediction.

```
Attribute Information:

# Attributes for both student-mat.csv (Math course) and student-por.csv (Portuguese language course) datasets:
1 school - student's school (binary: 'GP' - Gabriel Pereira or 'MS' - Mousinho da Silveira)
2 sex - student's sex (binary: 'F' - female or 'M' - male)
3 age - student's age (numeric: from 15 to 22)
4 address - student's home address type (binary: 'U' - urban or 'R' - rural)
5 famsize - family size (binary: 'LE3' - less or equal to 3 or 'GT3' - greater than 3)
6 Pstatus - parent's cohabitation status (binary: 'T' - living together or 'A' - apart)
7 Medu - mother's education (numeric: 0 - none, 1 - primary education (4th grade), 2 â€" 5th to 9th grade, 3 â€" secondary education or 4 â€" higher education)
8 Fedu - father's education (numeric: 0 - none, 1 - primary education (4th grade), 2 â€" 5th to 9th grade, 3 â€" secondary education or 4 â€" higher education)
9 Mjob - mother's job (nominal: 'teacher', 'health' care related, civil 'services' (e.g. administrative or police), 'at_home' or 'other')
10 Fjob - father's job (nominal: 'teacher', 'health' care related, civil 'services' (e.g. administrative or police), 'at_home' or 'other')
11 reason - reason to choose this school (nominal: close to 'home', school 'reputation', 'course' preference or 'other')
12 guardian - student's guardian (nominal: 'mother', 'father' or 'other')
13 traveltime - home to school travel time (numeric: 1 - <15 min., 2 - 15 to 30 min., 3 - 30 min. to 1 hour, or 4 - >1 hour)
14 studytime - weekly study time (numeric: 1 - <2 hours, 2 - 2 to 5 hours, 3 - 5 to 10 hours, or 4 - >10 hours)
15 failures - number of past class failures (numeric: n if 1<=n<3, else 4)
16 schoolsup - extra educational support (binary: yes or no)
17 famsup - family educational support (binary: yes or no)
18 paid - extra paid classes within the course subject (Math or Portuguese) (binary: yes or no)
19 activities - extra-curricular activities (binary: yes or no)
20 nursery - attended nursery school (binary: yes or no)
21 higher - wants to take higher education (binary: yes or no)
22 internet - Internet access at home (binary: yes or no)
23 romantic - with a romantic relationship (binary: yes or no)
24 famrel - quality of family relationships (numeric: from 1 - very bad to 5 - excellent)
25 freetime - free time after school (numeric: from 1 - very low to 5 - very high)
26 goout - going out with friends (numeric: from 1 - very low to 5 - very high)
27 Dalc - workday alcohol consumption (numeric: from 1 - very low to 5 - very high)
28 Walc - weekend alcohol consumption (numeric: from 1 - very low to 5 - very high)
29 health - current health status (numeric: from 1 - very bad to 5 - very good)
30 absences - number of school absences (numeric: from 0 to 93)

# these grades are related with the course subject, Math or Portuguese:
31 G1 - first period grade (numeric: from 0 to 20)
31 G2 - second period grade (numeric: from 0 to 20)
32 G3 - final grade (numeric: from 0 to 20, output target)
```

Fig. 4 Dataset

Comparing Approaches. In a comparison of models with the technique utilizing all evaluations, the two methodologies effectively affect the computations. RMSE esteem gives the standard deviation of the model forecast blunder. A littler worth shows better model execution. For this situation, models are viewed as dependent on RMSE esteem which is determined for multilinear regression, KNN, lasso, random forest and so forth.

Root Mean Square Error.v Root mean square is determined by distinguishing residuals and squaring, then finding the mean of the squared, and lastly, determining the base of the mean squared gives RMSE. By utilizing RMSE, we can recognize which calculations give high precision results for the given datasets. For this situation, we see that the direct model has the least RMSE.

$$\mathrm{RMSE} = \sqrt{\frac{\sum_{i=1}^{N} (\mathrm{Predicted}_i - \mathrm{Actual}_i)^2}{N}} \tag{9}$$

6.1 *Result Discussion*

Data visualization of bar graph on numerical valued to analyze the every individual feature. As in bar graph shows more relation with G3 in the Fig. 5. If the relation with G3 is high only be selected for predicting the student grade. As the less correlation will not help much to predict the grade, we can choose the features of 0, 8,10,11,15.

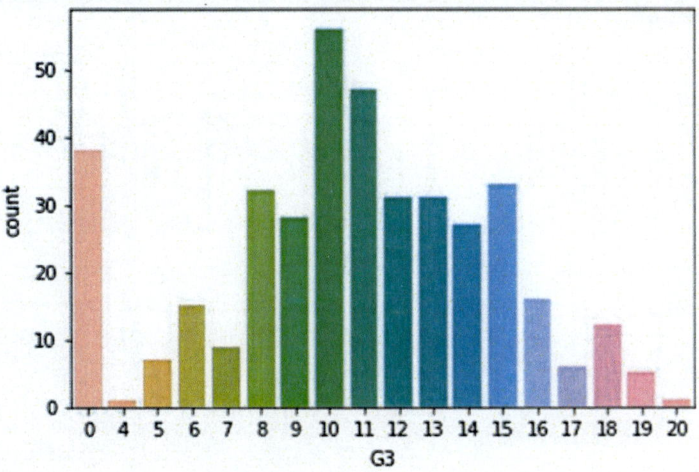

Fig. 5 Bar graph of independent features of G3

From the Fig. 6, the following can be concluded

- Male students did better than female
- Students living in urban areas had higher averages
- Students that intend on pursuing higher education have better scores.

Features used for prediction by analysis G1, G2, G3, Medu, Fedu, health, famrel, study time, absences and failures.

Bivariate analysis is an algorithm which compares the relation with every feature. Here, categorical features are compared to G3. By seeing the bar graphs, we can know which categorical variable shows high relation with the G3. So, the primer categorical features are taken for predicting the final grade.

Correlation matrix compares features to features as shown in Fig. 7, and all the features are compared to all the features. It is also multivariate analysis. All features mean categorical and numerical features also. When a feature is compared with the other feature, then if it gives positive which means highly correlated if the value is below 0 and negative then it is less correlated. Correlated means when the single feature depends on the other feature more that feature to be selected because related to that have more contribution is there [25]. So, with G3 highly correlated were chosen, and the rest can be avoided (Table 1).

When the model is trained by different regression algorithms than to choose the best among them, we have used root mean square error to find out the error rate among the algorithms.

Root mean square gives the error by comparison of predicted value and actual value [26]. As multilinear regression shows less error rate [27], we have chosen that particular model to predict the student grade (Fig. 8).

This is the feature selection process by data visualization of bar graph to analyze the features.

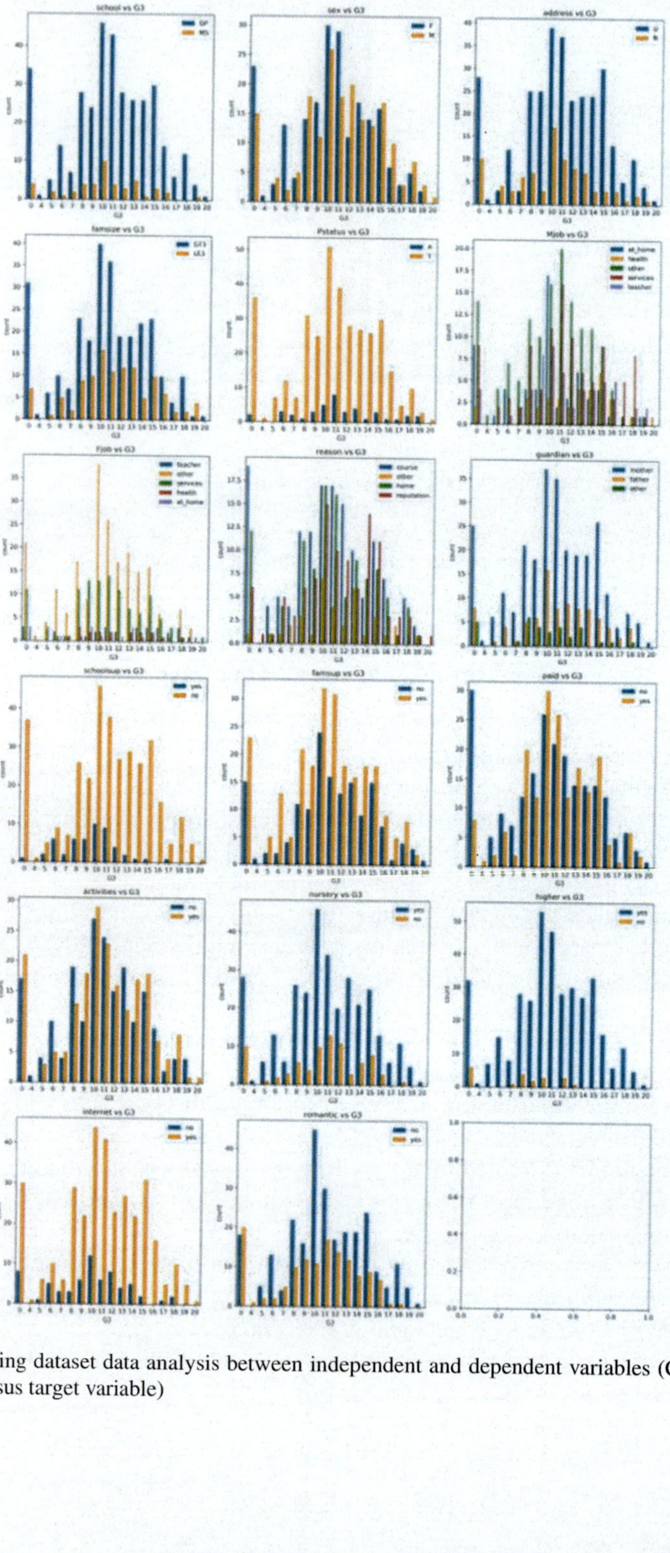

Fig. 6 Training dataset data analysis between independent and dependent variables (Categorical variables versus target variable)

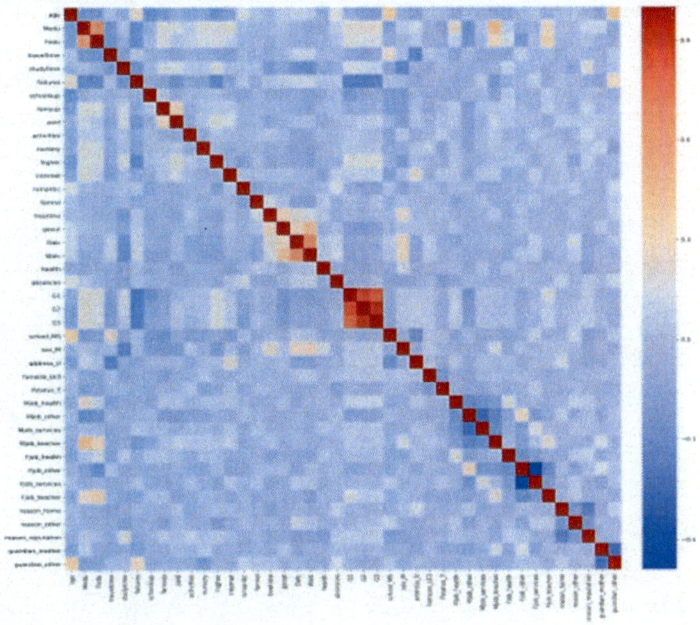

Fig. 7 Heatmap of features for feature selection

Table 1 Accuracy with root mean value

Algorithm	Accuracy with root mean value
Multilinear regression	2.018
Lasso regression	2.153
Rigid regression	2.166
ElasticNet regression	2.142
KNN	2.6908
Random forest	2.338

7 Conclusion

In this paper, we have used a different algorithm to predict student grade in the final examination. Prediction of student performance is one of the most important in research using machine learning techniques. In this, we have taken multilinear regression to find accuracy, predicted values, and using those values, we have found the final predicted value. In future, we wish to apply a different algorithm to predict the final grade of the student. This helps the student and teacher to make proper decisions to improve the performance of the weak students. The number of algorithms using we concluded, by comparison, finding the accuracy based on a study we have suggested using multilinear regression algorithm.

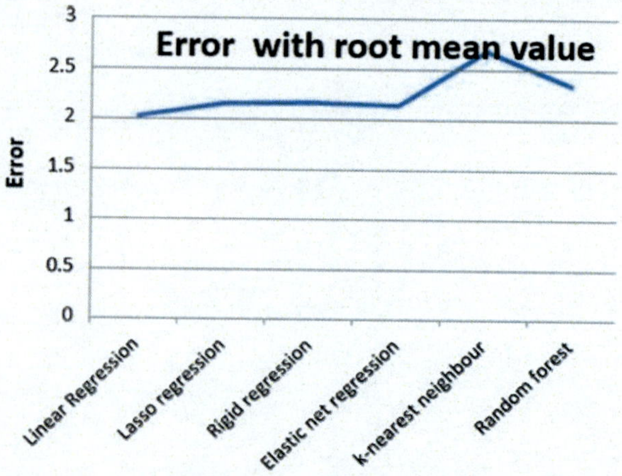

Fig. 8 Error with root mean value graph

7.1 *Future Work*

The project helps to predict the grades of a student. As we know the development of new technology, we need basic knowledge of particular subjects [28]. The dataset is consisting of every individual subject marks of student scores. These scores of a subject show that the student has learned. Based on this score, a new technology (subject) is suggested to them [29], and the best platform to learn with a brief description is explained.

Acknowledgements Illa Mani Krishna and R. Rahul Kumar, final year student of our department (Software Engineering, SRM Institute of Science and Technology), put their effort to complete this idea. I would like to thank to them for their dedicated support and their involvement. This work has been done at the laboratory of Software Engineering Division of SRMIST.

References

1. Iqbal Z, Junaid Q, Adnan NM, Faisal K (2017) Machine learning based student grade prediction: a case study. arXiv preprint arXiv, 1708.08744
2. Acharya A, Devadatta S (2014) Early prediction of students performance using machine learning techniques. Int J Comput Appl 107(1)
3. Baba SCHMH, Akhila G, Mani KSR, Somisetty VP (2014) Student performance analysis using ClassificatioTecniques. Int J Pure Appl Math 115(6):1–7
4. Pojon M (2017) Using machine learning to predict student performance. Master's thesis
5. Bendangnuksung PP (2018) Students' performance prediction using deep neural network. Int J Appl Eng Res 13(2):1171–1176
6. Oladokun VO, Adebanjo AT, Charles-Owaba OE (2008) Predicting students academic performance using artificial neural network: a case study of an engineering course
7. Sembiring S, Zarlis M, Hartama D, Ramliana S, Wani E (2011) Prediction of student academic performance by an application of data mining techniques. In: International conference on management and artificial intelligence IPEDR. vol 6. No 1, pp 110–114

8. Agrawal H, Mavani H (2015) Student performance prediction using machine learning. Int J Eng Res Technol 4(03):111–113
9. Yadav SK, Brijesh B, Saurabh P (2012) Mining education data to predict student's retention: a comparative study. arXiv preprint arXiv:1203.2987
10. Shmueli G, Otto RK (2011) Predictive analytics in information systems research. MIS Quarter 553–572
11. Quinlan JR (1987) Simplifying decision trees. Int J Man-Machine Stud 27(3):221–234
12. Kalles D, Pierrakeas C (2006) Analyzing student performance in distance learning with genetic algorithms and decision trees. Appl Artif Intell 20(8):655–674
13. Shaleena KP, Shaiju P (2015) Data mining techniques for predicting student performance. In: 2015 IEEE international conference on engineering and technology (ICETECH), IEEE, pp 1–3
14. Arsad PM, Norlida B (2013) A neural network students' performance prediction model (NNSPPM). In: 2013 IEEE International conference on smart instrumentation measurement and applications (ICSIMA), IEEE, pp 1–5
15. Ramesh VAMANAN, Parkavi P, Ramar K (2013) Predicting student performance: a statistical and data mining approach. Int J Comput Appl 63(8)
16. Jishan ST, Raisul IR, Naheena H, Rashedur MR (2015) Improving accuracy of students' final grade prediction model using optimal equal width binning and synthetic minority over-sampling technique. Dec Anal 2(1):1
17. Quadri MN, Kalyankar NV (2010) Drop out feature of student data for academic performance using decision tree techniques. Global J Comput Sci Technol
18. Osmanbegovic E, Suljic M (2012) Data mining approach for predicting student performance. Econ Rev: J Econ Bus 10(1):3–12
19. Angeline DMD (2013) Association rule generation for student performance analysis using apriori algorithm. The SIJ Trans Comput Sci Eng Appl (CSEA) 1(1):12
20. Bin M, Usamah NB, Pauziah MA, Rosni AK (2013) An overview of using academic analytics to predict and improve students' achievement: a proposed proactive intelligent intervention. In: 2013 IEEE 5th conference on engineering education (ICEED). IEEE, pp 126–130
21. Schmidhuber J (2015) Deep learning in neural networks: an overview. Neural Netw 61:85–117
22. Asogwa OC, Oladugba AV (2015) Of students academic performance rates using artificial neural networks (ANNs). Amer J Appl Math Stat 3(4):151–155
23. Oancea B, Raluca D, Stefan C (2013) Predicting students' results in higher education using a neural network. pp 190–193
24. Karamouzis ST, Andreas V (2008) An artificial neural network for predicting student graduation outcomes. In: Proceedings of the World congress on engineering and computer science, pp 991–994
25. Huang S, Fang N (2013) Predicting student academic performance in an engineering dynamics course: a comparison of four types of predictive mathematical models. Comput Educ 61:133–145
26. Arsad PM, Norlida B (2012) Neural network model to predict electrical students' academic performance. In: 2012 4th international congress on engineering education. IEEE pp 1–5
27. Shakil M (2006) A multiple linear regression model to predict the student's final grade in a mathematics class. Sam Houston State University
28. Hamsa H, Indiradevi S, Kizhakkethottam JJ (2016) Student academic performance prediction model using decision tree and fuzzy genetic algorithm. Procedia Technol 25:326–332
29. Vellingiri J, Kaliraj S, Satheeshkumar S, Parthiban T (2015) A novel approach for user navigation pattern discovery and analysis for web usage mining. J Comput Sci 11(2):372

A Graph Invariant-Based TGO Model for RailTel Optical Networks

Shanmuk Srinivas Amiripalli, T. Venkata Santosh Anish Dora, Ssvr Kumar Addagarla, Parvathaneni Naga Srinivasu, and G. S. Mallikarjuna Rao

Abstract Optimization methods are used to develop a model that can decrease production costs and maximize performance. There are several strategies for optimizing in soft computing techniques, such as scent marking optimization, ant colony optimization, and machine learning optimization. Optimizing the current network model is our issue. This paper chooses optimization based on graph theory because this approach has a solid theoretical history with a systematic approach and proof. RailTel is one of the country's largest neutral telecom infrastructure providers that owns an exclusive Right of Way (ROW) Pan-India optic fiber network along the railway track. We carefully researched the RailTel optical network's topology and noted that the current model can be optimized and improved in specific parameters during our study. We found a few invariant graph parameters such as number of nodes, number of edges, average degree, average clustering diameter, transitivity, connectivity edge, average length of path, and density. We made an effort to upgrade the current model. When we compared both models, we found that the proposed model had almost the same or better values for all parameters. We succeeded in having a diameter of half the current model, an optimized average path length, and transitivity improvement. The current model showed better optimized results compare to the existing method.

Keywords Optical network · Trimet graph · Topology · Optimization · Graph theory · NetworkX

S. S. Amiripalli · T. V. S. A. Dora · P. N. Srinivasu
Department of Computer Science and Engineering, GIT, GITAM University, Vizag, India

S. K. Addagarla (✉)
Department of Computer Science and Engineering, Gayatri Vidya Parishad College for Degree and PG Courses (A), Vizag, India

G. S. M. Rao
Department of Computer Applications, Gayatri Vidya Parishad College of Engineering(A), Vizag, India

S. S. Dash et al. (eds.), *Sixth International Conference on Intelligent Computing and Applications*, Advances in Intelligent Systems and Computing 1369,
https://doi.org/10.1007/978-981-16-1335-7_7

1 Introduction

A system or process is optimized when it is configured to function as effectively as possible. For all the possible solutions, the optimization challenge is seeking the right one. In other areas, such as mathematics, computer science, and economics, this can be used. The techniques of optimization are complexity, deterministic, multiobjective optimization of three forms. One of the oldest and best methods of optimization is graph theory [1]. This approach is the cornerstone of many fields of study since it is assisted by many mathematical principles that are the basis for many newer techniques. The optical network is a communication network that uses an optical fiber cable to transmit information [2, 3]. This data communication mode is considered to be one of the fastest ways of data communication available. It can hold data with greater capacity over longer distances. Topology is a way to structure a network, including the physical network's logical and physical topology, as the name implies that it refers to how the network nodes are linked, wiring, cables, etc. [4, 5]. Logical network topology refers to the philosophical understanding and techniques related to how and why the network is structured. There are several topologies available, but with promising results, choosing the right topology will help the company save a lot of money. Selecting an appropriate model, according to the specification, allows us to troubleshoot mistakes and error findings. There are many topology types, such as a ring, bus, point-point, star, mesh, tree, and hybrid. [6–9].

Figure 1 shows the RailTel optical network topology of India. Optimization is a process of solving a problem resulting in the objective function being optimally solved. An optimal solution guarantees feasibility along with higher cost-minimizing efficiency. Graph theory is a mathematical representation of a network that has been constructed [9–11]. The relationship between different points in the network (nodes) and the connection between them (edges) is defined here. This approach is used in all regions since this is one of the best types of representation. It is possible to understand the network very well as this gives many insights into how the graph is constructed and how it can be strengthened. NetworkX is a Python programming language software package that lets us design, visualize, and research several complicated networks. For working on Python language, data structures such as graph package are used for the development.

1.1 Trimet Graph

It is a graph that is mainly designed for the optimization of real-world problems. This graph focuses mostly on two properties like scalability and survivability. In the below figure, trimet with several nodes starting from 3 to 8 was displayed. It is a combination of the triangle, friendship graph, and new models of friendship graphs. This graph looks so simple, but it is an incremental model for star graph, so wherever the star graph is used, it can be replaced with trimet for better connectivity

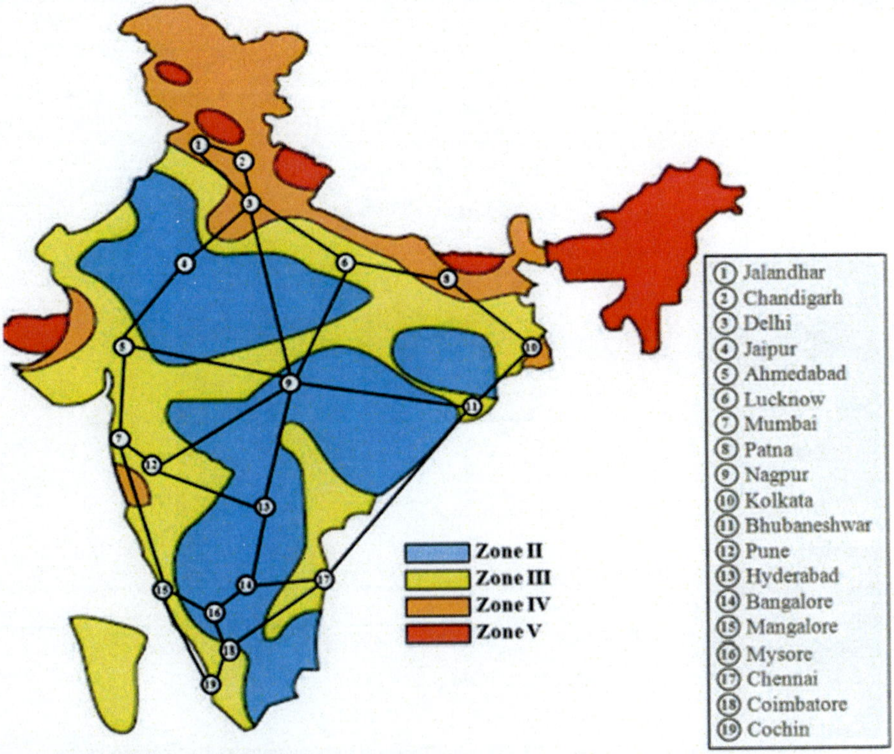

Fig. 1 RailTel optical network topology of India

and scalability. The model is a graph model, so it can be used in a wide variety of applications in wired and wireless networks. It is a two-edge survivable graph and one of the optimal graphs.

1.2 Parameters

The number of nodes is defined as the unique points available that represent each place in India. The number of edges represents the connection between any two locations. The average degree is the average number of links every node has. Diameter is the maximum distance among all the distances between a vertex to all the other vertices is considered the diameter of a Graph [12–14]. An efficient model has less diameter value because of the reduced path to be covered, resulting in faster delivery of data/information and the reduced cost for the usage of optical fiber. Density is the ratio of the number of edges to the total possible edges. The average path length is an idea in community topology defined by the average range of steps along the

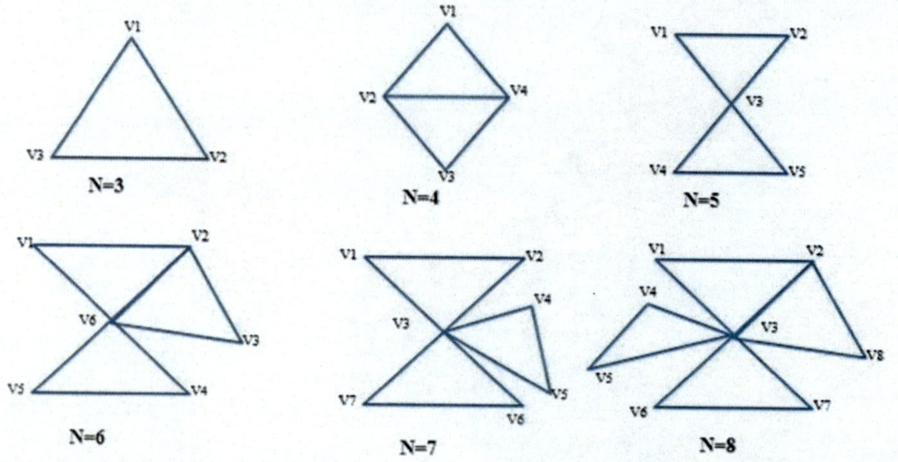

Fig. 2 Trimet graph model with 3–8 nodes

Table 1 Network requirements and graph invariants

Feature	Invariant
Cost	Order, size, average degree
Maximum latency	Diameter
Average latency	Average distance
Network congestion	Edge betweenness and degree variance
Survivability	Two connectivity

shortest paths for all possible pairs of community nodes. Transitivity is the general probability for the network to have adjoining nodes interconnected, consequently revealing tightly related groups' lifestyles. Edge connectivity is a parameter where it gives the survivability of edges; before disconnecting any graph, it will come to 1 edge. If a graph is a high connectivity, it is treated as a better model. The average clustering of the graph should be more for better connectivity. The trimet graph model for various nodes as shown in (Fig. 2) and network requirements are presented in (Table 1).

1.3 Optimization Techniques

CRO chemical reaction optimization is one of the soft computing techniques, and it was observed that in WSN, CRO model is used to solve routing and increase the efficiency of power. TGO optimization is one of the graph-based optimal models which is used for a wide range of real-time applications. Recently, this model was used in solving design problems in IoT. Similarly, we are inspired by other optimization techniques like scent marking optimization, ant colony optimization, swarm optimization method, honey bees inspired optimization method, animal migration optimization, etc.

2 Literature Survey

The author [1, 4] mentioned how disasters (natural calamities, weapons of mass destruction) can damage the optical backbone networks. The networks using optical backbone produce highly scalable connectivity even across greater distances. These networks make full use of optical technology to carry extensive data. If the backbone network is of high complexity and greater scale, there are chances for multiple failures, which results in end-to-end connection issues. These interruptions may affect many services or applications which are dependent on the optical layer. Having a model that can sustain natural calamities is essential, so it is vital to understand optical backbone networks' vulnerability. In [2, 15–17] describes the importance of network topology in modern wireless technologies. We have many basic topologies such as bus, ring, and mesh, which are lacking in fulfilling the requirements of contemporary problems. They have attempted to reduce the gap between the existing topologies and design problems. They have come up with a new trimet graph-based optimized topology that is highly scalable, reliable, secure, cost-efficient, and simple. This topology is continuously evaluated for various scenarios of wireless networks using the Cooja simulator. In [3, 18, 19], practically there could be situations where better connection results in better communication. Graph theory has anticipated the concept of the complete graph to tackle these problems. The author has come up with trimet derived from the optimized properties of the existing complete graph. The root purpose of using a trimet is to establish better communication with lesser infrastructure compared to the complete graph [4].Trimet oversees all the vertices with the help of the dominant vertex, which has a communication path from all vertices of the graph. In [5, 20, 21], there is a considerable gain in popularity for wireless networks in recent years growth in wireless, 5G cellular, etc. The problem that we face in the above technologies is survivability and scalability which will affect the performance. This development demands improvement of the existing topologies as they cannot be addressed by the traditional ring, mesh, star topologies due to a lack in scalability, survivability, and resilience properties. The author has used graph theory techniques as they are well equipped with mathematical techniques. The physical features like cost, latency, congestion, etc., can be correlated with graph invariants like nodes, edges, average distance, etc. The author in [6, 13] mentions the importance of having knowledge about the computer network topology and its development trend for the research of next-generation network design. It is essential to understand the current topology models to expand to meet the requirements of the next-generation networks. Here, the author analyzed the current situation of development. Knowing the advantages and disadvantages of every model will help us build a new model depending on the situation and the requirement. Communication has become very important in all the aspects of modern life, and thus, any disruption to this network can create a huge loss. As the network is not sufficiently resilient, dependable, and survivable, it has become a field of study where improvement is required. In [7–9], the author provides an architectural framework of how to build proper network infrastructure. A resilient strategy is presented to detect, defend, and remediate challenges. The set of rules to build such networks has been presented along with techniques used to describe and analyze network resilience.

3 Proposed Model

Firstly, we considered the paper where the author discussed network topology and optical topology (a reference to the RailTel paper). The proposed model is shown in Fig. 3. We thought that the model could be improved in a few ways in a detailed review, so we took the current model's network graph. We analyzed the number of nodes, edges, and tested whether there was a possibility of creating TGO. The current model was not a TGO model, so we tried to use TGO to construct the same graph, and then implemented both the generated graph models. Then, we compared all the invariant parameters of the graph and concluded that the proposed model is either equivalent or better in most of the parameters, making it a better model.

We have designed a model with the same number of nodes and edges to maintain similar conditions for the comparison. The existing model is one of the most optimized models designed by the experts that is why we got similar values as the existing model for some of the parameters. This shows that the existing model cannot be further optimized without degrading its strength in a few aspects as such. Even in improving the existing model resulted in few parameters with similar values that state that the model has achieved the optimal solution for those respective parameters; any further modifications may weaken the network, which has further implications on the other parameters. The average degree can be calculated by dividing the number of links between the cities divided by the number of cities. Edge connectivity is the minimum number of links between cities to be removed to break the entire network. Density is the ratio between total links to the total possible edges.

Algorithm for the proposed TGO model
Input: The number of nodes and edges Output: TGO model

(continued)

(continued)

Algorithm for the proposed TGO model
1. Start
2. Identify graph invariant parameters for the RailTel optical network
3. Analyse the existing RailTel optical network topology by using graph invariant technique
3. Design of trimer graph optimization (TGO)-based network model
4. Implement the TGO proposed model and generate results using the NetworkX package in Python
5. Compare the existing RailTel optical network topology with the TGO model by using a graph invariant technique
6. Stop

4 Results and Discussion

The issue we encountered during the development of the new model is to retain at least values closer to the new one. In Table 2, we have listed out various TGO parameters, but in four main parameters that can further optimize the model with cost and performance quality, we may manage to get better results. The enhanced parameter is that diameters can be defined as the maximum distance between a pair of cities. We have managed to get a value half of the current model that enhances the model by several folds. A lower value of the diameter means that it is simple to transfer the data/information. Transitivity is the total number of triangles to the number of connected triple nodes in the graph. We managed to get twice the current model's value, which indicates that the total number of triangles generated is higher, improving the connections and the network. The average length of the path is the shortest average path between any two cities. The lesser the path length, faster is the contact and the lesser the rate. Compared to the current model, we managed to achieve better value.

The average clustering of the proposed model is better since decentralized and centralized are balanced. As in the current model, they only concentrate on centralization. The proposed model has outshined the actual model in a few ways by comparing the two models, making it the best design for the given network. We managed to get twice the value of the current model, which indicates that the total number of triangles generated is higher, improving the connections and the network. The average length of the path is the shortest average path between any two cities.

The lesser the path length, faster is the contact and the lesser the rate. In Fig. 4, we showed the various performance measures using RailTel optical network. Compared to the current model, we managed to achieve better value. The average clustering of the proposed model is better since decentralized and centralized are balanced. As in the current model, they only concentrate on centralization. The proposed model has outshined the actual model in a few ways by comparing the two models, making it the best design for the given network.

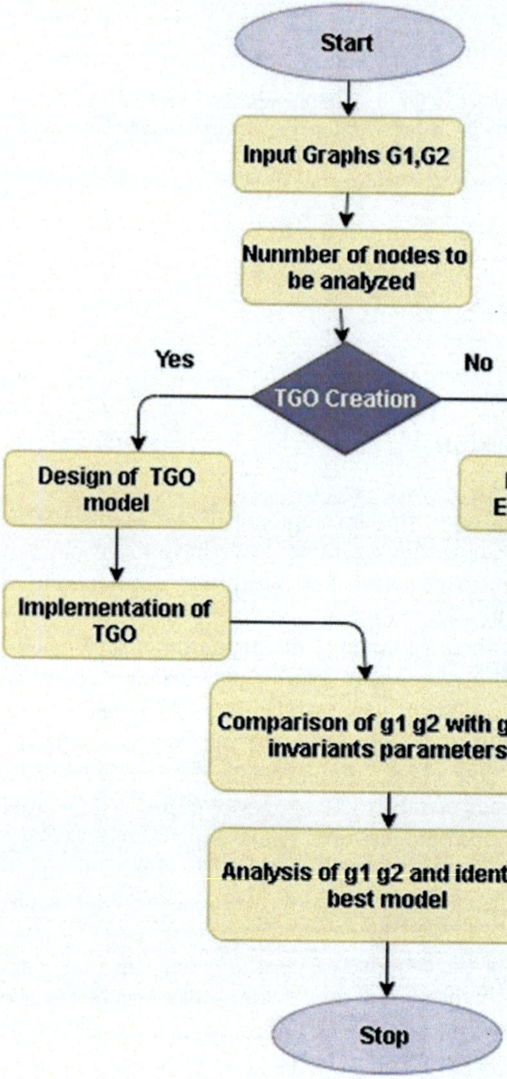

Fig. 3 Flowchart for the proposed TGO model

5 Conclusion

Survivability is an important component present in designing a complex network. We observed that survivability is optimal and two edge network problem during our existing study. After careful analysis, a novel two-edge survivable TGO model was identified. During the implementation, nine network science parameters are used for analysis. In this paper, both the existing and the proposed models are implemented

Table 2 TGO model parameters

Parameters	Existing model	Proposed model
Number of nodes	19	19
Number of edges	27	27
Average degree	2.8421	2.8421
Diameter	6	3
Average clustering	0.1754	0.869
Transitivity	0.15	0.2967
Edge_connectivity	2	2
Avg_Path_Length	2.812	2.3
Density	0.157	0.157

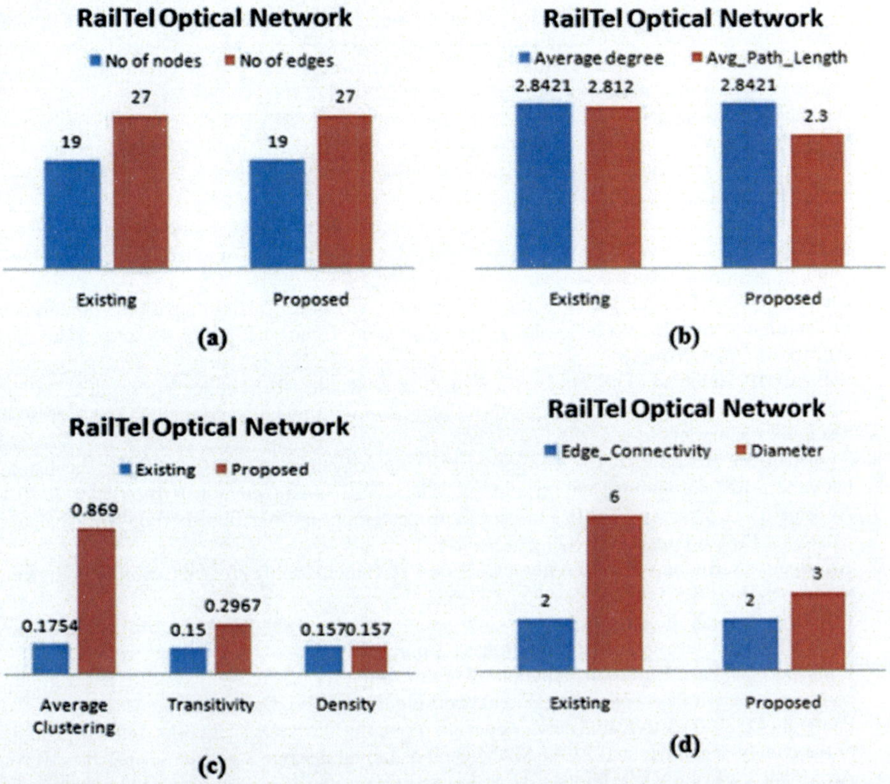

Fig. 4 Comparision of performance analysis for existing and current model is shown from **a** to **d**

using the Python language and NetworkX package. To maintain uniformity, a number of nodes and edges are taken equally. As the existing model is one of the optimal models, so we got similar values in average degree, edge connectivity, and density. The proposed model generates better results in diameter, average clustering, transitivity, and average path length. Finally, this paper's research contribution is to reduce the overall distance in a network, which leads to improve latency in a network using the invariant graph technique.

References

1. Habib MF, Tornatore M, Dikbiyik F, Mukherjee B (2013) Disaster survivability in optical communication networks. Comput Commun 36(6):630–644
2. Amiripalli SS, Bobba V, Potharaju SP (2019) A novel trimet graph optimization (TGO) topology for wireless networks. https://doi.org/10.1007/978-981-13-0617-4_8
3. Amiripalli SS, Kumar AK, Tulasi B (2016) Introduction to TRIMET along with its properties and scope. In: AIP Conference proceedings, AIP Publishing LLC, vol 1705. No. 1, pp 020032
4. Amiripalli SS, Bobba V (2018) Research on network design and analysis of TGO topology. Int J Netw Virtual Organ 19(1):72–86
5. Amiripalli SS, Bobba V (2019) Trimet graph optimization (TGO) based methodology for scalability and survivability in wireless networks. Int J Adv Trends Comput Sci Eng 8(6):3454–3460. https://doi.org/10.30534/ijatcse/2019/121862019
6. Liu , Liu Q (2014) A study on topology in computer network. In: 7th International conference on intelligent computation technology and automation, Changsha, pp 45–48. https://doi.org/10.1109/ICICTA.2014.18
7. Sterbenz JP, Hutchison D, Çetinkaya EK, Jabbar A, Rohrer JP, Schöller M, Smith P (2010) Resilience and survivability in communication networks: strategies, principles, and survey of disciplines. Comput Netw 54(8):1245–1265
8. Amiripalli SS, Bobba V (2019) An optimal TGO topology method for a scalable and survivable network in IOT communication technology. Wireless Personal Commun 107(2):1019–1040
9. Amiripalli SS, Bobba V (2019) Impact of trimet graph optimization topology on scalable networks. J Intell Fuzzy Syst 36(3):2431–2442
10. Amiripalli SS, Bobba V (2020) A fibonacci based TGO methodology for survivability in ZigBee topologies. Int J Sci Technol Res 9(2):878–881
11. Du H, Fan J, He X, Feldman MW (2018) A genetic simulated annealing algorithm to optimize the small-world network generating process. Complexity
12. Bannapure M, Patil VL (2014) Ant colony optimization for random network. In: IEEE global conference on wireless computing and networking (GCWCN), December, pp 41–46
13. Panigrahi D (2011) Survivable network design problems in wireless networks. In: Proceedings of the twenty-second annual ACM-SIAM symposium on discrete algorithms, pp 1014–1027
14. Amiripalli SS, Kollu VVR, Jaidhan BJ, Srinivasa L, Raju VA (2020) Performance improvement model for airlines connectivity system using network science. Int J Adv Trends Comput Sci Eng 9(1):789–792. https://doi.org/10.30534/ijatcse/2020/113912020
15. Amiripalli SS, Bobba V (2020) An optimal graph based zigbee mesh for smart homes. 79(4):318–322
16. Lloret J, Garcia M, Bri D, Diaz JR (2009) A cluster-based architecture to structure the topology of parallel wireless sensor networks. Sensors 9(12):10513–10544
17. Amiripalli SS, Surendra T, Venkata Dileep B, Priyadharshini EV, Lakshmi Charan Reddy V, Sateesh Kumar D (2020) Analysis of airline connectivity using network science. Int J Psychosoc Rehab 24(6):5229–5234

18. Ramiah Chowdary P, Challa Y, Jitendra MSNV (2019) Identification of MITM attack by utilizing artificial ıntelligence mechanism in cloud environments. J Phys: Conf Ser 1228(1):012044
19. Thota JR, Kothuru M, Shanmuk Srinivas A, Jitendra MSNV (2020) Monitoring diabetes occurrence probability using classification technique with a UI. Int J Sci Technol Res 9(4):38–41
20. Amiripalli SS, Venkatarao R, Jitendra MSNV, Mycherla NMJ (2020) Detecting emotions of student and assessing the performance by using deep learning. Int J Adv Trends Comput Sci Eng 9(2):1641-1645
21. Jitendra MSNV, Radhika Y (2020) A review: Music feature extraction from an audio signal. Int J Adv Trends Comput Sci Eng 9(2):973–980

Analysis of Intelligent Controllers for Juice Flow Process in a Sugar Industry

S. Ramesh, S. Anbarasi, M. Ettappan, and P. Chandrasekar

Abstract Control of juice flow process in a sugar industry is an important operating task in determining the quality of sugar production rate. This paper analyses the presence of uniform brix concentration by controlling the flow rate in a sugar industry through controllers tuned by evolutionary computing techniques. Based on the real-time data obtained from sugar industry, the mathematical model of system is evaluated with the help of various system identification models, and the best model is analysed through its fitness percentage, and its controller performance is initially analysed through the conventional Ziegler–Nicholas tuning. The performance of the controller is further analysed with popular evolutionary intelligent-based techniques. The effectiveness of intelligent techniques is demonstrated with respect to various time domain specifications and also being compared with conventional controllers. The results depict the role of efficient intelligent controller for juice flow process.

Keywords Juice flow process · System identification · PID controller tuning · Genetic algorithm · Particle swarm optimization · Self-adaptive differential evolution

1 Introduction

Generally, in the sugar factory various processes are carried out in the juice extraction process. At first, the raw sugar cane is crushed and the juice is extracted from the sugar cane. The extracted juice is then stored in the raw juice tank which contains raw juice. From the raw juice tank, the juice is transferred to the primary juice heater.

S. Ramesh (✉)
Department of EEE, Vel Tech Rangarajan Dr. Sagunthala R&D Institute of Science and Technology, Avadi, Chennai, Tamil Nadu, India

S. Anbarasi
Department of EEE, P.S.R. Engineering College, Sivakasi, Tamil Nadu, India

M. Ettappan · P. Chandrasekar
Department of EEE, Chennai Institute of Technology, Chennai, Tamil Nadu, India

S. S. Dash et al. (eds.), *Sixth International Conference on Intelligent Computing and Applications*, Advances in Intelligent Systems and Computing 1369,
https://doi.org/10.1007/978-981-16-1335-7_8

Here, the juice is heated up to 70 °C. The juice from the primary heater is then sent to the sulphitation tank. In sulphitation tank, the impurity of juice is cleared and the pH value of the juice is maintained by adding the milk of lime and sulphur dioxide. The juice processed in the sulphitation tank is normally known as treated juice. This treated juice is then transferred to the secondary juice heater and heated up to 103 °C. The heated juice is then transferred to the clarifier section. In the clarifier, the extra mud is taken out and filtered juice is sent to the pan section for recirculation process. The remaining juice in the clarifier maintained with the pH level of 7.1. The filtered juice of the clarifier section is termed as clear juice. This clear juice is sent to the heat exchanger and then to the evaporator. After the evaporator process, the juice will become syrup. This syrup is then deported to the dryer section, and the process is taken over for crystallization. Finally, the solid sugar is produced and dispatched with proper packing.

Few applications of intelligent-based controllers for sugar industry can be found in the literature. Lin et al. [1] discussed neural-network-based approximate dynamic programming model for the industrial sucrose crystalline process. Pitteea et al. [2] discussed about the application of intelligent-based controllers for multiple-effect evaporator process. The role of model predictive controllers for sugar industry is discussed in [3]. The tuning of parameters for proportional-, integral-, and derivative-based controller is being discussed in [4].

This study has been carried out with the real-time data collected from the sugar industry named E.I.D. Parry (India) Limited, Karur, India. In the factory, the compound effect evaporators (CEE) are used to extract the excess amount of water from the treated juice. The CEE consists of four evaporator vessels or five in some of the industries which are allied in sequence. Normally, the role of CEE is to increase the concentration of sugar cane juice syrup which is represented in terms of degrees of brix concentration. By definition, one degree of brix concentration is the presence of 1 gm of sucrose in 100 gm of solution. The system under study requires a brix concentration of 58 degree to be maintained for the quality product. With the help of system identification procedures [5–7], the mathematical model of the juice flow process needs to be represented in the form of transfer function for sugar industry. Based on the real-time data, various system identification models, such as ARX, ARMAX, Box–Jenkins and nonlinear ARX (NLARX), are evaluated for best fitness.

The objective of this research work is to maintain the brix concentration and to minimize the energy consumption by developing a prominent controller scheme for the CEE. In order to persuade the objective, a suitable control scheme is analysed using either proportional integral (PI) or proportional integral derivative (PID) controller. Initially, gain values of the PI and PID controller are tuned with conventional Ziegler–Nicholas (ZN) method, and the output results are compared. The comparative output responses obviously proved that the PI controller produces better results comparing to the PID controller.

To improve the system performance further, the gain parameters are then tuned with some of the evolutionary computation approaches like genetic algorithm technique (GA), particle swarm optimization technique (PSO), simulated annealing technique (SA), and self-adaptive differential evolution technique (SaDE).

2 Identification of Mathematical Model for CEE

The identification of mathematical model of CEE can be obtained with an investigational-based approach that considers input and output variables of the process such as input signal or control variable, output signal or forbidden variable, and even in some cases disturbances. In worldwide, this procedure comprises of: conducting experiment; acquiring data; model selection; model parameter estimation; and model validation.

Normally, two inputs have been considered in CEE. One is the steam flow rate, and the other one is the juice flow rate. In this research, for juice flow process, the juice flow rate (T/hr) is taken as input by assuming the steam flow rate as constant and the final stage of the evaporator is the syrup which is considered in terms of brix concentration which is considered as output. Many set of real-time data (3840 number of datasets) consist of input and output taken at the interval of 6 min from the sugar industry to analyse the obsolete mathematical representation of the system.

2.1 Models Structure Selection and Comparison

The selection of suitable model structure is highly essential which needs to be done through system identification procedures. Model complexity can influence the exact behaviour which exactly describes the real process. Proper selection of model structure needs to be done to determine the suitable parameters that resemble the model framework. A lot of different model structures are evaluated such as ARX, ARMAX, Box–Jenkins, and nonlinear ARX (NLARX). These structures are used to identify the system model. All the 3840 set of real-time data collected from the sugar industry are considered in the identification process. Evaporator also has the input as juice flow rate and output as brix concentration. The fitness parameter is the key factor considered for system identification. The fitness percentage of various models analysed are illustrated in Fig. 1, and the respective fitness percentage of models are given Table 1.

Comparing the fitness percentage of all the linear models, the NLARX model is found to be better one with highest fitness percentage, and hence, this model is preferred for the test system under consideration. The selection of suitable controller for NLARX model is a challenging task, and that is discussed in the subsequent session of this research paper.

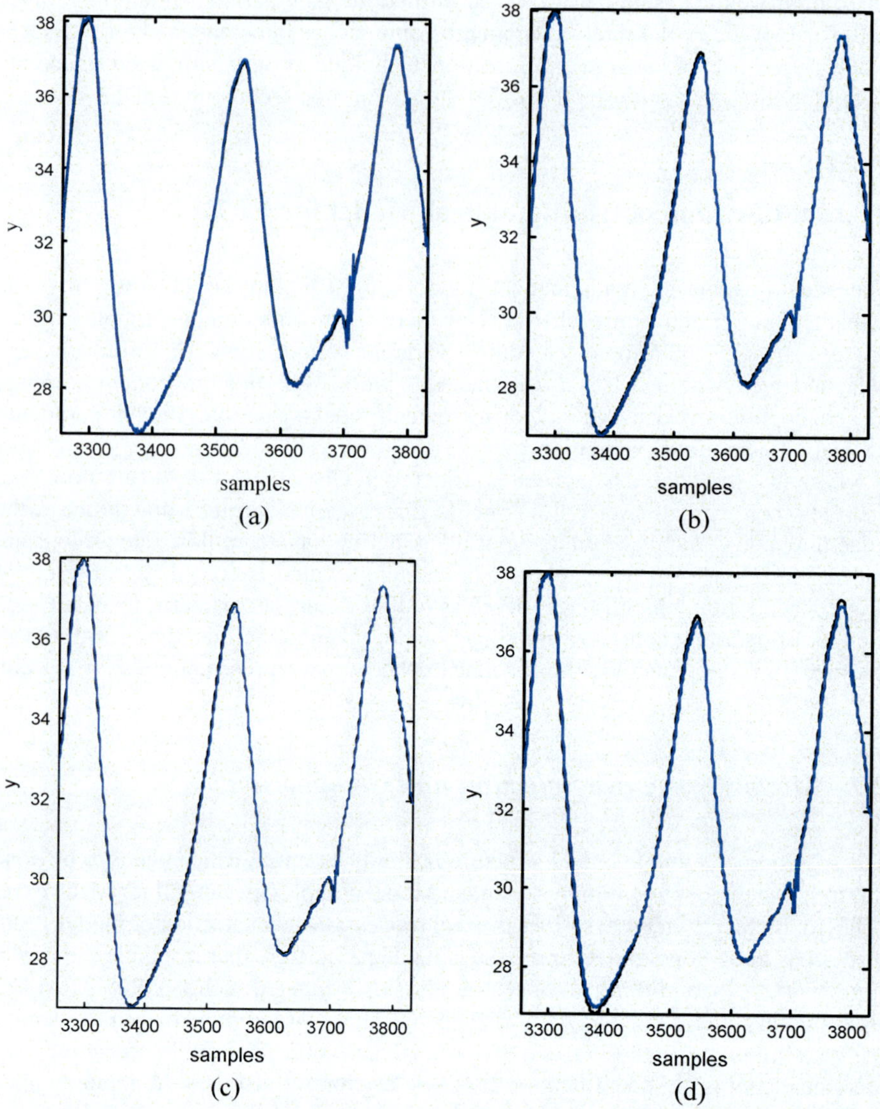

Fig. 1 Illustration of fitness for different models: **a** NLARX fitness **b** ARMAX fitness **c** ARX fitness **d** BJ fitness

Table 1 Fitness percentage of models

MODELS	Fitness %
NLARX	96.62
ARMAX	95.97
ARX	95.86
Box-Jenkins	94.18

Table 2 Performance of conventional controllers

Controller	Settling time (sec)	Rise time (sec)	Peak time (sec)	Peak over shoot (%)	System stability
PI	42.5	27.6	1.0	1.1	stable
PID	43	27.6	1.01	1.4	stable

3 Controller Design

A suitable controller is necessary to be designed to enhance the output performance of juice flow process. The controller design needs the representation of mathematical model in the form of transfer function or state space model. From the real-time datasets, the transfer function of the juice flow process has been identified through NLARX model and is represented in Eq. (1)

$$G(s) = \frac{0.2181s^3 + 6.058s^2 + 44.11s + 129.6}{s^3 + 20.89s^2 + 42.6s + 6.786} \quad (1)$$

The majority of the industrial controllers are PID based which is due to its simple functional features and ease of implementation. For the juice flow process, the PI controller is preferred. The reason for the selection of PI controller without D mode is due to the presence of disturbances and noise which induces large transport delays in the system. The PI controller can be effectively used compared with PID controller in suppressing the noise, disturbances and transport delays in the system.

The performance index of PID and PI controller is shown in Table 2. Several approaches have been followed for tuning the PI controller parameters. The Ziegler–Nichols method is the most commonly used for tuning controllers. The set point is controlled at the brix concentration of 58%. For juice flow process, the respective time domain specifications are reported in Table 2.

4 Evolutionary Algorithms

The EA has a wide range of implementation in an engineering application all over the world. In this article, the evolutionary computing techniques such as GA, PSO, SA, and SaDE are attempted to optimize the parameters of conventional controller.

4.1 Genetic Algorithm (GA)

Genetic algorithms [7, 8] are stochastic numerical search procedures that deal with a set of solutions called population (candidate) that undergo stable modifications through genetic functions: selection based on tournament, crossover, and mutation. At the end of every iteration, the solutions are ranked according to their fitness with admiration to the objective function where the best fit individuals are preserved and progress to the next generation. According to the fitness value, the individuals (parents) are chosen for the next generation by exchanging genetic information with the help of crossover to form children population. The worst performing parents are replaced by the best children to keep a stable population size. At the end of every iteration, a new population of individuals with improved fitness is progressed as parents. The mutation has been used to prevent the convergence to a local optimum and create to visit new regions of solution space. As the number iteration increases, the chance of getting optimal solution may be more. By using this algorithm, the conventional parameters of PI controller may get optimized. Initial parameters of GA are: population size of 10, iteration size of 300, and crossover rate of 0.8. The arithmetic crossover, uniform mutation, and the selection based on ranking are used to evaluate the controller parameters.

The performance of the Ziegler–Nicholas (ZN) method is effectively compared with the genetic algorithm in terms of various time domain specifications for juice flow process.

4.2 *Simulated Annealing (SA)*

Simulated annealing [9, 10] is a probabilistic method for evaluating a cost function's global minimum, although it may have many local minimums. In this study, the sugar flow process is imitated as a initial high temperature substance and it is gradually allowed to cool down for implementing simulated annealing. This exhibits minimum energy configuration of the process when the temperature attains its final frozen value. In this study, it is necessary to adjust conventional parameters within a certain range. By using the simulated annealing, the local optimization is proposed and is compared with the results of conventional and some evolutionary techniques like GA and PSO.

The effectiveness of simulated annealing technique (SA) is analysed with the genetic algorithm in terms of time domain specifications. According to the settling time, the ZN method gives better performance compared with simulated annealing technique (SA). The simulated annealing technique (SA) performance does not reproduce the global optimum since the simulated annealing is struck at local optimum; it does not work constantly in this optimization with a wide range of parameter optimization.

4.3 *Particle Swarm Optimization (PSO)*

Dr. Russell C. Eberhartand and Dr. James Kennedy introduced particle swarm optimization in 1995 [11, 12]. The PSO algorithm is an adaptive algorithm based on

Table 3 Selection parameters of PSO

Parameters	Values
Population size	10
Maximum number of iteration	300
Iteration counter	1
Function evaluation counter	1
Inertia weight	1
Velocity constants C_1 and C_2	1

a social–psychological metaphor that adapts a population of people (referred to as particles) by returning to previously effective regions stochastically. Particle swarm performs two important functions, viz. velocity update and position update. Each particle is thrusting towards the particles prior to the best location during each generation. The updated velocity value of particle is calculated in every iteration based on its present velocity. The new velocity value is then used to calculate the updated position of the particle for next iteration in the search space. This process is followed for the subsequent number of iterations until it reaches the maximum iteration count or converge to a least error value of subsequent iterations. In this paper, PSO is proposed to tune the PI controller. The parameters of the controller are included in the optimization with the particular range to achieve the global optimum solution. The parameters of PSO are given in Table 3.

The performance of PSO is compared with the conventional method (ZN) in the case of time response curve. According to the settling time, the PSO has maintained better settling time, but still it does not maintain the successful global optimum solution among 100 trial runs compared with the conventional method (ZN).

4.4 Self-Adaptive Differential Evolution Technique (SaDE)

In original DE [13], the performance highly depends on the strategies and parameter settings. SaDE is the variant version of DE [14] algorithm. It may take a massive amount of computational time to find the most effective strategy and the corresponding control parameters for a particular problem. Also, different methods and different parameter settings for different global and local search capabilities may be favoured during different evolution phases. But, the learning methods and parameter settings during evolution can be adapted automatically by the SaDE algorithm. The main ideas of the SaDE algorithm are summarized into three features such as strategy adaptation, parameter adaptation, and local search. Table 4 shows the selection parameters of SaDE algorithm for the PI controller of juice flow process.

Table 4 Selection parameters of SaDE

Parameters	Values
Population size	10
Maximum number of iterations	300
Function evaluation	3000
Learning period	50
Cross over rate (CR)	0.8

5 Comparison of Evolutionary Algorithms Based PI Tuning

In this paper, the performance comparison of different evolutionary algorithms has been considered in the sense of settling time, rise time, peak time, peak over shoot, and success rate among 100 runs. In evolutionary computation, there are four methods used, namely GA, SA, PSO, and SaDE, and are effectively compared with conventional controller. While comparing the four techniques, only two methods have given the best optimum solution, and they are genetic algorithm and self-adaptive differential evolution technique (SaDE). The remaining algorithms simulated annealing, and particle swarm optimization does not maintain the consistency in terms of settling time even for 100 runs. According to the genetic algorithm and self-adaptive differential evolution (SaDE), among 100 runs, the self-adaptive differential evolution (SaDE) maintains the constant optimum solution in terms of settling time whereas the success rate of GA is only 48%. The performance comparison of the various controllers in terms of time domain specifications are given in Table 5, and the comparison of various evolutionary techniques is shown in the form of step response curve in Fig. 3.

Based on Fig. 3 and Table 5, it is observed that the performance of SaDE is better when compared with all other controllers.

Table 5 Performance comparison of controllers

Controllers	Settling time (sec)	Rise time (sec)	Peak time (sec)	Peak over shoot (%)	Success rate (%)
SaDE	**4.11**	**0.58**	**1.62**	**6.59**	**98**
GA	4.11	0.58	1.62	6.59	48
PSO	15.3	3.79	8.53	9.14	18
ZN	42.5	27.6	59.5	1.1	100
SA	50.2	2.55	7.08	59.1	97

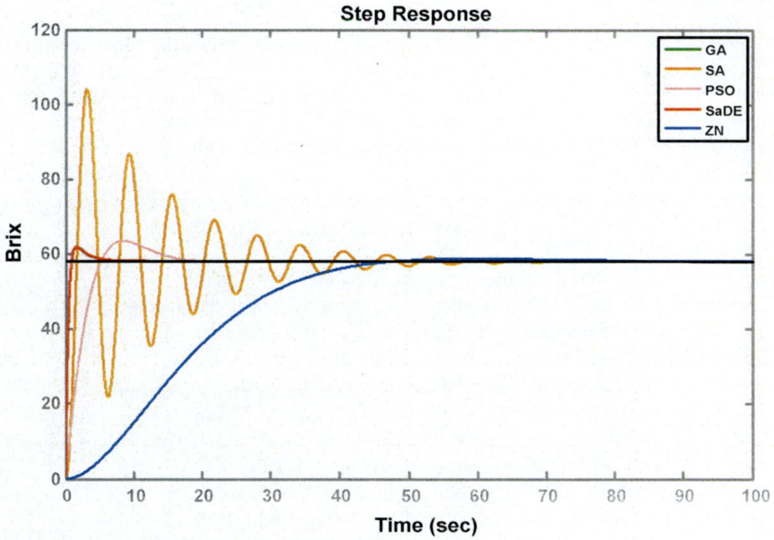

Fig. 3 Step response curve of GA, SA, PSO, SaDE, and ZN

6 Conclusion

In this article, for a sugar industry, the system identification using the linear and nonlinear models like ARMAX, ARX, Box–Jenkin and NLARX is designed for juice flow process. From various models, it is concluded that NLARX model gives more fitness (96.62%) compared with other models, and the system is concluded as a stable one. For the obtained model, optimal parameters of the controller are obtained using the evolutionary computing techniques such as GA, PSO, SA, and SaDE. The performance of various algorithms are compared with respect to settling time, peak time, rise time, peak over shoot, and success rate and is shown using MATLAB. After 100 runs, the SaDE has retained the global optimum solution with highest success rate among all algorithms reported. It is observed that SaDE can give best controlling action compared with other algorithms discussed in this work. The SaDE-based controller can be more suitable for the juice flow process in sugar industry to get the fine control action.

References

1. Xiaofeng L (2010) Optimal control for industrial sucrose crystallization with action dependent heuristic dynamic programming. In: Proceedings 8th World congress on international control and automation, Jinan, China
2. Pitteea V, Robert TF, Ah K, Harry CS, Rughooputh: intelligent controller for multiple-effect evaporator in the sugar industry. In: IEEE International conference industrial technology

3. Bordons C, Camacho EF (2000) Applications of model predictive controllers in a sugar factory. IFAC Proc 33(10):329–334
4. Dušan F, Iztok F, Iztok F, Riko Š (2016) Parameter tuning of PID controller with reactive nature-inspired algorithms. Robot Auton Syst 84:64–75
5. Ljung L (1999) System Identification–theory for the User, 2nd edn. PTR Prentice Hall, Upper Saddle River, NJ
6. Florakis A, Fassosis SD (2001) MIMO LMS–ARMAX identification of vibrating structures. Part II: a critical assessment. Mech Syst Signal Process 15(4):737–758
7. Valarmathi K, Devaraj D, Radhakrishnan TK (2009) Real-coded genetic algorithm for system identification and controller tuning. App Math Model 33:3392–3401
8. Jayachitra A, Vinodha R (2014) Genetic algorithm based PID controller tuning approach for continuous stirred tank reactor. In: Advances in artificial intelligence. vol 2014. Article ID 791230, pp 8. https://doi.org/10.1155/2014/791230
9. GirirajKumar SM, Bodla R, Anantharaman N (2010) Design of controller using simulated annealing for a real time process. Int J Comp App 6(2):0975–8887
10. Zhang Y, Hu Y (2008) On PID controllers based on simulated annealing algorithm. In: Proceedings of the 27th Chinese control conference. Yunnan, China
11. Kennedy P, Eberhart R (1995) Particle swarm optimization. In: Proceedings of IEEE international conference on neural networks. vol 4. pp 1942–1948
12. Maiti D, Biswas S, Konar A (2008) Design of a fractional order PID controller using particle swarm optimization technique. In: 2nd National conference on recent trends in information systems,
13. Suganthan PN (2009) Differential evolution algorithm with strategy adaptation for global numerical optimization. IEEE Trans Evol Comp 13(2)
14. Suganthan PN (2006) Self-adaptive differential evolution algorithm for constrained real-parameter optimization, In: IEEE Congress on evolution computation. Vancouver, BC, Canada

Performance Analysis of Three-Phase Cuk Inverter with PWM and SPWM Modulation Based on Power Quality

D. Venkata Ramana, P. Chandrasekar, R. Sreedhar, K. Karunanithi, K. Aruna Rani, and A. Vijayakumar

Abstract The objective of the paper is to analyse and compare the performance of three-phase Cuk-based inverter with PWM and SPWM techniques in the aspects of power quality using MATLAB/Simulink. In three-phase inverters configuration, a distinct method known as single-stage inverters, which has less control circuit complexity and requires minimum number of switching components, is implemented for this comparative analysis. Three-phase Cuk inverter is one of the capable methods to implement in grid-integrated and stand-alone applications. The normal PWM technique involves modulation of 50 Hz square wave with high frequency switching pulses, which was introduced originally for the inverters. Here, 125 kHz pulses are selected as high frequency switching pulses. In SPWM technique, a 50 Hz sine wave is modulated by high frequency switching pulses. Since sinusoidal signal is selected as modulating signal in SPWM, the harmonics are reduced when compared to other PWM control techniques.

Keywords MATLAB/Simulink · Pulse width modulation (PWM) · Sinusoidal pulse width modulation (SPWM) · Three-phase Cuk inverter (TCI) · Total harmonic distortion (THD)

1 Introduction

Power quality is an important aspect in power supply distribution system in order to reduce losses and maintain the stability of the system. When inverters are employed in

D. Venkata Ramana · P. Chandrasekar · R. Sreedhar (✉) · K. Karunanithi · K. Aruna Rani · A. Vijayakumar
Department of EEE, Vel Tech Rangarajan Dr Sagunthala R&D Institute of Science and Technology, Chennai, India

P. Chandrasekar
e-mail: hodeee@veltech.edu.in

K. Karunanithi
e-mail: drkkarunanithi@veltech.edu.in

S. S. Dash et al. (eds.), *Sixth International Conference on Intelligent Computing and Applications*, Advances in Intelligent Systems and Computing 1369,
https://doi.org/10.1007/978-981-16-1335-7_9

power system, they introduce lot of harmonics due which power losses increase. Also, the inverters involve inductive devices which will consume more reactive power, leads to poor power factor and reducing the stability level of power system. Therefore, power quality analysis is very important in inverter circuits. Even though plenty of power quality issues like individual harmonics, voltage fluctuations, electromagnetic interference (EMI) affects the transmission and distribution system, total harmonic distortion (THD) will be one of the major issues in terms of inverters.

Though performance characteristics of inverters varies with number of factors like type of load, source, control technique and output level, a Cuk inverter plays a major role because of its better performance, less noise generation, high efficient method and its stability. The design analysis and implementation procedure of power factor correction (PFC) Cuk converter for SRM motor are discussed in [1]. The detailed design of a new Cuk-based Z-source inverter is investigated [2]. The higher levels of inverters are more complex to control and instability exists there. A dual-mode grid-connected Cuk converter for PV/TE applications is modelled and implemented here [3]. Repetitive controller for Cuk inverter in CCM mode with phase lead compensation is analysed [4].

Discontinuous modulation schemes are more popular in inverters because of its simple circuitry and high efficiency. A discontinuous switching technique is proposed to control power electronics switches in single- and three-phase differential mode Cuk inverter [5]. Though the type of inverter plays major role to convert DC–AC, the performance, stability, load independency and the purity in sinusoidal output with power quality will be maintained by the proper selection and implementation of control circuitry of the inverter. New and innovative PWM-based control methods for three-phase three-level inverters are investigated and compared in [6]. A digital control chip is designed to implement feed-forward and frequency control technique for three-phase full-bridge inverter and controlling of VSI using digital control devices for renewable energy sources are reviewed in [7].

Various harmonics reduction techniques are investigated, analysed and compared for three-phase bridge inverter [8]. A detailed comparative analysis is carried out using several PWM techniques for three-level diode-clamped VSI, and based on this study, a new carrier-based PWM technique is introduced for low output voltage THD for multilevel inverters (MLI) [9]. A comparison study is carried over different modulation schemes for modular multilevel converters [10].

The above study involves a concise and an organized review of various inverter topologies and salient features of Cuk converter and modulation techniques to enhance the power quality. Detailed study of power quality issues is also carried. Here, A three-phase Cuk inverter is connected across the positive terminal of each module which results in greatly reducing the individual harmonics and THD. The working of the three-phase Cuk inverter is carried out with normal PWM and SPWM techniques with RL load and its performance is compared based on the THD in the following chapters of this paper.

2 TCI with PWM Control

PWM technique involves modulation of 50 Hz square wave with high frequency switching pulses. Therefore, the 50 Hz square wave is phase shifted by 120° with respect to each other to provide ideal three-phase reference signal. The block diagram of TCI with PWM control module is shown in Fig. 1. TCI has three single-phase modules connected to common DC bus to form a three-phase inverter circuit. Here, the switching pulses to the TCI are generated by open-loop PWM control. The performance analysis of TCI is done with a simple RL load. The AC voltage and current measurement blocks will measure the output voltage and current from the TCI. From the measured output voltage and current values, the output power from the TCI is manipulated. This is also designed to measure real and reactive power consumption of the load.

The simulation of TCI with PWM control which is connected with RL Load is shown in Fig. 2. A 100 V DC is given as input to generate 230 V RMS voltage across each phase. The simulation work is carried out to measure all the parameters of an inverter like the output voltage (both phase voltage and line voltage), output current, output power, power factor, real power, reactive power consumption by the load. The output power is manipulated by the multiplication of output voltage and current as shown in the simulation. But this work is carried out to analyse the output voltage THD and current THD. The FFT analysis of load current with RL load and load voltage with RL load are shown in Figs. 3 and 4, respectively.

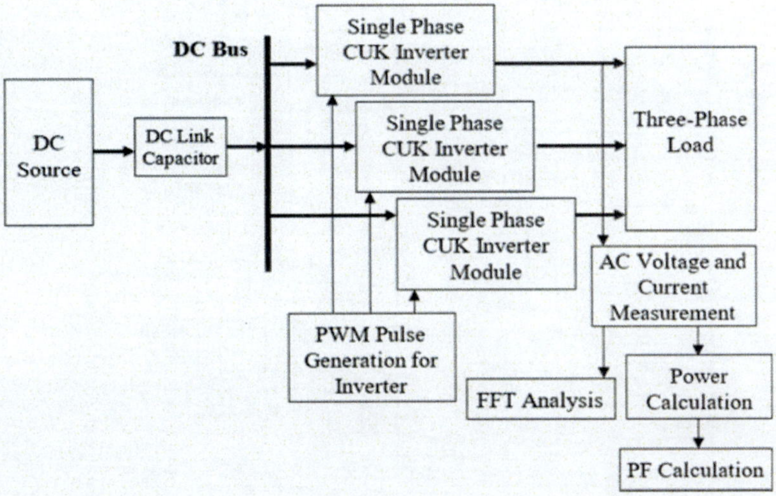

Fig. 1 Block diagram of TCI with PWM control

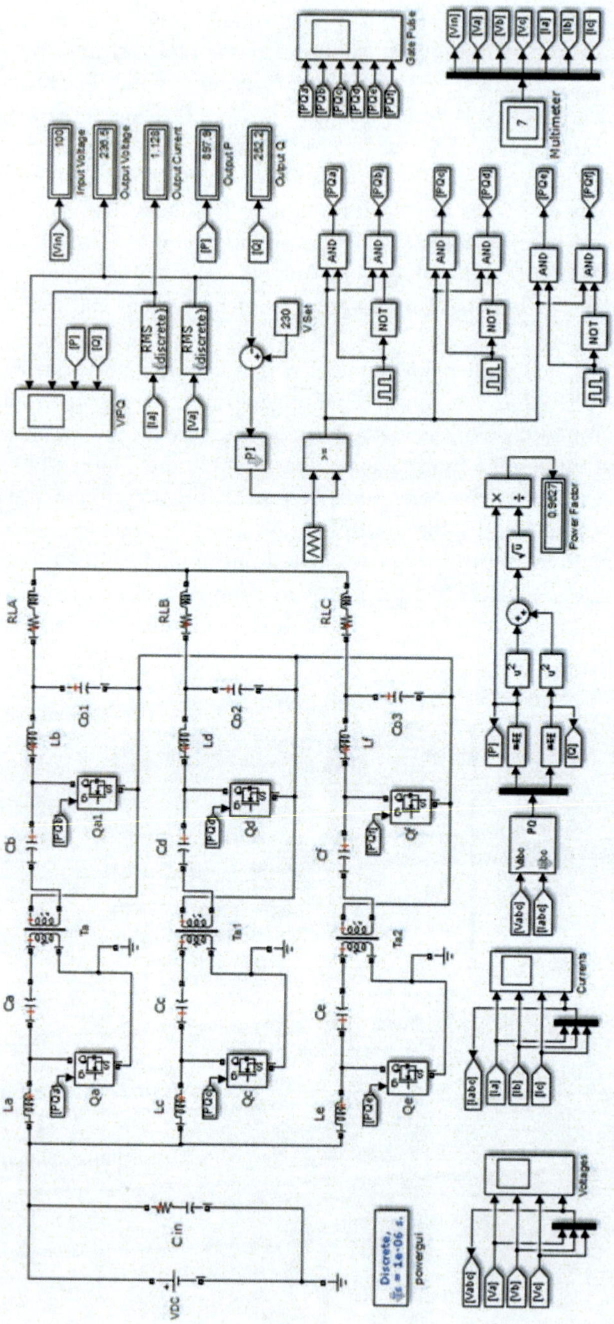

Fig. 2 Simulation circuit with RL load and PWM control

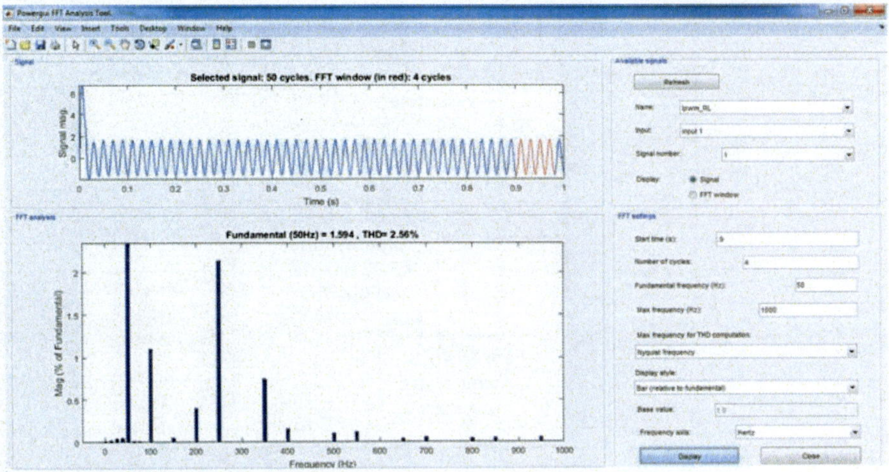

Fig. 3 FFT analysis of load current with RL load

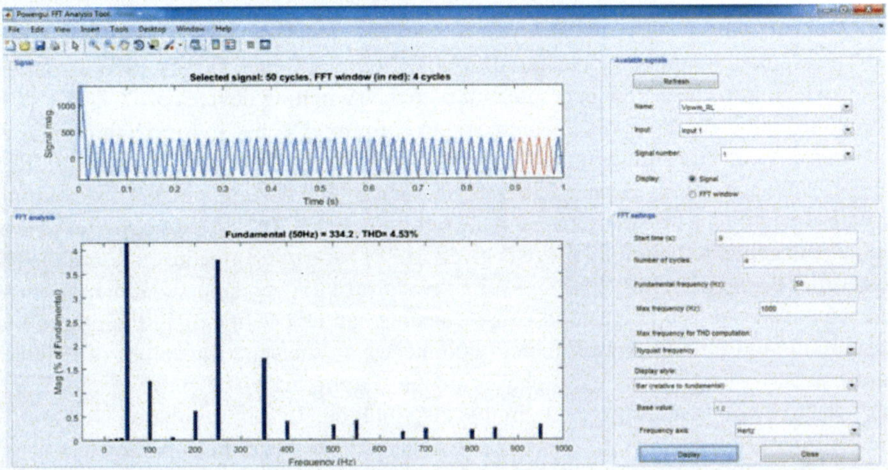

Fig. 4 FFT analysis of load voltage with RL load

3 TCI with SPWM Control

SPWM technique also follows like PWM method which involves modulation of 50 Hz sinusoidal wave with high frequency switching pulses. Therefore, the 50 Hz sine wave is phase shifted by 120° with respect to each other to provide ideal three-phase reference signal. The generated SPWM signals are used to control the switches in TCI to get required output AC voltage and current. The same value of RL load is connected as in PWM control.

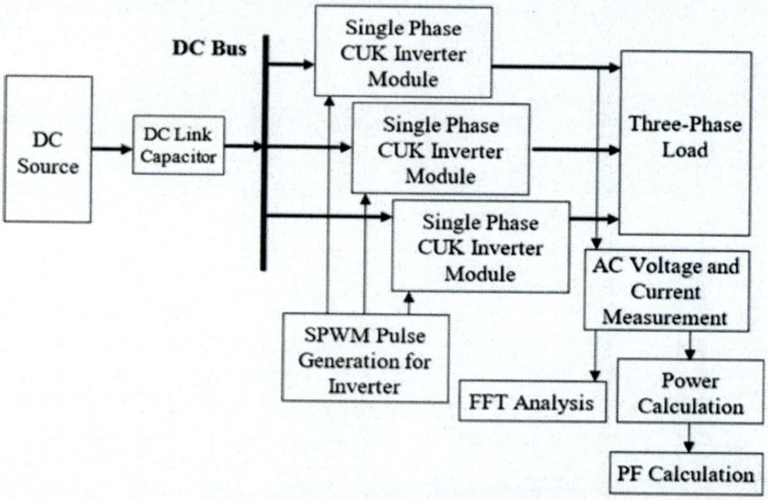

Fig. 5 Block diagram Cuk inverter with SPWM control

The block diagram of TCI using SPWM is shown in Fig. 5. Here, a SPWM-based control technique is used to generate pulses to the switching devices of TCI. The TCI has three single-phase modules connected to common DC bus to form a three-phase inverter circuit. The simulation of TCI with SPWM control with star connected RL load is shown in Fig. 6. Here, a 230 V RMS AC voltage is compared with the output voltage to avoid the drastic changes in the output voltage. The difference between the actual and generated output voltage is given to a PI controller, and the output of the PI controller is compared with the sinusoidal waveform to get a sinusoidal output from the TCI. The phase voltages and phase currents can be continuously monitored by this simulation circuit. A multimeter is connected to measure the values of voltage and current values across each phases of the output side. The power factor, real and reactive power measurement circuits also included in this simulation so that the power quality of this inverter can be also analysed based on these parameters. The FFT analysis of load current and load voltage is shown in Figs. 7 and 8, respectively.

The output of the Cuk inverter which was controlled by two different modulation techniques namely PWM and SPWM were analysed. The load current THD with RL load using PWM control is 2.56% and using SPWM control the load current THD is reduced to 2.23%. Similarly, the load voltage THD is around 4.53% when TCI is controlled by PWM technique, and it is reduced to 4.04% while using SPWM control.

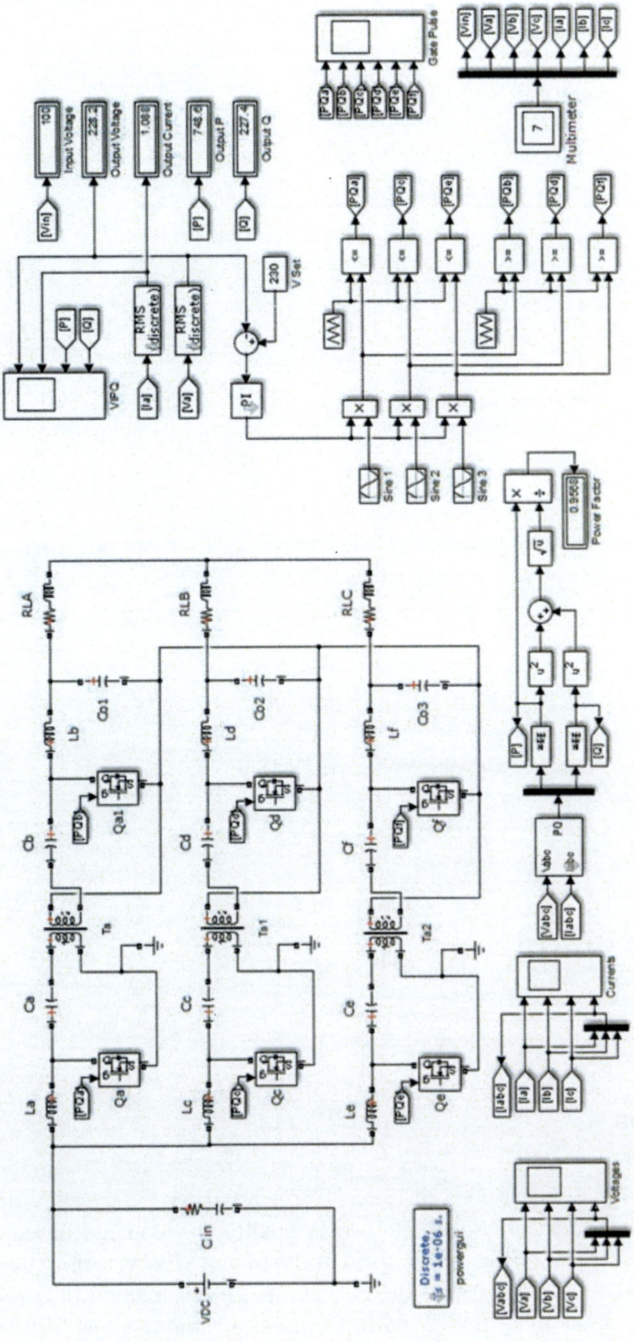

Fig. 6 Simulation circuit with RL load and SPWM control

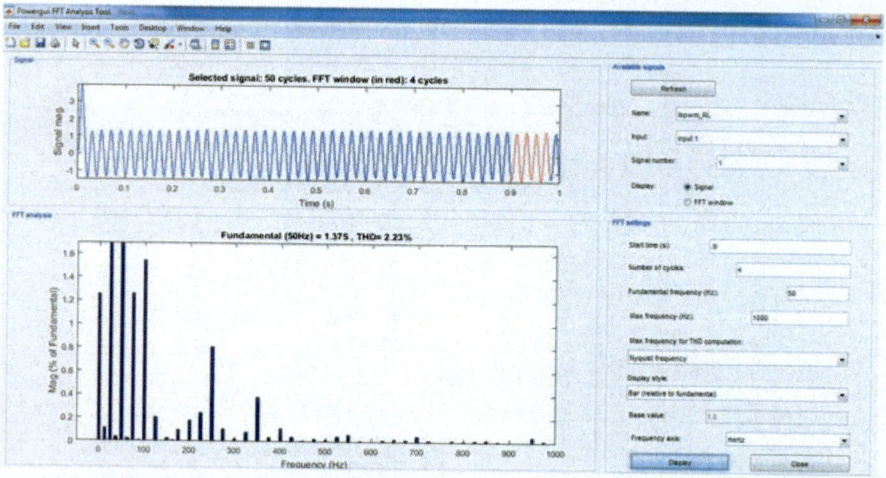

Fig. 7 FFT analysis of load current with RL load

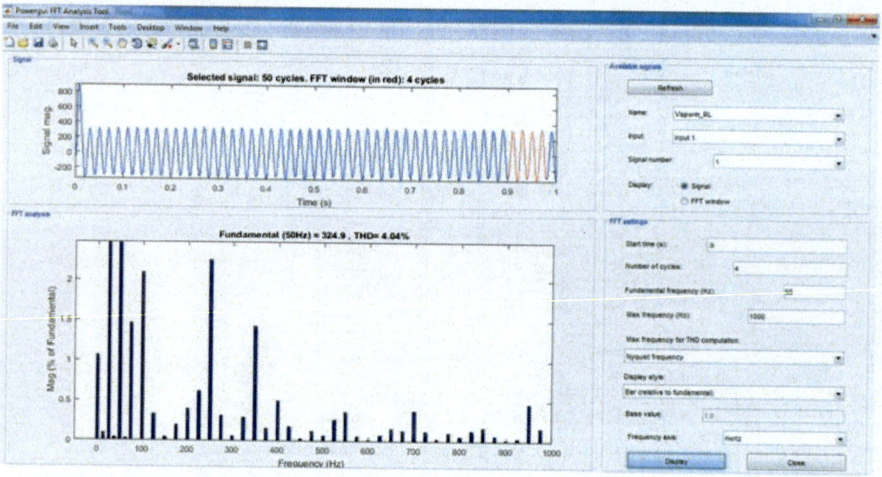

Fig. 8 FFT analysis of load voltage with RL load

4 Conclusion

The working of three-phase Cuk inverter (TCI) has been analysed with PWM and SPWM modulation techniques and power quality is compared based on total harmonic distortion (THD). The comparison of PWM and SPWM techniques reveals that SPWM modulation technique is better than the former method of control. The simulation results show that the SPWM modulation technique gives further reduction in THD value in both output voltage and output current. This inverter can be examined with other PWM techniques as well as the performance of this TCI can be further improved for the mitigation of different power quality issues with other commercial loads like motors.

5 Scope for Future Work

The scope for the future work is listed as follows.

1. Cuk inverter can be studied for other advanced PWM techniques.
2. Cuk inverter can be analysed with motor loads and time-varying heavy duty loads like electric vehicles.

References

1. Anand A, Singh B (2017) Design and implementation of PFC Cuk converter fed SRM drive. IET Power Electron 10(12):1539–1549
2. Wang B, Tang W (2018) A new CUK-based Z-source inverter. Electron 7(11):313
3. Han B, Lai J-S, Kim M (2017) Dynamic modeling and controller design of dual-mode Cuk inverter in grid-connected PV/TE applications. IEEE Trans Power Electron 33(10):8887–8904
4. Han B, Lee JS, Kim M (2017) Repetitive controller with phase-lead compensation for Cuk CCM inverter. IEEE Trans Indus Electron 65(3):2356–2367
5. Soni H, Mazumder SK, Gupta A, Chatterjee D, Kulkarni A (2017) Control of isolated differential-mode single-and three-phase Ćuk inverters at module level. IEEE Trans Power Electron 33(10):8872–8886
6. Chokkalingham B, Padmanaban S, Blaabjerg F (2018) Investigation and comparative analysis of advanced PWM techniques for three-phase three-level NPC-MLI drives. Electr Power Compon Syst 46(3):258–269
7. Tahir S, Wang J, Baloch MH, Kaloi GS (2018) Digital control techniques based on voltage source inverters in renewable energy applications: a review. Electron 7(2):18
8. Rana DR (2018) Analysis and comparison of harmonics reduction topologies for three phase bridge inverter . Int J Sci Res Develop 6(3):137–140
9. Noroozi M, Abrishamifar A, Rezaierad G (2020) A new carrier-based pulse width modulation scheme with reduced output voltage total harmonic distortion for multi-level inverters. IETE J Res: 1–12
10. Moranchel M, Huerta F, Sanz I, Bueno E, Rodríguez FJ (2016) A comparison of modulation techniques for modular multilevel converters. Energies 9(12):1091

A Pedagogical Approach on Computer Networking in Practice Using a Smartphone

Vemula Rajesh, K. R. Aravind Britto, P. Thanapal, G. Sasi, and V. Elamaran

Abstract A general opinion is that learning computer networking protocols is very burdensome for freshman students. The prime motive of this manuscript is to explore hands-on exercises to clinch the fundamentals behind computer networking. This article demonstrates ten significant hands-on exercises using smartphone iOS apps to raise the subject's interest level. Apps such as Depp Whois, wsFILTERS, Ping, Port Scan, and Trace Route are the first five practical experiments considered here. The remaining five hands-on problems are such as PathPing, Nslookup, OS fingerprinting, Telnet, and Address Resolution Protocol (ARP) for a better understanding of data communications on the Internet. This research article inspires both the teachers and students community with self-learning, which would help to proceed with further qualitative research on advanced topics such as network security, ethical hacking, and sensor networks.

Keywords Apps · Computer networks · Nmap · Ping · Port Scan · Trace Route · Wireshark

V. Rajesh
Department of ECE, School of EET, Kalasalingam Academy of Research and Education, Anand Nagar, Krishnankoil 626126, India

K. R. Aravind Britto
Department of ECE, PSNA College of Engineering and Technology, Dindigul 624622, India

P. Thanapal
School of Information Technology and Engineering, Vellore Institute of Technology, Vellore 632014, India
e-mail: thanapal.p@vit.ac.in

G. Sasi
Department of BME, Vel Tech Multi Tech Dr. Rangarajan Dr. Sakunthala Engineering College, Chennai, India

V. Elamaran (✉)
Department of ECE, School of EEE, SASTRA Deemed University, Thanjavur 613401, India
e-mail: elamaran@ece.sastra.edu

S. S. Dash et al. (eds.), *Sixth International Conference on Intelligent Computing and Applications*, Advances in Intelligent Systems and Computing 1369,
https://doi.org/10.1007/978-981-16-1335-7_10

1 Introduction

This article explores how university students and professors can explore e-learning (mobile/smart learning) with online engagement, active learning, and assessment tools in their classrooms. The elaborative list of such tools is shown in Fig. 1. Using such online tools in the classroom environment, student learning process, and individual and collaborative learning experiences can be improved effectively. These new teaching and learning methodologies can boost students' engagement, motivation, performance, and recall ability to sophisticated online supporting materials. Some tools would help to produce audience feedback instantly [1].

Moreover, in higher education, teachers, professors, and curriculum designers might effectively exploit such online platforms to speed up their typical workflow. Of course, engaging students digitally with such online tools might be more attractive to both the student and staff community. The article's motive is to create awareness of numerous types of innovative digital tools that are often used in the smart teaching and e-learning process. The approaches using innovative learning tools and techniques involve both the faculty-centered teaching and assessment and student-centered learning to create a real impact on any subject [2, 3].

To create and present a useful online lecture, an efficient and exciting way of teaching approaches, slide organization, and design skills, and innovative methods of asking quizzes are needed extensively. At present, during the corona pandemic, there is a considerable demand to maximize the active and student-centered learning experiences via online classrooms. The emotional intelligence of both faculty and student aid to form a bridge to active learning. Nowadays, tools such as Kahoot, Nearpod, Pear deck, Quizizz, and Socrative become popular game-based technology

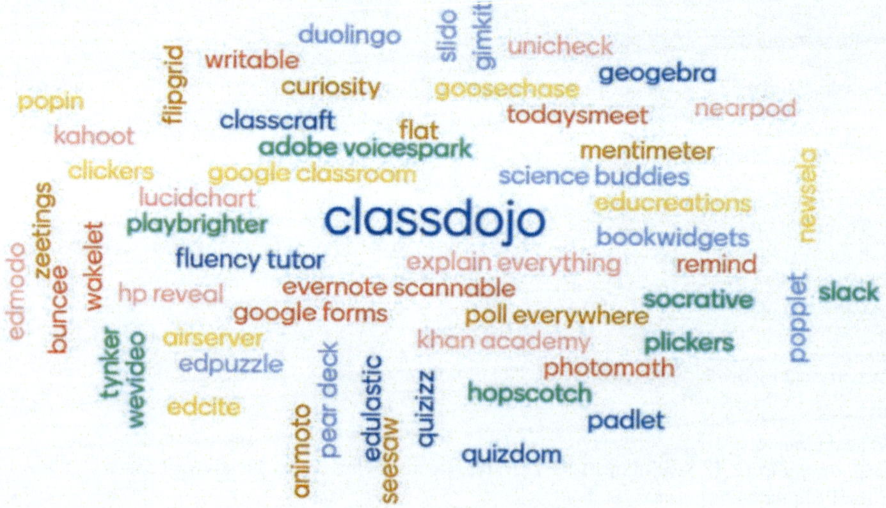

Fig. 1 Engagement, active learning, and assessment tools (online)

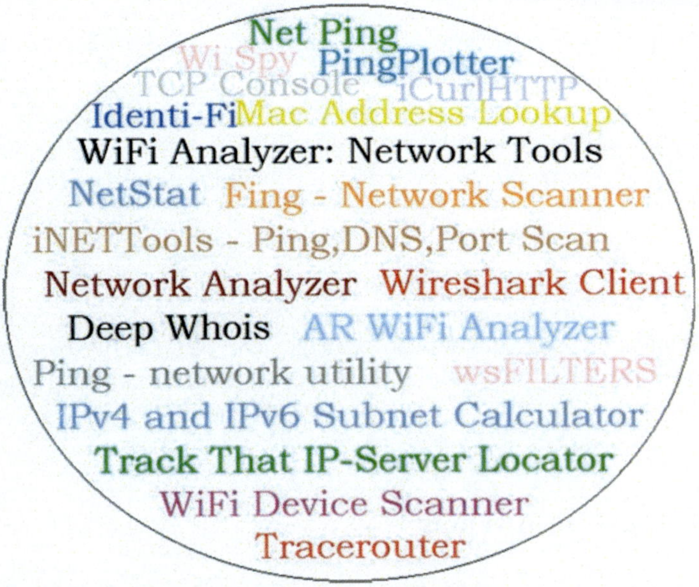

Fig. 2 Various iOS apps focusing on the computer networking field

solutions for online classrooms. Through such online gamification tools, faculty might observe the tremendous growth in student participation and performance. Also, the online feedback assessments become handier to the faculty at the cost of zero time [4, 5].

At present, survival in academia is a big challenge for teachers. The challenges for professors may arise through various forms, such as effective teaching, smart teaching, writing proposals for research funds, quality publications, and patents. To raise the level of interest in classes, students can exploit plenty of tools such as online educational apps, learning animations, and virtual laboratory sessions. A few impressive smartphone (iOS) apps related to the computer networking domain are shown in Fig. 2. Only Apple devices such as iPhones, iPads, iPods, iPad Pros, and iPad-mini can be used to run such iOS apps. Some of them are commercial, and few others are for free use. For instance, iOS apps such as AR WiFi Analyzer, WiFi Device Scanner, Wireshark Client, Nice Trace—Trace Route, and IPv4 and IPv6 Subnet Calculator are paid apps from the Appstore (Apple). Most of the other iOS apps in Fig. 2 are free to use.

2 Materials and Methods

The activity or hands-on-based computer networking course is designed to motivate the students and encourage them to research the allied subjects of computer

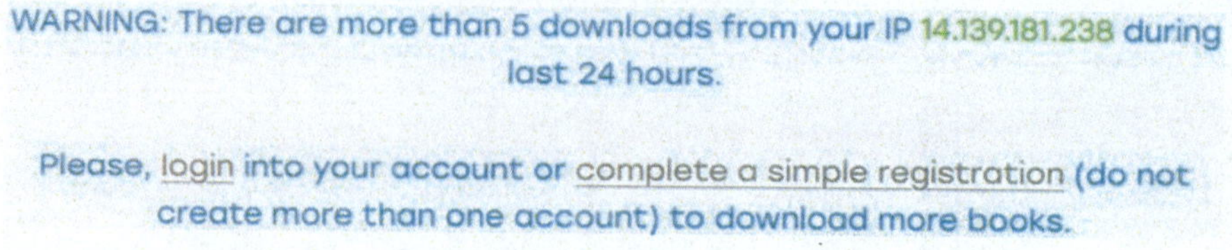

Fig. 3 Public address is seen by the private network

networking such as cryptography and network security, wireless networks, sensor networks, ethical hacking, and network management. Few activity-based exercises (experiments) and hands-on examples to explore computer networking concepts are demonstrated in this section.

2.1 *The App*—Deep Whois

Public and Private Addresses. During the development of IPv4, the scientists thought that the count of IP addresses would be enough to serve the Internet. Before the invention of IPv6, two temporary solutions such as classless IP addressing and private addressing were implemented with Network Address Translation (NAT) protocol. While the people are working inside the university campus or an organization, the LAN in which their systems are connected are not seen by the outside Internet; means their IP addresses are private, not public. But, the organization (or university) should have a public address that is seen by the Internet, means by everyone around the globe [6].

For instance, while the users connect to the domain "www.z-lib.org" to download the ebooks, a warning message as illustrated in Fig. 3, indicates that the IP address (public address of the university) exceeds the daily limit. The domain would offer free use without registration for the guests. The public address of an organization is seen from Fig. 4 using an iOS app (Deep Whois).

Here follows another possible exercise to know the public address. If the public address exploits "sci-hub" as a proxy to access the articles from the publisher, say for instance from the American Association of Physics Teachers, the publisher might block that address. Figure 5 pinpoints this scenario, i.e. the reason behind blocking an IP address [7].

2.2 *Filters Used in Wireshark—WsFILTERS*

The utility of Packet Sniffers. Nowadays, the packet sniffer tool such as Wireshark can be used in cyber-forensics due to its capability to reveal the source of the

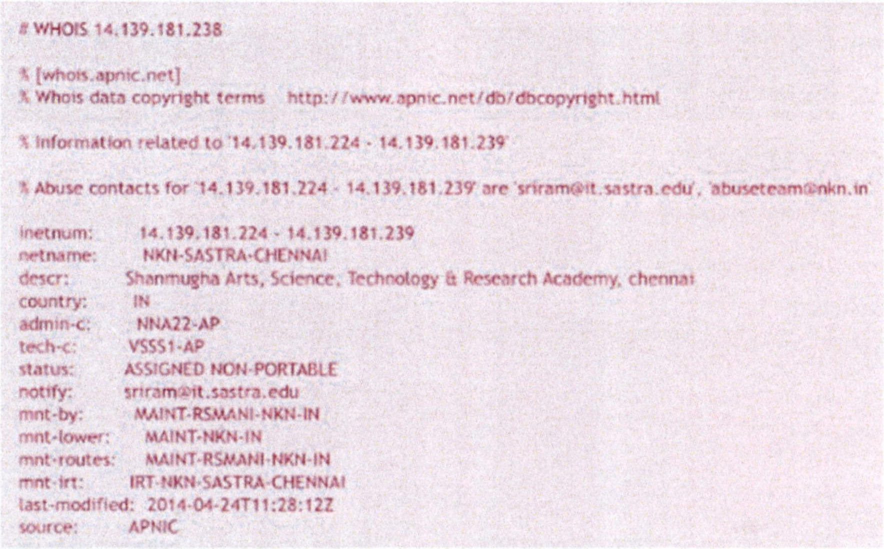

```
# WHOIS 14.139.181.238

% [whois.apnic.net]
% Whois data copyright terms    http://www.apnic.net/db/dbcopyright.html

% Information related to '14.139.181.224 - 14.139.181.239'

% Abuse contacts for '14.139.181.224 - 14.139.181.239' are 'sriram@it.sastra.edu', 'abuseteam@nkn.in'

inetnum:        14.139.181.224 - 14.139.181.239
netname:        NKN-SASTRA-CHENNAI
descr:          Shanmugha Arts, Science, Technology & Research Academy, chennai
country:        IN
admin-c:        NNA22-AP
tech-c:         VSSS1-AP
status:         ASSIGNED NON-PORTABLE
notify:         sriram@it.sastra.edu
mnt-by:         MAINT-RSMANI-NKN-IN
mnt-lower:      MAINT-NKN-IN
mnt-routes:     MAINT-RSMANI-NKN-IN
mnt-irt:        IRT-NKN-SASTRA-CHENNAI
last-modified:  2014-04-24T11:28:12Z
source:         APNIC
```

Fig. 4 Public address seen by the iOS app "Deep Whois"

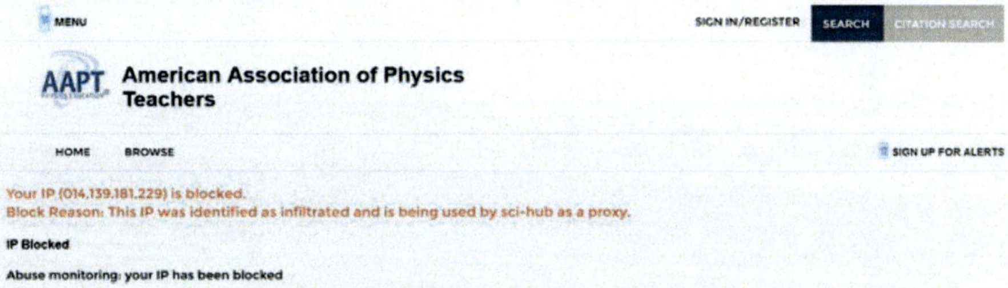

Fig. 5 Status of a public address while accessing the publisher "AAPT"

attack, time, and duration of the attack, protocols and ports involved, and data transmitted during the attack. Wireshark is a powerful tool having more than 1,40,000 filters applied for the packet analysis during the traffic. The packet sniffer tool's sole purpose is to interpret and analyze the packets during the network traffic. Mainly, two categories of filters, such as capture filters and display filters, are used during the analysis of captured packets [8].

One such exercise to analyze the DNS query and response packets for the domain "www.iitd.ac.in" is explored here. The relevant filter from the DNS protocol is selected from an iOS app, as illustrated in Fig. 6. The resulting filter is applied in the Wireshark tool for further packet analysis, as shown in Fig. 7.

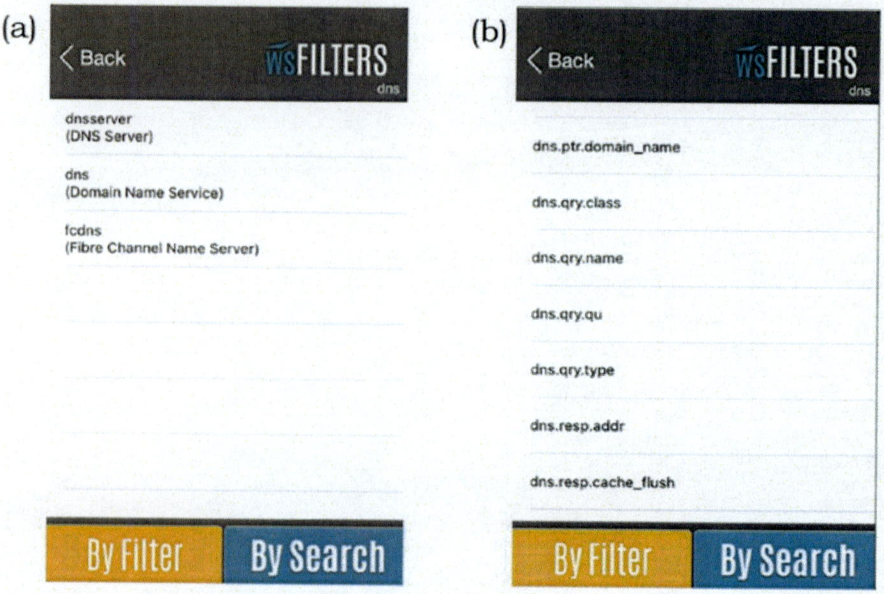

Fig. 6 Exploring Wireshark filters using iOS app "wsFILTERS": **a** search on "DNS"; **b** various options of filters involved in "DNS"

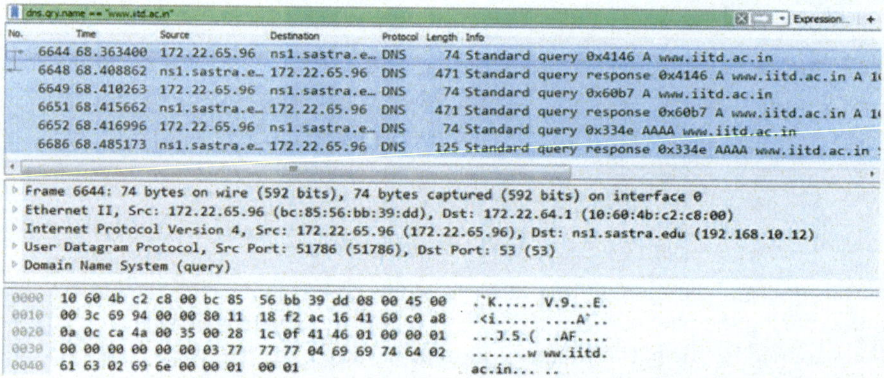

Fig. 7 Packets captured by Wireshark using the filter "dns.qry.name = = "www.iitd.ac.in"

2.3 *Network Diagnose Tools*—iNetTools—Ping, DNS, Port Scan

The Ping Command. The utility of the "ping" command is to find the reachability of the destination node. If it is connected to the network, means if it is reachable, the remote system would respond with an "echo-reply" for the "echo-request". The Internet Control Message Protocol (ICMP) plays a vital role during the ping activity.

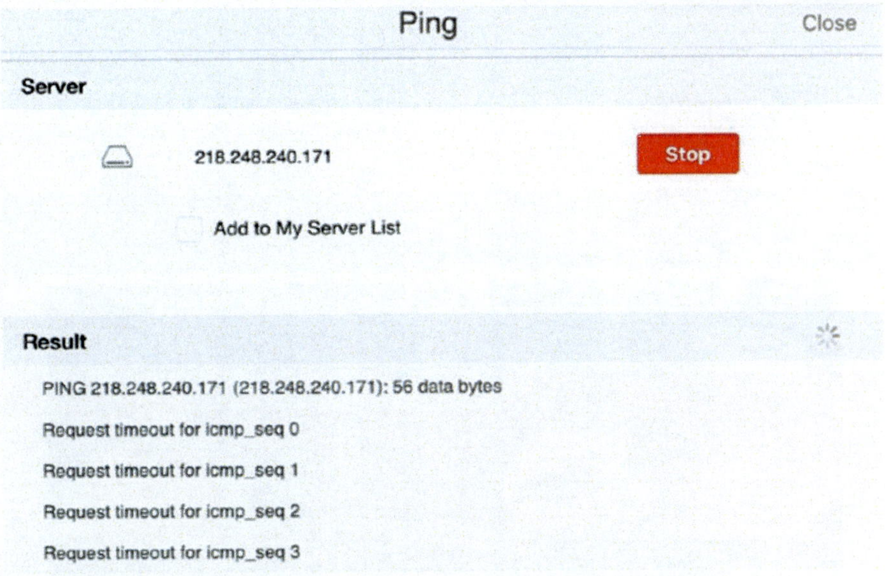

Fig. 8 Denial of ICMP packets by a firewall server from the portal "bel-india.in"

Due to the security issues such as hacking and attacks by intruders, some critical portals do not accept such ICMP echo packets. One such example is explored in Fig. 8, but the portal's home page of the portal is active, which can be seen in Fig. 9.

2.4 *Port Scan*—iNetTools—Ping, DNS, Port Scan

Port Scan. The sole objective of port scanning by the Nmap tool is to determine which ports are closed, opened, and filtered. For instance, if the ports fall under the filtered category, the packets under the specified ports will be filtered by firewall's rules that resides in the server. Similarly, if the ports fall under the open category, the server might accept the predefined services under the specified ports. One such exercise is illustrated in Fig. 10 to detect the available open ports on the portal "www.iiitb.ac.in" using an iOS app [9]. The results are closer to the findings obtained using the Nmap, a network scanner tool shown in Fig. 11.

2.5 *Trace Route*—iNetTools—Ping, DNS, Port Scan

Trace Route. Like the "ping" command, Trace Route is another simple application that helps provide useful troubleshooting information during the connectivity issues

Fig. 9 Secured portal "bel-india.in"

```
C:\>nmap 52.172.39.79
Starting Nmap 7.70 ( https://nmap.org ) at 2020-11-15 10:37 India Standard Time
Nmap scan report for 52.172.39.79
Host is up (0.047s latency).
Not shown: 995 filtered ports
PORT      STATE SERVICE
80/tcp    open  http
443/tcp   open  https
990/tcp   open  ftps
2121/tcp open   ccproxy-ftp
3389/tcp open   ms-wbt-server

Nmap done: 1 IP address (1 host up) scanned in 14.09 seconds
```

Fig. 10 Port scanning using "Nmap" tool

that arise between the hosts. Using Trace Route, the complete path between the source host and the destination host can be displayed; the "*" indicates the reachability failure of nearby switching node or router. One such example is illustrated in Fig. 12 using DOS command mode, and the equivalent steps are done using an iOS app, as shown in Fig. 13. From the results, it is evident that almost both display the same kind of network path with the intermediate nodes (IP addresses).

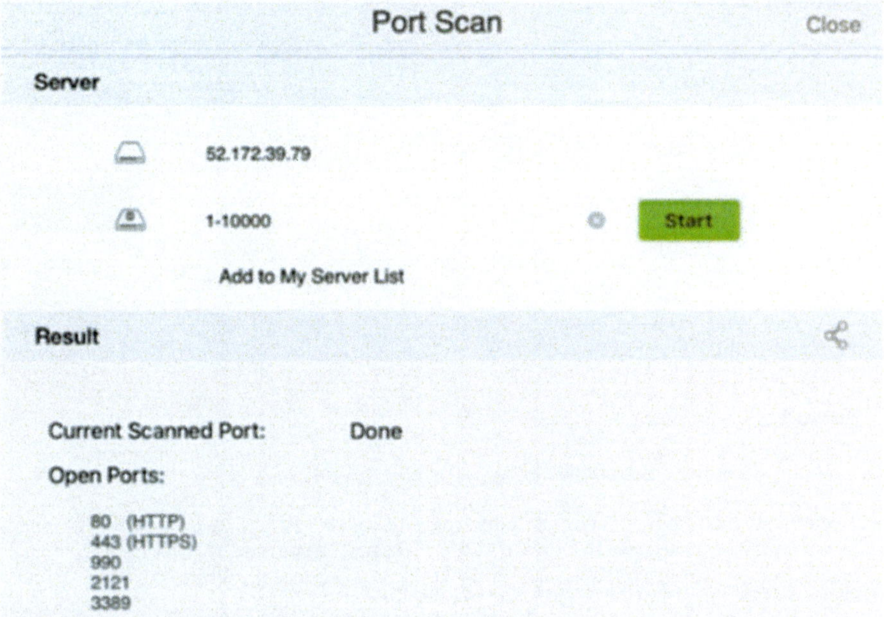

Fig. 11 Port scanning using an iOS app

```
C:\>tracert sastra.edu

Tracing route to sastra.edu [115.240.199.108]
over a maximum of 30 hops:

  1     *        *        *     Request timed out.
  2     *        *        *     Request timed out.
  3    59 ms    46 ms    49 ms  10.71.201.130
  4    83 ms    75 ms    52 ms  192.168.72.255
  5    71 ms    60 ms    59 ms  192.168.71.140
  6    89 ms    52 ms    49 ms  172.26.74.199
  7    72 ms    49 ms    89 ms  172.26.75.83
  8    62 ms    66 ms    71 ms  192.168.68.130
  9    88 ms    53 ms    88 ms  192.168.68.131
 10     *        *        *     Request timed out.
 11    78 ms    55 ms    69 ms  sastra.edu [115.240.199.108]

Trace complete.
```

Fig. 12 Traceroute of the portal "sastra.edu"

2.6 *Other Experiments*

The following exercises (experiments) are also explored during the course to gain further knowledge on computer networking.

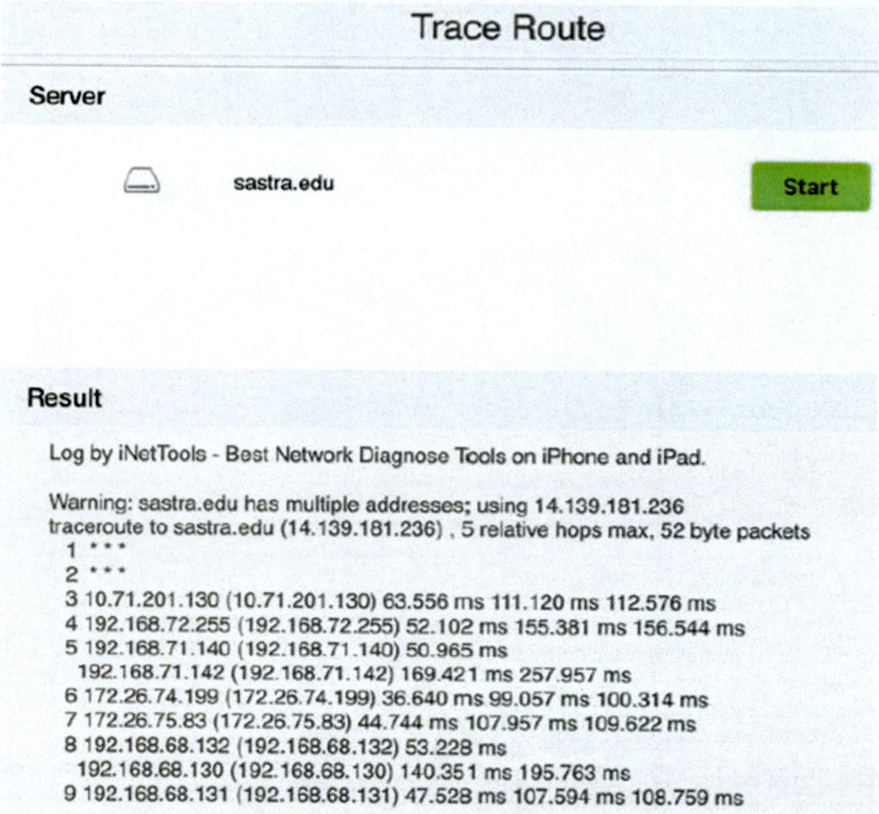

Fig. 13 Similar path (almost) by an iOS app using traceroute

- PathPing
- Nslookup
- OS fingerprinting
- Telnet
- ARP.

3 Survey Results

The sixty students of the short-term computer networking course were inspected to resolve which experiments they noticed most thought-provoking with a scale of 1 to 5. Answer one was "not attractive" and five was "most attractive" [10]. For the five experiments presented here, the average score (out of 60 students) for each experiment is shown in Fig. 14. For the remaining five experiments, Fig. 15 shows the survey results.

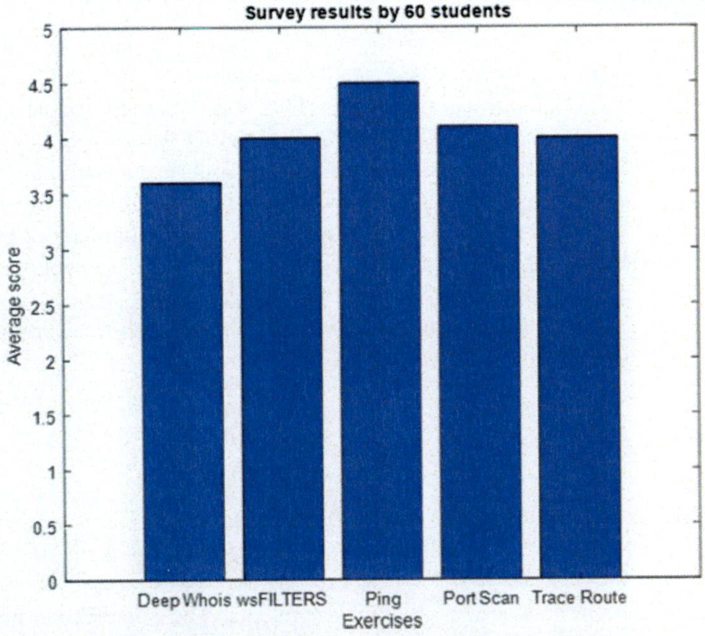

Fig. 14 Survey results for the first five experiments

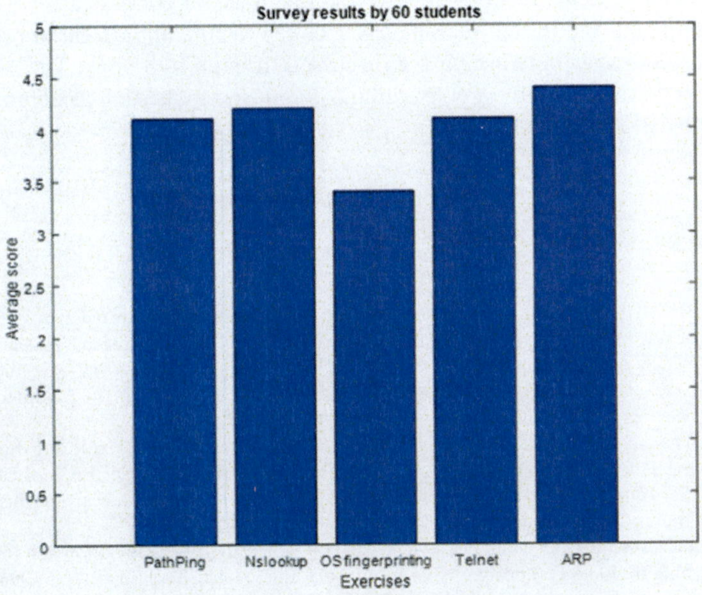

Fig. 15 Survey results for the remaining five experiments

A real-time demonstration with computer networking, for instance, can be more effective because the students can lively see their output results instantly. The survey has shown that most of them learn by doing. This is dramatically rising due to the current generation full of messages (WhatsApp), tweeting, e-mails, and texting. The iOS app allows users to easily view the results for such tasks such as traceroute, ping, and port scanning. The apps used in this article are available for free to use on the app store. Professors and students can use these apps for homework problems, in-class assignments, or an in-class demonstration. Thus, the smartphone or iPad provides an opportunity for students and teachers, especially in engineering, to create applications, or "apps" to aid the learners to gain knowledge on difficult subjects [11].

4 Conclusion

The computer-based exercises are often used in teaching subjects like computer networking and network security to better understand with hands-on examples. Wireshark, the most appropriate tool, can be used to explore the basics of computer networking protocols and network security concepts. The smartphone apps (iOS apps) are used in this article through hands-on examples to cover the fundamentals of computer networking. Computer networking is the most crucial subject in the computer science field. Troubleshooting and diagnosis of computer network nodes play a significant role in the Internet era. The key factors and essential procedures involved in computer networking are discussed through iOS apps. The results are compared with the standard or conventional diagnostic tools such as command-line tools, Wireshark, and Nmap tool.

References

1. An Y (2018) The effects of an online professional development course on teachers' perceptions, attitudes, self-efficacy, and behavioral intentions regarding digital game-based learning. Edu Tech Res Dev 66:1505–1527
2. Goksun DO, Gursoy G (2019) Comparing success and engagement in gamified learning experiences via Kahoot and Quizizz. Comput Educ 135:15–29
3. Sasi G, Thanapal P, Arvind Shriram RK, Balaji VS, Elamaran V (2020) Exploring digital signal processing concepts using on-line graphical DSP simulator. Int J Adv Sci Technol 29:410–421
4. Lopez CE, Tucker CS (2019) The effects of player type on performance; a gamification case study. Comput Hum Behav 91:333–345
5. Elamaran V, Jayapriya D, Hemavathy R, Upadhyay HN (2015) A handy approach for teaching and learning digital VLSI design using EDA tools. Int J Pharm Technol 7:9935–9944
6. Sathishkumar BR, Sundaravadivazhagan B, Martin B, Sasi G, Chandrasekar M, Rakesh Kumar S, Elamaran V, Balaji VS, Arunkumar N (2020) Revisiting computer networking protocols by wireless sniffing on brain signal/image portals. Neural Comput Appl 32:11097–11109
7. Tang T, Zhou MC, Quan Y, Guo JL, Balaji VS, Gomathi V, Elamaran V (2020) Penetration testing and security assessment of healthcare records on hospital websites. J Med Imag Health Inf 10:2242–2246

8. Sasi G, Thanapal P, Balaji VS, Venkat Babu G, Elamaran V (2020) A handy approach for teaching and learning computer networks using Wireshark. In: Proceedings of the fourth international conference on inventive systems and control (ICISC 2020), IEEE Press, Coimbatore, India, pp 456–461
9. Bagyalakshmi G, Rajkumar G, Arunkumar N, Easwaran M, Narasimhan K, Elamaran V, Solarte M, Hernandez I, Ramirez-Gonzalez G (2018) Network vulnerability analysis on brain signal/image databases using Nmap and Wireshark tools. IEEE Access 6:57144–57151
10. Hoffbeck JP (2015) Real-time digital signal processing demonstration platform. In: 122nd ASEE annual conference and exposition. American Society for Engineering Education, Seattle, Washington, pp 1–11
11. Keith JM, Nelson GC, Thompson A, Gazzini JL, Sprabery R (2013) A heat conduction iPhone and iPad app for engineering education. In: 120th ASEE annual conference and exposition. American Society for Engineering Education, Atlanta, Georgia, pp 1–11

A Novel Monitoring Arrangement for Single and Multiple Power Quality Occasions Calculation and Classification in Supply System: A GSA-FUZZY Strategy

K. Karthikumar, V. Senthil Kumar, M. Karuppiah, N. Udhaya Raj, A. Arunbalaj, and S. C. Vijayakumar

Abstract Expanded use of non-direct loads and deficiency occasion on the force framework has developed decrease in the nature of intensity gave to the clients. It is essential to perceive and recognize the force quality (PQ) aggravations in the conveyance framework. In order to perceive and recognize the PQ unsettling influences, it requires the advancement of profoundly assurance plans. This paper presents an ideal insurance plot for PQ occasions expectation and characterization in the circulation framework. The proposed insurance conspire is the consolidated presentation of both the gravity search algorithm (GSA) and fuzzy framework. Here, GSA is used in two stages with the target capacity of expectation and grouping of the PQ occasions. The standard fuzzy first stage is used for perceiving the sound or unfortunate state of the framework under different circumstances. Fluffy is used for see the framework signal solid or undesirable condition under various conditions. In second stage, fluffy plays out the characterization of the unfortunate signs to perceive the privilege PQ occasion for affirmation. Here, the second stage fuzzy learning technique is improved by using the SSO dependent on the base mistake target work. These proposed techniques play an evaluation methodology to guarantee the framework and organize the ideal PQ occasion which happens in the dispersion framework. By then, the recomended work is carryout in MATLAB/Simulink software, and the exhibitions of the recomended framework are contrasted and distinctive existing strategies like MUSIC-ANN, GA-ANN. Examination result exhibits prevalence of GSA-fuzzy method, affirm its capability of PQ occasions gauge and course of action.

K. Karthikumar (✉) · M. Karuppiah · N. U. Raj · A. Arunbalaj · S. C. Vijayakumar
Department of Electrical and Electronics Engineering, Vel Tech Rangarajan Dr Sagunthala R & D Institute of Science and Technology, Avadi, Chennai, India

S. C. Vijayakumar
e-mail: drkarthikumark@veltech.edu.in

V. S. Kumar
College of Engineering Guindy, Anna University, Chennai, India
e-mail: sensang@annauniv.edu

S. S. Dash et al. (eds.), *Sixth International Conference on Intelligent Computing and Applications*, Advances in Intelligent Systems and Computing 1369,
https://doi.org/10.1007/978-981-16-1335-7_11

Keywords Occosions and monitoring scheme · Fuzzy · GSA · XCF · Power quality · Feature removal · Calculation · Discrimination

1 Introduction

Power quality (PQ) is a significant aspect for electric force values, and its customers accordingly investigation happening insurance conspire is expanding more enthusiasm for late years. Voltage hang, swell, interference, flash, indent, drifters and music are the kinds of electrical cable occasions. These occasions are the genuine causes which downgrade the nature of electric force in the framework [1]. These occasions may in like manner prompt glitches, hazards, decreased lifetime, disappointment of electrical gear, etc. These occasions are required to recognize earlier the assurance change can be made to confine PQ occasions [2]. In order to choose the sources and purposes behind occasions, PQ occasions are perceived/ portrayed into different sorts which are the critical issue in the security conspire ask about. The genuine necessities in affirmation plot ask about the ability to perform customized PQ checking and information examination [3]. In such way, the significant job shows in the power quality occasion grouping framework is the component removing and order for which the power qualiy event gauge having the element taking away from the unsettling influences [4]. Consequently, ghastly investigation utilizing discrete Fourier transform and fast Fourier transform [5, 6] has applied. In any case, because of the non-fixed landscape of the PQ occasions, such changes are not ground-breaking in distinctive the unsettling influence graphs. Defeat the downsides of together DFT furthermore, FFT, the wavelet transform has been extensively used. So, a wavelet change be able to dismember diverse power quality occasions simultaneously in both time and recurrence spaces [7, 8]. All things considered, it shows a couple of issues like over the top calculation, affectability to clamor level and the reliance of its exactness on the chose wavelet.

Concerning occasion and portrayal of such aggravation [9–11], in numerous papers, fluffy standards are used to choose. In such procedures, a broad number of commitments to the fluffy situation increase the privilege recognizing confirmation pace of the unsettling influences. Regardless, the impediment is that it moreover fabricates the technique capriciousness and decreases its speed. In any case, for the estimate of numerous PQ occasions, no thought has been paid. Clearly, the various methodologies, for instance, S-change [12], discrete wavelet change and fake neural organization with fluffy rationale [13], Hilbert et al. change [1], S-change, TT-change [14], multilevel wavelet change found neural organization [15], S-change and fluffy master framework [16], altered S-change and molecule swarm improvement [17], wavelet bundle change [18] have been presented for the forecast plus order for PQ occasions. The whole recently referenced frameworks be able to recognize the power quality events anyway it needs broad number of tests plus hereafter the unconventionality of calculation be adequate [19]. Expanded use of non-straight loads and issue occasion on the force framework has advanced decrease in the nature of intensity gave to the clients. It is basic to perceive and recognize the PQ unsettling influences in the circulation framework. To perceive and recognize the PQ unsettling influences, it requires the advancement of exceptionally insurance plans. This paper presents an ideal novel insurance conspire for power quality (PQ) occasions expectation and grouping in the dispersion framework. The proposed insurance conspire is the consolidated exhibition of both the gravity search algorithm (GSA) fuzzy. The proposed procedure is obviously depicted in detail. Remainder of this paper is coordinated as ongoing test, and the found of the exploration work is briefed about in Sect. 2. The current method intensive clarification is clarified in Sect. 3 and 4. The modified strategy complement results and connected conversations are wrote in segment 5, and the article is ended in Segment 6.

2 Resent Research Work: A Review

Different examination works have been just existed in the writing who relied upon forecast arrangement to intensity quantity in the conveyance framework using numerous strategies—alternate points of view. A bit of the works is evaluated and shown a novel way to deal with manage ideal element choice for the grouping of PQDs to perceive the PQ occasions. For the ideal component choice, their methodology contains the discrete wavelet trasform (DWT) and stastic neural organization-based counterfeit honey bee province (PNN-ABC). The element be separated beginning the aggravations using discrete wavelet transform through multi-goal examination (MGE). Representing the grouping of PQDs, the PNN ranger was used. DWT-MSD-based forecast and order of different power quality occasions be shown with Kubendran et al. [12]. For removing the highlights of various unsettling influences, the DWT coefficients-based methodology for the energy substance in the distinctive recurrence zone and the coefficients on every level were used.

It takes model prediction plus allows checking: Power quality distrubance contacts quality, relations throughout lines, identify power flow apparatus and correlate action between monitoring nodes. To assess an electrical system, the exhibited smart sensor network was a capable device which conceivably can identify disturbances in the network. A sorting procedure in view of MST and ELM has recommended with Zhang et al. [21] in favour of classifying PQ disturbances. For improving the fulfilment of point function, in MST, predictable parameters were generated that are more important to discover solutions to time and frequency. Nonsettling reaction in the power system to track and organize control quality control, Kanirajan et al. [22] have suggested a new method that uses RBFNN. A RBFNNN was defined by the extracted function threw the wavelet for the characterization of points.

In combination with a mathematical model, improving the organized procedure was introduced by Rodriguez–Guerrero et al. [23] suggested that waveforms consisting of synchronous PQ disturbances be represented. Because of the way in which a large number of synchronous PQDs can be reproduced from a single descriptive word, it was detailed. Ahila et al. [24] have introduced a fast classification technique for multi-class power system disturbances called an ELM algorithm. The ELM can conduct a multi-category classification without any variation. As far as classification accuracy is concerned, the execution of the ELM classifiers has been enhanced by the displayed approach by defining the best subset of accessible features and solving the uncertain problem of model determination.

P. The fuzzy-flower pollination algorithm (FFPA) hybrid for the performance improvement of the radial distribution system (RDS) was detected by Mariaraja et al. [25]. Under regular and abnormal working conditions, the main objective of this work was to update the re-configuration algorithm of the distribution system. For a continuously stirred tank reactor and pH neutralization mechanism, the double-layer and nonlinear model predictive control method was developed by Cai et al. [26]. The nonlinear structures can be represented as a Hammerstein–Wiener model. It is possible to portray the nonlinear systems as a Hammerstein–Wiener model. Mouzakitis [27] has discovered what modes of fault diagnosis occur and what are their main benefits. In three major classes, namely model-based, hardware-based and history-based fault diagnoses, the fault diagnosis classification has been studied. Recent work by Karthikumar et al. [28] on SSO-detection ANNs of PQ incidents.

2.1 Background of Research Work

In the appropriation context, the fundamental contributing variable is the existence of the electrical force which is recognized by incessant force notification efficiency. This reporting system is used to distinguish, document and ensure the consistency of force issues. In spite of this, the diffusion system gives them instructions on PQ occasions, such as voltage list, swell symphonic bending, move, gleam and drifters. Voltage swells and drops that occur due to electrical drives that start up, similar to circuit failures or malfunctions that can lead to power interruptions. In addition, heaters, the exchange of capacitors, the use of nonlinear loads and lightning strikes are the other agitating PQ impacts. These agitating impacts condense the life of the system and cause electrical types of gear to be affected. Thus, pros use a range of techniques, such as Fourier transform, S-change, H Huang transform, wavelet transform, etc., which are used for the identification of PQ occasions within the force structure.

3 Proposed Delivery Device Security Mechanism for One and Many PQ Incidents

Figure 1 demonstrates expectation and characterization of the proposed PQ insurance scheme in the dispersion context. The information power signal from the circulation system is disassembled in the proposed structure as expected. From the outset, in the pre-preparation stage, standardization and divisional steps have been finished in the

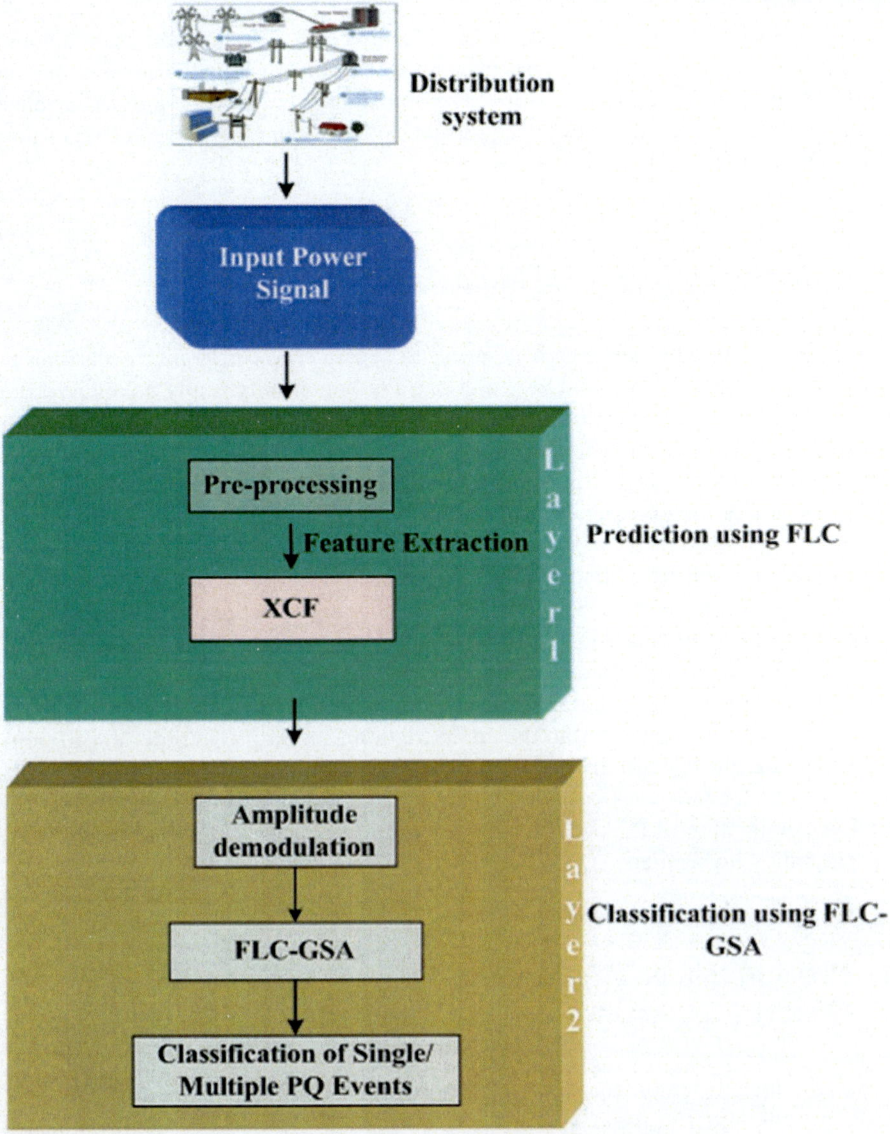

Fig. 1 Proposed assurance plan of PQ occasions forecast and arrangement in the appropriation framework

light of one or thrice) stage occasional details. Perceived information is transferred to the extraction process of the component, which recognizes the unmistakable highlights of the occasion signal. At this point, the fuzzy reasoning controller (FLC) is used to predict PQ opportunities, taking into account the extricated highlights. Here, the single and various PQ occasions are perceived dependent on the cross-connection work (XCF). Perceived PQ opportunities are evaluated for arrangement in the second FLC era. Enough de-modulation has been used here for detecting the cover of person plus various power quality occasions. The adjusting sign can have high recurrence sections, so as to recognize elevated recurrence portions; the gravity search algorithm (GSA) figuring is used.

3.1 Generated Single and Multiple PQ Events

PQ resources may be considered as one and separate power quality occasions in the sharing context [9, 19, and 28]. The one PQ opportunity imply a solitary kind from PQ opportunity inside a balanced symbol, for example, list, swell, music, indent, commotion, etc. These occasions predominantly caused as a result of the programmed detachment of burden, stumbling of insurance gadgets, protection disappointment, lightning and protector flashover. Of course, various PQ occasions infer that more than one sort of PQ occasion, for example, droop with music, swell with music, hang With passing, swelling with music, etc. These occasions are caused by non-straight loads, movements, extra energy misfortunes plus high heating. A few dangerous results of reduced PQ occasions involve high heating of rails, unfavourable production of hardware and appliances, displacement of defence equipment, engine disillusionments, mixed metering and impedance with correspondence circuits. These are the opportunities provided in the appropriation framework. In order to differentiate between the PQ and the category of occasions, the proposed calculation was attempted on a continuous basis to display prevalence. The forecast and characterization examination of PQ occasions by the proposed method is portrayed in detail in the going with segment.

4 Presumption and Taxonomy of Fuzzy-Based Power Quality Measurements

In the anticipated method, fuzzy is used here two steps among the purpose of the standards plus the order of the PQ event. In stage one, the typical fuzzy logic controler is employed for seeing a strong or unfortunate framework signal beneath dissimilar circumstances. In stage two, FLC examines the arrangement of unfortunate signs seeing perfect disturbunce potential for assurance. Here, the second stage FLC learning technique is upgraded by using the GSA calculation in context of the base mistake target work. The proposed assurance plot plays an evaluation strategy to make sure about the framework and characterize the ideal PQ occasion which happens in the dispersion framework.

4.1 *Phase 1: FLC Prediction*

Fluffy rationale is commonly used in machine control. The expression "fluffy" implies the rationale that can oversee thoughts cannot be imparted as "valid" or "bogus", however, rather as "To some degree valid". Disregarding the way that in the plan of the regulator, elective systems, for instance, GA and neural networks (NNs) can play out the fluffy reasoning is a lot of the time. Fluffy rationale has the favored angle in that phrasing that human administrators can see, empowering the utilization of their experience. Starting at now, this observing tasks that are performed adequately by people are made easier

The FLC is used for the recognition and classification of the PQ occasions appeared in the sign produced using the circulated framework. In stage 1, utilization of the FLC classifier helps confinement of the sign got from the conveyance framework as solid or undesirable sign. In stage 2, the FLC classifier the privilege PQ occasions are seen for the framework observing. In the subsequent stage, target work GSA is used to revive the FLC learning system considering the setting of the base blunder. Figure 2 shows the proposed fuzzy rationale controller.

4.1.1 Modelling of Fuzzy Logic Controller

Three important steps for the fuzzy system are

(a) Fuzification
(b) Fuzzy inference engine
(c) De-Fuzification

This is briefly shown in the figure above,

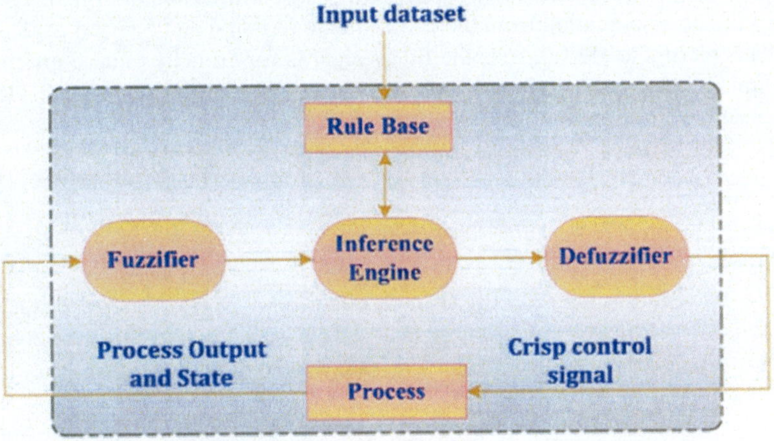

Fig. 2 Proposed assurance plan of PQ occasions forecast and arrangement in the appropriation framework

(a) **Fuzification:** In the fluffy information base, the new commitment to a semantic variable is changed in this technique with the participation work gathered. The fuzzification cycle from the information power signal, the single (or) three stages is utilized in instating PQ events in the expectation stage. Here, by separating the second voltage esteems the different voltage level PQ occasions values with the best voltage estimate in the fuzzification cycle are scaled to per unit in midst of occasion.

(b) **Fuzzy inference engine:** Here, the framework without is shortcoming or under the flaw that exists, and the other conveying the conditions, the principles shape the conditions. The participation capacity can be remembered for both layers of fuzzification and de-fuzzification. On the other hand, the fuzzification and de-fuzzification layers, the fluffy etymological situations to the non-fluffy data concerning the enrollment work all around are pushed out to list the phonetic word.

In different frameworks like three-sided, Gaussian, trapezoidal and so on, the participation capacity can be depicted.

Decision Making: The fluffy standards can be used for every part approach for making impelled fuzzification method. The choice made is based on the norms set out above. The overall methodology for the fluffy norm is In case A, at that point B" (Melin and Castillo 2014). By then, the deficiency is available on the off chance that An is positive and B is negative, and by then, the structure is no shortcoming if An is zero and B is zero. The versatile guidelines are in the framework,

$$\text{If } c_r \text{ is } x_i \text{ and } ec_r \text{ is } y_i, \text{ THEN } z_i^f(ti) \text{ is } F_i(ti) \tag{1}$$

where fuzzy subsets are x_i and y_i, the singleton value of the fuzzy system represents.

(c) **De-Fuzification:** The fluffy regards are gotten while encountering the system of the de-Fuzification from the choice made (Abonyi and Szeifert et al. 2003). This technique de-fuzzifies the fluffy regards for crunchy yield. Participation limit of the yield variable is the reason of the de-fuzzification. It yields a structure information base.

$$F_i(ti) = z_i^f(ti) \tag{2}$$

where target solution after fuzzification is denoted by $z_i^f(ti)$. Fuzzification is the basic objective of the level of support assessed for each person. With the enhanced intrigue work, definitively the fuzzy system predicts the outcomes.

4.2 Phase 2: GSA- Fuzzy Classification

The preparation point for stage 2 is the understanding of single and different PQ opportunities within the circulation system. To evaluate the degree of similarity between the two force signals, XCF is used for this purpose. The category of PQ possibilities was recognized in terms of the limit value. The estimation of XCF coefficients so procured has been differentiated and a current edge esteem. This relationship selects and demands if the modified input sign comprises a single or a number of PQ occasions. Recognizing the ultimate aim of ordering the right PQ opportunity, the analysis is concluded between a single and a variety of occasions. Enough demodulation is used to differentiate between single and multiple PQ occasions. There may be a dose of high recurrence segments in the balanced sign. checking plan in the circulation framework. In the principal layer, FLC is used to see the framework signal solid or unfortunate condition. The second layer of the FLC is utilized for the characterization of the unfortunate signs. FLC learning technique is overhauled by GSA. The proposed method is accepted at the level of MATLAB/Simulink at that time. In different research, the prevalence of the suggested strategy is carried out, and the existing approaches are contrasted and diverse. The outcome of replication indicates that the suggested strategy is convincing in ensuring the PQ propagation system is highly accurate. In like way, the proposed procedure ensures the framework with a more modest multifaceted nature for the location and arrangement of the PQ occasions, and consequently, the classification exactness of the framework is raise.

4.2.1 Gravity Search Algorithm for Accurate Classification

In an arrangement of usage in science, designing, industry and the executives, heuristic calculations, for example, stochastic worldwide advancement strategies have been comprehensively associated starting late (Nezamabadi-Pour and Baranin.d). Inclination steadiness examination, work advancement, boundaries recognizable proof, highlight assurance, the estimate of energy interest, financial and emanation dispatch are a piece of the application comprehensively used here. Extensive exploration has been guided especially for the advancement issues to consider and upgrade its execution given by the heuristic calculation. In various fields of science, designing and business advancement issues are appeared. In this manner, dealing with these issues is unworkable using standard procedures. Thusly, heuristic calculations assembled gigantic lessening for dealing with streamlining issue [29]. Calculations normally fuse ant colony optimization (ACO) simulated annealing (SA), particle swarm optimization (PSO), genetic programming (GP), evolutionary strategy (ES), genetic algorithm (GA) evolutionary programming (EP), BEE's calculation and so on for a large portion of the heuristics. Heuristic calculations, having a decent harmony among investigation and abuse, are basic for achieve both profitable worldwide and neighbourhood looks. In the inquiry space, the function of researching is to find new and better arrangements, and for the worldwide

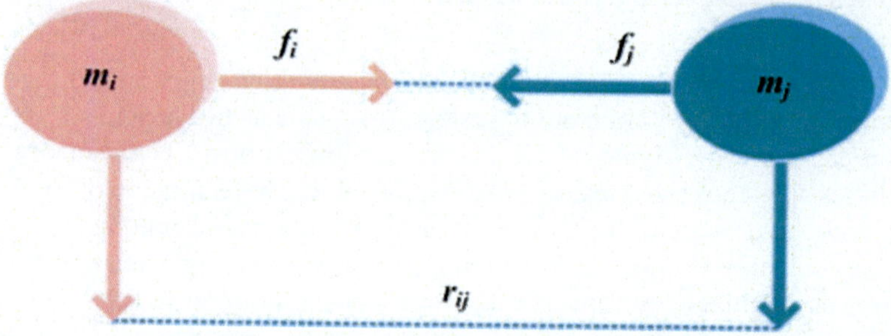

Fig. 3 Gravitational phenomena

best arrangement around the ideal of the neighbourhood search space, the limit of abusing is to look. The gravitational wonders are depicted in Fig. 3.

Their execution is estimated by their masses and agents that are considered as items in GSA. Any of these objects is pulled towards each other by force of gravity into objects with heavier masses. This force allows the creation of all to be equal. Heavier masses are related to better arrangements of the issue. Utilizing a fitness function, the agent position relates to an answer to the issue, and its mass is resolved. In the pursuit, space which would preferably show an optimum arrangement by pass of time, masses are pulled in by the heaviest mass. The GSA could be considered as an isolated system of masses [30]. Complying with the Newtonian laws of gravitation and motion, it resembles a little counterfeit universe of masses. Gravity force induces the worldwide creation of all objects to the largest mass as shown in Fig. 4. At the objects with bigger masses, this force causes a worldwide development of all objects the majority of the objects pull in one another by the gravity force. Great arrangements of the issue can be compared to the vast masses. Appropriate changes in the gravitational and inertial masses lead to each mass speaking to an answer, and the algorithm is explored [31]. In the inquiry space which speaks to the ideal arrangement, it is normal that the masses will be pulled in by the biggest mass with pass of time. By four parameters, for example, the position of the mass in dth dimension, inertia mass, active gravitational mass and passive gravitational mass each agent in GSA is determined. An answer of the issue whole the places of the mass of an agent and its monitoring from make its development moderate the inertial mass of an agent reflects [32, 33]. The velocities of an agent in indicated measurement are figured using the fitness evaluation of the issue and these control both the gravitational mass and the inertial mass. With each iteration, the positions of the agents in determined measurements are refreshed, and the best fitness is recorded alongside its comparing agent [34, 35]. The end state of the algorithm is characterized by a settled measure of iterations; naturally achieving which the calculation stops. At a predefined measurement of the relating specialist at final iteration turns into the global fitness for a specific issue and the situation of the mass turns into the worldwide arrangement of that issue after end of the algorithm, the recorded best fitness is assessed. Idea of gravitational forces between particles is appeared in Fig. 5.

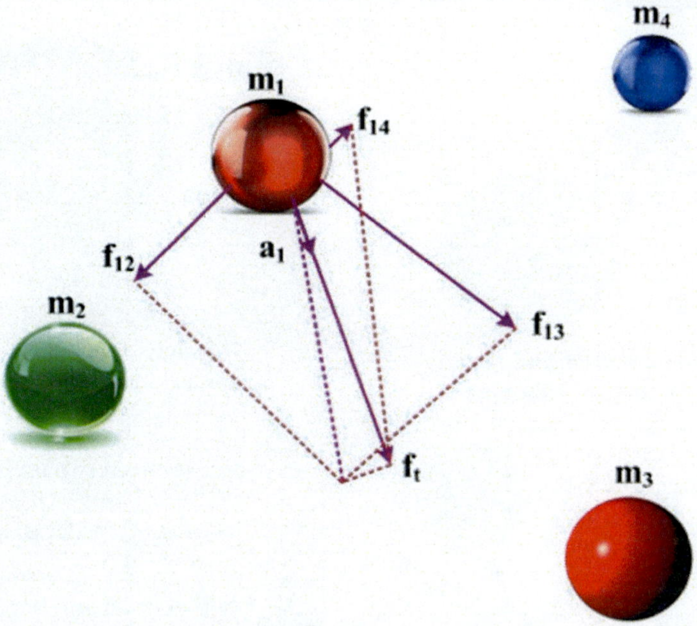

Fig. 4 Gravity power causes a worldwide development of all items towards the biggest mass

Step 1: Initialization of the agents

Initialize the random generation of agents positions inside the given search interval in this step as beneath:

$$Y_j = \left(y_j^l, \ldots, y_j^n, \ldots, y_j^R\right) \quad \text{for } j = 1, \ldots, n \tag{3}$$

where location of the j th mass in nth dimension as y_j^n as well as space dimension is the R.

Step 2: Fitness evaluation

In this progression, the target work is surveyed subject to determining the precision for each arrangement. The categorization accuracy (CA) is resolved as,

$$\text{CA}(\%) = \frac{T_P + T_N}{T_P + F_N + F_P + T_N} \times 100 \tag{4}$$

where T_P and T_N represent the number of accurate predictions with the actual class as true and false, the number of incorrect predictions with the actual class as true and false are F_P and F_N. Represents responsibly.

Step 3: Gravitational computation of constant

The gravitational constant $g(t)$ is computed utilizing the t iteration by Eq. (5).

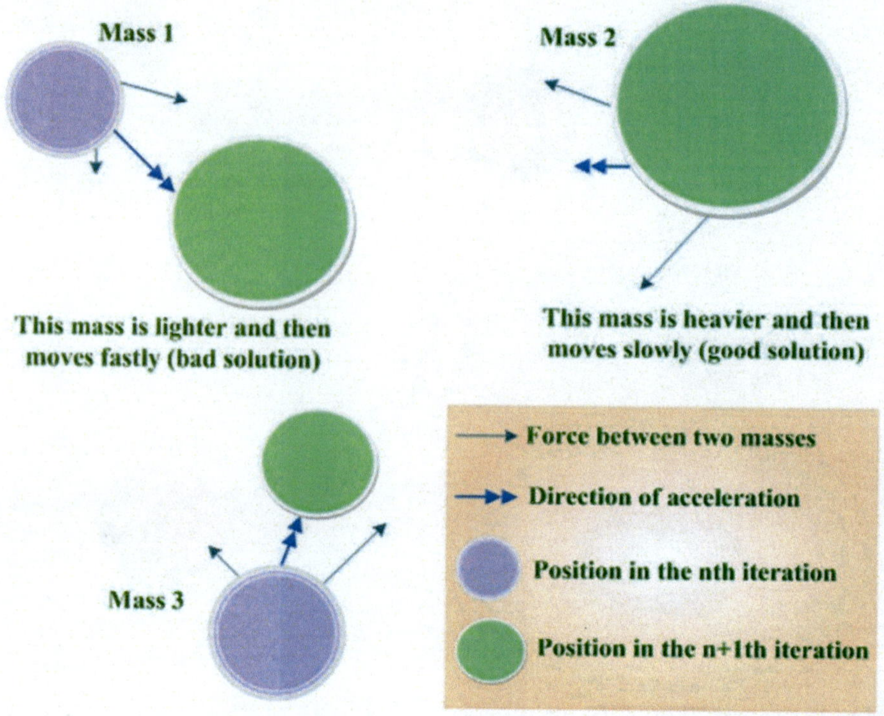

Fig. 5 Concept of gravitational forces between particles

$$g(t) = g_0 \exp\left(-\sigma \frac{t}{T}\right) \tag{5}$$

where g_0 is the randomly chosen gravitational constant, σ is the constant, cumulative iteration number represents T, and current period is the t.

Step 4: Revise the intertial weights

By subsequent iteration, the inertial weights and gravitational constants are modified.

$$x_j(t) = \frac{F_i(t) - w(t)}{b(t) - w(t)} \tag{6}$$

The weight of the jth agent is given below

$$X_j(t) = \frac{x_j(t)}{\sum_{i=1}^{n} x_i(t)} \tag{7}$$

Step 5: Total force calculation

At iteration t, the cumulative force acting on the jth agent shall be measured as follows.

$$F_i^{\mathrm{d}}(t) = \sum\nolimits_{i \in k\,\mathrm{best}\, i \neq j} R_i F_{ji}^{\mathrm{d}}(t) \tag{8}$$

where R_i is a random number between interval [0, 1]. The force acting on mass $(X_j(t))$ from mass $(X_i(t))$ is equated as below, according to the gravitational principle.

$$F_{ji}^{\mathrm{d}}(t) = g(t)\frac{X_j(t) \times X_i(t)}{r_{ji}(t) + \in}\left(y_i^{\mathrm{d}}(t) - y_j^{\mathrm{d}}(t)\right) \tag{9}$$

where constant is the $\in$ and $r_{ji}(t)$ are defined as the range of the Euclidian between jth and ith agents.

Step 6: Speeding up and rate measurement

The speeding up $A_j^{\mathrm{d}}(t)$ at iteration t and the rate of the jth negotiator at next iteration $(t + 1)$ in d th measurement is adapted by the law of enormity and law of wave as succeeding equation below.

$$A_j^{\mathrm{d}}(t) = \frac{F_j^{\mathrm{d}}(t)}{X_j^{\mathrm{d}}(t)} \tag{10}$$

$$V_j^{\mathrm{d}}(t+1) = R_j \times V_j^{\mathrm{d}}(t) + A_j^{\mathrm{d}}(t) \tag{11}$$

Step 7: Update the location of the agent

Then, the next locations of jth agents in d th dimension are updated as follows.

$$y_j^{\mathrm{d}}(t+1) = y_j^{\mathrm{d}}(t) + V_j^{\mathrm{d}}(t+1) \tag{12}$$

Step 8: Best return solution

The means from two to seven are repeated until the emphasis accomplishes its most extreme cutoff. At conclusive emphasis, the best arrangements of calculation are enlisted as a worldwide wellness capacity of the issue. Figure 6 is the flowchart of proposed GSA.

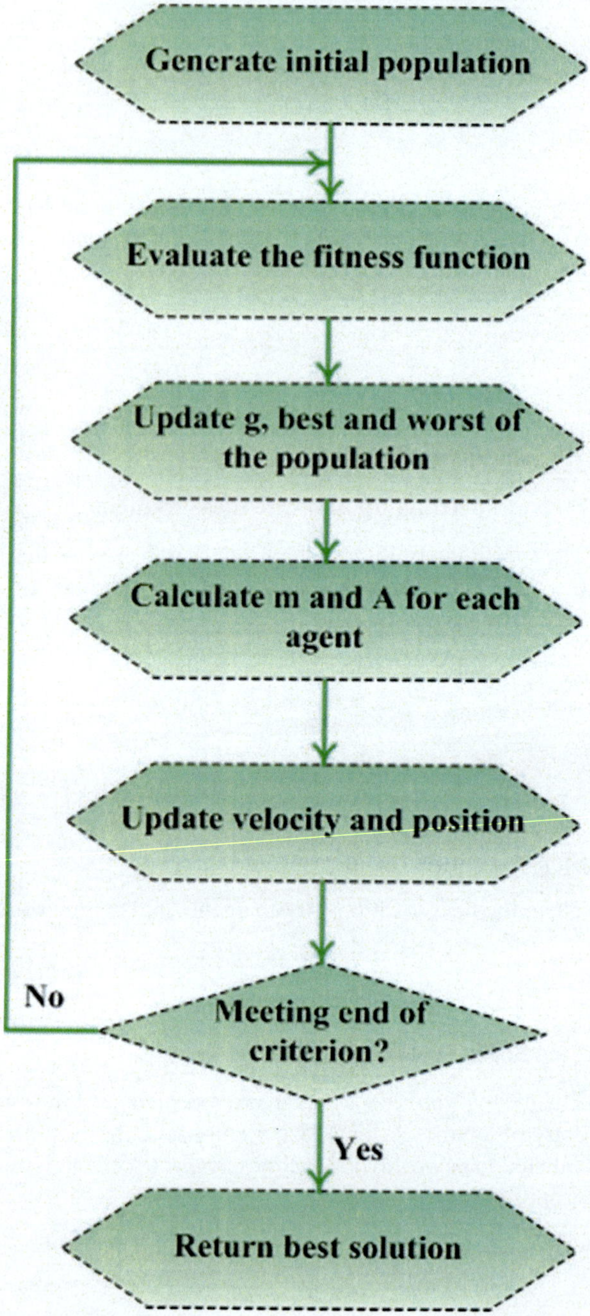

Fig. 6 Flowchart of gravity search algorithm

5 Results and Discussions

The proposed procedure for the location and arrangement of single and different PQ opportunities in the circulation system is applied in this section. Proposed approach is completed at the working stage of MATLAB/Simulink. The PQ occasion shall be checked in the single and three-stage conveyance lines, and the occasion details shall be accumulated. Three single- and three-stage tests are used with a view to implementing the proposed protocol.

Here, the proposed strategy recognizes and arranges for the PQ to accept the XCC of the strength signal. XCC isolates the power signal considering the limit esteem. The decision between for single and different PQ occasions is made in the light of the edge estimation of the breaking point. The edge estimation of XCC under the kind of no aggravations the XCC range is ≥ 0.95. The limit estimation of XCC under the sort of single *PQ* aggravation the XCC range is (0.95, 0.5). The edge estimation of XCC under the sort of various PQ unsettling influences the XCC range is (0.55, 0.10). The limit estimation of XCC under the sort of interference the XCC range is (0.10, 0.01).

5.1 Test Case 1: Normal Case

In this experiment, the proposed framework under ordinary condition is shown. Here, the ordinary state of the single stage current and voltage, genuine (*P*) and responsive force (*Q*) are examined utilizing the proposed method. Figure 7a portrays the presentation of single stage current versus time (sec). As observed from the figure, the single stage current in typical state offers a uniform current benefit of 1 A. Figure 7b shows the presentation of single-stage voltage versus time (sec). As

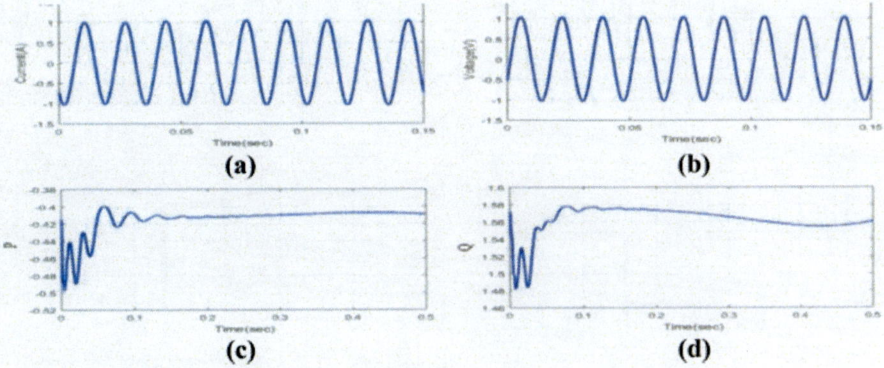

Fig. 7 Performance of **a** single-phase current **b** Single phase voltage **c** real power (*P*) **d** reactive power (*Q*) in normal condition

observed from the figure, the single-stage voltage in the ordinary state gives the uniform voltage estimation of 1 V. Figure 7c shows the presentation of P versus time (sec). As observed from the figure, the P in typical state. The presentation of Q versus time (sec) is portrayed in Fig. 7d. As observed from the figure, the Q in ordinary state.

5.2 Test Case 2: Analysis of Single Pq Events

In this segment, the examination of single PQ occasions is made and analyzed. Single PQ occasions are, for example, list, swell, homeless people, music, etc. Figure 8a shows the presentation of current list versus time. The figure shows the unsettling influence of current gets because of the hang condition at the waveform in the time moment of t = 0.2–0.32 s. Figure 8b shows the presentation of voltage list versus time. Figure shows the voltage getting upset because of the droop condition at the waveform in the time moment of t = 0.2–0.3 s. Figure 8c shows the presentation of P droop versus time. Figure shows P getting upset because of the droop condition at the waveform in the time moment of t = 0.2–0.3 s. Figure 8d shows the exhibition of Q droop versus time. The figure uncovers the Q getting upset because of the droop condition at the waveform in the time moment of t = 0.2–0.3 s.

Execution of current swell versus time is portrayed in Fig. 9a. It is seen from the figure that at the waveform in the time moment of t = 0.21–0.3 s, the current

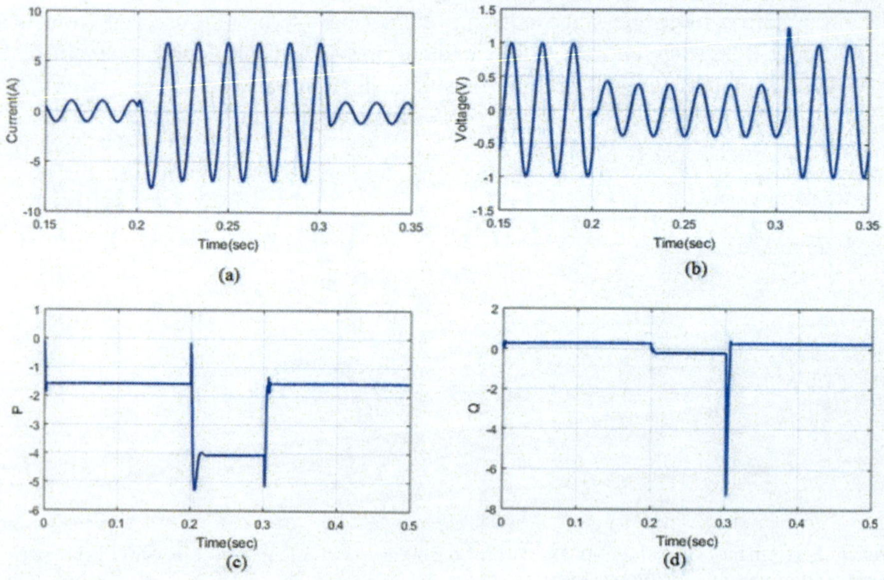

Fig. 8 Performance of sag **a** Current **b** Voltage **c** P **d** Q

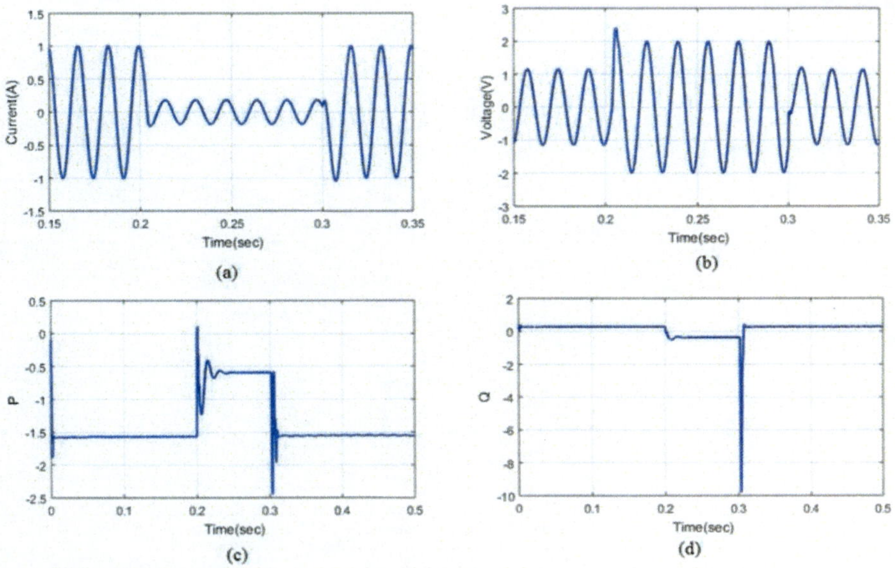

Fig. 9 Performance of swell **a** Current **b** Voltage **c** *P* **d** *Q*

gets upset because of the swell condition. Execution of voltage swell versus time is portrayed in Fig. 9b. It is seen from the figure that at the waveform in the time moment of t = 0.21–0.3 s, the current gets upset because of the swell condition. Execution of *P* swell versus time is portrayed in Fig. 9c. It is seen from the figure that at the waveform in the time moment of t = 0.2–0.3 s, the current gets upset because of the swell condition. Execution of *Q* swell versus time is portrayed in Fig. 9d. It is seen from the figure that at the waveform in the time moment of t = 0.2–0.31 s, the current gets upset because of the swell condition. Figure 10a shows the diagram of symphonious current versus time. At the time moment of t = 0.03–0.15 s, the waveform gets contorted because of the symphonious condition in the current stream. Figure 10b shows the chart of symphonious voltage versus time. At the time moment of t = 0.025–0.15 s, the waveform gets twisted because of the symphonious condition in the voltage stream. Figure 10c shows the diagram of symphonious *P* versus time. At the time moment of t = 0.03–0.15 s, the waveform gets contorted because of the symphonious condition in the P stream. Figure 10d shows the diagram of consonant *Q* versus time. At the time moment of t = 0.015–0.15 s, the waveform gets mutilated because of the consonant condition in the *Q* stream.

Figure 11 is the presentation chart of 10% hang state of framework boundary, for example, current, voltage, *P* and *Q*. Figure 11a is the diagram of 10% current hang versus time which gives the perception of the waveform of current stream in the framework getting twisted because of the 10% list condition at the time moment of t = 0.2–0.32 s. Figure 11b is the diagram of 10% voltage droop versus time which shows the waveform of voltage stream in the framework getting misshaped because of the 10% list condition at the time moment of t = 0.2–0.3 s. Figure 11c is the chart

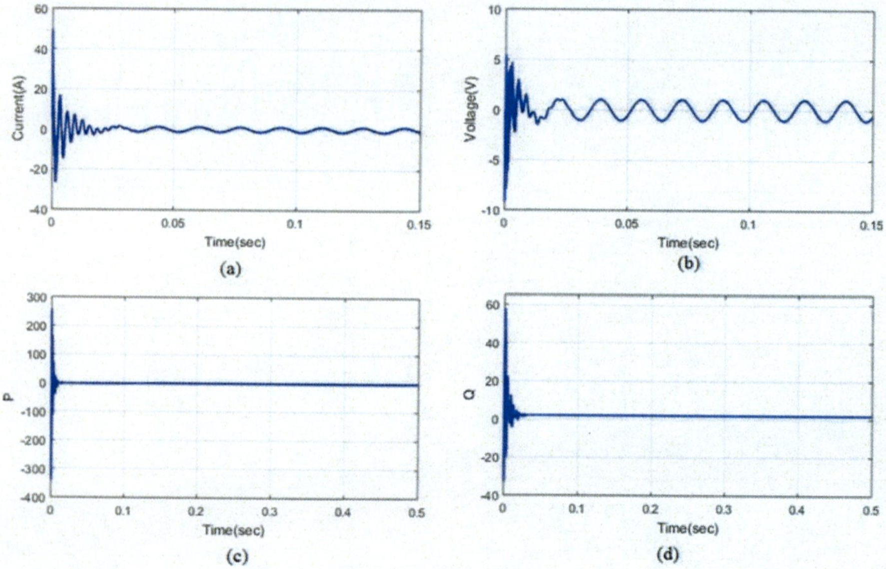

Fig. 10 Harmonic output of **a** Current **b** Voltage **c** P **d** Q

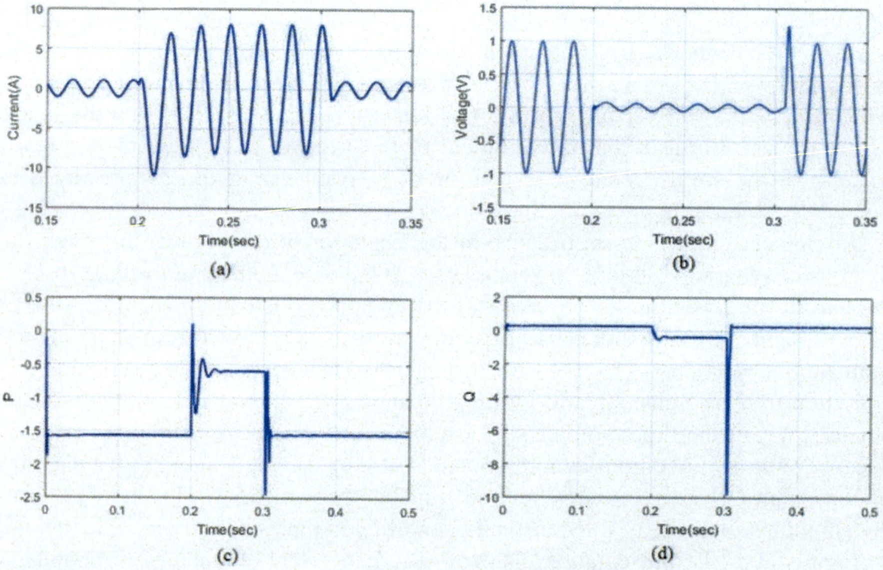

Fig. 11 Performance of 10% Sag **a** Current **b** Voltage **c** P **d** Q

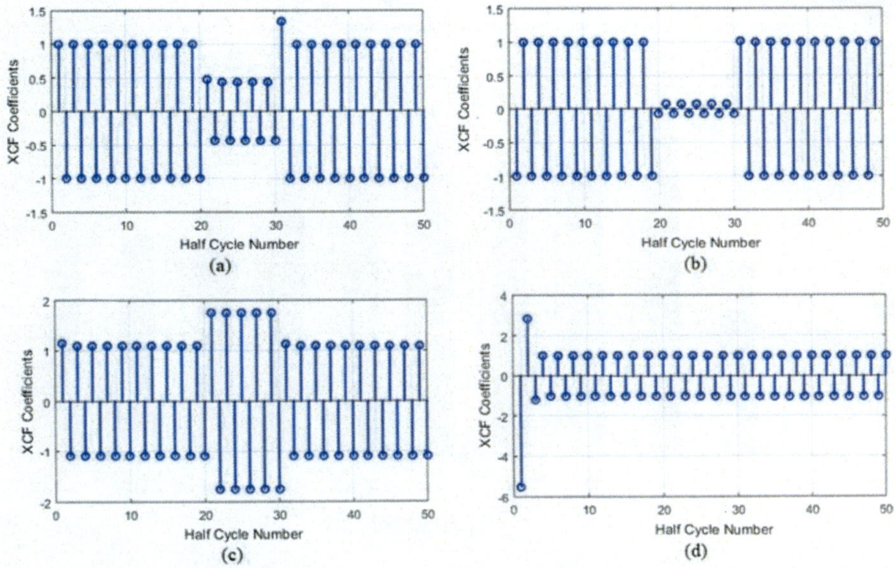

Fig. 12 Performance analysis of XCC **a** droop potential **b** 10% drooppotential **c** Swell potential **d** Harmonic potential

of 10% list *P* versus time which shows the waveform of P stream in the framework getting contorted because of the 10% list condition at the time moment of $t = 0.2$ to 0.3 s. Figure 11d is the diagram of 10% hang *Q* versus time which unmistakably shows that the waveform of *Q* stream in the framework getting contorted because of the 10% list condition at the time moment of $t = 0.2–0.305$ s.

Figure 12 shows the exhibition examination of XCC of droop and swell state of framework boundaries. Figure 12a is the XCC diagram of hang voltage versus time. The x-hub of the chart shows half cycle number, and the y-pivot of the diagram shows XCC coefficients. The XCC of the hang state of the voltage gets upset in the half cycle number of 20–29 with the XCC coefficient of roughly 0.5. Figure 12b is the XCC diagram of 10% list voltage versus time. The XCC of the 10% hang state of the voltage gets upset in the half cycle number of 20–30 with the XCC coefficient of around 0.1. Figure 12c is the XCC diagram of swell voltage versus time. The XCC of the swell state of the voltage gets upset in the half cycle number of 22–30 with the XCC coefficient of roughly 1.8. Figure 12d is the XCC chart of symphonious voltage versus time. The XCC of the consonant state of the voltage gets upset in the half cycle number of 3–50 with the XCC coefficient of around 1.

Figure 13 portrays the correlation of characterization precision of proposed and existing strategies, for example, GA-ANN and MUSIC-ANN. From the figure, it is seen that the proposed procedure is preferable in order precision over the current methods.

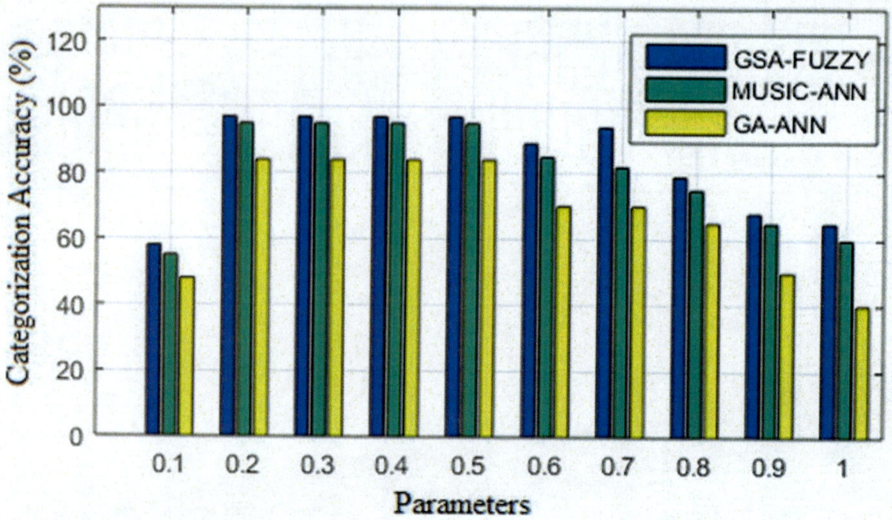

Fig. 13 Comparison of classification accuracy of proposed and existing technique

6 Conclusion

A half and half methodology is proposed for PQ occasions discovery and order which invigorates the observing plan in the circulation framework. In the principal layer, FLC is used to see the framework signal sound or undesirable condition. The second layer of the FLC is utilized for the arrangement of the unfortunate signs. FLC learning method is updated by GSA. With the aim of point, the planned method be accepted at the MATLAB/Simulink level. The commonality of the new approach is tried on the basis of various studies, and the current methods are contrasted and distinctive. The replication consequences how so as to planned policy is persuasive inside ensuring a high accuracy of PQ actions in the conveyance scheme. Similarly, the suggested protocol guarantees a system with a more modest intricacy for the identification and ordering of PQ opportunities, and henceforth, the arrangement precision of the framework is the framework is raise.

References

1. Smith Jayasree T, Devaraj D, Sukanesh R (2010) Power quality disturbance classification using Hilbert transform and RBF networks. Neurocomputing 73(7–9):1451–1456
2. Mishra S, Bhende C, Panigrahi B (2008) Detection and classification of power quality disturbances using S-Transform and probabilistic neural network. IEEE Trans Power Deliv 23(1):280–287
3. Singh U, Singh S (2017) Detection and classification of power quality disturbances based on time–frequency-scale transform. IET Sci Measurem Technol 11(6):802–810

4. Erişti H, Demir Y (2010) A new algorithm for automatic classification of power quality events based on wavelet transform and SVM. Expert Syst Appl 37(6):4094–4102
5. Wang H, Wang P, Liu T (2017) Power quality disturbance classification using the s-transform and probabilistic neural network. Energies 10(1):107
6. Babu P, Dash P, Swain S, Sivanagaraju S (2013) A new fast discrete S-transform and decision tree for the classification and monitoring of power quality disturbance waveforms. Int Trans Electri Energy Syst 24(9):1279–1300
7. Biswal B, Mishra S (2014) Power signal disturbance identification and classification using a modified frequency slice wavelet transform. IET Gener Transm Distrib 8(2):353–362
8. Masoum M, Jamali S, Ghaffarzadeh N (2010) Detection and classification of power quality disturbances using discrete wavelet transform and wavelet networks. IET Sci Measurem Technol 4(4):193–205
9. Hooshmand R, Enshaee A (2010) Detection and classification of single and combined power quality disturbances using fuzzy systems oriented by particle swarm optimization algorithm. Electric Power Syst Res 80(12):1552–1561
10. Kanirajan P, Kumar V (2015) Wavelet-based power quality disturbances detection and classification using RBFNN and fuzzy logic. Int J Fuzzy Syst 17(4):623–634
11. Kapoor R, Gupta R (2011) Fuzzy lattice based technique for classification of power quality disturbances. European Trans Electri Power 22(8):1053–1064
12. Puliyadi Kubendran A, Loganathan A (2016) Detection and classification of complex power quality disturbances using S-transform amplitude matrix-based decision tree for different noise levels. Int Trans Electri Energy Syst 27(4):
13. Sridhar S, Rao K, Sukrutha Jade S (2016) Detection and classification of power quality disturbances in the supply to induction motor using wavelet transform and neural networks. Balkan J Electri Comput Eng 4(1):37–45
14. Jashfar S, Esmaeili S, Zareian-Jahromi M, Rahmanian M (2013) Classification of power quality disturbances using S-transform and TT-transform based on the artificial neural network. Turkish J Electri Eng Comput Sci 21(6):1528–1538
15. Singh K (2015) Classification of power quality disturbances using wavelet transform and neural network. Int J Eng Res 4(5):423–1427
16. Sundaram P, Neela R (2016) Characterization and localization of power quality disturbances based on s-transform and fuzzy expert system. IOSR J Electri Electron Eng 11(4):42–53
17. Ahila R, Sadasivam V, Manimala K (2012) Particle swarm optimization-based feature selection and parameter optimization for power system disturbances classification. Appl Artif Intell 26(9):832–861
18. Liu Z, Cui Y, Li W (2015) Combined power quality disturbances recognition using wavelet packet entropies and s-transform. Entropy 17(8):5811–5828
19. Kapoor R, Saini M (2011) Hybrid demodulation concept and harmonic analysis for single/multiple power quality events detection and classification. Int J Electri Power Energy Syst 33(10):1608–1622
20. Morales-Velazquez L, Romero-Troncoso R, Herrera-Ruiz G, Morinigo-Sotelo D, Osornio-Rios R (2017) Smart sensor network for power quality monitoring in electrical installations. Measurement 103:133–142
21. Zhang S, Li P, Zhang L, Li H, Jiang W, Hu Y (2016) Modified S transform and ELM algorithms and their applications in power quality analysis. Neurocomputing 185:231–241
22. Kanirajan P, Suresh Kumar V (2015) Power quality disturbance detection and classification using wavelet and RBFNN. Appl Soft Comput 35:470–481
23. Rodriguez-Guerrero M, Carranza-Lopez-Padilla R, Osornio-Rios R, de J Romero-Troncoso R (2017) A novel methodology for modeling waveforms for power quality disturbance analysis. Electric Power Syst Res 143:14–24
24. Ahila R, Sadasivam V, Manimala K (2015) An integrated PSO for parameter determination and feature selection of ELM and its application in the classification of power system disturbances. Appl Soft Comput 32:23–37

25. Mariaraja P, Manigandan T, Thiruvenkadam S (2018) An expert system for distribution system reconfiguration through fuzzy logic and flower pollination algorithm. Measurem Control 51(9–10):371–382
26. Cai H, Li P, Su C, Cao J (2018) Double-layered nonlinear model predictive control based on Hammerstein-Wiener model with disturbance rejection. Measurem Control 51(7–8):260–275
27. Mouzakitis A (2013) Classification of fault diagnosis methods for control systems. Measurem Control 46(10):303–308
28. Karthikumar K, Senthil Kumar V, Karuppiah M (2019) A new protection scheme for pq events prediction and classification in distribution system: a sso-ann strategy. Measurement Control 52(5–6):449–461
29. Khokhar S, Mohd Zin A, Memon A, Mokhtar A (2017) A new optimal feature selection algorithm for classification of power quality disturbances using discrete wavelet transform and probabilistic neural network. Measurement 95:246–259
30. Manikandan M, Samantaray S, Kamwa I (2015) Detection and classification of power quality disturbances using sparse signal decomposition on hybrid dictionaries. IEEE Trans Instrument Measurem 64(10):27–38
31. Rostek K, Morytko Ł, Jankowska A (2015) Early detection and prediction of leaks in fluidized-bed boilers using artificial neural networks. Energy 89:914–923
32. Wang G, Hao J, Ma J, Huang L (2010) A new approach to intrusion detection using artificial neural networks and fuzzy clustering. Expert Syst Appl 37(9):225–6232
33. Samanta B, Al-Balushi K, Al-Araimi S (2003) Artificial neural networks and support vector machines with genetic algorithm for bearing fault detection. Eng Appl Artif Intell 16(7–8):657–665
34. Mirjalili S, Gandomi A, Mirjalili S, Saremi S, Faris H, Mirjalili S (2017) Salp Swarm Algorithm: A bio-inspired optimizer for engineering design problem. Adv Eng Softw 114:163–191
35. El-Fergany A (2018) Extracting optimal parameters of PEM fuel cells using Salp Swarm Optimizer. Renew Energy 119:641–648

Adaptive Active Contour Segmentation for Melanoma Diagnosis from Dermoscopy Images

K. S. Abhilash, P. S. Shijin Kumar, Zina Ravindran, and A. Nisam Rahman

Abstract Melanoma is a type of skin cancer having high survival rate with early stage detection. Dermoscopy images are used for the detection of melanoma. This work is planned to develop a novel technique for segmenting skin lesions from the dermoscopy images is analyzed and implemented. The method implemented is adaptive active contour segmentation. The results obtained by this method are analyzed by using various metrics. From this, the adaptive active contour-based segmentation provides a better segmentation results when compared with other methods with average error rate of 0.13. So, the adaptive active contour segmentation method is used along with hair removal technique, and this segmented image will be used for the classification, i.e., whether melanoma is present or not.

Keywords Melanoma · Dermoscopy · Adaptive active contour · Region growing · Segmentation

1 Introduction

One of the prominent reasons for skin cancer is the exposure to ultraviolet (UV) radiation. There are different stages in the development of melanoma. In stage 0, melanoma affects the top most layer which is termed as melanoma in situ. During stage 1, the tumor surface may appear broken down and have 1 mm thickness. In stage

K. S. Abhilash
Jai Bharath College of Management and Engineering Technology, Perumbavoor, Kerala, India

P. S. S. Kumar
Department of Electronics and Communication Engineering, Marri Laxman Reddy Institute of Technology and Management, Hyderabad, India

Z. Ravindran
Department of Electronics and Communication Engineering, Noorul Islam Centre for Higher Education, Kanyakumari, Tamilnadu, India

A. N. Rahman (✉)
Innovation and Research Society, Mavelipuram, Kerala, India

S. S. Dash et al. (eds.), *Sixth International Conference on Intelligent Computing and Applications*, Advances in Intelligent Systems and Computing 1369,
https://doi.org/10.1007/978-981-16-1335-7_12

2, tumor is between 1 and 2 mm thick, and surface may appear broken down. In stage 3, the cells having melanoma can spread nearer to different lymph nodes. In stage 4, the cells having melanoma can expand to different parts like liver, lung, tissues under the skin, lymph nodes, etc. Dermoscopy is a non-insertion method for a diagnostic visualization of pigmented lesions present in the skin. An example for clinical and dermoscopy image is shown in Fig. 1. Through visual diagnosis, experts are capable of identifying melanoma in 65–80% of cases based on experience. But here, the clinicians may improve their diagnostic performance only if trained formally. Hence, computerized image analysis techniques are in demand. For the identification of melanoma from dermoscopy images, the lesion area should be separated from the background skin image. This work concentrates on the segmentation of lesion border from the dermoscopy image.

Various existing methods for the segmentation of lesion from the dermoscopy images are explained below. Region growing [1] can assign initial seed points automatically for the segmentation of lesion from dermoscopy image. Here, the problem of over segmentation is reduced through optimal selection of seed points. An improved watershed-based algorithm [2] can reduce the border detection error which occurs due to the implementation of pre-processing steps. Active contour via partial differential [3] equation was used to define geodesic curve. The geodesic active contour approach detects spurious objects for noisy images. The geodesic edge tracing approach requires user feedback for marking initial points. Omar et al. [4, 5] used Otsu's method for computing global threshold, after applying Gaussian filter in gray-scale dermoscopy image and segmented the lesion using active contour-based segmentation. The major drawback of the above methods is the availability of good thresholding which results in the presence of high contrast between skin and lesion. The region-based approaches lead to over-segmentation and are time consuming.

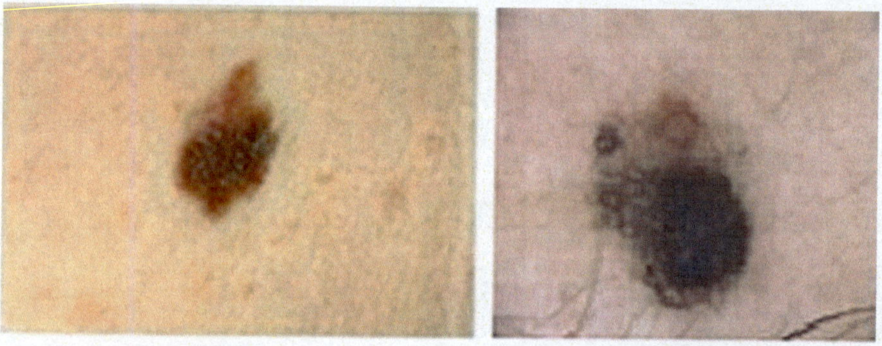

(a) (b)

Fig. 1 Sample images, **a** clinical image, **b** dermoscopy image

2 Methodology

Segmentation is a major step in any detection algorithms because it influences the overall efficacy of the framework. Yet, segmentation is tedious task due to diversity in size, color, and shapes of the lesions. Also, some lesions have uneven boundaries with inseparable surrounding skin. Dark hair surrounding the lesions and glare will delimit the segmentation performance. Hence, hair removal technique is important in accurate skin lesion segmentation. The block diagram of the prospective lesion segmentation system is depicted in Fig. 2.

2.1 Hair Removal

Presence of hairs in the dermoscopy images decreases the results of the accurate segmentation. For removing hairs, the RGB dermoscopy image is converted into gray-scale image, and the closing operation is performed on gray-scale image. Closing is a dilation followed by erosion. While performing dilation process, the structuring element s is placed over the image F and slides it throughout the image. This operation will turn the pixels to '*black*.' The dilation of input image F using s is described as $H(x, y)$.

$$I(x, y) = \begin{cases} 1, \text{ if } s \cap F \\ 0, \text{ otherwise} \end{cases} \quad (1)$$

Erosion is done for reducing the thickening of the regions obtained from dilation. The dilation operation is same as that of erosion, but this operation will turn the pixels to '*white*.' The erosion of input image I using s is described as $H(x, y)$.

$$H(x, y) = \begin{cases} 1, \text{ if } s \text{ fits } I \\ 0, \text{ otherwise} \end{cases} \quad (2)$$

The resulting image is subtracted with the gray-scale image and thresholding is done to obtain a mask. This will help to increase the segmentation results.

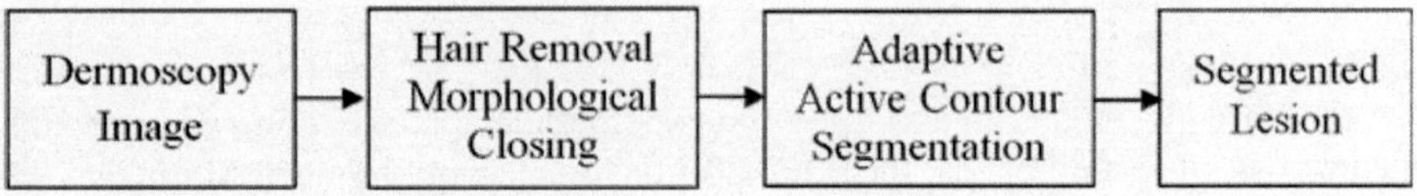

Fig. 2 Block diagram of lesion segmentation method

2.2 *Adaptive Active Contour Segmentation*

Adaptive active contour utilizes the initial boundaries and iterative steps to achieve segmentation. It dependent on a bunch of measures. The objective is to begin with an underlying shape near an ideal area and afterward utilize the versatile dynamic form of computation to derive a final desired segment. The model of the system is developed using a discrete set of points indicated by Eq. 3.

$$V(i) = (x_i, y_i) \text{ for } i = 1, 2, \ldots, n \tag{3}$$

The energy at each pixel in sliding window is calculated and the point of contour i is moved toward the pixel having least energy. Internal energy is defined as a square of the distance between points given in Eq. 4. α is the user defined parameter.

$$E_{\text{internal}} = (x_i - x_{i+1})^2 + (y_i - y_{i+1})^2 \tag{4}$$

$$E_{\text{internal}} = \alpha \sum_{i=0}^{n-1} (x_i - x_{i+1})^2 + (y_i - y_{i+1})^2 \tag{5}$$

External energy is defined as the negative of the brightness and is represented in Eq. 6.

$$E_{\text{external}} = -I(x_i, y_i) \tag{6}$$

$I(x_i, y_i)$ represents intensity of the limage. $V(i)$ is the contour. The energy present in the active contour can be represented as,

$$E[V(i)] = \int_{i=0}^{n-1} E_{\text{internal}} + E_{\text{external}} \tag{7}$$

3 Results and Discussion

The dermoscopy image dataset used for experimental process comprises of 200 images. Out of these images, there are 80 images of common nevi, 80 images of atypical nevi, and 40 images of melanoma. All RGB images have 8-bit depth and having resolution of 768 × 560. Each image in the database was evaluated by an expert dermatologist. İmage samples from the PH2 database are given in Fig. 3.

The active contour method uses the contour which helps to find the desired location of the lesion. The output images obtained through this method are depicted in Fig. 4.

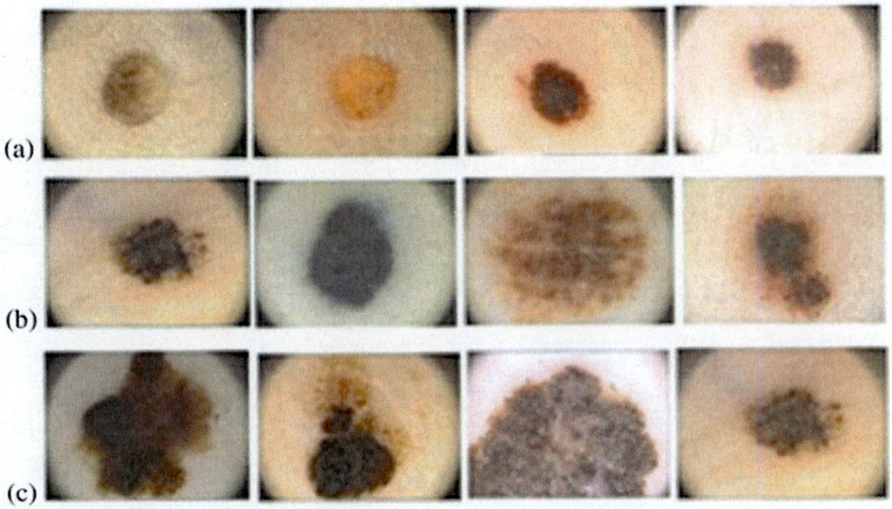

Fig. 3 Image samples from the PH2 database, **a** common nevi, **b** atypical nevi, **c** melanoma

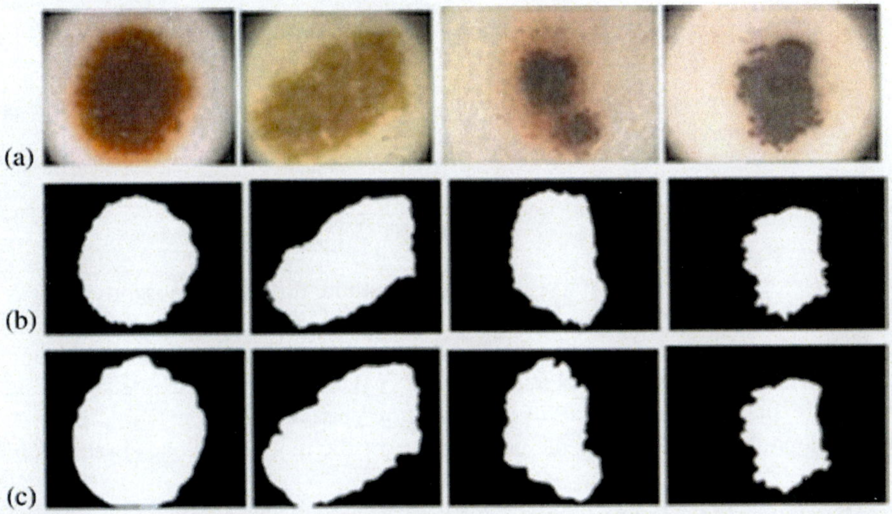

Fig. 4 Segmentation outputs **a** input dermoscopy image, **b** segmented output obtained by adaptive active contour method, **c** ground truth for the corresponding input image

The efficacy of proposed segmentation technique is assessed by utilizing various metrics such as measure of overlap (MOL), measure of under segmentation (MUS), measure of over-segmentation (MOS), dice similarity measure (DSM), and XOR measure. MOL is the area overlapped when the segmented image (A) is placed over the ground truth image (B). It is also known as Jaccard Index (J) and is expressed in Eq. 8.

$$J = \frac{|A \cap B|}{|A \cup B|} \tag{8}$$

MUS is the percentage of area which is not segmented compared to the ground truth area. It is expressed as U and is given by Eq. 9.

$$U = \frac{|B - (A \cap B)|}{|B|} \tag{9}$$

MOS is the percentage of lesion area in the segmented image that exceeds the ground truth area. It is expressed as V and is given by Eq. 10.

$$V = \frac{|A - (A \cap B)|}{|B|} \tag{10}$$

DSM is a measure of similarity between segmented area of lesion and the area of ground truth. The dice similarity coefficient (D) is given in terms of Jaccard Index (J) in Eq. (11).

$$D = \frac{2J}{1 + J} \tag{11}$$

XOR measure quantifies the percentage of border detection error (e) and is expressed in Eq. 12.

$$e = \frac{\text{area}(A \oplus B)}{\text{area}(B)} \tag{12}$$

The performance of AAC segmentation method for 10 dermoscopy images evaluated using above metrics is given in Table 1.

The efficiency of segmentation methods is superior, when the values of MOL and DSM are high. For good segmentation, MOS, MUS, and the error rate should be low. An exhaustive comparison of segmentation performance is given in Table 2. The column chart analysis for the performance of the three techniques is shown in Fig. 5.

The average performance measure of ten segmented images is compared for three techniques. From the above comparison, the measure of overlap for active contour method is 18.6% higher than Otsu's method and 32.48% higher than region growing. The dice similarity measure for active contour method is 10.88% higher than Otsu's method and 25.9% higher than region growing. The measure of under segmentation for active contour method is 48.9% lower than Otsu's method and 73.22% lower than region growing. The measure of over-segmentation for active contour method is 61.42% lower than Otsu's method and 44.74% higher than region growing. The error rate for active contour method is 53.8% lower than Otsu's method and 61.6% lower than region growing. From this, the active contour segmentation technique is

Table 1 Performance of AAC segmentation method

Image	MOL	MUS	MOS	DSM	Error
01	0.8636	0.1302	0.0071	0. 9268	0.1373
02	0.9165	0.0461	0.0408	0.9564	0.0869
03	0.9190	0.0072	0.0803	0.9578	0.0875
04	0.8186	0.1415	0.0487	0.9003	0.1902
05	0.9313	0.0448	0.0256	0.9644	0.0705
06	0.8082	0.1874	0.0054	0.8940	0.1928
07	0.8951	0.0300	0.0837	0.9447	0.1137
08	0.7973	0.1496	0.0665	0.8872	0.2162
09	0.8600	0.0935	0.0541	0.9247	0.1476
10	0.9327	0.0511	0.0173	0.9652	0.0684
Average	0.8742	0.0881	0.0429	0.9321	0.1311

Table 2 Performance comparison

Method	MOL	MUS	MOS	DSM	Error
Otsu Threshold	0.7142	0.1324	0.0692	0.8307	0.2016
Region Growing	0.6012	0.1526	0.0621	0.6907	0.2118
Proposed	0.8742	0.0881	0.0429	0.9321	0.1311

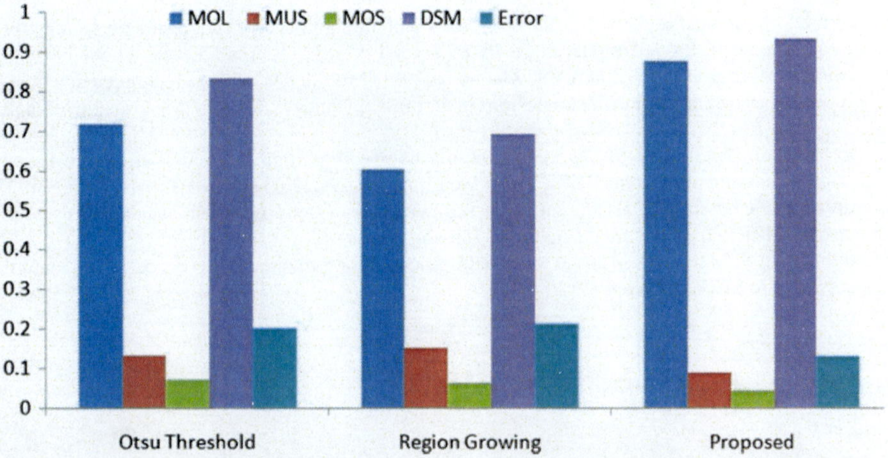

Fig. 5 Performance comparison

capable of achieving good segmentation performance with average error rate of only 0.13.

4 Conclusion

In this paper, the existing methods for segmenting skin lesions from the dermoscopy images are analyzed and implemented. The various methods implemented are Otsu's thresholding, region growing, and adaptive active contour segmentation. The results obtained by these methods are analyzed by using various metrics. From this, the adaptive active contour-based segmentation provides a better segmentation results when compared with other methods. But in dermoscopy images that contain hairs and bubbles, the segmentation will not hold good. So the proposed work removes the hairs by using morphological operations. After the removal of hair, the lesion from the image is segmented using adaptive active contour method. This segmented image will be used to detect the presence of melanoma.

References

1. Guru DS, Mahamad S, Sumithra R (2015) Segmentation and classification of skin lesions for disease diagnosis. Proc Comput Sci 45:76–85
2. Ashfaq AM, Emre CM, William VS, Xiaohe C (2011) Modified watershed technique and post-processing for segmentation of skin lesions in dermoscopy images. Comput Med Imaging Graph 35(2):116–120
3. Do HC, Guillermo S (2000) Segmenting skin lesions with partial-differential-equations-based image processing algorithms. IEEE Trans Med Imaging 19(7):763–767
4. Amir RS, Farzaneh K, Maryam Z (2013) Segmentation of dermoscopy images using wavelet networks. IEEE Trans Biomed Eng 60(4):763–767
5. Amr S, Mai S, Mariam AS (2012) Automatic detection of melanoma skin cancer using texture analysis'. Int J Comput Appl 42(20):22–26

Low-Power IoT-Based Smart Ingredient Measurement System

Bhavik Chauhan, T. M. Navamani, and Vatsal Soni

Abstract The world is progressing with the rapid germination of technology. In this IoT era, all technologies are automated and smart. Home automation is one of them, and the kitchen is an essential part of home automation. Nowadays, it is an area of the reformation in the intelligent and automated system in the kitchen. Nowadays, users are facing lots of issues regarding monitoring and remembering ingredients left in the containers. Hence, we have designed an android application for the ingredient measurement system as part of home automation. In the proposed system, BLE technology is used for data communication so that the power requirement is significantly less (a small battery can last long around five years). A sensor is installed into each container so that we know how much ingredients are left in the containers. Using this application, users can check the measurement and monitor ingredients with low-power BLE technology.

Keywords Internet of things (IoT) · Bluetooth Low Energy (BLE) · Generic Attribute Profile (GATT) · Raspberry Pi · Ultrasonic sensor (HC-SR04) · Android application

1 Introduction

Currently, demand is high for wireless communications. The Internet of things (IoT) is a recent technology for developing various useful applications in wireless communication. IoT is a network that can share data via Bluetooth technology or Internet-based and all physical objects linked to the Internet. It is a brilliant and innovative

B. Chauhan · T. M. Navamani (✉) · V. Soni
VIT University, Vellore, Tamil Nadu, India
e-mail: navamani.tm@vit.ac.in

B. Chauhan
e-mail: bhavik.chauhan2020@vitstudent.ac.in

V. Soni
e-mail: vatsal.yogeshbhai2020@vitstudent.ac.in

S. S. Dash et al. (eds.), *Sixth International Conference on Intelligent Computing and Applications*, Advances in Intelligent Systems and Computing 1369,
https://doi.org/10.1007/978-981-16-1335-7_13

technology that eliminates human effort and convenient access to physical equipment. It has an independent control feature that allows the device to control without social interaction. Nowadays, home automation using IoT is one of the essential things that all researchers can research. With a home automation system, devices are designed with Bluetooth Low Energy (BLE) chip to reduce power consumption compared to other wireless communication technologies. BLE technology allows users to develop viable products that can communicate to any modern mobile platform using chips, tools, and easy access standards. The kitchen is the core part of the home; it is possible to perform many activities, such as preparing meals and storing the required ingredients in separate containers. For making a meal, he/she takes an ingredient, according to the total members at home. No facility nowadays offers us the level of ingredients in containers or vessels that are available. This work provides an automatic update level of an ingredient in a container that the user can quickly check using an android application.

There is no such mechanism to measure the level of ingredients in the containers in the current scenario. We have to check manually how many ingredients are left in all the containers in the kitchen. In the literature, some researchers have proposed a system to measure the ingredients in the vessels using the android application and Wi-Fi module (for transmitting data from the sensor to the android app). If the Wi-Fi module is used, the users could plug in the device every time. It is not convenient for all the users, and some of the researchers incorporated the Zigbee communication technology for the data transmission of sensor data to the android application. However, both these technologies, such as Wi-Fi and Zigbee, consume more power, not suitable for the particular use case. The cost of the device is very high. So, we have proposed a system that contains a Bluetooth Low Energy (BLE) chip, which is of low cost for transmitting data to an android app, which is also BLE-enabled. This BLE-enabled chip and the android app reduce the power consumption than the Wi-Fi and Zigbee communication technology.

This paper intends to present an overview of the ingredient measurement system using low-power BLE technology achieved in the IoT. The structure of this paper is organized as follows. Section 2 exhibits the analysis and study of several related articles. In Sect. 3, we have proposed an overview of the system for achieving the ingredient measurement system using low power consumption. Section 4 exhibits the result and analysis of the system. Section 5 concludes this paper.

2 Literature Survey

In recent years, most of the researchers put their contributions in designing smarter home automation systems. The kitchen is one of the essential places in the house. Azran et al. [1] have proposed a method to measure the kitchen ingredients in an automated manner using the sensors and a Wi-Fi shield connected to the Arduino board and data represented in the android application. However, in this work the authors have incorporated a Wi-Fi module that drains more energy. Hence to overcome that, we have proposed BLE technology to diminish energy consumption. E-kitchen is proposed to execute everyday life more comfortable, vitality-efficient, usable, and safe. This whole design and implementation of home automation are based on Zigbee communication technology [2]. However, in this work authors have incorporated Zigbee technology that nowadays not compatible with android and other operating systems. Also, the smart kitchen using an intelligent screen to provide all the needed information concerning the list of menu, a record of ingredients and nutrition control measures, etc., is proposed by Mogali [3]. Damodhar et al. [4] have suggested a system that monitors and controls basic parameters like light intensity, apartment temperature, and heat. It also sends a message to the concerned person using the Wi-Fi module. Device-to-device communication communicates to all the devices autonomously without centralized control and collaboration to gather relevant information in real time [5]. Edward et al. [6] have discussed the smart refrigerator system developed on Raspberry Pi 3 and control by an android application. The various features introduced include knowing the abundance of a constituent in the fridge and notifying the user—automatic and smart kitchen safety monitoring system using DTH11, IR flame, and MQ-3 sensors. Arduino Uno does the interfacing of sensors and the controlling network handled by relay. It depends on kitchen safety like fire, gas leakage, and temperature [7]. Hadjakos et al. [8] have proposed an information acquisition system (IAS) for smart kitchen infrastructure to use new augmented kitchen apparatuses to sense the intercommunication that the user performs. Leung et al. [9] have suggested a system for cooking over electric devices instead of gas. Also, this work addresses electric load balancing, reducing power consumption, power consumption statistics, appliances' usage statistics, and optimized use of electric machines. Detection of the gas leakage, water inundation, and container superintendence system in a kitchen using the ATmega32 microcontroller and GSM module to send the alert message to the android mobile app is proposed by Chatterjee et al. [10]. Arya et al. [11] have proposed a system that contains a safety sensor to measure the distance of an ingredient into the container and send the data over the Wi-Fi to the android application. Sinha et al. [12] have proposed a system which provides smart communication between various appliances in building and home using NodeMCU. Rezwan et al. [13] have proposed a system that maintains the restaurant inventory in the kitchen more productive and unproblematic. This system notifies the users about the current stock and automatically orders new items if the quantity decreases. Hong et al. [14] have proposed an automated grocery ordering system using RFID. Patil1 et al. [15] have suggested that the calculation of distance using an ultrasonic sensor and sensor's data will be sent to the data cloud server and display it to the android application.

From the existing literature, we observe that most of the literatures use the safety sensors to measure an ingredient in the container. They also used Wi-Fi module and Zigbee technology for the data transmission from device to device. Hence, we have proposed a smart ingredient measurement system for home automation using low-power BLE technology.

3 Proposed System

The essential characteristics and architectural fundamentals of the proposed system are present in this section.

The proposed system's design embodies an ultrasonic sensor (HC-SR04), Raspberry Pi, android application, and Bluetooth Low Energy (transmitting data from Raspberry Pi to android application). In this system, the container acts as a peripheral, and android application acts as a central. Peripheral (container) always makes the advertisement. Central (android application) scans peripheral (BLE container), and it always looks for the advertisement. For our system, the container is always advertising, and the android app that scans this device and connection is established. After establishing the connection, they start moving Generic Attribute Profile (GATT) metadata to one another. Depending on the type of data transmission, one or other serves as the host. First, the ultrasonic sensor measures the level of an ingredient. After measuring an ingredient's level, it will send the level of an ingredient to the android application over the BLE. Once the data is transmitted, the user can see an ingredient's level, as shown in Fig. 1. Every five seconds, containers will check the level of an ingredient and update it to get an updated value every time.

The design of the offered system consists of two phases, such as hardware design and software design.

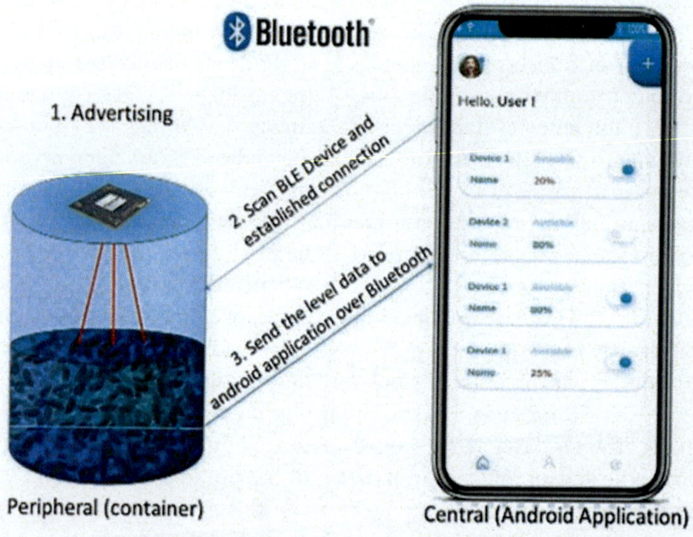

Fig. 1 Design of the proposed system

3.1 Hardware Design

3.1.1 Raspberry Pi

The Raspberry Pi is a credit card-size single-board computer developed in the UK by the Foundation. In this work, we have used the Raspberry Pi 3 Model B. It has a 1.2 GHz Broadcom 64-bit CPU with 1 GB of RAM and also supports Bluetooth Low Energy (BLE) and numerous other functionalities [16].

3.1.2 Ultrasonic Sensor (HC-SR04)

Two ultrasound transducers are composed of the ultrasonic distance sensor HC-SR04—the transmitter functions as an electrical signal translating into ultrasonic sound pulses of 40 kHz. A 4-pin module with the pin names Vcc, activate, echo, and ground, is the ultrasonic (US) sensor. In many applications, this sensor is used where distance measurement or sensing artifacts are required.

The sensor is regulated through +5 V power and ground pins of the sensor. The sensor's current consumption is less than 15 mA, maximum range is up to 4 m, and its accuracy is 3 mm. The trigger pin must be high for 10 μS and then off to start the measurement. This action triggers an ultrasonic wave from the transmitter at a frequency of 40 Hz, and the receiver waits for the tide to return. Distance is computed as the product of time value and speed of sound. The speed of the sound is 340 m/s. We must transform the speed into cm/μs to measure distance. A sensor is fitted on top of the container to measure the ingredient [17].

3.1.3 Bluetooth Low Energy (BLE)

Bluetooth Low Energy (BLE) is an intellectual, cost-effective paraphrase of Bluetooth wireless technology. Although Bluetooth Low Energy's energy efficiency makes it ideal for devices that need to run on small batteries for long periods, the magic of Bluetooth Low Energy lies in its ability to work with apps on smartphones or tablets. So, Bluetooth enables original device designers and manufacturers to develop solutions that will work with the billions of Bluetooth devices on the market [18].

3.2 *Software Design*

3.2.1 Python

Python is a high-level, general-purpose interpretation programming language. Python's design philosophy emphasizes code reading through the meaningful use of large spaces. Compared to any other language like C programming, the code line is shorter. Many functions are also supported with built-in library functions. It runs the code line by line [19].

3.2.2 Android

Android is a mobile operating system developed specifically for mobile trackpads such as cell phones and tablets and built on the Linux kernel and other open-source applications' updated variant. Android supports BLE, which helps the creation of a framework [20].

When an android app communicates with a BLE device, here are the functions and responsibilities that apply:

- **Central versus peripheral**: This applies to the BLE connection itself. Devices in the central role search for advertisements, and devices in peripheral roles create advertisements [20].
- **GATT server vs. GATT client**: After establishing this connection, it determines how the two devices will communicate with each other [20].

3.2.3 Generic Attribute Profile (GATT)

Generic Attribute Profile (GATT) describes how all profiles and user data are shared over a BLE connection. GATT only addresses actual processes and formats for data transfer. The GATT uses the Attribute Protocol (ATT) for transmitting the data between devices. These data hierarchically structure into sections called facilities, which group user data parts called features. A UUID is a 128-bit (16 bytes) number guaranteed to be globally unique (or has a high probability). UUIDs can be used in many non-Bluetooth protocols and applications and are specific in ITU-T Rec for their format, utilization, and generation. ISO/IEC 9834-8:2005 is often referring to as X.667. Two more UUID formats can be added: 16-bit and 32-bit UUID [21] (Fig. 2).

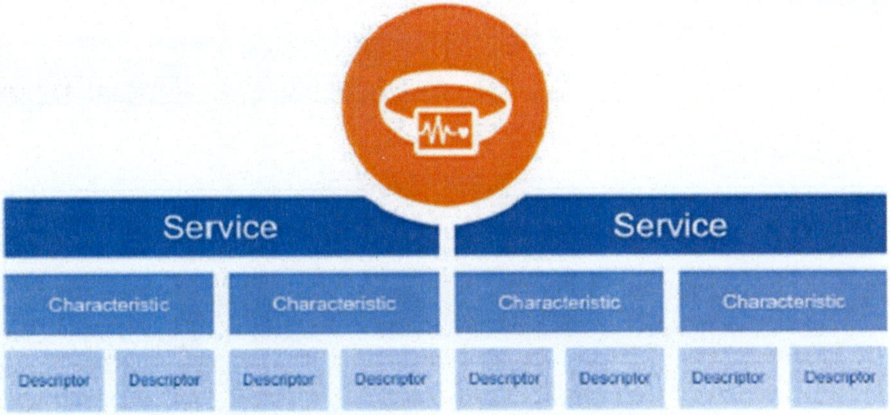

Fig. 2 Services, characteristics, and descriptors

4 Results and Analysis

In the proposed system, the sensors and chips can continuously monitor the ingredient levels in the containers, and the android application first connects to the device/container over the Bluetooth. After the connection establishment, the sensor sends the data to the android application with low power consumption. The user can see the threshold value of the ingredient stored in the containers.

The ultrasonic sensor is calculating the distance of the ingredient into the container. This sensor is connected to the Raspberry Pi using jumper wires as shown in Fig. 3. The Raspberry Pi is enabled with the Bluetooth Low Energy (BLE), and it always advertises for establishing a connection to another BLE device. Our android application scans all the nearby devices, as shown in Fig. 4. After the connection establishment, the sensor sends an ingredient's level, as shown in Fig. 5.

Fig. 3 Connection of ultrasonic sensor to Raspberry Pi

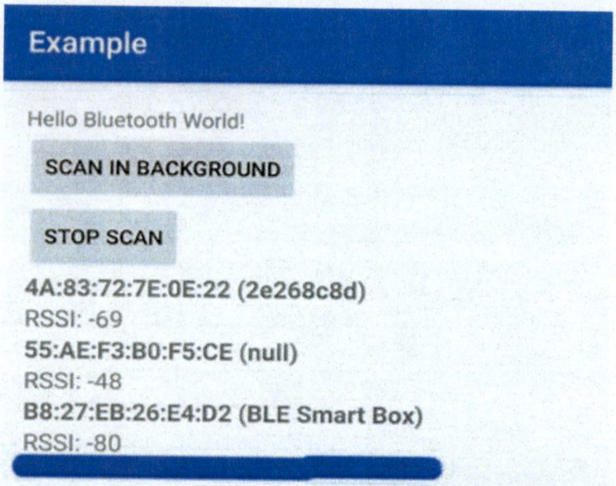

Fig. 4 Scanning BLE device

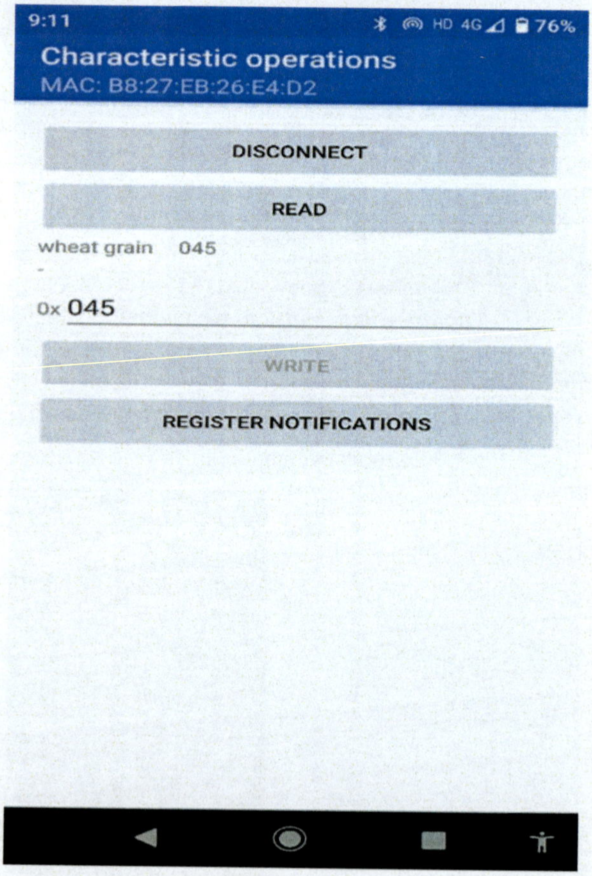

Fig. 5 Monitoring ingredient quantity

Table 1 Comparison between Bluetooth, Wi-Fi and Zigbee technologies [22]

Wireless technologies			
Parameter	Bluetooth	Wi-Fi	Zigbee
Frequency band	2.4 GHz	2.4 GHz	2.4 GHz
Range	77 m	75 to 90 m	Indoors—30 m
Power consumption	60 mA	400 mA	35 mA
Protocol stack size	250 KB	1 MB	32 KB
Maximum bandwidth required	15 MHz (dynamic)	22 MHz (static)	3 MHz (static)

In general, for the wireless data transmission Bluetooth, Wi-Fi, and Zigbee technologies are most commonly used. Table 1 shows the benefits and barriers of BLE over Zigbee and Wi-Fi [22].

Table 1 shows some of the advantages and disadvantages of all three wireless technologies. In our system, we have used BLE technology because they consume less power than the Wi-Fi. Also, Wi-Fi has a high range, but in our use case, we do not need higher coverage, so the BLE is more suitable for our use case. For power consumption, Fig. 6 shows the comparison and the advantage of Bluetooth over the Zigbee and Wi-Fi [22].

It is observed that Bluetooth technology consumes less power, and its cost is less than the Wi-Fi-enabled device. Hence, in our system, we have used BLE chip for

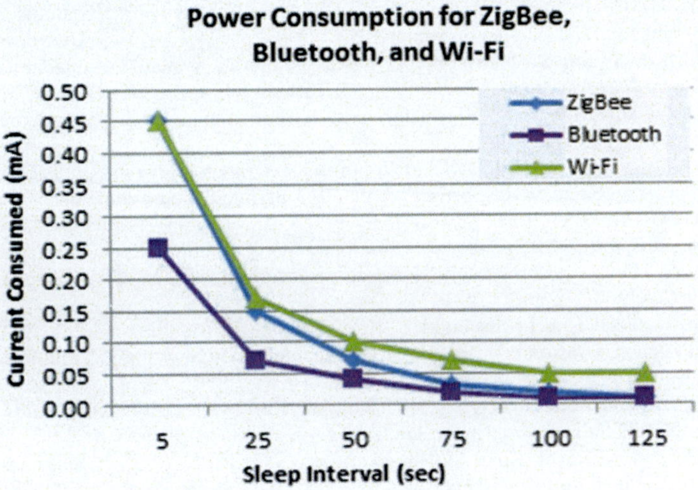

Fig. 6 Power consumption for Bluetooth, Wi-Fi, and Zigbee

the less power consumption for which battery power remains approximately for 2 to 4 years, and the cost of the product is also low.

5 Conclusion and Future Work

The composition and evolution of an interactive ingredient measurement system with the ultrasonic sensor, Raspberry Pi, Bluetooth Low Energy, and affordable user interface of android application are presented here. Using BLE technology is the main benefit of our use case because it consumes less power than the Wi-Fi module (NodeMCU) and Zigbee communication technology. It consumes only 60 mA and gives 2 to 4 years of battery life of the BLE-enabled device, and the user can easily monitor the level of an ingredient using the android application. In the future work, we have planned to include a module according to the level of an ingredient value; users will receive the nearest located store notification based on the users' current location and then recommend different online Web sites to buy online or offline.

References

1. Azran MAM, Zaid MFM, Sakeeb MSM, Althaff MMM, Fernando SGS (2017). Smart kitchen—a measurement system. Int J Comput Appl (0975–8887), 179(2)
2. Ceccacci S, Menghi R, Germani M (2015) Example of a new smart kitchen model for energy efficiency and usability. In: Fourth international conference on smart systems, devices and technologies, pp 12–18
3. Mogali SS (2015) Internet of Things and its role in Smart Kitchen. In: 4th National conferrence of Scientometrics and Internet of Things
4. Damodhar J, Swathi S (2016) Internet based monitoring system for smart kitchen using embedded web server architecture. Int J Innov Technol Res 4(4):3306–3311
5. Bello O, Zeadally S (2014) Intelligent device-to-device communication in the internet of things. IEEE Syst J 10(3):1172–1182
6. Edward M, Karyono K, Meidia H (2017, November). Smart fridge design using NodeMCU and home server based on Raspberry Pi 3. In: 2017 4th international conference on new media studies (CONMEDIA). IEEE, pp 148–151
7. Pudugosula H (2019, May) Automatic smart and safety monitoring system for kitchen using Internet of Things. In: 2019 international conference on Intelligent Computing and Control Systems (ICCS). IEEE, pp 1174–1177
8. Ständer M, Hadjakos A, Lochschmidt N, Klos C, Renner B, Mühlhäuser M (2012, December) A smart kitchen infrastructure. In: 2012 IEEE international symposium on multimedia. IEEE, pp 96–99
9. Leung YW, Tsang KF, Chui KT, Chu HY, Chow WH (2012) The energy profile study of electric kitchen utensils for residential smart kitchen
10. Chatterjee JM, Kumar R, Khari M, Hung DT, Le DN (2018) Internet of Things based system for Smart Kitchen. Int J Eng Manuf 8(4):29
11. Arya A, Taliyan A, Chauhan P, Gautam A (2019, April) Smart Kitchen with new measurement, web and application based with affordable design. In: 2019 4th international conference on Internet of Things: Smart Innovation and Usages (IoT-SIU). IEEE, pp 1–6

12. Sinha A, Sharma S, Mahboob MR (2017, May) An Internet of Things based prototype for smart appliance control. In: 2017 International Conference on Computing, Communication and Automation (ICCCA). IEEE, pp 1358–1363
13. Rezwan S, Ahmed W, Mahia MA, Islam MR (2018, October) IoT based smart inventory management system for kitchen using weight sensors, LDR, LED, Arduino Mega and NodeMCU (ESP8266) Wi-Fi Module with Website and App. In: 2018 fourth International Conference on Advances in Computing, Communication & Automation (ICACCA). IEEE, pp 1–6
14. Hong KS, Kim HJ, Lee C (2007, December) Automated grocery ordering systems for smart home. In: Future Generation Communication and Networking (FGCN 2007). IEEE, vol 2, pp 87–92
15. Patil CS, Pawar KN (2016) Smart grocery management system using Internet of Things. Int J Res Eng Technol
16. Online Web resource. https://www.raspberrypi.org/
17. Online Web resource. https://lastminuteengineers.com/
18. Online Web resource. https://www.bluetooth.com/
19. Online Web resource. https://en.wikipedia.org/wiki/
20. Online Web resource. https://developer.android.com/
21. Online Web resource. https://www.oreilly.com/
22. Asaduzzaman A, Chidella KK, Mridha MF (2015, April) A time and energy efficient parking system using zigbee communication protocol. In: SoutheastCon 2015. IEEE, pp 1–5

Implementation of Security Orchestration, Automation and Response (SOAR) in Smart Grid-Based SCADA Systems

Abdul Wahid Mir and Ramkumar Ketti Ramachandran

Abstract Critical infrastructures have come a long way in adoption of latest technologies to enhance the operational processes, automation and overall efficiency. The security of these vital and critical services-related infrastructures is on the top of agenda for any country as any security incident related will have disastrous consequences both socially and economically. Since there is an evolution of security-related risks and threats, the equal development in security-related technologies has also occurred. The automation of security is slowly and steadily taking place with new product offerings by the vendors in this space. The emergence of artificial intelligence and machine learning has simplified the automation process of security-related operations that seemed impossible in the past. With the implementation of smart grid-based supervisory control and data acquisition (SCADA) systems in the critical infrastructures, there are inevitable security-related challenges which needs to be addressed to safeguard the vital services reliant on these systems. The security operations face several challenges that include resource shortage, endless product development, huge volume of security alerts, static security controls with no orchestration, timely detection and response of security-related incidents, and cost of implementing security solutions. The security orchestration, automation and response (SOAR) implementation in smart grid-based SCADA systems offers an ideal solution for organizations to address the security challenges as SOAR utilizes technologies such as artificial intelligence, machine learning, deep learning, automation, threat intelligence and orchestration.

Keywords SOAR · Critical infrastructure · Security orchestration · Smart grid · SCADA · Industrial control system (ICS) · Operational technology (OT)

A. W. Mir · R. K. Ramachandran (✉)
Chitkara University Institute of Engineering and Technology, Chitkara University, Punjab 140401, India
e-mail: k.ramkumar@chitkara.edu.in

A. W. Mir
e-mail: wahids@live.com

S. S. Dash et al. (eds.), *Sixth International Conference on Intelligent Computing and Applications*, Advances in Intelligent Systems and Computing 1369,
https://doi.org/10.1007/978-981-16-1335-7_14

1 Introduction

Supervisory control and data acquisition (SCADA) systems are integral part of the critical infrastructures and industrial control system (ICS) such as electricity services, nuclear plants, transportation services, water management services and waste management services [1]. In particular, the electricity sector has adopted smart grid-based solutions and technologies that has transformed this sector with better control, efficient operations and cost-effective supply. It has enabled a two-way communication between the service providers and the customers with better control over service utilization and transparent consumption cost for the customers. Since the smart grid-based electricity services form the broad network of interconnected parties that include the service providers, service consumers, integrators, vast communication mediums/networks, etc., the security threat attack surface is not limited to electricity service providers only but to all the integrated ecosystems. Since the threat value is very high and critical for electricity services due to being high-value targets, the security of such systems is very vital [2].

Security of critical infrastructure-based systems has become the top priority for the organizations as the threats are real and the damages it has caused can be catastrophic. The threat landscape is so dynamic that the professionals in this area are finding it very difficult to keep up with the challenges of evolving threats on daily basis. There are several challenges being faced by the security professionals that include the overload of alert management, disparate tools, repetitive tasks and shortage of skilled professionals. Security of the critical infrastructures and industrial control systems bring in the significant challenges related to both operations and technology [3]. The protection of digital assets demands the appropriate suite of tools and processes for the security professionals to manage this critical job competently. Since the volume and the seriousness of the threat landscape continue to grow, it poses the greater challenges for the organizations. The workload of the security professionals continues to burden them with manual management of security events, alerts and disparate security systems.

The security orchestration, automation and response (SOAR) provides the platform that enables to connect and synchronize technologies/solutions through automation. Gartner describes SOAR as "technologies that enable organizations to collect security threat data and alerts from different sources, where incident analysis and triage can be performed using a combination of human and machine power to help, define, prioritize and drive standardized incident response activities according to a standard workflow" [4]. The monitoring of automation is overseen by human authority. It enables the security teams to efficiently operate the security operations and respond to the threats in timely and efficient manner. The implementation of SOAR helps to streamline and improve the daily tasks for the security teams by deploying security orchestration, automation and response tools. The mean time to resolution (MTTR) can be greatly improved with the help of SOAR platform.

SOAR solutions support security teams and professionals to manage mounting volume of events, alerts and incidents competently by automating manual response processes and workflows. It offers a complete function set in a solo platform for the organizations to manage the full security life cycle typically based on API's for simple integration with other security tools with option of extensibility and scalability features. The better usability of the SOAR solution for the security teams and security professionals with efficient user experience will enable the maximum productivity. The existing manual workflows, processes and security management tasks are automated with the combination of built-in processes and the technical capabilities of the SOAR solutions. The integration with a wide range of third-party disparate tools that is through two-way integration mechanism enables centralized security operational management by orchestrating the incident response by the SOAR platform. With the help of SOAR platform, the security teams or professionals can have a single dashboard to view and manage the full set of disparate tools that may include UEBA's, SIEM's, emails, threat intelligence systems, ticketing, endpoint detection and response system, firewalls, IPS/IDS systems, etc. (Fig. 1).

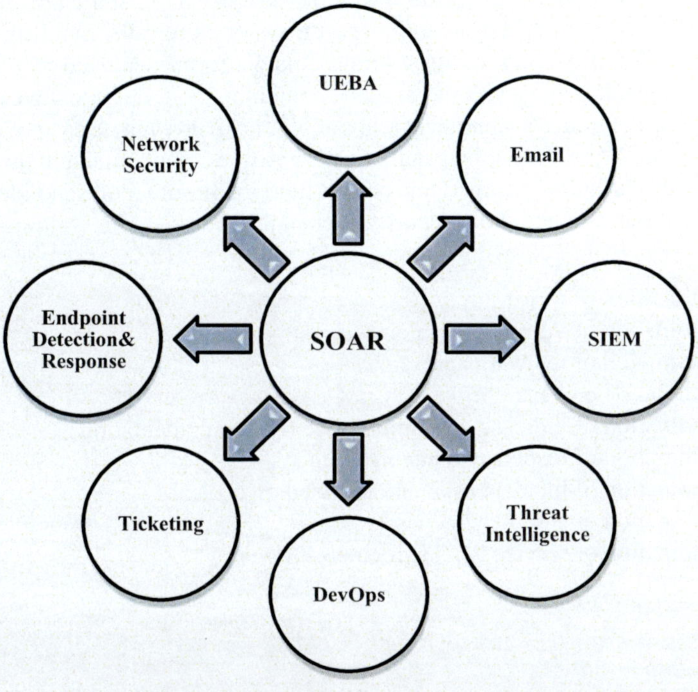

Fig. 1 Example of SOAR integration platforms

2 Functional Components of SOAR

A SOAR solution consists of the following key components that perform various functions and activities within the security operation life cycle (Fig. 2):

2.1 *Orchestration*

Security orchestration is the process of integrating disparate and a wide range of tools and technologies that are security-specific and non-security-specific working together to enhance the incident response in a timely manner. The alerts are ingested and analyzed from multiple sources of disparate tools within the organization that may include SIEM system, user and entity behavior analytics (UEBA), threat intelligence platforms/sources, intrusion detection system (IDS), intrusion prevention system (IPS), endpoint detection and response system, firewalls, incident response systems, etc. The disparate security solutions and platforms deployed by the organizations often result in huge volume of alerts containing both false positives and true positives. It also requires a significant amount of time to investigate each of the alerts by security teams or professionals that makes human element crucial in the security orchestration. The automated systems cannot function in solo mode and identify the subtle indications of a particular security eventuality. The security orchestration use cases may include the following:

- The automation of threat-hunting
- The automation of vulnerability management
- The automation of malware analysis
- The automation of VPN checks
- The automation of indicator of compromise (IOC) augmentation
- The automation of response to any incident
- The automation of incident assignment based on criticality
- The automation of case management
- The automation of service level agreement (SLA).

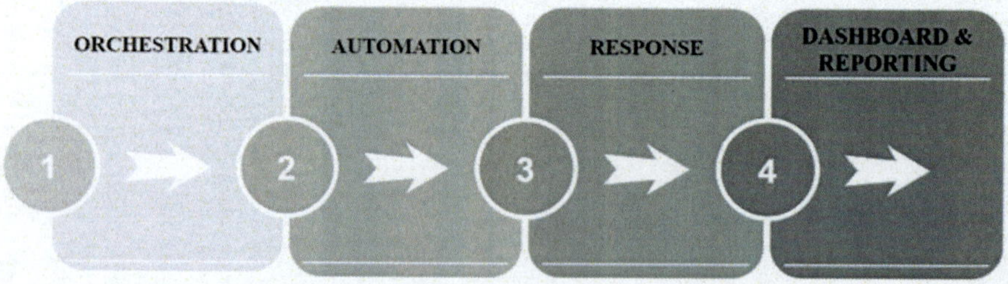

Fig. 2 SOAR functional components

2.2 Automation

The process of the machine-driven execution of actions is known as automation. The actions are performed on IT systems and security tools as a part of incident response mechanism. The detection and response to security incidents is a daunting and time-consuming task as the volume of alerts is really huge and automation helps to address each of them in due time with established action plan. It also helps to save lot of time on repetitive alerts and standardized steps or actions to be followed. It supports either of the security approaches, i.e., proactive security measures or reactive security measures. In case of proactive security measures, the automation playbook can be used to deploy measures like security operations and threat-hunting which can support security teams or professionals to identify vulnerabilities and threats in advance before they occur. In case of reactive security measures, the automation playbook can be used to deploy incident response and case management processes. The whole idea of automation is to automate the tasks which can be handled automatically by the machine itself in the form of standardized automation steps, workflows, actions, checking of status, auditing, verification, etc. It enables the security teams or professionals to focus on more productive tasks that machine cannot handle on itself, i.e., actual security incidents and their resolutions.

2.3 Response

This enables security teams or security professionals to manage the security incidents, collaborate with stakeholders, share information among the teams for incident resolution and devise the action plan for the future preventions for similar type of incidents. The response module is responsible to handle alert triage and processing phase in the SOAR platform. The case management is also handled here that includes collaboration and communication with appropriate stakeholders and task management with the security operation teams. The SOAR uses threat intelligence to gather information with respect to the incidents raised and recommends the appropriate actions to be taken. The remediation processes also kick in during this phase to resolve the raised incidents.

2.4 Dashboard and Reporting

The SOAR must offer the dashboard and reporting capabilities to generate reports for various audiences that include security analysts, security team managers, the chief information security officer (CISO) and other security experts who are related and responsible for security operations and processes within the organizations. The main

Table 1 SOAR solution summary

#	Product name	Vendor name	Launch year
1	ATAR	ATAR Labs	2017
2	Ayehu NG Platform	Ayehu	2007
3	SOC 3D	Cyberbit	2015
4	CyOPs	CyberSponse	2011
5	D3 SOAR	D3 Security	2002
6	Demisto Enterprise	Demisto	2015
7	IncMan	DFLabs	2013
8	EclecticIQ Platform	EclecticIQ	2014
9	Resilient	IBM	2010
10	InsightConnect	Rapid7	2000
11	Resolve	Resolve	2014
12	Security Operations	ServiceNow	-
13	Siemplify	Siemplify	2015
14	Phantom	Splunk	2014
15	Swimlane	Swimlane	2014
16	IR Flow	Syncurity	2014
17	ThreatConnect	ThreatConnect	2011
18	ThreatQ	ThreatQuotient	2013

aim of dashboard and reporting capabilities is to provide centralized reporting platform for better security intelligence and lessons learned from the previous incident reports so that the security posture can be improved.

3 Review of Soar Solutions

The summary of SOAR solutions is presently in the tables below consisting of product name, vendor name, launch year, vendor-specific solution features [5] (Tables 1 and 2):

4 Security Challenges for the Smart Grid-Based SCADA Systems

The digitization of smart grid operational technology (OT) environments has posed multitude of challenges due to disconnect and isolation of environments between information technology (IT) and operational technology (OT) [6]. The main gaps are related to security and optimal performance disruption. The history of working in

Table 2 SOAR solution—vendor-specific feature summary

Vendor-specific solution features	Vendors # (as in Table 1)
Audit Trail Generation	2
Automated Kill Chain Playbooks	5
Automation	1, 2, 3, 4, 5, 6, 7, 8, 9, 10, 11, 12, 13, 14, 15, 16, 17, 18
Automation-Assisted Patching and Threat Containment	10
Big Data Investigation	3
Case Management	4, 9, 11, 12, 13, 14
Computer Security Incident Response Team (CSIRT)	7
Configuration Compliance	12
Context-Driven Investigation	13, 16
Contextual Recommendation	17
Dashboard and Reporting	1, 12, 14, 16
Fraud Automation	4
Governance Risk and Compliance	12
Human- and Machine-Based Intelligence	9
Incident and Response	1, 4, 6, 7, 8, 9, 10, 11, 13, 16, 18
Incident Response and Threat Intelligence	7, 8, 11, 15
Integrations with Other Security Products	1, 2, 3, 4, 5, 6, 8, 10, 14, 17, 18
Leverage Intelligence Led Techniques	8,
Log Evidence and Audit Trails	7
Machine Learning	6, 7, 12, 13
Monitor KPI's	1
Orchestration	1, 2, 3, 4, 5, 6, 7, 8, 9, 10, 11, 12, 13, 14, 15, 16, 17, 18
OT Visibility	3,
Out-of-Box Connectors	4, 5, 6
Playbooks	1, 2, 3, 4, 6, 15
Playbooks on NISTSP 800–61 Rev 2	11
Rapid Response Runbooks	7

(continued)

Table 2 (continued)

Vendor-specific solution features	Vendors # (as in Table 1)
RBAC and Chain-of-Custody Features	5
Role-Based Access to Workflows	2, 16
Rollback of Changes to Workflows	2
Secure Provisioning	11
Security Operations Metrics	12
SOC Analytics	1
Training and Simulation	3
Vulnerability Management	4, 7, 10, 12
War Room for Analysts	6

isolation model for IT and OT has been there with different departmental ownerships and least dependence on each other within an organization. The critical infrastructural-based smart grid organizations have adopted a paradigm shift over the past decade that converged both information technology (IT) and operational technology (OT) into unified entity. This was due to the result of benefits of convergence and unification of services that could save organizations considerable cost and time. Although the goal was to enhance the productivity, there are still core challenges that arose due to this convergence. The core challenges include security and communication between IT and OT departments [7]. The plan to have unified network for IT and OT will increase the operational efficiency, but there are several underlying challenges that include the following:

- The OT-based industrial control systems (ICSs) are not designed keeping in mind the security-related challenges.
- The systems used in OT are still using the old technologies that were developed long time before the invent of present-day security protocols and platforms.
- The life cycle of OT-based systems and devices is designed for longer life; thus, upgrades are accordingly planned by both vendors and organizations.
- The updates and patches for OT systems are not done on regular basis which leaves them highly vulnerable.
- Even though the patches for some systems may be available, e.g., operating system patches, it is not possible to patch them due to other related system dependences.
- The OT and IT systems are often connected using unsecured devices and even using Wi-Fi networks that are not encrypted that makes them vulnerable to the attacks.
- The typical issue with the industrial control system (ICS) is the lack of network and system visibility.

Since OT is deployed for critical infrastructures such as electricity services, the attacks are growing in OT networks as the magnitude of damages could lead to massive loss to the organizations on both social and economic fronts. The primary goal of OT vendors is to build systems that will help in increased efficiency, increased productivity and increased product life or system life, and there is a need to converge them so that IT capabilities support OT to cater to security-related challenges. There is no doubt that OT and IT operational and departmental-level ownership cannot be unified but the convergence between them will improve the operational performance, reduce the costs, enhance the communication between both the environments and increase the overall productivity [8].

In the present era, the organizations are relying on security operation center (SOC) to cater to the growing cyber security-related challenges. In such scenario, having two separate SOCs will not only increase the cost but also affect the efficiency and productivity. The security team members in unified SOC can cater to both OT and IT as the employees with diverse skills can be employed. The SOC needs to provide the technologies that can cater to both OT and IT sectors [9]. The implementation of SOAR platform that offers orchestration capabilities will enable both the OT and IT to fill the gaps especially in the areas of communication and response among different team members.

5 Advantages of Using SOAR for Smart Grid-Based SCADA Systems

The organizations are facing surge in security challenges, and the SOAR platform is one of the ideal solutions which can be considered for both IT and OT systems. It is leveraging on latest and modern technologies that include machine learning, artificial intelligence, automation, threat intelligence and orchestration which are really helpful in managing the complex systems in OT specifically [10]. The following are some of the advantages which can be considered by the organizations who wish to implement the SOAR solutions:

- It can help to integrate disparate set of systems, applications and tools used by an organization for security management and enable the security teams to automate incident response process.
- It enables to collect alerts from all the integrated platforms into a centralized repository, thus making the investigation process simple.
- The cases are stored in a centralized location, thus offering a case knowledgebase for the security team to leverage from.
- The integrations enable high level of automation in processes and workflows which helps in reducing the time to address the incidents, faster resolutions and adaptive enhanced line of defense for the organizations.
- Use of playbooks for response to threats can help in achieving higher and faster level of compliance to security metrics or service level agreements.

- It can help organizations to reduce considerable budget restraints as one platform can be used for both IT and OT.
- It can help organizations to improve time management, productivity and efficiency among the security teams.
- It can enable organizational vide collaborative centralized platform with dashboards and reporting capabilities.
- It provides opportunity to optimize the entire security operation and management processes.
- It can support to extend the security operation center competencies to the OT network.
- It can enable improvement and better control of standard operating procedure (SOP) for industrial control systems.
- The responses and measures applied for incidents can be based on best practices defined by the organization.
- It can provide management of incidents which are similar in nature to avoid repetition of remediation measures.
- The OT departments can get better visibility of the critical systems with access to relevant data, key performance indicators and potential risks.
- The incident response and reporting time can be faster and more efficient which is very critical in industrial control systems.
- It provides efficient mechanism for handling escalation matrix.

6 SOAR Use Case Implementation Scenarios

The security teams are challenged with enormous pressure of broad and challenging attack surface which is evolving every single day. SOAR helps security teams to improve their security maturity and defenses with efficient automation, orchestration and response mechanisms. With the help of automation of the security-related tasks, the security operation teams can handle more tasks with shorter mean time to resolutions (MTTR's). The SOAR use cases depend upon various factors that include organizational internal security controls and environment, the line of business for an organization, legal and regulatory compliance that organization needs to comply with, etc. The frequent and common three use cases for SOAR are discussed below (Tables 3, 4 and 5):

7 Conclusion

There are unavoidable challenges that need to be addressed if the convergence between OT and IT networks is goal of the organizations. The OT network such as smart grid-based SCADA system running in critical industrial control systems of electricity business cannot afford failure due to any security threats and breaches.

Table 3 Use case—phishing email attack

<table>
<tr><td>Attack type</td><td>Phishing emails attack: Millions of phishing emails are sent to the users over the internet on daily basis and it is one of the most common type of attack we can see</td></tr>
<tr><td>Challenges</td><td>• Volume of phishing emails sent is to too large to handle on daily basis by the security teams
• Need for multiple security tools for investigation
• Manual investigation of each threat can be a time-consuming task
• Typically, organization does not have enough resources to handle volume of phishing attacks on daily basis
• The slow incident response can cause potential damages and increased risk to the organizations</td></tr>
<tr><td>Solution</td><td>Process automation: Automating the investigation and quarantine of the phishing emails detected by the SOAR</td></tr>
<tr><td>Automation workflow</td><td>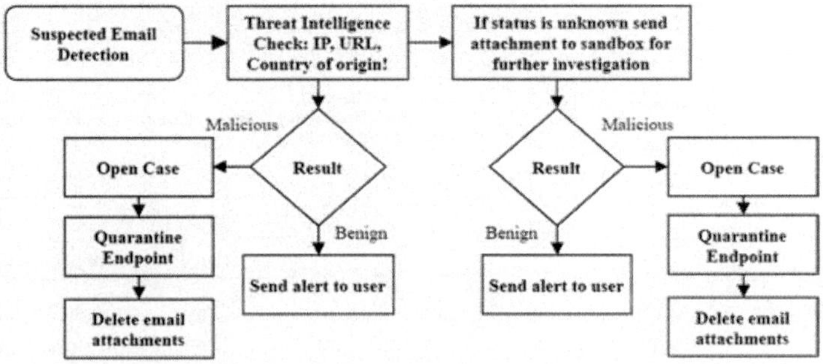
</td></tr>
<tr><td>Advantages</td><td>• The identification and resolution of attacks can be faster and large volume of phishing emails can be handled by the security operation team
• Mean time to resolution (MTTR) can be reduced drastically
Improvising automated workflows is very easy and can be instantly done</td></tr>
<tr><td>Technology requirements</td><td>Email system, threat intelligence, sandboxing system, endpoint detection and recovery, case management system</td></tr>
</table>

SOAR is a platform which can help to integrate most of the compatible security stack solutions allowing organizations to have better control and visibility of security alerts, threats and remediation activities for both OT and IT. The main objective of using SOAR is to provide better visibility, process improvement and efficiency in security operations. SOAR solution should help organizations to enhance the efficiency and efficacy of the enterprise security platform. The return on investment (ROI) should be a key selection factor while evaluating any SOAR platform. The focus of automation and orchestration of the SOAR platform should be flexibility of operations or processes for the security teams and not the machine replacement of humans. SOAR helps to minimize the limitations of the IT-OT convergence and helps to enhance security efficiency, communication and discernibility without affecting the critical business processes in industrial control system (ICS).

Table 4 Use case—endpoint attack

Attack type	**Endpoint attack**: The endpoints are major points of the attack into an organization, and the volume of alerts is directly proportional to the size of an organization. The number of alerts generated can burden the security teams, thus preventing the on-time alert response. These attacks can be disastrous since the endpoints are already inside the network and any delay in action can affect the entire internal network
Challenges	• The number of endpoints in an organization can be in 100′s to 100,000′s • Manual investigation of each threat on each endpoint can be a time-consuming task • The slow response can cause potential damages and greater risk to entire network
Solution	**Process automation**: Automating the endpoint alert investigation and remediation action process
Automation workflow	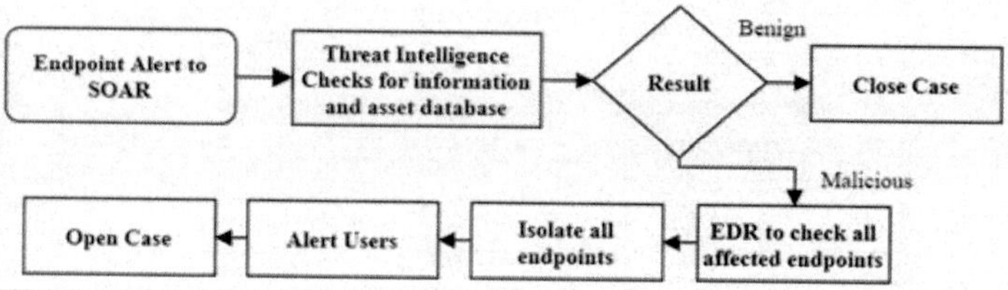
Advantages	• Automation and orchestration of process will ensure that any alert related to the endpoints is address and immediate action like isolation of endpoint or process kill can be initiated immediately • Mean time to resolution (MTTR) can be reduced drastically
Technology requirements	Security information and event management (SIEM) solution, endpoint detection and recovery, case management system

Table 5 Use case—forensic investigation

Attack type	**Forensic investigation**: In the event of security compromise, the data collection from affected endpoints as well as disparate tools needs to be collected as evidence for further investigation purposes
Challenges	• The data collection being post-incident task is cumbersome activity • Third-party system evidence is usually required by the auditors • The source of evidence is usually not a single source
Solution	**Process automation**: Automating the forensic data collection process
Automation workflow	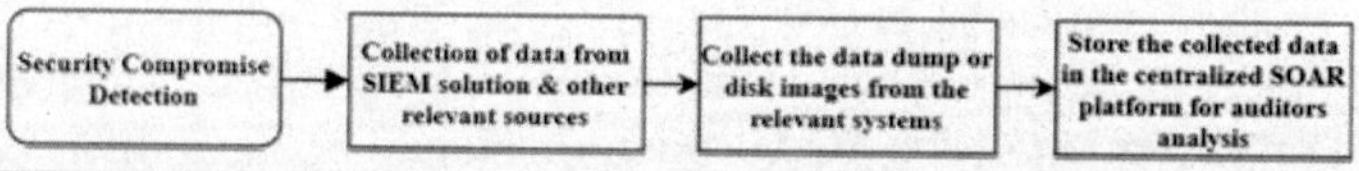
Advantages	• Automation process will ensure most of the evidence data is collected automatically which will be later used by the forensic teams
Technology requirements	Security information and event management (SIEM) solution and forensic software management system

References

1. Mir A, Ketti Ramachandran R (2019) Security gaps assessment of smart grid-based SCADA systems. Information Comput Security 27(3): 434–452
2. Paverd A, Martin A, Brown I (2014) Security and privacy in smart grid demand response systems. In: International workshop on smart grid security. Springer, Cham, pp 1–15
3. Mir A, Ketti Ramachandran R (2019) A survey on security challenges and research opportunities in Smart Grid based SCADA systems. Int J Comput Sci Eng 7(3): 734–755
4. Goel S, Hong Y (2015) Security challenges in smart grid implementation. Smart Grid Security. Springer, London, pp 1–39
5. Gartner: Fueling the Future of Business. https://www.gartner.com/doc/reprints?id=1-1OA93LZ7&ct=190716&st=sb. Last accessed 2020/06/01
6. Kumar P, Lin Y, Bai G, Paverd A, Dong JS, Martin A (2019) Smart grid metering networks: a survey on security, privacy and open research issues. IEEE Commun Surv Tutorials 21(3):2886–2927
7. Brewer R (2015) Cyber threats: reducing the time to detection and response. Netw Security 2015(5):5–8
8. Architecture and design requirements for Enterprise Security Monitoring Platform: Addressing security monitoring challenges in the financial services industry. https://www.diva-portal.org/smash/record.jsf?pid=diva2%3A1215841&dswid=440. Last accessed 2020/06/01
9. Inayat Z, Gani A, Anuar NB, Khan MK, Anwar S (2016) Intrusion response systems: foundations, design, and challenges. J Netw Comput Appl 62:53–74
10. El Mrabet Z, Kaabouch N, El Ghazi H, El Ghazi H (2018) Cyber-security in smart grid: survey and challenges. Comput Electr Eng 67:469–482

IOT-Based Wearable Assistive Device for Visually Impaired People

S. P. Rajamohana, Zbigniew M. Wawrzyniak, E. G. Radhika, S. Priya, and G. Sindhu

Abstract In this paper, we present a sensor-based wearable assistive gadget to help visually impaired individuals stroll without anyone's help through the roads. The principle segments of the framework are microcontroller board, ultrasonic sensors, flame sensor, and GSM and GPS modules. The proposed framework utilizes a lot of sensors to follow the way and alert the user of impediments before them. The user is alerted by vibrations on different fingers based on the direction of the obstacle, which is useful when the user is in the noisy environment and has hearing misfortune. In addition, the flame sensor is utilized to distinguish the fire in the environment and caution the visually impaired individual through vibration. When the emergency button is pressed, the system alerts the blind individual's concerned persons with the system location that is sent as a message. The prototype model developed for the proposed framework has shown its usefulness and viability. The proposed framework has a larger number of highlights than other comparative systems and anticipates that it should be a valuable device to improve the personal satisfaction of visually impaired people.

Keywords Visual impairment · Prototype model · Wearable assistive gadget

S. P. Rajamohana (✉) · E. G. Radhika · S. Priya · G. Sindhu
Department of Information Technology, PSG College of Technology, Coimbatore, Tamil Nadu, India
e-mail: spr.it@psgtech.ac.in

E. G. Radhika
e-mail: egr.it@psgtech.ac.in

S. Priya
e-mail: spa.it@psgtech.ac.in

G. Sindhu
e-mail: gsu.it@psgtech.ac.in

Z. M. Wawrzyniak
Faculty of Electronics and Information Technology, Warsaw University of Technology, Warsaw, Poland
e-mail: z.wawrzyniak@ise.pw.edu.pl

S. S. Dash et al. (eds.), *Sixth International Conference on Intelligent Computing and Applications*, Advances in Intelligent Systems and Computing 1369,
https://doi.org/10.1007/978-981-16-1335-7_15

1 Introduction

According to the recent survey made by World Health Organization (WHO), there are around 285,000,000 visually affected individuals in the earth. In that, there are 39 million who are visually impaired in the world. Over the previous years, visual impairment that is brought about by ailments has diminished because of the accomplishment of public health actions. However, the count of visually impaired individuals has been expanded by 50,000 by year. Lamentably, every one of these numbers is assessed to be doubled by 2020. The requirement for a gadget for way finding and flexibility has been elevated. The least complex and the most inexpensive way of finding and accessible are pets which were trained for such cause and smart cane. These equipments could help the visually impaired individuals just partially in light of the fact that they cannot give all data and features for secured mobility. To guarantee the total security of the visually impaired individuals, obstacle detection alone is not adequate. Besides, the visually impaired individuals ought to be cautioned regarding those hindrances which ensure their safety and secured movement in unfamiliar environments. To achieve this, obstacles are identified and limited at first and after that the data of those obstacles will be sent to the visually impaired people by utilizing different modalities such as voice and vibration [1]. The proposed framework has a few highlights to give outwardly debilitated people methods for sheltered and free portability and nonstop contact with their families and guardians, who can follow their location [2]. The framework depends on sensors to recognize objects at some separation before the user, alert the family and parental figures when the client bumbles over, and alert the family, guardians, and encompassing individuals when the client needs support.

2 Assistive Technology

Assistive innovation is a term that incorporates all gadgets and appliances which are used by optically provoked people to help in their step-by-step lives, make their activities easier, and give a protected versatility. The visual assistive technology is isolated into 3 groups: vision enhancement, vision substitution, and vision replacement [3]. This supportive mechanization abets users with recognition of the objects and discerns external environment using sensors that have become usable for the visually impeded individuals through electronic gadgets [4]. The vision replacement model is more intricate than the other two models. It manages medicinal and innovative controversies. Vision replacement incorporates showing data legitimately to the visual cortex of the cerebrum or through a ocular nerve. In any case, the postulation of vision enhancement and vision substitution is analogous. The differentiation is that in vision enhancement, the camera input is handled and after that the results will be outwardly displayed.

Vision substitution is like vision enhancement, yet the result builds up non-visual display, which can be vibration, sound-related, or both dependent on the touch and hearing senses that can be successfully controlled and felt by the outwardly debilitated users [5]. The proposed venture is based on visual substitution. The vision substitution classification incorporates three subcategories. They are electronic travel aids (ETAs), electronic orientation aids (EOAs), and position locator devices (PLDs). Each one of the three classifications tries to improve the visually impaired people's mobility with slight contrasts.

2.1 Types of Assistive Technology

2.2 Electronic Travel Aids (ETAs))

These gadgets exchange data to the user about encompassing environment through sensor cameras, sonar, or laser scanners. The principles of ETAs as demonstrated by the National Research Council seem to be:

- Ascertain snags starting from the head to ground throughout the user body.
- Rendering a few guidelines to the user about the gesture surface that comprises holes or fissures.
- Uncovering things encompassing the impediments.
- Dispensing data to the user regarding the distance of the obstacle with advancing directions.
- Proposing remarkable sight areas.
- The preparedness of self-orientation and intellectual guide of the environment are granted through information.

2.3 Electronic Orientation Aids (EOAs)

These are gadgets that furnish people on foot with directions in new places. The rules of EOAs are

- Characterizing the course to pick the best way.
- Tracing the way to generally register the zone of the user.
- Providing movability rules and way signs to manage the user and develop her/his cerebrum about condition.

2.4 Position Locator Devices (PLDs)

These gadgets compute the exact locale of its possessor, for example, gadgets utilizing GPS module. Hurdle spotting and GPS components are the centered substructure in the propounded framework.

2.5 Most Suggestive Electronic Gadgets

Most electronic contraptions that give administrations to outwardly impeded people depend upon the information assembled from the encompassing condition and transmitted to the user by methods for either tactile, sound or both.

2.6 Smart Cane

The smart cane is used for identifying the objects and to produce exact directions for navigation. It is a convenient device that is outfitted with a sensor framework. The components used in the smart cane comprise ultrasonic sensors, microcontroller, vibrator, buzzer, and water detector so as to guide outwardly impeded individuals. The system uses servo engines, ultrasonic sensors, and fuzzy controller to perceive the obstruction in front of the user and after that give guidance through voice messages or hand vibration. Obstacles are identified using ultrasonic sensors [6]. Consequently, the fuzzy controller can give the careful decisions dependent on the information got from the servo motors and ultrasonic sensors to explore the user. Smart cane results rely upon congregating the data to produce audio notification to the user via speaker. Also, visually impeded individuals are provided with special vibrator gloves that give particular vibration for each finger and each one with a different meaning. The smart cane has accomplished its objectives in identifying the objects and obstructions, delivering the required feedback [7]. The drawbacks of smart cane are that the water sensor will not distinguish water except if it is 0.5 cm or more and the signal of water indicator will not stop before it is dried or cleaned. The future work that can be improved in the smart cane is that so as to screen the power status, it is smarter to have a power supply meter being introduced and to add a signal clock to decide the period to illuminate the buzzer's issue as well.

2.7 Vision Substitution

The embedded device is utilized as an eye substitution that helps the blind people in directions and navigation [1]. The vision substitution device is a TI MSP 430G2553 microscale controller. The vision substitution device is linked with android application by making use of proposed algorithms [8]. The job of this android application is to make use of GPS, GSM, and GPRS to extract the location of the individual and navigate better directions in order to help the blind people with good facilities. The vision substitution device comprises 2 ultrasonic sensors and 3 vibrator motors [9]. The design of the device is very light and helpful to all visually impaired people. The downside of this eye substitution is that the structure could be better if the authors did not utilize the wood establishment that will be carried by the user more often than not. The framework is not reliable and is restricted to android devices [10].

3 Related Works

Author proposed an electronic travel aid (ETA) that is fixed to the shoe and helps the outwardly weakened people in navigation. Authors have actualized the techniques like Zigbee module interface, vibrator sensor, microcontroller Arduino, ultrasonic separate estimation module, gyroscope interface, and power supply. The ultrasonic sensor utilized starts the vibration on the heel if the article is behind and front. The alarms happen likewise through voice message [11].

Author proposed a snag acknowledgment framework for the visually impaired individuals utilizing the radio frequency identification (RFID) [8]. Authors have introduced a RFID tag on every one of the regions that should have been recognized. These labels will fill in as a milestone to the individual utilizing the stick. Each label will be furnished with as much data as expected to plainly characterize the area of that exact tag. It has a RFID label reader on the stick, and a Braille interfaces of the visually impaired individual can utilize it. Also, authors proposed the fuse of a mechanical battery that will empower the stick to dependably have a full battery which will stay away from its clients to come up short on battery while utilizing it. The RFID reader identifies the label stage and reads all the data put away in the label utilizing remote association, the data is then read and examined, and the information recovered is transmitted to a control unit that makes an interpretation of the data into Braille code. This data that is currently in Braille is shown on a Braille interface. The impediment of the framework is that the stick wiring confounded, and changing the battery can be troublesome particularly on the off chance that it kicks the bucket at a basic minute.

Author [1] propounded canny framework for managing people who are visually impaired. Authors felt that the conventional stick could not assist with detecting risks from all dimensions of obstructions. The framework has been utilized to get the information from the detecting gadgets and direct the GSM module. The GSM module will coordinate with ultrasonic, accelerometer sensors to identify snags thereby increases point by point information for visually impaired people. Assessments of the framework created have been directed by joining the model to the handle of the white stick. The trial results have demonstrated the handiness of the framework in enabling visually impaired individuals to move freely, securely and rapidly among deterrents and dangerous places. But the challenge is that could not be able to discover an answer for broadening the battery life and utilizing GSM module which expanded the framework cost.

In this work [12], author developed a obstruction recognition and area following framework, which involves two sections, one to identify deterrents utilizing a ultrasonic sensor and caution the client through a signal, and the other to follow the client area utilizing a GPS module and send SMSs with area data to enrolled gadgets by means of a GSM module.

Author has implemented a remote indoor situating framework as a component of an answer for autonomous versatility of VIPs [13]. The framework is supplied with an ear by situating server, a discretionary worldwide situating server, a neighborhood database server, and a cell phone to work in an alternate remote system as reference terminal. In particular, the terminal estimates the strength of the flag transferred from a reference station inside a structure and passes the outcome to then ear by situating server. This system means to store data about the region it serves. At that point, the put away reference estimations are utilized to appraise the client area based on the got flag quality pointer, consequently conveying one of the principal fruitful indoor situating frameworks. Moreover, the terminal is in charge of the correspondence with the worldwide restriction server.

In this work [14], author proposed a savvy stick for visually impaired which can give early cautioning of a hindrance utilizing infrared (IR) sensors. Subsequent to distinguishing the impediments, the stick is made to caution the outwardly disabled individuals utilizing vibration signals but neglected to make the shrewd stick for aiding crisis required by the visually impaired yet concentrate just for snag discovery. Furthermore, the IR sensors are not by any stretch of the imagination sufficiently effective on the grounds that it can identify just the closest deterrent in short separation.

In this work [15], author proposed a framework that utilizes a shrewd stick utilizing laser sensors to recognize the impediments and down controls. They signalized the deterrent location by a high pitch "Blare" utilizing a mouthpiece. The plan of the laser stick is straightforward and instinctive. Be that as it may, their stick can just distinguish obstruction, yet cannot give intellectual and mental help. There exists just signal sound that is activated by any deterrent, and there is no any help to guide them.

In this work [16], author presented a device for the blind people that provide navigation and real time monitoring. The key idea is to use device for remote assistance by monitoring video transmitted by a RF camera. Author designed ultrasonic assistive headset for VIP. It employs ultrasonic sensors, microcontroller circuit, and solar cells [10]. All the components deployed are inexpensive and easy to use. The ultrasonic assistive headset aids the person by determining the location and distance of the obstacle and vocalizes to the blind individual where the obstacle is present. It paves the way for indoor and outdoor mobility. Overall, the proposed framework has obstacle detective highlights; besides, the system permits consistent correspondence between the user and his family members, with location detection. Moreover, the flame sensor is incorporated into the gadget that detects the fire in the surrounding and alerts the blind person through vibration.

4 Proposed System

The proposed system provides a smart electronic wearable aid for visually challenged people. This wearable device helps the blind to walk easily in urban areas that indicates obstacle data to the blind person through vibration by using ultrasonic sensors. Fire in the surroundings can be detected using flame sensor that alerts the blind person using vibrator. And when a person is in a trapped location, he/she can press the emergency button and get the latitude and longitude position of the blind person through GPS. And the location can be sent to the relatives of the blind person through GSM.

4.1 Arduino Controller

Arduino Uno can be programmed with the product called Arduino Software IDE. The Arduino Uno board can be controlled utilizing the USB association or with an outside power supply. Arduino controller has 32 kB memory out of which 0.5 kB is involved by the boot loader. It additionally has 2 kB and 1 kB of SRAM and EEPROM separately [17]. A library called Software Serial Library permits the sequential correspondence on any of the Uno's computerized pins (Fig. 1).

4.2 Flame Sensor

The flame sensor identifies then earness off lame dependent on the infrared (IR) wavelength radiated by the fire. Arduino Uno checks the rationale level on the yield stick of the sensor, and as a reaction the vibrator engine begins pivoting. Fire sensor has three pins, VCC, GND, and D0. The D0 is associated with digital *I/O* stick of Arduino. With no flame, the yield from the flame sensor is HIGH. At the point when the sensor distinguishes any flame in its surrounding, its yield turns out to be LOW. Arduino distinguishes this LOW flag on its information stick and takes suitable activities. The 10*k* potentiometer on the board can be utilized to change the affect ability of the sensor.

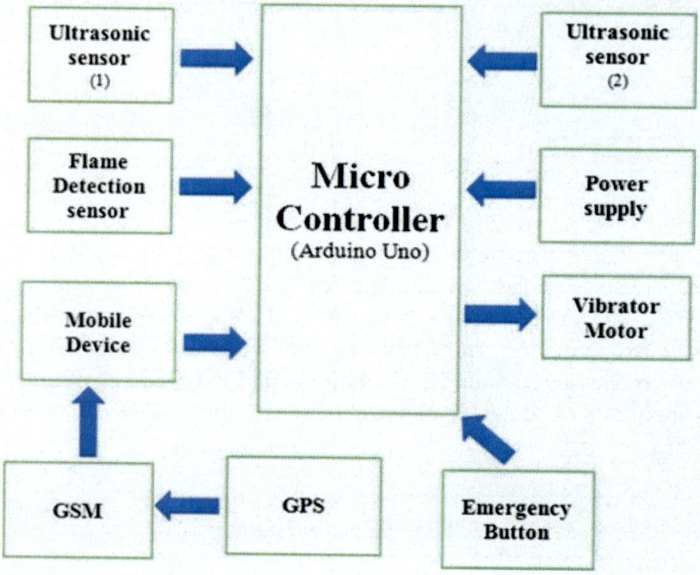

Fig. 1 Design of the obstacle assistive model

4.3 Ultrasonic Sensor

An ultrasonic sensor gadget estimates separation to an article with the assistance of sound waves. It works by conveying a sound wave at ultrasonic recurrence and hangs tight for it to resound once more from the item. At the point when an electric beat of high voltage is connected to the ultrasonic transducer, it vibrates over a particular range of frequencies and produces a burst of sound waves (Khatri (2014)). At whatever point any snag comes in front of the ultrasonic sensor, the sound waves will reflect as reverberation and produce an electric pulse. The trig stick sends a flag—a HIGH recurrence sound. At the point when a flag finds an item, it is reflected and the echo stick gets it. The sensor computes the time taken by the distinction between sending sound waves and accepting reverberation. The echo patterns are compared with the patterns of sound waves to determine detected signal conditions. The time between the transmission and the gathering of the flag enables us to know the distance of an item [18].

4.4 GPS Module

The GPS recipient Neo6M gets the information of scope and longitude from the satellite. The data will be utilized to share the area of visually impaired individual to others. It has an outside receiving wire and implicit EEPROM. It works with standard NMEA sentences. It speaks with Arduino by means of serial correspondence. To get crude GPS information, the sequential correspondence must be begun with the GPS module utilizing Software Serial. When correspondence is begun, the information will be passed from the module in NMEA position [19]. NMEA represents National Marine Electronics Association which is the standard information design bolstered by GPS makers. This crude information can be changed over into intelligible configuration with the assistance of TinyGPS++ library.

4.5 GSM Module

The GSM shield outfitted with an Arduino board sends a SMS to the predetermined individual. At first, embed the SIM card into the SIM card holder. Ensure the radio wire is very much associated. On the sequential port select, ensure the jumper top is associated with use programming sequential. The shield can be controlled utilizing an outside 5 V power supply. Make a point to choose the outside power source with the flip change alongside the DC jack as shown in Fig. 1. To control up or down the shield, press the power key for around 2 seconds. At that point, the status LED will illuminate and the NetLight LED will squint for each 800 milliseconds until it finds the system. When it finds the system, the NetLight LED will begin squinting for at regular intervals. It tends to be tried whether the shield is working appropriately or not by sending AT commands from the Arduino [20].

4.6 Arduino IDE

Arduino has a hardware programmable microcontroller circuit board and Arduino Software IDE (Integrated Development Environment) that is used to compose and transfer PC code to the physical board. The Arduino IDE is a cross-stage application that is written in the programming language Java. It is utilized to compose and transfer projects to Arduino compatible boards. TinyGPS++ is a new Arduino library for parsing NMEA information streams given by GPS modules. Like its forerunner, TinyGPS, this library gives minimized and simple-to-utilize techniques for separating position, date, time, height, speed, and course from shopper GPS gadgets. TinyGPS++'s developer interface is significantly easier to use than TinyGPS, and the new library can remove subjective information from any of the heap NMEA sentences out there, even exclusive ones.

5 Test and Results

5.1 Wearable Gloves with Obstacle Detective Embedded Device

This is the overall output of the prototype model embedded with ultrasonic sensors, flame sensor, GPS, GSM, vibrator motor, and emergency button shown in Fig. 2.

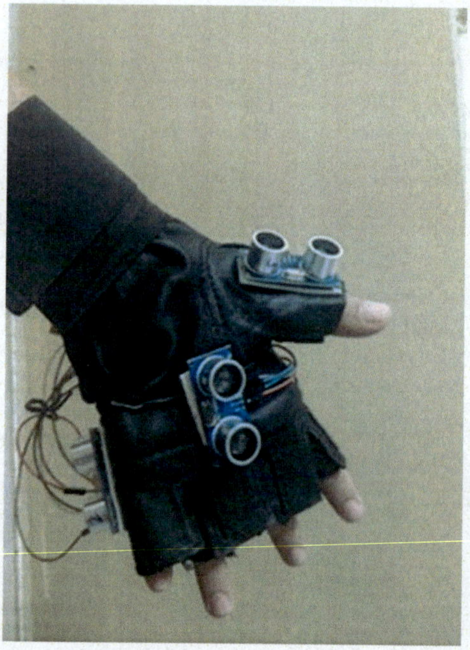

Fig. 2 Wearable gloves with embedded device

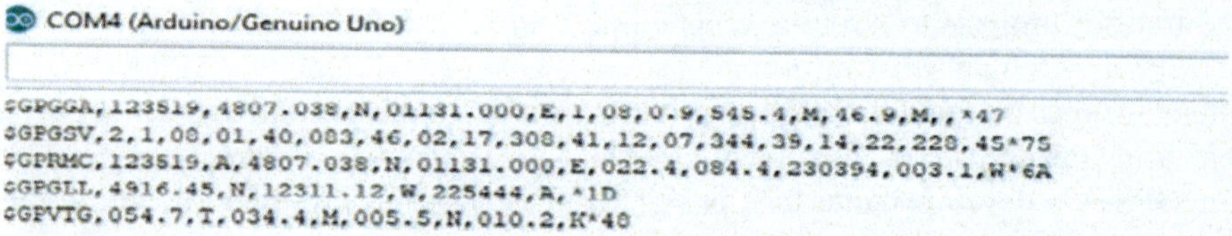

Fig. 3 NMEA format

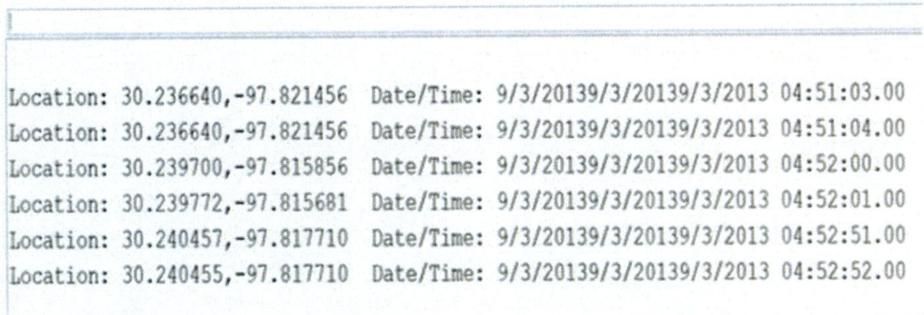

Fig. 4 GPS readable format

5.2 *NMEA Format*

The NMEA stands for National Marine Electronic Association which is the standard data format supported by all GPS manufacturers as shown in Fig. 3.

5.3 *Results of GPS Module*

When the emergency button is pressed, the location of the blind person is displayed immediately on the Arduino IDE screen. GPS module extracts the latitude, longitude, and date/time and displays in the Arduino IDE screen. There are two output formats in GPS. They are NMEA format and human readable format as shown in Fig. 4.

5.4 *Results of Ultrasonic Sensor*

When the ultrasonic sensor detects an obstacle, the range of detection is displayed in the serial monitor of Arduino IDE as shown in Fig. 5. The range of detection is fixed as 15 cm. When the obstacle is within 15 cm, the range will be displayed and the vibrator motor rotates (Fig. 5).

5.5 *Results of Flame Sensor*

When the flame is detected within a range of 750–1100 nm, then the output pin is set high and vibrator motor rotates (Figs. 6 and 7).

If the flame is detected, then the serial monitor of Arduino IDE displays FIRE. If the flame is not detected, then the serial monitor displays NO FIRE.

Fig. 5 Results of ultrasonic sensor

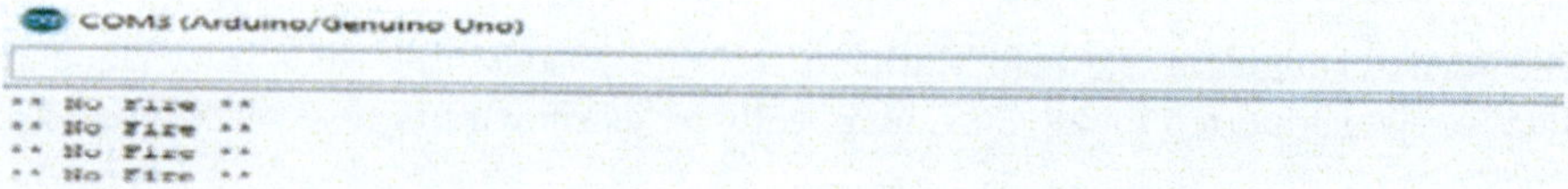

Fig. 6 Results of flame sensor

Fig. 7 Results of flame sensor

6 Conclusion

In this paper, a wearable assistive gadget is utilized as help for VIPs. The principle preferred standpoint of this assistive model is that the gadget is wearable, and an extra element incorporated into this wearable aid is flame sensor which recognizes the fire in the environment and alerts through vibration. In spite of other wearable devices effectively existing in the market, this gadget is portable and compact since the segments are inserted in a hand glove. The vibrator engines utilized are of extremely little size which makes the gadget exceptionally light and simple to use. This component will serve an extraordinary help to the user when they are struck some place as emergency button is added to the gadget which when pressed extracts the latitude and longitude area from the GPS module and sends this coordinates to the phone number indicated in the source code by means of the GSM module. In contrast to certain current frameworks existing in the literature survey, there are no required conditions for the gadget to work. The gadget can work any place whenever. Lifetime of the battery is a detriment of this framework. Another disservice is that the GPS module cannot extricate the coordinates except if it is in an open space. Future works should be possible to expand the lifetime of the battery by expanding the voltage and to utilize a GPS module which works under all conditions.

References

1. Al-Fahoum, Amjed S, Al-Hmoud HB, Al-Fraihat AA (2013) A smart infrared microcontroller-based blind guidance system. Active and Passive Electronic Components 2013
2. Amutha B, Ponnavaikko M (2009) Location update accuracy in human tracking system using zigbee modules. arXiv preprint arXiv:0912.1019
3. Sangami A, Kavithra M, Rubina K, Sivaprakasam S (2015) Obstacle detection and location finding for blind people. Int J Innov Res Comput Commun Eng 3(2):119–123
4. Kumar MN, Usha K (2013) Voice based guidance and location indication system for the blind using GSM, GPS and optical device indicator. Int J Eng Trends Technol (IJETT) 4(7): 3083–3085
5. Rao BS, Deepa K, Prasanth H, Vivek S, Nanda Kumar S, Rajendhiran A, Saravana J (2012) Indoor navigation system for visually impaired person using gps. Int J Adv Eng Technol 3(2): 40–43
6. Dambhare S, Sakhare A (2011) Smart stick for blind: obstacle detection, artificial vision and real-time assistance via GPS. In: Proceedings of the 2nd National Conference on Information and Communication Technology, Chennai, India, vol. 2324
7. Gayathri G, Vishnupriya M, Nandhini R, Banupriya M (2014) Smart walking stick for visually impaired. Int J Eng Comput Sci 3(03)
8. Nassih M, Cherradi I, Maghous Y, Ouriaghli B, Salih-Alj Y (2012) Obstacles recognition system for the blind people using RFID. In: 2012 sixth international conference on next generation mobile applications, services and technologies. IEEE, pp 60–63
9. Ramadhan AJ (2018) Wearable smart system for visually impaired people. Sensors 18(3):843
10. Aymaz Ş, Çavdar T (2016) Ultrasonic assistive headset for visually impaired people. In: 2016 39th international conference on Telecommunications and Signal Processing (TSP). IEEE, pp 388–391
11. Khatri A (2014) Assistive vision for the blind. Int J Eng Sci Invent 3:16–19
12. Nalavade KC, Bharmal F, Deore T, Patil A (2014) Use of ultrasonic sensors, GPS and GSM technology to implement alert and tracking system for blind man. In: Proceedings of the International Conference of Advance Research and Innovation (ICARI), pp 272–274
13. Wawrzyniak P, Korbel P (2013) Wireless indoor positioning system for the visually impaired. In: 2013 federated conference on computer science and information systems. IEEE, pp 907–910
14. Alshbatat N, Ilah A (2013) Automated mobility and orientation system for blind or partially sighted people. Int J Smart Sensing Intelligent Syst 6(2)
15. Sharma P, Shimi SL, Chatterji S. A review on obstacle detection and vision. Int J Eng Sci Res Technol
16. Chandana K, Hemantha GR (2014) Navigation for the blind using GPS along with portable camera based real time monitoring. SSRG Int J Electron Commun Eng 1:46–50
17. Jain D (2014) Path-guided indoor navigation for the visually impaired using minimal building retrofitting. In: Proceedings of the 16th international ACM SIGACCESS conference on Computers & accessibility, pp 225–232

18. Jothi R, Kayalvizhi M (2017) Smart walking stick for visually challenged people. Asian J Appl Sci Technol (AJAST) 1:274–276
19. Gawari H, Bakuli M (2014) Voice and GPS based navigation system for visually impaired. Int J Eng Res Appl 4:48–51
20. Kher Chaitrali S, Dabhade Yogita A, Kadam Snehal K, Dhamdhere Swati D, Deshpande Aarti V (2015) An intelligent walking stick for the blind. Int J Eng Res Gen Sci 3: 1057–1062

Cybersecurity Awareness on Video Conferencing Service During Coronavirus Pandemic

Venkata Ramana Ayinavalli, L. Kavisankar, K. Ravichandran, and S. Balasubramani

Abstract The entire world is looking for a solution that helps us to come out of the current pandemic situation caused by the novel coronavirus. It has not only caused lakhs of death but also the biggest economic shutdown. In this scenario, with the need for social distancing the world needs to have a certain strategy to sustain. Even though online e-commerce has hit new highs, now it is growing even faster with video conferencing service; here the need for awareness of cybersecurity comes into action. Since the entire population is depended on the information technology, all age people without awareness of securing themselves from cyberattack use the technology which creates lots of problem for the end user and the provider of that service. This paper considers the statistics on the cybersecurity awareness on video conferencing service from the different level of users of different age group.

Keywords Cybersecurity · Awareness · Video Conferencing · Coronavirus pandemic · COVID-19

V. R. Ayinavalli (✉)
Department of Information Technology, GMR Institute of Technology, Rajam, Andhra Pradesh, India
e-mail: ramana.av@gmrit.edu.in

L. Kavisankar
School of Computing, College of Engineering and Technology, SRM Institute of Science and Technology, Kattankulathur, Tamilnadu 603203, India
e-mail: kavisanl@srmist.edu.in

K. Ravichandran
Department of Visual Communication, Dr. MGR Educational and Research Institute, Chennai, Tamilnnadu, India

S. Balasubramani
Department of Computer Science and Engineering, Hindustan Institute of Technology and Science, Chennai, Tamilnadu, India

S. S. Dash et al. (eds.), *Sixth International Conference on Intelligent Computing and Applications*, Advances in Intelligent Systems and Computing 1369,
https://doi.org/10.1007/978-981-16-1335-7_16

1 Introduction

As huge quantities of representatives adjust to telecommuting, video conferencing has given an open and moderate approach to empower the cutting edge working environment. In any case, the expanded dependence on video conferencing innovation has likewise given the ideal occasion to cybercriminals to listen in on private discussions, send parody messages, capture screen controls and dispatch a variety of vindictive assaults.

As of late, it has been uncovered that Zoom has become an objective for programmers because of a genuine weakness in its work area conferencing application. Indeed, programmers around the globe are effectively occupied with bug abundance chasing, looking for likely weaknesses in Zoom's innovation.

2 Literature Review

Videoconference as assistance [1] proposes an approach to offer videoconference as a Web administration over an interface which can be utilized by outsiders to advance their applications. This interface incorporates a security instrument supporting assigned approval to permit coordination into outsider's surroundings. The public and private cloud video conferencing implementation [2] is done using various scalable advancements in the cloud service with essential parameters like scalability, accessibility and pervasiveness.

To guarantee the transmission security of the data of the video conferencing framework P2P-based [3], here this paper reports a bunch of new security engineering and sets up the relating security the board technique by presenting security levels, encryption calculations and the methodology tree.

In this examination, the investigation is done on the protection given that might be in danger by going to virtual gatherings [4]. Its extricated private data from montage pictures of meeting members that are openly posted on the Web. It utilizes the picture prepared, text acknowledgment devices, just as interpersonal organization investigation to investigate the web creeping curated dataset of more than 15,700 composition pictures, and more than 142,000 face pictures of meeting members.

As of now, the online video gathering framework normally needs to introduce this application programming. Hence, the product should be created copy adaptation for various working frameworks (OS), e.g., Android, Windows and MacOS. Notwithstanding, the online video gathering framework is OS free in order to spare advancement cost [5]. In the paper, we present execution of an electronic video meeting arrangement by utilizing WebRTC [6]. It very well may be utilized for meeting and talk with a program. Rather than customer worker brought together plan, in a digital video meeting, every part sends the media stream to one another shared. Moreover, the media stream is scrambled, and an administration system is created to deal with the passage of gathering room staff and accomplish a protected multi-party video meeting. We likewise conquer shared (P2P) web NAT crossing of media transmission. Our answer covers four primary highlights: P2P Web NAT crossing, dynamic join and leave, access control, dynamic video goal change.

Multi-client video conferencing [6] is a universal innovation. Progressively end-has in a meeting is helped by cloud-based workers that improve the nature of involvement for end clients. This paper assesses the effect of methodologies for the situation of such workers on client experience and organization cost. We consider situations dependent on the Amazon EC2 foundation just as future situations in which cloud occurrences can be situated at a bigger number of potential locales over the planet. We analyze various potential systems for picking which cloud areas should have administrations and how traffic should course through them. Our examination is driven by genuine information to spur interest situations with sensible topographical client conveyances and diurnal conduct. We infer that on the EC2 foundation a very much picked static determination of workers performs well however as more cloud areas are accessible a unique selection of workers gets significant.

Our exploration shows that video conferencing frameworks are inclined to various assaults which can bring about the revelation of delicate data [7]. This paper features a portion of those issues just as giving suggestions to make sure about these gadgets.

This exploration shows [8] that the advancement of private cloud asset pools can conquer the restrictions of customary video conferencing frameworks, with points of interest of improved asset usage, self-administration and higher framework soundness.

Cloud-based provisioning of the conferencing administrations [9] on which these applications depend will bring benefits, for example, simple assistance provisioning and versatile adaptability. Nonetheless, it stays a major test. This paper proposes a PaaS for conferencing administration provisioning. The proposed PaaS depends on a plan of action from the best in class. It depends on conferencing IaaSs that, rather than VMs, offer conferencing substrates (e.g., dial-in flagging, video blender and sound blender). The PaaS empowers piece of new meetings from substrates on the fly. This has been prototyped in this paper and, to assess it, a conferencing IaaS is likewise executed. Execution estimations are additionally made.

With the ever-expanding requests on interactive media applications, distributed computing, because of its prudent however ground-breaking assets, is turning into a characteristic stage to measure, store and offer sight and sound substance. Be that as it may, the work of distributed computing additionally brings new security and protection issues as not many public cloud workers can be completely trusted by clients. In this paper [10], we center around how to safely share video substance to a specific gathering of individuals during a specific time-frame in cloud-based mixed media frameworks and propose a cryptographic methodology, a provably secure time–space characteristic-based admittance control (TAAC) plot, to make sure about the cloud-based video content sharing.

In this paper, a web-RTC-based sound/video conferencing framework on the virtualized cloud has been executed [11]. Through this framework, clients can encounter sound/video meeting with just an Internet browser. Additionally, this framework gives adaptable capacities through the versatile assets the executives from the virtualized cloud.

This paper proposes [12] online protection relief activities that can make sure about WFH stages, working from home and video conferencing applications, and make them more secure choices to coordinate correspondences both now and possibly in the post COVID-19 time.

3 Secure Video Conferencing

3.1 *Tips for Secure Video Conferencing*

Tips for Secure Video Conferencing

The accompanying tips are imparted to the number of inhabitants in all age gathering.

•When you get a gathering greeting, confirm that it's from a known, confided the in sender. Try not to click dubious connections which may attempt to fool you into entering your login subtleties into a satire phishing site.
•Do not offer a connection to a video chat via web-based media. Just give the private connection straightforwardly to explicit individuals who ought to partake in the gathering.
•Check that whiteboards, reports and different materials which may contain delicate data are not considering the camera.
•Ensure you are utilizing the most recent adaptation of the video conferencing programming, as security weaknesses are probably going to be misused in more established forms.
•Avoid unveiling gatherings. For additional security, make your gathering settings private and add a gathering secret key to control the permission of visitors. In the event that the administration lets you make a secret phrase for the gathering, use secret word creation best practices.
•Make utilization of sitting area includes in conferencing programming. Such highlights put members in a different virtual room before the gathering and permit the host to concede just the individuals who should be in the room.
•Manage your screen sharing choices. Set the host to oversee screen sharing of course, at that point the host can permit explicit members to share when proper.
•Have a video conferencing strategy setup. Like other significant authoritative strategies, a video conferencing strategy empowers you to define clear limits and safe practices for clients.
•Don't record gatherings except if you have to. In the event that you do record a gathering, ensure all members are educated and the account is spared in a free from any danger area.
•Educate all representatives about the dangers of video conferencing and the particular advances they should take to guarantee their gatherings are secure.

The results were analyzed, nonparametric models. In total, 243 public participated in the study. Following knowledge screening, a final sample of three hundred was achieved. Out of the three hundred responses, there have been 101 females (42%) and 142 males (58%). People are in age below 30 30.3%, between 31–35 years 40.3%, between 41–45 years 22.6% and on top of forty-five years 6.8%. The survey is done with students, IT employees, bank employees, academicians, doctors, cameramans, business and home makers. The various video conference tools like Google Hangouts, Skype, MS Teams, Zoom, Jitsi, Go-to-meeting, Hangouts meet, Big blue Button, Life Size, Cisco Webex meetings were analyzed with their features like security, cloud security, admit people only those who supposed to attend, password protected, sharing video conference links increases number of threats, turn off video when not required, use latest version, eject participants, educate all employees, balancing security with ease of use, encrypting all chat conversations, firewall/NAT traversal, service availability and data centers.

The cybersecurity awareness for the user of video conference is provided, and the level of understanding of the security need and implementation is analyzed based on the statics gathered in this paper. The result clearly shows the level of awareness of the video conferencing servicing users.

Among the various questionnaires related to securing video conference service, one of the questions is found with very less significance value (Fig. 1) of 0.002 and

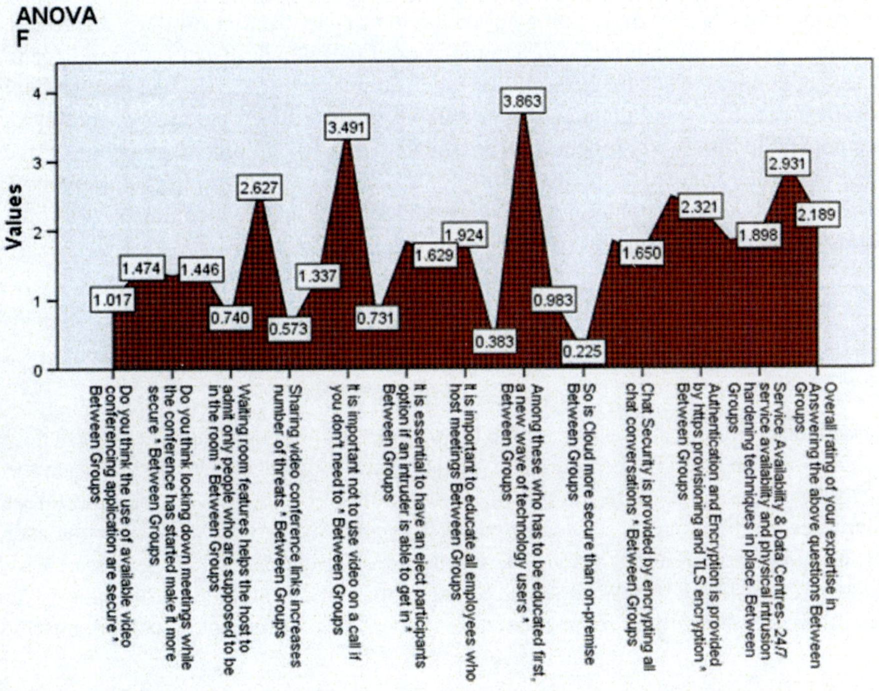

Fig. 1 ANOVA F values

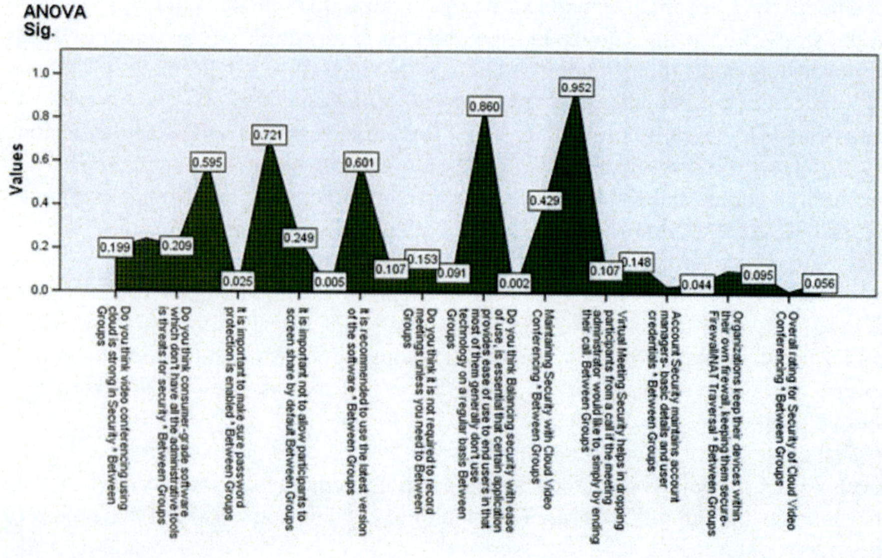

Fig. 2 ANOVA significance

ANOVA F value (Fig. 2) of 3.863 which clearly shows that more number of people agree that a new wave of technology users has to be educated first similarly it is found, it is not very much important to use video on a call which doesn't demand for it. The stakeholders of secure cloud-based video conferencing service feel that the 24/7 service availability and the data center and physical intrusion hardening techniques are not stable; its clearly reflecting from the data that 10.67 percentage people agree they have they very good infrastructure for the availability and intrusion prevision. 28.9 percentage people neither agree nor disagree. Around 60.9 percentage of people totally support that sufficient infrastructure is not in place.

4 Conclusion

The video conferencing service which the people thought would use very rarely now has become the need of the hour. The technology always takes a big stride when the need for it increases. With the use of the technology by the naive users, the threat increases for the user and service provider. The cybersecurity awareness for the user of the video conference is provided, and the level of understanding of the security need and implementation is analyzed based on the statics gathered in this paper. The results clearly show the level of awareness of the video conferencing servicing users.

References

1. Rodríguez P, Gallego D, Cerviño J, Escribano F, Quemada J, Salvachúa J (2009, November) Vaas: Videoconference as a service. In: 2009 5th international conference on collaborative computing: networking, applications and worksharing. IEEE, pp 1–11
2. Gavish A, Curtis M, Wolf S (2010) Video conferencing in the cloud. Conference: Fall 2010 Internet2 Member Meeting
3. Wu W, Bai E (2011, July) P2P-based video conferencing security management strategy. In: 2011 international conference on multimedia technology. IEEE, pp 614–617
4. Kagan D, Alpert G, Fire M (2020) Zooming into video conferencing privacy and security threats. arXiv preprint arXiv:2007.01059
5. Lin CHR, Zhang H, Liu JS, Chen SH (2020, June) Implementation of secure web conferencing. In: 2020 2nd International Conference on Computer Communication and the Internet (ICCCI). IEEE, pp 12–17
6. Clegg RG, Landa R, Griffin D, Rio M, Hughes P, Kegel I, Williams D (2017) Faces in the clouds: long-duration, multi-user, cloud-assisted video conferencing. IEEE Trans Cloud Comput 7(3):756–769
7. Jam N, Proebstel E (2006) Work in progress: who is watching you? Video conferencing security. In: 2006 Securecomm and workshops. IEEE, pp 1–2
8. Feng F, Wang W (2015, December) Video conferencing system using private cloud computing technology. In: 2015 international conference on Computational Intelligence and Communication Networks (CICN). IEEE, pp 960–963
9. Alam AF, Soltanian A, Yangui S, Salahuddin MA, Glitho R, Elbiaze H (2016, June) A cloud platform-as-a-service for multimedia conferencing service provisioning. In: 2016 IEEE Symposium on Computers and Communication (ISCC). IEEE, pp 289–294
10. Yang K, Liu Z, Jia X, Shen XS (2016) Time-domain attribute-based access control for cloud-based video content sharing: a cryptographic approach. IEEE Trans Multimedia 18(5):940–950
11. Yoon S, Na T, Ryu HY (2016, January) An implementation of Web-RTC based audio/video conferencing system on virtualized cloud. In: 2016 IEEE International Conference on Consumer Electronics (ICCE). IEEE, pp 133–134
12. Okereafor K, Manny P (2020) Solving cybersecurity challenges of telecommuting and video conferencing applications in the COVID-19 pandemic. Int J IT Eng (IJITE) 8(6): 24–32. June 2020 ISSN: 2321-1776

Development of an Analytical Model for Channel Potential in Junction-Less Double-Gate FETs

D. Sudha and G. Amarnath

Abstract An analytical model for the potential distribution in channel region of junctionless double-gate FETs by considering Gaussian-like vertical doping is presented in this paper. To calculate the channel potential distribution in the channel region, the two-dimensional Poisson's equation has been derived by including the evanescent mode technique. The channel conduction path position also has been modelled to derive the channel potential at various points in the channel conduction path. The channel potential considered to be the addition of the conventional device potential obtained by Poisson's solution and the potential function calculated by Laplace equation. The validity of the developed model is verified by comparing with physics-based numerical simulation results obtained by commercially available ATLAS TCAD tool and found in good agreement.

Keywords Channel potential · Double-gate · JLFET · Modelling · TCAD

1 Introduction

The "Junction-Less Metal–Oxide–Semiconductor Field Effect Transistors" or "JLFETs" are considerable devices for development of digital memory circuits [1] and analog RF applications [2] due to their superior device characteristics [3, 4]. Additionally, these devices gives excellent better carrier transit time/speed and better scaling properties in the nanometre regime due to less scattering effects at surface [4]. An extensive study and research have been reported on the device physics and fundamental device working principle of JLFETs [4–8] and also on physics-based analytical models [9–12]. In [8], Boltzmann statistics is used for realizing the threshold voltage change with respect to channel doping and oxide thickness. In

D. Sudha (✉)
Department of ECE, CMR College of Engineering and Technology, Hyderabad, India

G. Amarnath
Department of ECE, Marri Laxman Reddy Institute of Technology and Management, Hyderabad, India

S. S. Dash et al. (eds.), *Sixth International Conference on Intelligent Computing and Applications*, Advances in Intelligent Systems and Computing 1369,
https://doi.org/10.1007/978-981-16-1335-7_17

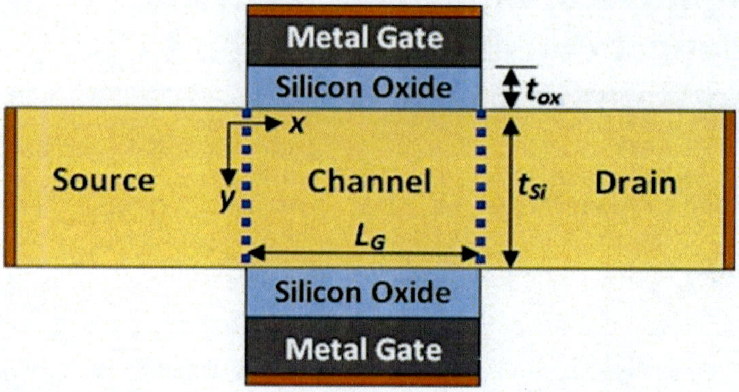

Fig. 1 A two-dimensional schematic of junctionless double-gate FETs

[6], bulk conduction method based quasi threshold voltage has been developed for double-gate (DG) JLFETs. The sub-threshold current and threshold voltage model have been developed by calculating channel potential with Poisson's equation solution in [9]. Sub-threshold current and sub-threshold voltage model have been developed by Poisson's equation solution with superposition method. In [10], a potential model is proposed for the DG-JLFET by considering depletion regions of source and drain. These mentioned studies have considered a uniformly doped channel. Fabrication steps like ion-implantation and diffusion take different doping concentrations for different regions. So it is difficult to organize the fabrication steps with uniform doping as well as threshold voltage adjustment [11]. The construction of JLFET of graded channel with minimum doping on the channel bottom surface and maximum doping on the channel top surface is reported in [12] results in Gaussian-like doping profile. The distinct doping profiles can be extracted from the general doping profile by changing the Gaussian function parameters in a required manner [12]. The improvement is shown in off-state current with a Gaussian-like vertical doping in the channel [13]. So, this paper presents a physics-based analytical model for the channel potential with a Gaussian-like vertical doping profile of DL-JLFET. The model outcomes are related to 2-D numerical simulations for the validity of the developed model results. In order to validate the numerical simulation results, the drain current results obtained by ATLAS TCAD [14] are compared with the experimental results available in the literature [4]. The validity of the developed model outcomes is tested by relating with numerical simulation results acquired by TCAD tool.

2 Model Development

The considered schematic of the device structure is presented in Fig. 1, where t_{ox}, t_{si}, and L_{g} are thickness of oxide, thickness of channel and length of the channel respectively with drain-to-source, V_{ds} and gate-to-source, V_{gs} voltages.

For easy calculations, the drain and source terminals are considered to be of no thickness as mentioned in [6] which are located at the end of each side. The x- and y-coordinates are shown along thickness of the silicon and interface of source-to-channel, respectively. The Gaussian-like doped structure considered for numerical simulations is presented in Fig. 2.

2.1 *Channel Potential*

There is no need to let the channel region of the doping profile $N_{\mathrm{GL}}(x)N_{\mathrm{GL}}(x)$ is considered in the vertical direction and is represented as [11]

$$N_{\mathrm{GL}}(x) = cN_{pk}\big((a + 2baX)^2 - 2b\big)\exp\big(-a\alpha X - bX^2\big) \tag{1}$$

where $a = 0.646$, $b = 1.786$, and $c = 0.55$ are the appropriate parameters; $\alpha = -1$ for $x < 0$ and $\alpha = +1$ for $x > 0$; $X = \left(x - \frac{R_p}{\sqrt{2}\sigma_p}\right)$ where σ_p and R_p are the straggle and projected range of the Gaussian-doping profile; and N_{pk} is peak-doping

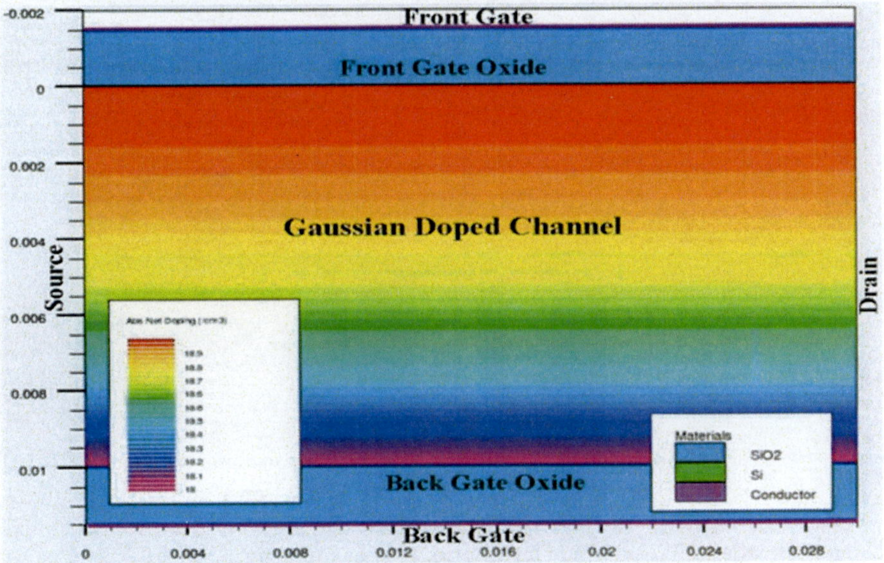

Fig. 2 A 2-D view TCAD-based simulated structure of Gaussian-like doped profile

at $x = R_p$. The two-dimensional channel potential $\Psi(x, y)$ can be calculated by solving Poisson's equation:

$$\frac{\partial^2 \Psi(x, y)}{\partial x^2} + \frac{\partial^2 \Psi(x, y)}{\partial y^2} = -\frac{q}{\varepsilon_{si}} N_{\mathrm{GL}}(x) \tag{2}$$

To solve Poisson's equation, following boundary conditions are used [10, 14]

$$\Psi(x, 0) = V_{bi}(x) \tag{3}$$

$$\Psi(x, L) = V_{bi}(x) + V_{ds} \tag{4}$$

$$V_{gs} - V_{fb} - \Psi(t_{si}, y) = \frac{\varepsilon_{si}}{C_{0x}} \frac{\partial \Psi(x, y)}{\partial x} |_{x=t_{si}} \tag{5}$$

$$V_{gs} - V_{ff} - \Psi(0, y) = -\frac{\varepsilon_{si}}{C_{0x}} \frac{\partial \Psi(x, y)}{\partial x} |_{x=0} \tag{6}$$

where V_{fb} and V_{ff} are the back and front flat-band voltages at silicon–oxide interfaces; $C_{0x} = \varepsilon_{ox}/t_{ox}$ is oxide capacitance at gate region; ε_{ox} and ε_{si} are the permittivity of oxide and silicon, respectively; t_{ox} is thickness of oxide; and $V_{bi}(x)$ is the built-in potential calculated by considering the band gap narrowing due to heavily doped channel region. Since the changes occurred in electron affinity, energy bandgap, and intrinsic carrier concentration due to heavily doping are need to incorporate in the model development [15–17].

$$V_{bi}(x) = V_T \ln\left(\frac{N_{GL}(x)}{n_{ieff}}\right) \tag{7}$$

$$n_{i\mathrm{eff}} = \sqrt{n_i^2 \exp\left(\frac{\Delta E_g}{kT}\right)} \text{ and } \chi_{\mathrm{eff}} = \chi + 0.5(\Delta E_g) \tag{8}$$

$$\Delta E_g = BGN \cdot E \left\{ \ln\left(\frac{N_{\mathrm{GL}(x)}}{BGN \cdot N}\right) + \left[\ln\left(\frac{N_{GL(x)}}{BGN \cdot N}\right)^2 + BGN \cdot C \right]^{\frac{1}{2}} \right\}$$

and

$$E_{g\mathrm{eff}} = E_g - \Delta E_g \tag{9}$$

where $BGN \cdot N$, $BGN \cdot C$ and $BGN \cdot E$ are empirical constants with values of 1.0 $\times 10^{17}$ cm^{-3}, 0.5 and 9.0 $\times 10^{-3}$ eV; χ_{eff} = effective electron affinity; $n_{i\mathrm{eff}}$ = effective intrinsic concentration; V_T = thermal voltage; the solution of two-dimensional channel potential $\Psi(x, y)$ of (2) by considering evanescent-mode analysis can be written as [13, 18]

$$\Psi(x, y) = U(x) + V(x, y) \tag{10}$$

$U(x)$ and $V(x, y)$ are the potential function accountable for the conventional long-channel characteristics and short-channel effects, respectively.

$$\frac{d^2U(x)}{dx^2} = -\frac{q}{\varepsilon_{si}} N_{\mathrm{GL}}(x) \tag{11}$$

$$\frac{\partial^2 V(x, y)}{\partial x^2} + \frac{\partial^2 V(x, y)}{\partial y^2} = 0 \tag{12}$$

The solution of (11) is $U(x)$ can be written as

$$U(x) = -\frac{qx_b^2}{\varepsilon_{si}}\left[cN_{pk}\exp\left(-a\alpha X - bX^2\right) + YX + Z\right] \tag{13}$$

where Y and Z are the constants and are calculated with following boundary conditions:

$$V_{gs} - V_{fb} - U(t_{si}) = \frac{\varepsilon_{si}}{C_{0x}}\frac{\delta U(x, y)}{\delta x}|_{x=t_{si}} \tag{14}$$

$$V_{gs} - V_{ff} - U(0) = -\frac{\varepsilon_{si}}{C_{0x}}\frac{\delta U(x, y)}{\delta x}|_{x=0} \tag{15}$$

From Eqs. (14) and (15), the constants Y and Z can be written as

$$Y = \frac{\left(V_{\mathrm{beff}} - V_{f\mathrm{eff}}\right) - \frac{qx_b^2}{\varepsilon_{si}}\left(E_f - E_b\right) - \frac{qx_b}{C_{ox}}\left(E_{df} + E_{db}\right)}{\frac{qx_b^2}{\varepsilon_{si}}\left(X_f - X_b\right) - \frac{2qxb}{C_{ox}}} \tag{16}$$

$$Z = -\frac{\varepsilon_{si}}{2qx_b^2}\left[\left(V_{f\mathrm{eff}} + V_{\mathrm{beff}}\right) + \frac{qx_b}{C_{ox}}\left(E_{df} - E_{db}\right) + \frac{qx_b^2}{\varepsilon_{si}}\left(E_f + E_b\right) + \frac{qx_b^2}{\varepsilon_{si}}Y\left(X_f + X_b\right)\right] \tag{17}$$

The solution of (12) after using the boundary conditions can be written as [14]

$$V(x, y) = \sum_{n=1}^{\infty}\frac{\cos(\eta x)}{\sinh(\eta L)}\left[V_n \sinh(\eta L - \eta y) + V_n^{'} \sinh(\eta y)\right] \tag{18}$$

where $\eta = \left(\frac{\pi}{\lambda}\right)$ with $\lambda = \left[t_{si} + \left(\frac{2\varepsilon_{si}t_{ox}}{\varepsilon_{ox}}\right)\right]$.

$V(x, y)$ can be estimated by considering the lowest value ($n = 1$) and can be expressed as [14]

$$V(x, y) \approx \frac{\cos(\eta x)}{\sinh(\eta L)}\Big[V_1 \sinh(\eta L - \eta y) + V_1' \sinh(\eta y)\Big] \quad (19)$$

3 Results and Discussion

The channel potential distribution model is developed for the double-gate junctionless transistor (DG-JLFET) by considering Gaussian-like doping profile in vertical direction. The model results are compared with 2-D numerical simulations for the validity of the developed model results. The different parameter values incorporated for computations and simulations are $t_{ox} = 1.5$ nm, $W \cdot F = 5.1$ eV, and $N_{pk} = 1 \times 10^{19}$ cm^{-3} The ATLAS-based numerical simulations are realized by using the fldmob, conmob, auger, consrh, bgn and prpmob models for the lateral electric field-dependent mobility, concentration-dependent mobility, Auger-recombination at large carrier density, concentration-dependent lifetime with Shockley–Read–Hall recombination, bandgap narrowing to add the effects occurred due to heavily doping on channel and the electric field-dependent mobility in perpendicular, respectively. To ignore the quantum effects in device characteristics and model simplification, the thickness of the channel is considered to be more than 7 nm.

In order to validate the numerical simulation results, the drain current results obtained by ATLAS TCAD are compared with the experimental results available in the works [4] and presented in Fig. 3. The remarkable matching gives the strength of the proposed device simulation results.

Figure 4a presents the potential at a position in the channel conduction path model results for variant values of straggle parameter, σ_p with Gaussian-like doping profile in comparison with numerical simulation results. It is observed from the results that

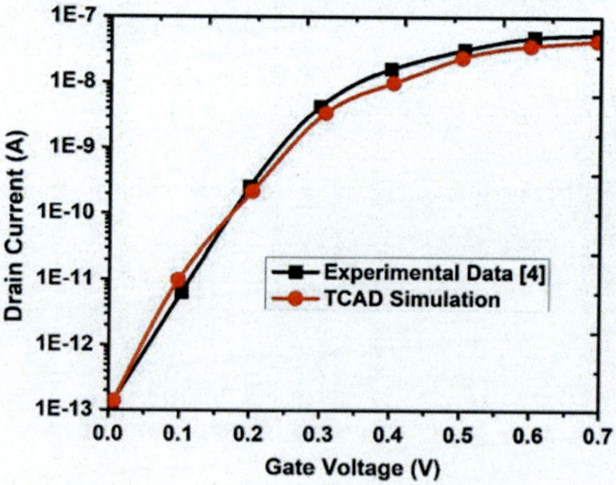

Fig. 3 Simulation model estimation against experimental nonplanar JLFET $I_{DS} - V_{GS}$ data from [4]

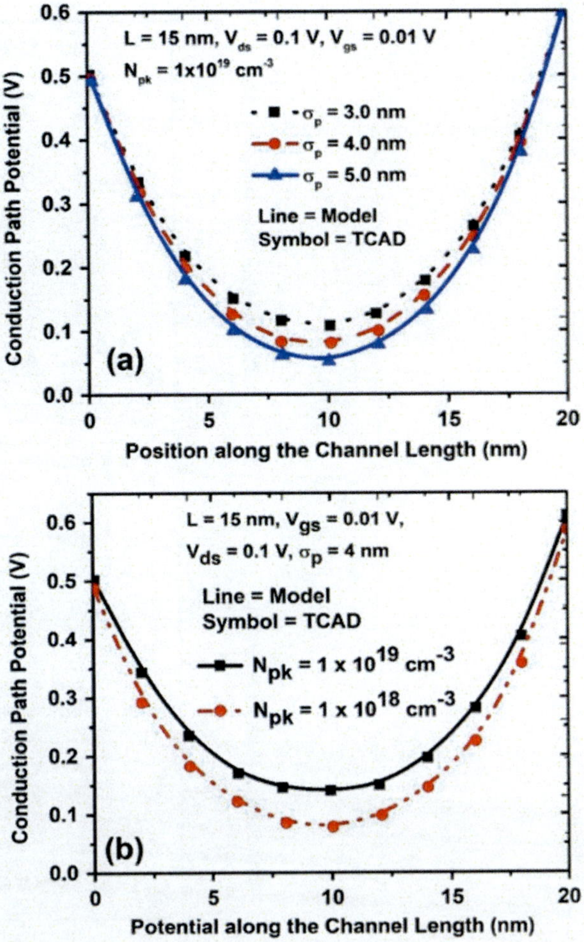

Fig. 4 Conduction path channel potential alteration with channel length for **a** Different straggles and **b** different doping concentrations

there is an increment in the minimum channel potential with σ_p shows a decrement in barrier potential from the source to the channel.

Figure 4b shows the conduction path potential variation in the channel for different values of maximum doping concentration. It is observed that the decrease in the barrier potential from source to the channel due to drain-induced barrier lowering (DIBL) and short-channel effects (SCEs) can be enhanced by heavily doping in the channel.

The change in channel potential alongside the conduction path for altered gate lengths and drain–source voltages has been shown in Fig. 5a, b. In Fig. 5a, a remarkable reduction in the potential barrier at source-to-channel is observed with increase in drain-to-source voltage for 15-nm gate length due to DIBL effect and almost remains unaffected for 65 nm gate length as shown in Fig. 5b.

Fig. 5 Conduction path channel potential variation at different drain voltages for **a** channel length at 15 nm and **b** channel length at 65 nm

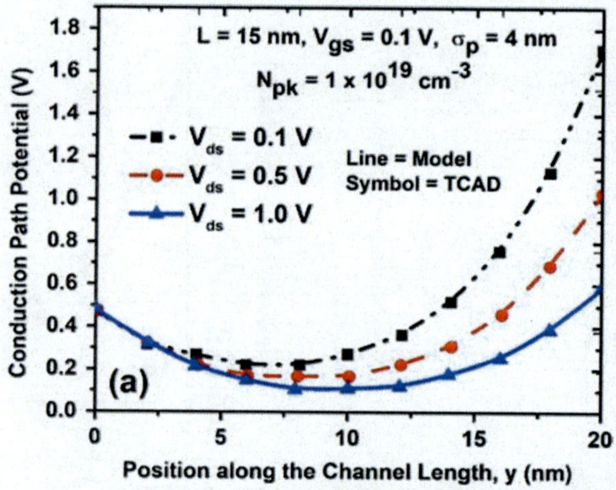

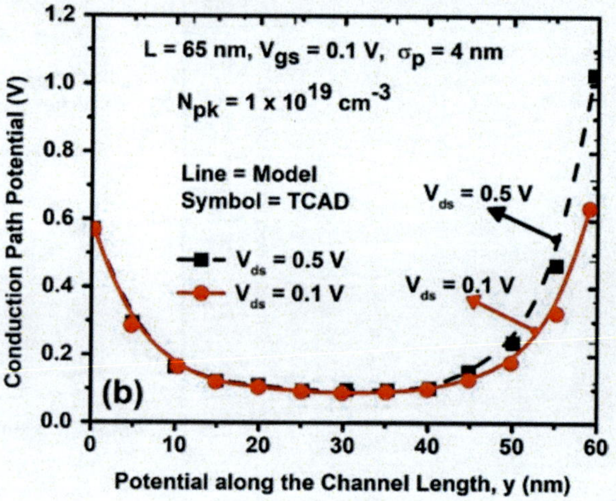

4 Conclusion

A physics-based analytical model for the channel potential in double-gate junctionless transistor has been developed and presented in this paper. A Gaussian-like vertical doping profile is considered, and the two-dimensional Poisson's equation has been solved to estimate potential field with charge by incorporating superposition method. The potential function at channel region has been estimated by calculating 1-D Poisson's equation for long-channel device and 2-D Laplace equation for short-channel 2-D potential function. The validity of the developed model outcomes is confirmed by relating with numerical simulation outcomes attained by TCAD tool. From the results, it is perceived that the channel potential is reduced by reducing the peak doping and channel length for gate lengths below ~65 nm. From these observations, it can be said that the straggle parameter can be used for the SCEs and DIBL optimization. Hence, an established model outcome is verified with the device physics and numerical simulations and found to be in good agreement.

References

1. Smith TF, Waterman MS (1981) Identification of common molecular subsequences. J Mol Biol 147:195–197
2. Choi S-J et al (2011) Nonvolatile memory by all-around-gate junctionless transistor composed of silicon nanowire on bulk substrate. IEEE Electron Device Lett 32:602–604
3. Doria RT et al (2011) Junctionless multiple-gate transistors for analog applications. IEEE Trans Electron Devices 58:2511–2519
4. Lee C-W et al (2009) Junctionless multigate field-effect transistor. Appl Phys Lett 94:053511
5. Gupta SK, Rawat AS, Verma YK, Mishra V (2019) Linearity distortion analysis of junctionless quadruple gate MOSFETs for analog applications. Silicon 11:257–265
6. Chiang T-K (2012) A quasi-two-dimensional threshold voltage model for short-channel junctionless double-gate MOSFETs. IEEE Trans Electron Devices 59:2284–2289
7. Manikandan S, Balamurugan NB, Nirmal D (2020) Analytical model of double gate stacked oxide junctionless transistor considering source/drain depletion effects for CMOS low power applications. Silicon 12:2053–2063
8. Dasgupta A, Lahiri SK (1986) A novel analytical threshold voltage model of MOSFETs with implanted channels. Int J Electron 61:655–669
9. Baidya A, Lenka TR, Baishya S (2016) Mixed-mode simulation and analysis of 3D double gate junctionless nanowire transistor for CMOS circuit applications. Superlattices Microstruct 100:14–23
10. Vanlalawmpuia K, Bhowmick B (2020) Linearity performance analysis due to lateral straggle variation in hetero-stacked TFET. Silicon 12:955–961
11. Jin X et al (2012) Modelling of the nanoscale channel length effect on the subthreshold characteristics of junctionless field-effect transistors with a symmetric double-gate structure. J Phys D Appl Phys 45:375102
12. Gnudi A, Reggiani S, Gnani E, Baccarani G (2013) Semianalytical model of the subthreshold current in short-channel junctionless symmetric double-gate field-effect transistors. IEEE Trans Electron Devices 60:1342–1348
13. Dubey S, Tiwari PK, Jit S (2010) A two-dimensional model for the potential distribution and threshold voltage of short-channel double-gate metal-oxide-semiconductor field-effect transistors with a vertical Gaussian-like doping profile. J Appl Phys 108:034518
14. ATLAS Device Simulation Software, Silvaco Int., Santa Clara, CA, USA, (2015)
15. Dixit VK, Gupta R, Purwar V, Srinivas PSTN, Dubey S (2020) Effect of Substrate Induced Surface Potential (SISP) on threshold voltage of SOI Junction-Less Field Effect Transistor (JLFET). Silicon 12:921–926
16. Mondal P, Ghosh B, Bal P (2013) Planar junctionless transistor with non-uniform channel doping. Appl Phys Lett 102:133505
17. Bavir M, Abbasi A, Orouji AA (2020) A simulation study of junctionless double-gate metal-oxide-semiconductor field-effect transistor with symmetrical side gates. Silicon 12:1593–1602
18. Jeon D-Y et al (2013) Revisited parameter extraction methodology for electrical characterization of junctionless transistors. Solid-State Electronics 90:86–93

Mobile Sink-Based Data Aggregation Protocol Using Genetic Algorithm Employing Binary String (MSDAP-GABS) for Extended Network Lifetime in Wireless Sensor Networks

Amiya Bhusan Bagjadab and Santosh Majhi

Abstract Out of various challenges encountered in wireless sensor networks, data aggregation by the sink is an important issue of consideration. Generally, in static sink-based networks (SWSN), the sink node is fixed at a particular position. Further, all the sensors send their sensed data through multi-hop forwarding. However, the loophole of such network is that the nearest neighboring node to the sink expend its energy comparatively faster than rest of the nodes. Thus, if the nearest node to the sink fails, the entire network becomes disconnected. To overcome such situation sometimes, a mobile sink is deployed to collect the sensed data. In this article, we have devised a mobile-sink-based data aggregation protocol using genetic algorithm that employs binary string for overcoming the drawback of nearest node failure so as to preserve the connectivity in the entire network. In this proposal, among the sensors some of the sensors are chosen as the cluster heads those are liable to gather the sensed data from own neighbors. Moreover, the deployed mobile sink node moves through the cluster heads so as to collect the overall sensed data. The cluster heads are so chosen that each of the sensor node has at least one neighboring cluster head as well as the length of the cycle formed by the cluster heads is minimized. Results obtained from extensive experimental evaluation and comparative performance analysis justify the superiority of our proffered approach.

Keywords Wireless Sensor Network (WSN) · Data Aggregation · Mobile Sink · Static Sink

1 Introduction

In this modern era of technological progress, the adoration of sensors is predominantly due to their multifold uses in several domains of life. Generally, wireless sensor

---

A. B. Bagjadab (✉) · S. Majhi
Veer Surendra Sai University of Technology, Burla, India

S. Majhi
e-mail: smajhi_cse@vssut.ac.in

S. S. Dash et al. (eds.), *Sixth International Conference on Intelligent Computing and Applications*, Advances in Intelligent Systems and Computing 1369,
https://doi.org/10.1007/978-981-16-1335-7_18

network (WSN) is comprised of huge no. of sensors having processing, communication as well as measuring capabilities. The sensors are basically used to measure the peripheral states, viz. sound, light, pressure, temperature, etc. and finally communicate their sensed information to the base station, i.e., the ultimate destination. The location information of sensors needs to be predetermined for sensing the environmental conditions prevailing in the region to be speculated. The prime goal of the WSN is sensing and reporting the data that is sensed to the concerned user. Be that as it may, the necessity is to delay the life span of the corresponding network. Since the appliances are deployed with finite bulk of battery power, the proper usage of energy becomes essential. From various experimental investigations, it is completely famed that communication of information takes bulk quantity of energy. Therefore, efficient data aggregation along with transmission lowers significant bulk of energy consumption. Various types of energy-effective data aggregation approaches exist like as cluster-based approach, chain-based approach, etc. A group of nodes is used to form the cluster, and one among the nodes out of individual cluster becomes the cluster head (CH) in act of cluster-based method. Moreover, all the nodes within the cluster communicate the information to the CH. Thereafter, the CH transmits the communicated data to the sink. A different node becomes the CH in every round. This results in variable communication distance. The point to be noted is that the sensors consume larger quantity of energy during this method. However, in course of chain-based method, the nodes transmit the sensed information to their neighboring nodes and likewise, it frames a chain.

The chain leader further transmits the information to the intended base station. Moreover, arbitrarily the chain leader is adopted. However, the issue is that the communication range stands as additional and variant for expanded network in dispersed manner leading to more amount of energy expenditure. In both the approaches, the sensors send information to the base station through other sensors. Moreover, the communicating range endures variant so that, the energy consumption becomes higher. This results in rapid dissipation of energy. However, while sending the information to the sink, the nearest neighboring node to the sink that conveys its data to it becomes overloaded with accumulated information from other nodes. Sometimes, it so happens that in case of failure of this node, the entire network becomes disconnected as well as the desired data cannot reach the sink node. To overcome such hurdle, we have developed a novel method called "Mobile Sink-based Data Aggregation Protocol using Genetic Algorithm Employing Binary string (MSDAP-GABS)" that uses genetic algorithm for extending the lifetime of the network while conjointly overcoming the issue of failure of the entire network in case of failure of the nearest neighboring node to the sink due to overloading of larger amount of data.

2 Related Work

In this portion, we have elaborated few of the related works regarding sink-based data gathering protocol in case of wireless sensor networks with static and mobile sinks.

As discussed by Khan et al. [1], WSN consists of static sensor nodes and a static sink positioned in the middle of the observed area of consideration and the nodes monitor several events in the area of interest. Further, a static sink node is deployed for collecting information in wireless sensor network through multi-hop forwarding that leads to greater amount of energy expenditure by nearby nodes surrounding the base station. In real-life applications, multi-hop transmission is essential for reporting the information from sources toward the sinks. Therefore, the energy expenditure relies upon the corresponding transmission distance. Again, these static sinks division the wireless network into still smaller sub-fields while each of them having one static sink. However, the benefit of number of static sinks for energy effectiveness will also be accomplished along a static sink by rationally divisioning the respective sensing field at particular level or through hierarchical arrangement. Such divisioning can either be static or dynamic, as well as it can either be self-organized or predetermined. Further, for averting the expiring of nodes closer to the desired sink, the divisioning of the field into sub-regions has been done within every cluster, with a determined CH to that the local nodes communicate their information and the CHS bear to attain greater capability than the normal nodes. Again they are liable for forwarding the gathered information to the destined sink through either single hop or multiple hops. Further, the formation or construction of cluster as well as the cluster head selection is accomplished so that the amusement of energy during routing will be lowered. This technique also may be increased multilevel hierarchies wherever clusters as well as the structures may either be decisive statically or modified dynamically. A self-organizing method may be needed to define a cluster, where every sensor node independently chooses if it has to become the CH or not, or whether a rigid normal framework of the clusters is necessary during the starting of the whole process. In this context, while considering the idea of multilevel hierarchies, the network lifetime may further be improved, employing desired number of master aggregators. Again (operation dependent) data collection is done at every cluster head prior to data transmission to the desired sink. In many works as suggested by Wankhade et al. [2], it becomes quite hectic to recharge these sensors or to deploy new nodes to replace the dead nodes. Again, the WSN may be needed to provide multiple missions under several conditions, and these requirements may not be easily gratified by deploying large number of sensor nodes.

Another method as discussed by Das et al. [3] is that another method for data gathering of the nodes closer to the desired sink is the appropriate usage of the mobile sink. Further, the mobile sink node is there to gather the information from sensor nodes as well as to store it to the corresponding base station. However, because of the failure of this node the whole method fails since just one mobile sink node is there in use. To run-over the lacuna marked for static sink, the utilization of a mobile sink has been depicted. Further, a mobile sink will use many sorts of quality patterns like random mobility, controlled mobility, and fixed path mobility that have consequences according to the energy efficiency furthermore as information gathering ways in which the random mobility, during this category of the sink uses a random path within the sensor field. Generally, for gathering information from the sensor nodes the sink adopts the pull strategy. In pull strategy, a node sends its information only if the corresponding sink sends an invitation for this and in case of push strategy, a node proactively transmits its information to the sink node. Moreover, single hop information assortment ends up in the strongest lowering of energy expenditure, since no information passing on the sensor nodes is there. This may result in incomplete operation from WSN, sort of a random-quality pattern there's no such warranty that either the sink can influence all the nodes at intervals the sensor field or this might presumably take an excessive quantity of time to do and do that. But if the time needed for entire coverage of the field is even lower, then the sink is computed to gather the concerned information from all the nodes that are inside the maximum no. of hops greater than the one. There is a crucial compromise between inclusion time of the WSN as well as the dissipation of energy. But, the inclusion time may further be lowered if numbers of mobile sinks randomly get off in the sensor field in a efficient manner. If the information gathering isn't provoked by the concerned sink, however it pursues a push strategy, another main overhead in the course of random sink mobility takes place due to the adversity of tracking the recent locate of the sink as well as acclimating the routing paths to concerned sink in multi-hop communications that lead to enhanced energy dissipation.

When a neighbor sink-node makes out such information, it amends the sink region in afterward sent out packet. Each node that gets the packet from neighboring nodes of the sink updates the placement coordinates of the sink; thus ultimately all nodes get updated the placement-coordinates of the sink for geographic routing. Likewise, He et al. [4], Rahmaan and Narendran [5], Isik et al. [6], Xu et al. [7] have proposed and used the framework of multi-agent based on data routing as well as mobile sink tracking approach to privilege multiple intermediate nodes between a source node as well as the concerned sink that are considered as agents. Moreover, these agents (specifically the one highest to the sink) are those which are liable of chase region of the mobile sink thereby lowering the trail change value that eventually curtails the overhead in chase the situation of the every which way moving sink node. However, mounted quality, during this category of models, the sink is computed to pursue a mounted path during a round-robin manner. Further, at runtime this mounted path is preset. This cannot be influenced by the behavior of the network. Coverage of the sensor field is to be warranted by an efficient strategy for choosing the routing path for the respective information packets. Moreover, sink mobility is planned in order that the accomplish sensor field will be traversed in lowest possible time. In case the sink is ready to conclude its future positions, this will communicate this information to a node positioned within the locality of its involved future position. This node is accountable for aggregation the sensor information in its locality in order that once sink truly reaches this position, it shouldn't have to be compelled to watch for the information and it had been thought-about that the info rate is exact for all the sensors. Just in case of variable information rates across sensors, the energy dissipation will be balanced by the partitioning of the nodes in teams in such a fashion that every cluster has just about the equal total rate. Associate degree approach that denotes the cluster heads as sub-sinks that are utilized in the involved device field. Every of the sensor node is related to one in all the sub-sinks. Further, the association paradigm is predicated on the number of your time the sink puts in with every sub-sink. However, if a sub-sink has the mobile sink in its locality for a extended length of your time, then bigger variety of device nodes are related to it and the other way around, which boosts the out-turn of the sensor field. Controlled mobility suggests that the ways wherever sink mobility is controlled or target-hunting supported the parameter of interest, like residual energy of nodes or predefined objective perform or predefined noticeable events. Similarly, a mobile-sink-based methodology, wherever the sink tries to stay off from the nodes with lesser residual energy and tries to be within the locality of these nodes that have bigger remaining energy. It helps in equalization the energy dissipation from the corresponding nodes.

3 Proposed Approach

3.1 Motivation

Many researchers have been conducted in *Static Wireless Sensor Networks* (*SWSN*), and so many methods, models as well as algorithms have been developed. Most of the existing studies consider that the sink is static, which works as a gateway between the sensor network and users. Further, all the sensing data from the sensors are relayed to it through multi-hop. Therefore, the sensors near to the sink node become the bottleneck. As they have to relay the data of other nodes, this expends greater amount of energy. Once they die, the sink gets disconnected from the rest of the network, whereas the rest of sensors are still fully operational having sufficient amount of residual energy. This is regarded as static sink neighborhood problem. To overcome such problem, we have proposed a mobile-sink-based data gathering protocol employing genetic algorithm.

3.2 Problem Statement

A set of sensors $p = p_1, p_2, p_3, p_4, p_5 \ldots, p_n$ with uniform communication range cr and locations $(x_1\ y_1), (x_2\ y_2)\ (x_n\ y_n)$ are taken. Two sensors p_i and p_j are neighbor of each other if Euclidean distance between them is lesser than or equal to cr. A subset ch, known as cluster heads of p (ch_ p), is to be selected so that each sensor in p–ch has at least one neighbor in ch and the whole distance traversed to form a cycle on the nodes in ch is the minimal.

A scenario of considered problem is portrayed in Fig. 1. Set of sensors $P = P_1$, $P_2, P_3, P_4, P_5, P_6, P_7$; Cluster heads CH $= P_1, P_3, P_6$. We have given the best cost of the diagram. Further, three cluster heads are given and their coordinates. They are P_1 (0.4, 2), P_3 (2.6, 3.5), P_6 (2, 0.5).

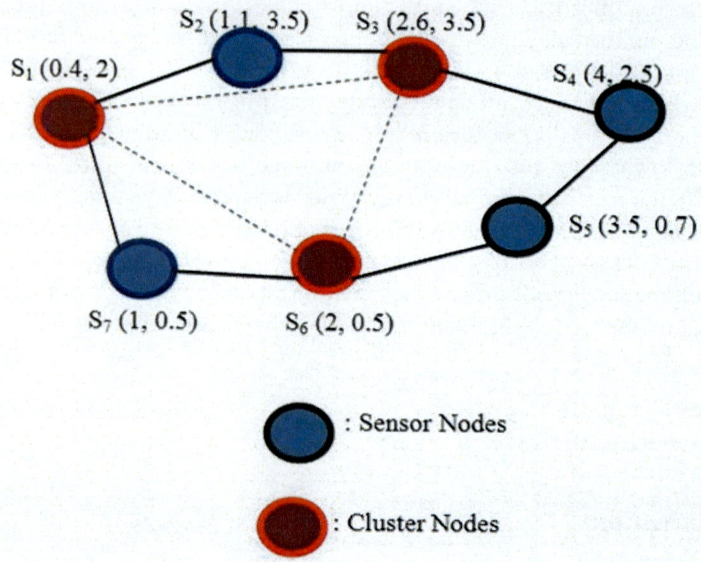

Fig. 1 Example of problem statement

Distance from $P1$ to $P3$ is given as follows:

$$\text{dist1}(p_1, p_3) = \sqrt{(x_3 - x_1)^2 + (y_3 - y_1)^2} = 2.64$$

Distance from P3 to P6 is given as follows

$$\text{dist2}(p_3, p_6) = \sqrt{(x_6 - x_3)^2 + (y_6 - y_3)^2} = 3.05$$

Distance from P6 to P1 is given as follows:

$$\text{dist3}(p_6, p1) = \sqrt{(x_6 - x_1)^2 + (y_6 - y_1)^2} = 2.19$$

$$\text{Cost} = \text{dist1}(p1, p3) + \text{dist2}(p3, p6) + \text{dist3}(p6, p1) = 2.64 + 3.05 + 2.1 = 7.88$$

Here the best cost is 7.88.

3.3 *Solution*

We have solved the problem using genetic algorithm in data gathering (chromosome of binary string where genes are either 0 or 1).

3.3.1 Genetic Algorithm in Data Gathering (Chromosome of Binary String Where Genes Are Either 0 or 1)

In this scheme, genetic algorithm (GA) is deployed to construct a mobile sink-based data gathering protocol in WSNs. The major components of GA include (i) crossover, (ii) population, (iii) fitness function, (iv) cross-over, (v) repair-function and (vi) mutation.

(i) Chromosome: A chromosome is a collection of genes, and genes denote the nodes for a given network. Every chromosome has fixed length size decisive by the number of nodes. Gene values 1 and 0 represent cluster nodes and nodes, respectively in the network.

(ii) Population: A population is generally a set of chromosomes. A new population is generated in two manners particularly (1) steady-state GA and (2) generational GA. Just in case of steady-state GA, a number of chromosomes area unit replaced in each generation and therefore the generational GA replaces all the present chromosomes in every of the generations. Moreover, a variety of generational GA is considered ideology that replicas a definite range of the most effective chromosomes from the present generation to the new generation. Additionally, a replacement generation is made using crossover moreover as mutation operations.

(iii) Fitness function: For the purpose of finding the fitness value, we have applied Euclidean Minimum Spanning Tree (EMST) as well as depth first search (DFS). EMST is the minimum spanning tree of a set of n points in the concerned plane, where the weight of the edge between every pair of points is the Euclidean distance between the two points. Euclidean minimum spanning trees is to search out the links price a set quantity per unit length. Afterward, we've got conjointly applied DFS for visiting all the nodes during a graph during a specific fashion, whereas change similarly as checking their values on the means. We've to realize the EMST during a plane is to notice that it's a sub-graph of every Delaunay triangulation of n points, a much-reduced set of edges.

(iv) Selection: The selection method chooses that chromosomes can mate (crossover) to supply a new chromosome. Further, a chromosome with higher fitness value encompasses a better likelihood of choice for the aim of mating. Various selection strategies exist to work out chromosomes for mating like—Roulette-wheel selection, rank selection and tournament selection. We've got applied tournament choice, wherever a group of chromosomes is chosen every which way from the present population, wherever best two chromosomes that have most fitness values.

(v) Crossover: Crossover is named because the genetic operator engaged on a pair of chromosomes. This generates a brand new chromosome called offspring. Besides, crossover results depend upon the selection method conducted from the population. A crossover operation wherever the crossover purpose is chosen because the best two chromosomes of the cluster, that has best fitness value and therefore the gene values of taking part parents flipped to make a pair of child chromosomes.

(vi) Repair function: A repair function is employed to spot in addition as repair or prevent the inclusion of invalid chromosomes into the new generation. For checking validity of the chromosome, we have to find the uncovered sets of the chromosomes.

Uncover sets = Total nodes or genes of the chromosome-(cluster heads + neighbor of cluster heads).

(vii) Mutation: Mutation adds some variation within the next generation. In mutation, a node is haphazardly picked from chromosome and therefore the parent is modified haphazardly. Like crossover, the mutation operation could generate associate invalid chromosome, that's conjointly mounted using the repair perform.

Description of work done:

In this approach, a special node called mobile sink moves in the certain zone with minimum cost and collects data from the concerned sensors.

Here, I have taken some assumptions:

a. All sensors are having equal capability.
b. The sensors are static in nature and have limited amount of power.
c. The sink is a special node having infinite power that has mobility power
d. Sink will move across cluster heads for trapping data.

Our proposed Genetic-Algorithm works as follows:

1. Generate a random population.
2. Choose a small subset of the population and make a group.
3. Include the best chromosome in this group.
4. Crossover the best two chromosomes of the group and build two offsprings.
5. Apply mutation with very low probability.
6. Evaluate the validity of the offspring.
7. If it is found invalid then repair it and make it a valid chromosome.
8. Repeat the above process till convergence criteria gets satisfied.

4 Simulations and Performance Assessment

Using MATLAB, we have conducted extensive experimental study of our proposed protocol. In the result analysis, we have dispersed 200 sensor nodes randomly in a network field having communication range 100 m (Table 1).

Table 1 Parameters maintained

Parameters	Definition	Values
n	No of sensors	200
CR	Communication range	100 m
Length	Length of the region	500
Width	Width of the region	500
Pop_size	Population size	500
Group_size	Group size	10

4.1 Simulation Setup

All the sensors are distributed randomly in a region having length and width of 500 m each. Every sensor is assumed to communicate within a fixed unit range. It is considered that the communication range of sensor is 100 m. If Euclidean distance between them is $\leq Cr;$, then the two sensors Si and Sj are neighbor of each other. One cluster node must be at least one of the neighboring nodes. Further, the cycle is formed using cluster heads for the purpose of covering the entire network area. The mobile sink moves to the cycle for collecting the desired data with minimal distance. Keeping simulation in mind, we have assumed size of group as 10 and size of the population as 500. Moreover, *x*, *y* and dist depends on number of sensor nodes.

4.2 Result Analysis

We have done simulation analysis for analyzing the performance of the proposed protocol. For this, 200 nodes were densely deployed in the sensor field having length of the region and width of the region as 500 m each. Since the communication-range of the sensor node is fixed and in every round of data gathering process, every node will expend a certain amount of battery power. Further, they are static in nature. Again, the sink may be a special node having unlimited power that has capability to move. Mobile-sink moves within the such region with minimum cost and gathers information from the sensors. Therefore, the mobile-sink travels through the cycle the nodes in cluster heads. The red line as portrayed in Fig. 2 indicates the total distance traversed to make a cycle on the nodes in cluster heads; which is found to be the minimal.

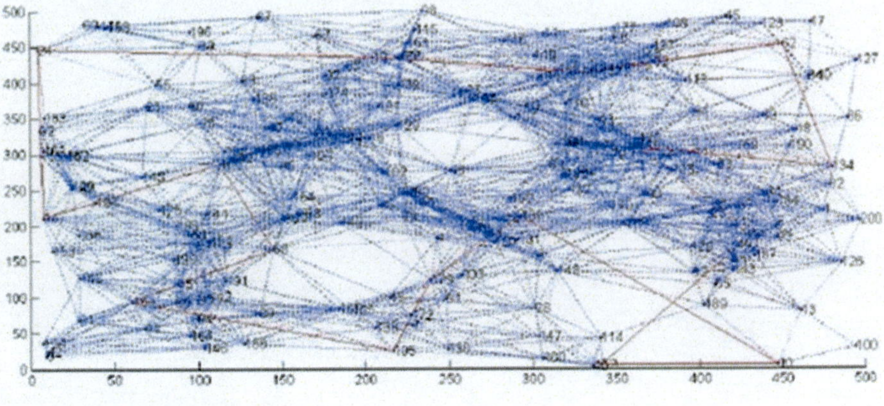

Fig. 2 Proposed_Graph

5 Conclusion

In this paper, we have presented a mobile-sink-based data gathering protocol in case of WSNs deploying genetic algorithm at the same time keeping sensor nodes design as simple as feasible. Further, our proposed protocol enhances the lifetime of the network. Since the sensor nodes transmit with the mobile sink node through cluster nodes, they expend limited amount of battery power that gives longer lifetime to the desired network. The simulation outcomes clearly demonstrate that the entire distance traversed to form a cycle on the nodes in cluster heads is the least. In future, we are planning to measure the performance using various novel algorithms for conducting the comparative performance analysis.

References

1. Khan MI, Gansterer HG, Wilfried N (2013) Static-sink vs. mobile-sink: the influence of basic parameters on energy efficiency in wireless-sensor-networks. Comput-Commun 36(9): 965–978
2. Wankhade SR, Chavhan NA (2013) A review on data collection method with sink node in wireless sensor network. Int J Distributed Parallel Syst (IJDPS) 4(1): 67–74
3. Das SK, Anastasi G, Di Francesco M (2011) Data collection in wireless sensor networks with mobile elements: a survey. ACM Trans Sensor Netw (TOSN) 8(1)
4. Guru SK, Dhamdhere SD (2014) Robust data collection in wireless sensor networks with mobile sinks. Int J Comput Sci Information Technol (IJCSIT) 5(4): 4999–5002
5. Nair NM, Felix Arokya JA (2013) Survey on data collection methods in wireless sensor networks. Int J Eng Res Technol (IJERT) 2(12): 2220–2225
6. He L, Pan J, Xu J (2011) Data gathering protocols for wireless sensor networks with mobile sinks. In: IEEE Global Telecommunications Conference (GLOBECOM 2011). IEEE, pp 1–5
7. Narendran M, Rahmaan K (2012) Enabling energy efficient sensory data-collection with mobile sink. Int J Emerging Trends Technol Comput Sci (IJETTCS) 1(2): 134–140

Mobile Application Integrated Intelligent Energy Audit System for Smart Home

Manoviraj Singh, Riya Desai, and Ganesan Subramanian

Abstract As the years passed by, the conception of embedded network and persuasive computing evolved into the most paramount technology of this era, kenned as Internet of things (IoT). This technology has been put forward by sundry companies so that the public can reap the benefits of this technology. There is no doubt that today's astute homes system is one of the most prosperous technologies and yet only 7.7% of the homes are perspicacious. This research illustrates a methodology for a low-cost astute home utilizing wireless fidelity (Wi-Fi). The users can exercise seamless control over the contrivances in their astute home through an Android/iOS application-predicated graphic utilizer interface (GUI) on a smart phone. This system is not only accessible utilizing the GUI but withal utilizing a voice command system. Utilizing this application, users will additionally be able to perform astute metering by the integration of one gettable component. This could be an accessible system with a possible solution to the challenges that will be discussed in this research as an immensely colossal-scale implementation of this system will cost approximately AED 150.

Keywords Blynk · Home automation · Internet of things · Astute metering · Wi-Fi

1 Introduction

Natural resources are scarce and take a few million years to develop. We require to conserve these resources as they provide for our quotidian needs. If we do not find alternatives to non-renewable energy sources, we will not only harm the ecosystem but adscititiously ourselves. In today's world, astute habitation is one of the best economical solutions for preserving natural resources. Home automation is a mundane term when it comes to enhancing our home's comfort, productivity, sociability, and efficiency. While most of the population is environmentally cognizant and possesses erudition about automation and its usefulness, many still do not fixate on

M. Singh · R. Desai · G. Subramanian (✉)
Manipal Academy of Higher Education Dubai, Dubai, United Arab Emirates
e-mail: ganesan.subramanian@manipaldubai.com

S. S. Dash et al. (eds.), *Sixth International Conference on Intelligent Computing and Applications*, Advances in Intelligent Systems and Computing 1369,
https://doi.org/10.1007/978-981-16-1335-7_19

implementing these facts in genuine time. There are many reasons why the public does not implement home automation technology [1]. Our research topic targets to engender transparency among people who do not optate to deal with IoT-predicated home automation systems due to their misinterpretation of it being intricate systems or systems which are accessible only to the opulent and technically sound groups. By studying the current IoT-predicated home automation systems and obtaining technical erudition, we wish to utilize some gettable hardware to virtually automate our own home [2]. All around the world, the idea of IoT has been put forward by sundry companies, and there are many people who have implemented this as well, but for the rest of the mob, IoT has remained a conception. This conception is just an accumulation of the theoretical cognizance about IoT and its applications, whereas our research will show how this information can be assimilated and implemented into their own homes. First, let us investigate the reasons for which this technology has not been virtually applied in every household. Unreliable technology due to utilization of more frugal third-party products in the market.

IoT is not only a few sensors and contrivances connected to each other utilizing a wired or wireless communication network, but it is supplementally an integration of the genuine and virtual world. The term astute home refers to a modern home environment otherwise kenned as ambience perspicacity which provides a better quality of life utilizing controlled security and accommodations. This system is sensitive and adaptive to modern homeowner as their needs like the automation of domestic tasks, facile communication, and security can be enhanced [3]. Astute homeowners can perform tasks and engage in activities very effortlessly compared to their anterior lifestyle. Many of these elementary task include automation: the capability to install contrivances that are automatic and carry out automatic functions; multifunctionality: the capability to execute multiple functions; adaptability: the capacity to adjust to meet the desiderata of users; interactivity: the possibility to sanction for interaction among users; and efficiency: the facility of performing functions to cut back time and preserve cost in a congruous way [4]. The necessity to install a perspicacious system in every domicile is not only due to the astuteness of the technology but due to its numerous advantages [5]. Today, this technology is not only applied in homes but been implemented in offices, hospitals, shopping malls, etc. It is one of the leading technologies in the world. Being very flexible, every utilizer can reap the multiple benefits of this technology in their own home.

The organization of the thesis is as follows: Chapter "A Novel Network Learning for Image Compressive Sensing" discusses the challenges of the different IoT-predicated systems utilized for keenly intellective homes. These challenges have come up due to the misconceptions instilled in commoners. Here, each of the challenge is explained in detail, and we have also mentioned the reasons due to which each issue has been raised. Chapter "Performance of Photovoltaic based ZETA Converter Water Pumping Application" discusses on hardware and software equipments utilized. We have explained the hardware and software to be used and given a detailed explication of how this system can be installed in individual homes. Chapter "Performance Analysis of Radial Distribution System by Optimal Deployment of

DG and DSTATCOM Considering Network Reconfiguration using a SAR Algorithm" gives the conclusion that summarizes the experiences we have drawn from this project and a future scope for the better results.

2 Challenges with Existing Smart Home Technology

After understanding the consequentiality of IoT and its applications, the users need to consider the challenges as well. The challenge reviewed here is the interoperability of IoT contrivances. In the IoT ecosystem, every day 5.5 million devices get connected [6]. The rate at which the developments are taking place in IoT and the solutions to interoperability between IoT contrivances are not impending at the same rate. Interoperability is essential to promote the growth secure systems at a frugal cost. Under this reviewed topic, we discuss in detail the issues of interoperability in IoT and mechanisms habituated to resolve the issues. The interoperability issue at the physical level has been minimized by utilizing Wi-Fi as the communication medium. The first issue to be considered is the incompatibility of the hardware and software of the devices due to the difference in protocols utilized for communicating [7]. The format of the message used to communicate can be another issue in the operation of contrivances. If quotidian standards and units are not used between sensors and IoT devices, then perception of data is arduous, for example, if the temperature sensor is calibrated according to Fahrenheit and the system is calibrated according to Celsius, then there will be interference in communication [8].

We need to overcome interoperability of IoT devices as it proves to be an enormous hurdle on our journey to a smart future. The organizations like Google and Microsoft are working on constricting the interoperability of IoT devices. Linux substructure is also building an open framework known as Edge Foundry to accommodate as an ecosystem for interoperable devices. Government and non-governmental organizations have funded the design and development of prevalent layer and system layer for application interface (API) management for more cost effective and efficient APIs. Device manufacturers will build a product platform utilizing one standard which will facilitate the interoperability of contrivances [9]. Last but not the least, the Internet community needs an immensely colossal heart to develop open APIs. Thankfully, even more companies are investing in open-source solutions and APIs that would resolve interoperability.

The share of 60 + people will rise from 11% in 1950 to 22% in 2050. With the incrementation in life expectancy and magnification of elderly population, there is an exordium to evolving models for the positive aging of older people such that they can lead a consummated life and habituate to the degenerative changes to maintain a good worthy life. Creating awareness among the senior citizens can influence the betterment of environment and society. There are many issues that the older people must deal with while living independently. While coping with their rudimentary health, they need to obviate more prevalent risks like falls, sensory, impairment, immobility, and isolation. One of the leading causes of deaths in older adults is a fall with one of three adults being a victim of a fall. According to a research, 80% of older adults have fair to adequate vision, 36.8% a disorder that extirpates the central vision, and 2% are consummately blind. Around 30% of older adults, mostly consisting of women, are in peril of becoming socially isolated because they live alone. Then, one-third of older adults have some condition inhibiting mobility. Globally 60% of older adults are victims of counterfeit medicines, use of those medicines could lead to plethora of damage [10]. In order to move further with this study, an initiative is placed in the framework of Ageing in Place, an incipient model for long-term care of adults. In this model, clients approve the timing, intensity of health, and personal care accommodations. Many such projects have been initiated in both the USA and Europe. The Ageing in Place project includes Tiger Place, a 34,000 square foot facility in Columbia, Missouri, developed by the University of Missouri Columbia [11]. This study is implemented to unravel the reactions of senior denizens to devices and sensors that can be installed in their homes which will avail to address the desiderata and concern of older adults. Initially, the senior denizens were introduced to the purport of the study. Questions that were asked at the focus group included experience of the participants with the current technology and their insight of the usefulness of this technology concretely for health-cognate issues. Further, different questions were asked predicated on areas where the older adults might need assistance, for example, privacy infringement and confidentiality. A total of 15 adults including seven males and eight females volunteered and were divided into three sessions, each of them were over the age of 65, and they were an active part of this session for an hour [12]. The assistance that can be provided utilizing the astute home technology is emergency avail, assistance with aurally perceiving and visual impairment, aversion and detection of falls, temperature monitoring, automatic lighting, monitoring of blood pressure and glucose levels, property security, reminder system promulgating upcoming appointments or events, and timely and accurate information on adverse drug events.

3 Materials and Methods

3.1 Hardware and Software Used

The four main hardware components used in this research are the NodeMCU, 4-channel relay module, ACS712 current sensor and Arduino Mega. The **NodeMCU** comes with a Wi-Fi integrated module, and the deep sleep operating features make it suitable for IoT applications [13]. As the module already has Wi-Fi, there is no need for any extra modules to be attached as we would have to do in case of Arduino. Further, it consists of a micro-USB port, and using that, we can connect it to the PC for programming and power supply. **Relay module** is an electromechanical switching device. The product applied in this research is a four-channel relay, and each channel needs a 15–20 mA driver current. **ACS712** is a Hall effect-based sensor integrated

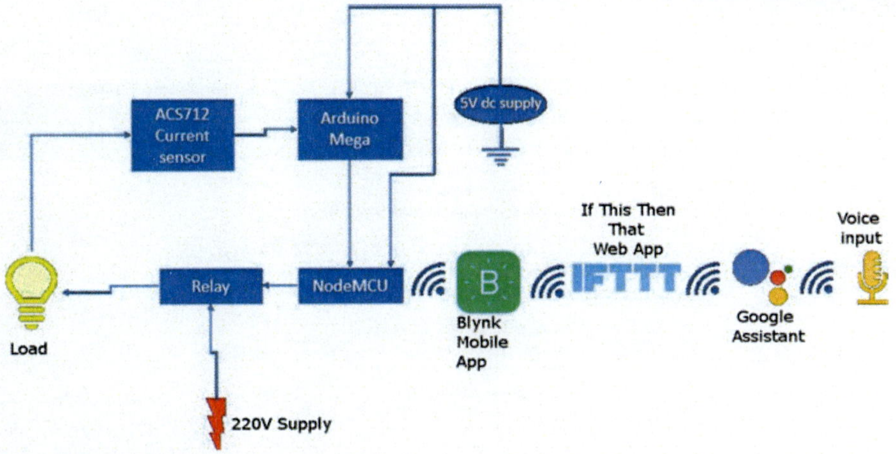

Fig. 1 Block diagram

with a low resistance current conductor. This conductor in the current sensor is used to calculate and measure the amount of current consumed by the load (Fig. 1).

The software components include: **Arduino Mega** is used in this design to take a stable ampere reading from the current sensor, since they both have the same operating voltage value. Both microcontrollers work on the same open-source software, i.e., Arduino IDE.

The **Arduino Integrated Development Environment** is a cross-platform application that is written in functions from C and C+ + . This open-source application is used to write, compile, and upload codes to Arduino compatible boards such as Raspberry Pi, Arduino series, and NodeMCU. **Blynk** [14] is an iOS and Android mobile phone application which can be used to control Arduino, Raspberry Pi, and NodeMCU over the Internet [15]. Within the app, we can build our own graphic user interface using Blynk's own widgets such as buttons, sliders, and meters by simply dragging and dropping them (Fig. 2).

Google Assistant is an artificial intelligence-powered application developed by Google. The applications of Google Assistant are immeasurable as it is capable of two-way conversations [16]. It can understand whatever vocal commands are said to it and reply or perform actions accordingly. This is useful in our design as it is a ready-made voice control application that is freely available to the public and budding IoT developers (Figs. 3 and 4).

IFTTT Another Web-based service application that needed to be employed to successfully complete our voice-controlled smart home application was IFTTT. It is an online platform that helps us connect all our different applications. The features of IFTTT Web services allow us to connect Google Assistant and Blynk, thus forming as a link in between our graphic user interface and voice control application [17]. Through the free service is provided by IFTTT, we can create applets that allow us to

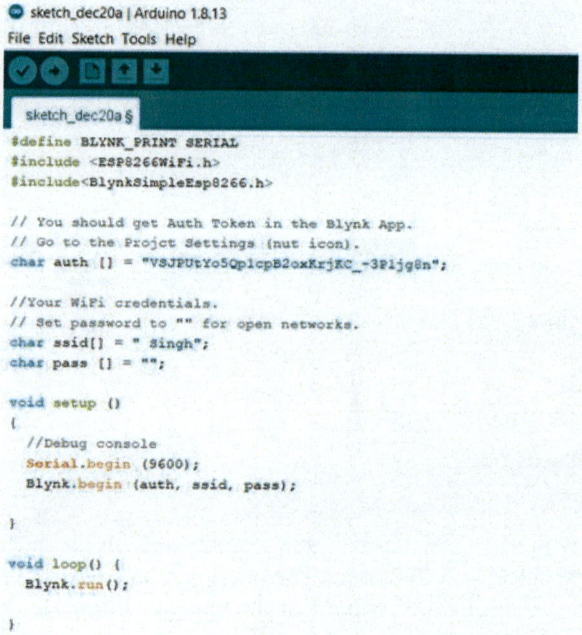

Fig. 2 Established connection between Arduino and Blynk app

Fig. 3 Google Assistant used for switching lights

connect various apps together, and these applets let us create instructions in between the two apps, so that they can communicate with each other [18].

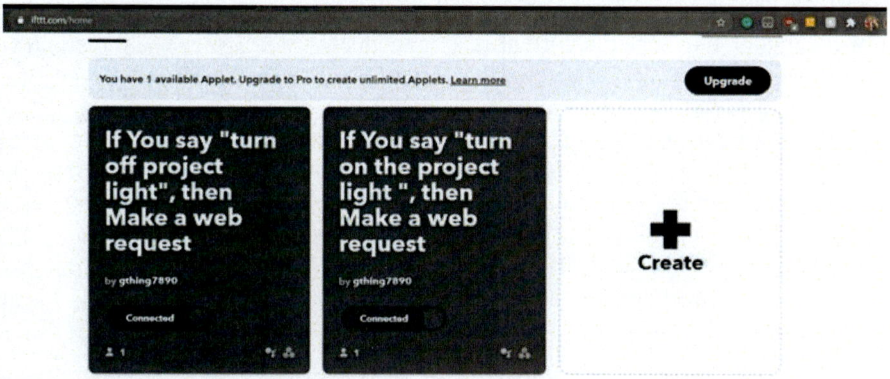

Fig. 4 Applets configuration on IFTTT.com

3.2 *Experimental Setup*

The ACS712 is connected to the live wire of any load to read a raw data value from it and provides raw data value to the Arduino Mega. The Arduino Mega is used in this project because it operates at a 5v level, and the output from the ACS712 sensor is received at 5v level. If the ACS712 sensor was connected directly to the NodeMCU, then there would be random current values showing because the NodeMCU operates at a 3.3 v level (Fig. 5).

The Arduino Mega reads the raw data values from the ACS712 sensor using the “analogRead” function. Then, it further uses mathematical equations to convert this

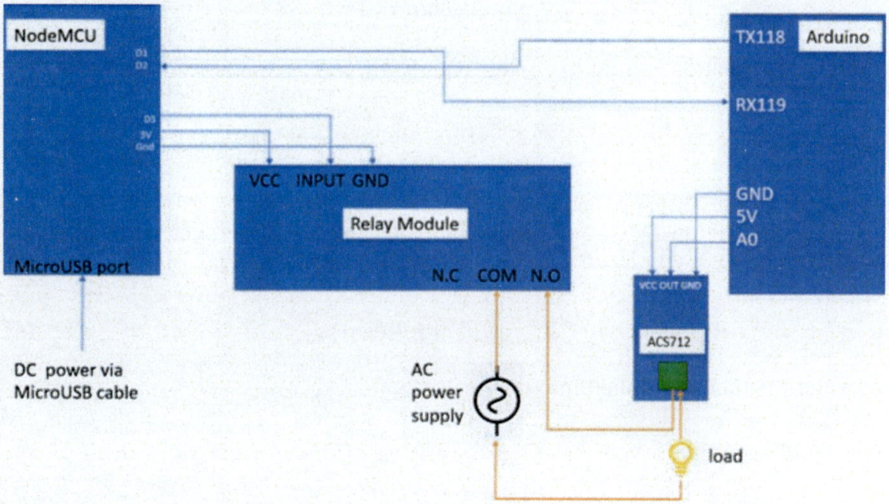

Fig. 5 Circuit diagram

raw data value into sensible current readings which have been defined as the float integer value “amperage”. The average of 5000 samples of raw data is taken from the ACS712 sensor to achieve higher accuracy and reduce fluctuations [19].

SoftwareSerial Library is included in the code so that the amperage value can be serially transmitted to NodeMCU via jumper cables. Ports TX1 18, RX1 19 on the Arduino Mega are set for communicating data to the NodeMCU ports D1, D2. Once data communication initializes, the “amperage” data is further used to calculate other data such as “Wattage” and “Tariff bill” using simple mathematical equations. All the values calculated are displayed on the user’s mobile phone through the Blynk application [20].

3.3 Energy Calculations

The below-stated example will occur when the smart system is connected to a 200 mA, 22 watt/h load:

A value in the range of 0–1024 is taken into the Arduino from ACS712 sensor, using analog read function. The average of 5000 values is taken every 3 s. The first part occurring in the Arduino Mega is

$$\text{svamplesum} = (\text{analogread}(\text{acs712}) - 511)^2 \tag{1}$$

“samplesum” is squared to avoid negative values. Assume that the “analogread (ACS712)” raw value reads 515 and our offset for the sensor is 511.

$$\therefore \text{samplesum} = 16 \tag{2}$$

The next equation is used to get the average of 5000 samples:

$$\text{Meanof} = \text{samplesum/samplecount} = 16 \tag{3}$$

“samplecount” increases by one every time a sample is taken, which is every 3 s. The next equation is used to eliminate the squaring operation that had been done while calculating “samplesum”.

$$\text{Value} = \text{sqrt(meanof)} = 4 \tag{4}$$

“Value” is then multiplied into the live voltage “vpc”

$$\text{mV} = \text{value} * \text{vpc} = 4 * 4.882 = 19.528(5) * \text{vpc} = 4.882 \text{ volts} \tag{5}$$

Further, we calculate the amperage using

$$\text{amperage} = \text{mV}/100 = 0.195 = 0.200 \text{ amps} \tag{6}$$

This final amperage value is sent to the NodeMCU via serial communication. The next part occurs in the NodeMCU:

$$\begin{aligned}\text{Watt} = &\text{amperage} * Vp = 0.200 * 110 = 22 \ \text{watts} \\ &* \text{Voltage peak}(Vp) = 110 \text{ volts}\end{aligned} \tag{7}$$

Then, we have a live updating energy meter which updates every 3 s by using the formula:

$$\begin{aligned}\text{Wh} = &\text{watt} * 0.00083 = 0.01826 \\ &* 0.00083 = 3/3600\end{aligned} \tag{8}$$

0.00083 is a value we get when we divide 180 ms (3 s) by 216,000 ms (1 h). This is done because a value is taken from the ACS712 sensor for every 3 s, and the value to be shown in the meter at the end of one hour is the watt/hour value.

Basically, 0.018296 W/hour is added to itself every 3 s when a load of 22 watts is on. Then, for the first 30 min, we will get an output of 11 W/hour. Finally, for first hour, we will get an output of 22 W/h.

To get the final bill value as per Dubai tariff rates:

$$\text{kWh} = \text{Wh}/1000 = 0.021912 \text{ kW} \tag{9}$$

$$\text{Bill} = 0.269 * \text{kWh} = 0.00589 \text{ Dirhams} \tag{10}$$

Depending on various loads, we will receive various fairly accurate values. But keep in mind that the loads connected cannot cross 20 amperes value. The ACS712 is also available in maximum 30 amperage rating. Virtual ports are set to communicate this data to the Blynk app over Wi-Fi medium. Virtual port V7 is for amperage. Virtual port V8 is for watt/hour. Virtual port V5 is for Kilowatt/hour. Virtual port V9 is for bill. Memory locations in the NodeMCU EEPROM are set so that the previous values can be stored, and new values can be added to them. So that even when the load is switched off, the meters on Blynk do not go to zero, and the users can see how much power they have used at any time. After uploading the code, the live values can be observed in the Arduino IDE application via serial monitor option. The final step is to add new widgets on the Blynk application. For our graphical user interface, we have chosen gauges to represent the live ammeter, watt-hour meter, and an LCD screen to show the Kilowatt hour value and bill value in Dirhams.

Setting up and implementing the voice-controlled load part of the project was simple, as the codes and data were available online. After the installation of the voice-controlled load, running and testing the system were also easy. The real challenge was faced when we tried to come up with the code for the metering application, even though it seems that it would be easy to take readings from a sensor. We faced many embedded issues while working with the ACS712 current sensor, such as the offset value jumps around randomly, thus providing unstable readings. We could have gone for a more stable current sensor device, but the ACS712 was commonly available. This issue was resolved by adding the Arduino Mega to take the readings. The components chosen for the project did not showcase any issues, other than the ACS712 sensor. After getting the readings from the ACS712 under control, sending the data to the NodeMCU was the next challenge. This was overcome using serial communication via jumper cables (Figs. 6 and 7).

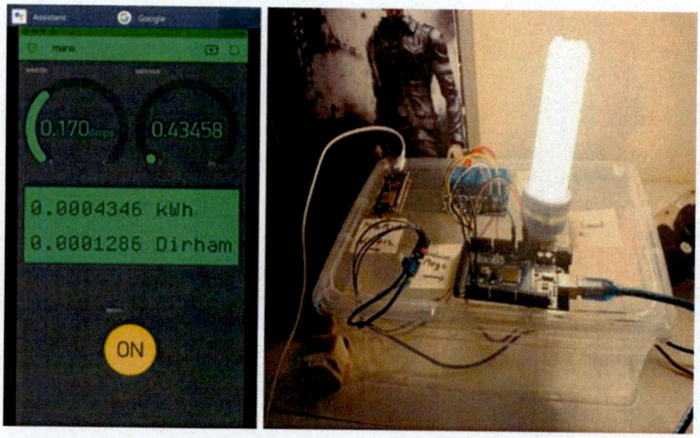

Fig. 6 Final hardware and app showing output for 1-min runtime of a 170 mA, 18 watt load

Fig. 7 Final hardware and app showing output for 1-h runtime of a 170 mA, 18 watt load

4 Conclusion

From there on out, it was simple to complete the project because the data just had to be sent to Blynk with a simple timer to keep transmission and reception in sync. The hardware setup for the project can be done in a neater and smaller manner with the use of printed circuit boards to reduce wiring. Each of these components work based on an open-source software available online, and the softwares used can be put to auto update. Hence, the user does not have to depend on any third party for administration of the system, saving the user from any kind of excessive labor charges, making installation of these devices easy. Security of the network is taken care by multiple factors such as email ID password, Wi-Fi password, Blynk verification codes, and without that data, no one can access the project. The components used in this project are easily accessible to any layman. With the use of IFTTT, we were able to overcome interoperabiltiy between Google Assistant and Blynk.

Besides overcoming the challenges, smart home applications can be very effective in preserving the environment. Users can check the amount of energy consumed by the load using the Blynk app. Thus, the user can exercise control over the energy usage by remotely controlling the system, supporting sustainable use of non-renewable sources.

This system is great for home energy auditing. User can identify cost-effective ways to improve comfort and efficiency of buildings. In our current output, the user can only access the monitored energy and tariff value by opening the Blynk application, but this system can be made more efficient. The Blynk application that we have used in our project is a flexible application that can be shaped by the users according to their desire. The user can receive regular pop ups/ notifications alerting them about the appliances being switched on during their absence in the house. They can also receive a notification alert for high consumption of energy. By setting limits on the energy consumption per day, they will not only save money but also contribute to sustainable future.

References

1. Mocriia D, Chenb Y, Musilekac P (2018) IoT-based smart homes: a review of system architecture, software, communications, privacy and security. Internet of Things 1–2:81–98
2. Pătru I-I, Carabaş M, Bărbulescu M, Gheorghe L (2016) Smart home IoT system. In: 15th RoEduNet conference: networking in education and research
3. Stefanov DH, Bien Z, Bang W-C (2004) The smart house for old persons and persons with physical disabilities: structure, technology perspective and arrangements. IEEE Trans Neural Syst Rehabil Eng
4. Lê Q, Nguyen HB (2012) Tony barnett smart homes for older people: positive aging in a digital world. Future Internet
5. Blue speed AV (2012) The 7 greatest advantages of smart home automation
6. Pokomi S (2019) Reliability and availability of internet of things. Military technical courier
7. Banafa A (2017) 3 Major challenges IoT is facing

8. Choi J, Li S, Wang X, Ha JA (2012) General distributed consensus algorithm for wireless sensor networks
9. Al Hinai1 S, Vikram Singh A (2017) Internet of things: architecture, security challenges and solutions. In: International Conference on Infocom Technologies and Unmanned Systems
10. Demiris G. Rantz M, Aud M, Marek K, Tyrer H, Skubic M, Hussam A (2004) Older Adults' attitudes towards and perceptions of "Smart Home" Technologies: a pilot study
11. Balta-Ozkana N Davidson R, Bicket M, Whitemarsh L (2013) Social barriers to the adoption of smart homes
12. Rantz MJ (2003) Aging in place. Nurseweek, Midwest/Heartland Edition 4:7
13. Blynk (2019) Home. https://blynk.io/
14. Content Development Team (2019) Robindia. Blynk for NodeMCU-Introduction. https://robo india.com/tutorials/blynk-introduction-nodemcu/
15. Yuan M (2017) Getting to know NodeMCU and its DEVKIT board. Available: https://developer.ibm.com/technologies/iot/tutorials/iot-nodemcu-open-why-use/
16. Kevan T (2018) Smart home pursuit of interoperability
17. Alam MR. Reaz MBI, Ali MAM (2012) A review of smart homes-past, present, and future. IEEE Trans Syst Man Cybern Part C (Appl Rev) 42(6):1190–1203 https://doi.org/10.1109/TSMCC.2012.2189204
18. Ricquebourg V, Menga D, Durand D, Marhic B, Delahoche L, Loge C (2006) The smart home concept: our immediate future. In: 1ST IEEE international conference on e-learning in industrial electronics
19. Demiris G, Rantz M, Aud M, Marek K, Tyrer H, Skubic M (2017) Challenges and solutions of interoperability on IoT. In: International conference on smart technologies for smart nation
20. Wang X, McGill TJ (2018) I want it any way: perception of smart home devices. J Comput Inf Syst

PSO with Moderate Random Search Strategy (MRSPSO)-Based PI/PID Controller in AVR System

Sangram Keshori Mohapatra, Asini Kumar Baliar Singh, and Nanda Kishore Ray

Abstract In this paper, the effect of high gain fast response on decreases the power system stability with enhance the transient stability in automatic voltage regulator (AVR). The most controller tuning in plant is PID controller used in the purposed system and tuning of controller parameter can be obtained by applying moderate random search strategy (MRSPSO) with PSO optimization technique. The ITAE objective function has been applied to obtain the parameters of the controller under different value of generator gain and generator time constant to increases the performance of the controller in term of settling time and overshoot.

Keywords Automatic voltage regulator (AVR) · Proportional–integral–derivative (PID) controller · PSO with moderate random search strategy (MRSPSO)

1 Introduction

The power system is the dynamic and most complex network. Each of the components which are connected to the system has its rating i.e. the range of voltage under which it can perform as desired. Due to the dynamic behaviour of the connected load to the system there may be sudden rise or fall in the level of voltage. So we must develop a system which can accommodate the sudden change and maintain a steady level of voltage at the output. The AVR is used for this purpose in the power system. It always takes the fluctuating voltage and gives a steady output voltage. The voltage consistency is very much important in power system as it has many

S. K. Mohapatra (✉)
Department of Electrical Engineering, Government College of Engineering, Keonjhar, Odisha, India

A. K. Baliar Singh
Department of Electrical Engineering, Government College of Engineering, Kalahandi, Odisha, India

N. K. Ray
Department of Electrical and Electronics Engineering, Centurion University of Technology and Management, Sitapur, Odisha, India

S. S. Dash et al. (eds.), *Sixth International Conference on Intelligent Computing and Applications*, Advances in Intelligent Systems and Computing 1369,
https://doi.org/10.1007/978-981-16-1335-7_20

impacts. If it is maintained within the level the equipments life span increases and the performance also improves [1]. This also operates the machine close to the steady state stability. Due to the AVR the reactive power also distributed between the interconnected alternators as a result the losses also reduced. If there is a fault in the system AVR increases the excitation system due to which maximum synchronising power can be fed to the system. It is a closed loop control system with some controllers that maintain the output at the desired level. Here PID controller is used for better performance. So in order to get the robust performance of AVR tuning of the PID controller plays a major role [2]. Unless we have a dynamic fine tuning it will have a huge impact to the system. In earlier days there are various conventional methods were available to determine the gains of PID controller [3]. But all those conventional methods have many disadvantages when the system is complex and non linear in nature. In order to overcome this and to get a fine tuning of PID controller there are many algorithms are developed at present. All these algorithms suggest the best solutions from a set of alternatives without explicitly enumerating and evaluating all possible alternatives. This is so called optimization techniques, which is now a days are the mostly used to find the best solutions for non linear systems. There are many such optimization techniques are available such as Tabu search, variable neighborhood search, simulated annealing, Population based approaches like genetic algorithm, evolutionary programming and ant colony Optimization, etc. [4–9]. Particle Swarm Optimization algorithm is found to be the most efficient from all the proposed methods. It gives the most accurate optimized parameter [10]. There are a lot of similarity among PSO and GA but yet PSO can be distinguished. PSO is inspired by social behavior of birds and fishes. It uses a number of particles that constitute a swarm moving around in the search space looking for the best solution. Each particle in the search space adjusts its flying according to its own flying experience as well as the flying experience of the other particles. PSO is easy to implement as there are very parameters to adjust. It gives more accurate solution with less convergence time [11]. The moderate random search swarm optimization (MRSPSO) was attempted in the literature for economic load dispatch which is an improved particle swarm optimizer which enhance the capability of swarms for building the solution space further fruitful by escalating their rate of convergence [12]. The main purpose of MRSPSO optimization algorithm for tuning of PI/PID controller parameters for the improvement of the performance of transient stability. The purposed algorithm employed for the PID controller design by choosing the best objective function of Integral Time Absolute Error (ITAE). The minimum value of objective function of controller parameters which give less settling timeless peak overshoot rise time and peak time of AVR system. It is also analyzed that performance of PID controller compared PI controller in the purposed system.

2 Modeling of AVR System

The objective of an AVR is to hold the terminal voltage magnitude of a synchronous generator at a specified level. Figure 1 shows that a simple AVR system consists of four main components, i.e., amplifier, exciter, generator, and sensor. The amplifier has a time delay represented by a time constant TA. Similarly, the time constant for the exciter and generator model is given as Te and Tg.. The time constants TA will be in the range of 0.02–0.1 s, Te is in the range of 0.4–1.0 s while Tg is in the range 1–2 s. The transfer function model of the AVR loop is shown in Fig. 1. The transfer function is given as (Fig. 2)

$$\frac{\Delta V_t(s)}{\Delta V_{\text{ref}}(s)} = \frac{0.1\,\text{s} + 10}{0.0004\,\text{s}^4 + 0.045\,\text{s}^3 + 0.555\,\text{s}^2 + 1.51\,\text{s} + 11} \tag{1}$$

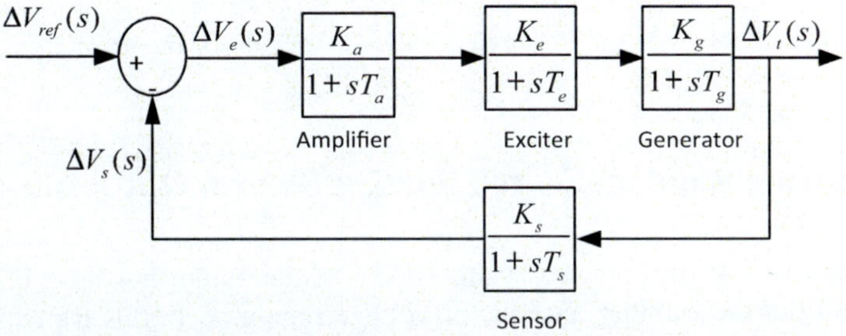

Fig. 1 Transfer function block diagram of AVR loop

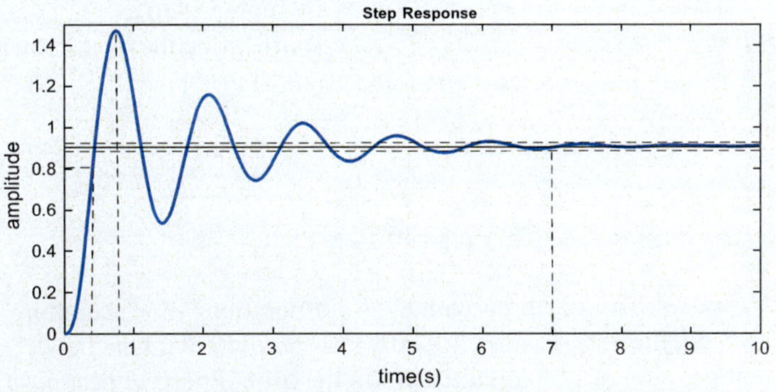

Fig. 2 Behavior of the AVR system without PID controller

2.1 PID Controller

To reduce the steady-state error and improve the dynamic response, we use PID controller. The derivative controller improves transient response. The integral controller reduces the steady-state error due to a step function to zero. The transfer function of PID controller is given as

$$\mathrm{TF_{PID}} = K_p + \frac{K_i}{s} + K_d s \tag{2}$$

The PID controller output can be obtained by using Eq. (3)

$$u(t) = K_p \cdot e(t) + K_i \int_0^t e(t) \cdot dt + K_d \frac{\mathrm{d}e(t)}{\mathrm{d}t} \tag{3}$$

where $u(t)$ denotes the control signal and $e(t)$ speaks for the error signal.

3 Moderate Random Search Particle Swarm Optimization

The tuning of AVR in the power system most of the author has used PSO strategies to find out the stability analysis of voltage regulator. In this paper, moderate search particle swarm optimization (MRSPSO) algorithm is applied to tuning the PI/PID controller. The standard PSO is a population-based stochastic algorithm of the velocity and position update in heuristic manner. The best-known portion and velocity of ith particle in d-dimensional space are represented as

$P_i = (p_{i1}, p_{i2} \ldots p_{id})$ and $V_i = (v_{i1}, v_{i2} \ldots v_{id},)$. If 'g' be the best-known position of the entire swarm, then $(Pg)_{\text{best}} = (p_{g1}, p_{g2}, \ldots, p_{gd})$.

$$V_{i(K+1)=} W V_{iK} + c_1 \text{Rand}_1 \times (P_{\text{besti}} - S_{ik}) + c_2 \text{Rand}_2() \times (g_{\text{besti}} - S_{iK}) \tag{4}$$

$$S_{i\,(K+1)} = S_{iK} + V_{iK+1} \tag{5}$$

where V_{ik} = velocity of ith individual at kth iteration, W = weighing factor,c_1, $c_2$acceleration coefficients, Rand_1 (), Rand_2 () = random numbers between 0 and 1, Sik = current position of ith individual at kth iteration, Pbesti = best position of ith individual,gbest = best position of the group. The acceleration coefficients c_1 and c_2 drag all particles toward pbest and gbest positions. The PSO is widely used to solve many optimization problems but it has drawbacks of premature convergence and low accuracy. So an improved convergence particle, i.e., moderate random search and particle swarm optimization, is used which has the flexibility in algorithm parameter update the randomness in particle position only. This proposed method improves the convergence rate with higher accuracy.

$$S_{i\,(K+1)} = P_d + \propto \lambda\,(m_{\text{besti}} - S_{Ik}) \tag{6}$$

$$m_{\text{besti}} = \sum_{i=1}^{s} \frac{P_{\text{best}}}{s} \tag{7}$$

where S represents the population size of MRSPSO, P_d = the attractor moving direction of particles given as

$$P_d = \text{rand0}\,P_{\text{best}} + (1 - \text{rand0})\,g_{\text{best}} \tag{8}$$

4 Results and Analysis

In this paper, PI/PID controller parameters can be found by using the MRSPSO optimization technique by taking the boundary value of KP, KI, and KD with different values of Kg. The best value of objective function as per past analysis, integral time absolute error (ITAE), is taken in the purposed analysis which is given by

$$\text{ITAE} = \int_0^t t\,|V_r - V_t|\,\text{d}t \tag{9}$$

In this paper, our main aim is to vary different parameters of the system in order to obtain an optimal value of the gains, KP, KI, and KD so as to improve all the aspects of the voltage response, that we are getting as output like maximum peak overshoot, rise time, settling time, etc. The PID gains obtained by using optimization techniques of MRSPSO in SIMULINK model developed in Fig. 3, and the optimal controller parameter can be taken by different value of objective function to run the system minimum 20 to 30 times. The best five cases are taken in each Kg value during simulation model running to optimization the parameters. Table 1 shows the optimal controller parameters with different value of Kg with different ranges of KP, KI, and KD taken during simulation. From Table 1, MATLAB/SIMULINK results compare characteristics of MRSPSO-based PID controller with different sets of minimum and maximum ranges of KP, KI, KD and also varying the value of Kg from 0.7 to 0.9; thus, we came to conclusion that in all case-3 is showing best response with minimum overshoot. The controller-based AVR system is taken as per range of 0.1 < KP < 50, 0.1 < KI < 2.5, 0.1 < KD < 2.5 at (Kg = 0.8). Here K_P is varied between 0.1 to 50, and K_I, K_D is varied between 0.1 to 2.5. We have taken different results. From this result, Fig. 5, it is clear that case-3 (blue line) is better response than other

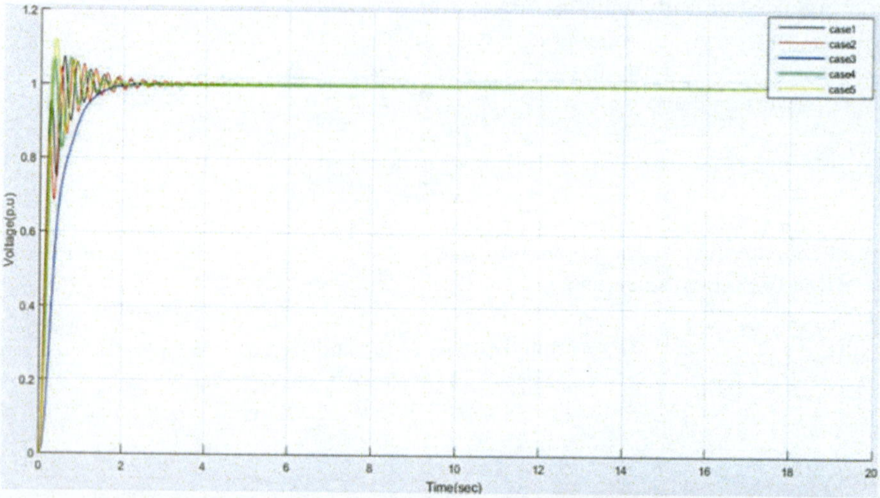

Fig. 3 Step response of incremental change in terminal voltage of MRSPSO-PID

Table 1 MRSPSO-based optimized PID gains

Kg	Ranges	Cases	KP	KI	KD	Obj
0.6	0.2 < KP, KI, KD < 2.0	Case 1	1.2239	1.0449	0.2214	0.3151
		Case 2	1.9713	0.5009	0.3912	0.2713
		Case 3	1.2442	0.2306	0.4175	0.184
		Case 4	1.3565	0.4659	0.2132	0.1904
		Case 5	1.4776	0.4999	0.2278	0.4784
0.7	0.1 < KP < 2.0 0.1 < KI, KD < 2.5	Case 1	5.1287	1.1129	0.5418	0.4397
		Case 2	10.163	0.6489	2.1021	0.376
		Case 3	3.4333	2.0222	0.442	0.2673
		Case 4	5.6959	0.2604	2.325	0.2626
		Case 5	4.2962	0.2844	1.0041	0.6491
0.8	0.1 < KP < 5.0 0.1 < KI, KD < 2.5	Case 1	7.826	0.4454	1.5543	0.18
		Case 2	5.786	0.4276	1.5645	0.3167
		Case 3	8.8444	0.5637	1.5794	0.1117
		Case 4	10.7394	0.5638	2.0986	0.2415
		Case 5	5.7795	0.4812	0.954	0.2344
0.9	0.1 < KP, KI, KD < 1.0	Case 1	0.5455	0.2328	0.1495	0.3371
		Case 2	0.7268	0.2128	0.2171	0.1651
		Case 3	0.6761	0.2626	0.1405	0.1447
		Case 4	0.5455	0.2328	0.1495	0.3371
		Case 5	0.7268	0.2128	0.2173	0.1651

four cases. Case-3 has minimum overshoot, and lesser settling time, it gives a much stable required output (Fig. 4).

The controller-based AVR system. 0.1 < KP < 1.0, 0.1 < KI < 1.0, 0.1 < KD < 1.0 at Kg = 0.9. Here KP, KI, KD are varied between 0.1 and 1.0. We have taken different results. From this result, we can conclude that Fig. 6 shows that the case-3 (blue line) is better response than other four cases. Case 3 has minimum overshoot and lesser settling time. It gives a much stable required output.

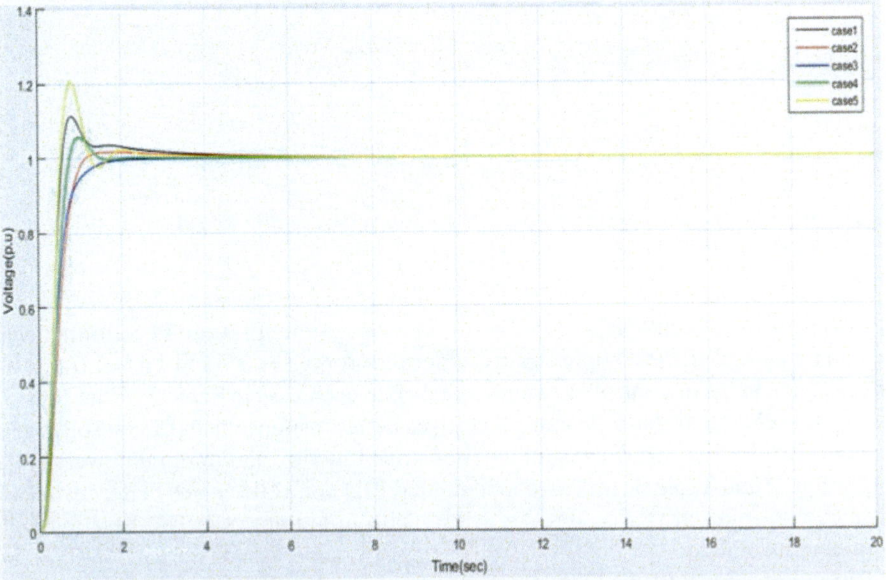

Fig. 4 Step response of incremental change in terminal voltage of MRSPSO-PID

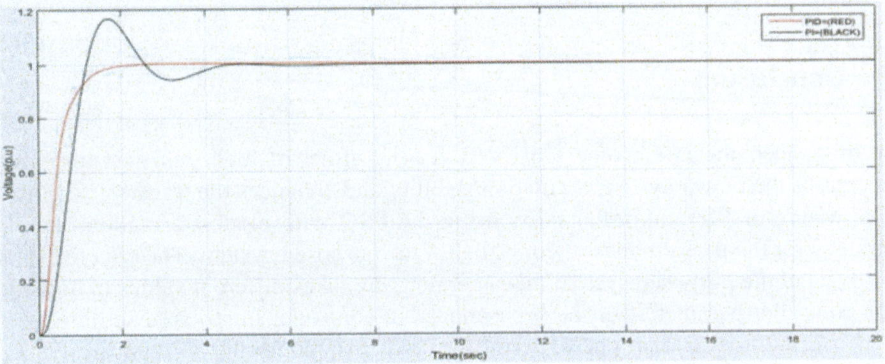

Fig. 5 Step response of incremental change in terminal voltage of PID- and PI-based AVR system (Kg = 0.7, $\zeta g = 1.0$)

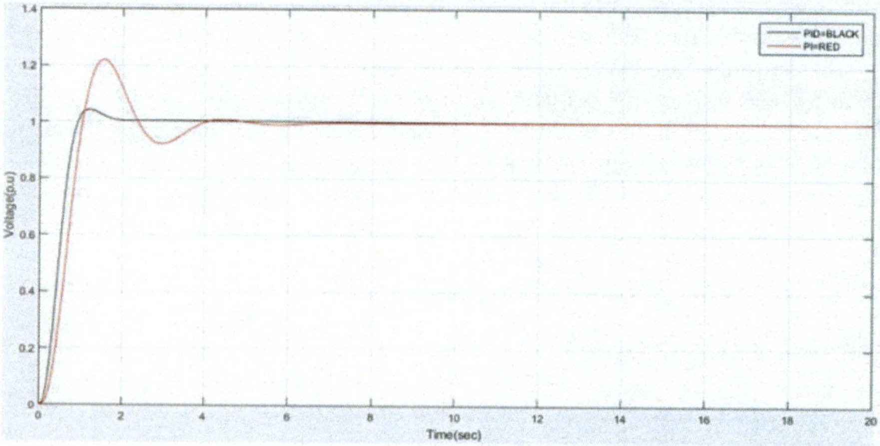

Fig. 6 Step response of incremental change in terminal voltage of PID- and PI-based AVR system ($Kg = 0.7$, $\zeta g = 2.0$)

To check the effectiveness of the purposed AVR system, PID and PI controller can obtain by using MRSPSO optimization algorithm for $Kg = 0.7$ and $\zeta g = 1.0$ & 2.0, the best five value of controller parameter taken in each case in Table 2. From Figs. 7 and 8, it is clear that that PID controller gives better response than PI controller at ($Kg = 0.7, \zeta g = 1.0, 2.0$) with minimum objective function and minimum overshoot. From Fig. 9 and Table 3, we can conclude that PID controller gives better response than PI controller at ($Kg = 0.8$, $\zeta g = 2.0$) with minimum objective function and minimum overshoot.

From Figs. 8 and 9 and Table 4, we can conclude that PID controller gives better response than PI controller at ($Kg = 0.9$, $\zeta g = 1,2$) with minimum objective function and minimum overshoot.

5 Conclusion

In this paper, the effectiveness of AVR for the effect of high gain fast response decreases the power system osculation stability and enhances the transient stability. The tuning of PID controller is by using the PSO with moderate random search strategy for the performance of controller of the purposed system. The effectiveness analysis of the purposed system the generator gain takes different values to find out the controller parameter for the performance of controller in terms of settling time and overshoot. It is also check performance of PID controller to compare the PI controller under different generator gain and generator time constant. It is show that PID controller gives better response than PI controller with minimum overshoot.

Table 2 Optimized PID gains using MRSPSO approach for AVR system with Kg = 0.7 and ζg = 1.0, 2.0

Kg	Controller	ζg	MRSPSO	Obj. function	KP	KI	KD
0.7	PID	1	1	0.4027	1.648	1.821	0.3413
			2	1.2445	0.5345	0.222	1.5579
			3	1.8563	0.8565	0.4453	1.3541
			4	1.7565	1.452	1.2203	0.1104
			5	0.3097	1.5639	1.6386	0.2218
	PI		1	1.15	0.5213	0.3231	0
			2	0.9455	0.306	0.1686	0
			3	0.9392	0.536	0.2625	0
			4	0.9996	0.6737	0.2703	0
			5	0.9056	0.3248	0.1741	0
	PID	2	1	0.4312	1.648	1.821	0.3413
			2	0.2403	1.9697	0.4176	0.3018
			3	0.4024	1.423	0.338	0.347
			4	0.4247	1.3162	0.4432	0.185
			5	0.5078	1.59	1.4406	0.3077
	PI		1	0.8453	0.5852	0.2685	0
			2	0.9301	0.719	0.2823	0
			3	1.0509	0.306	0.1686	0
			4	0.8494	0.536	0.2625	0
			5	0.904	0.6737	0.2703	0

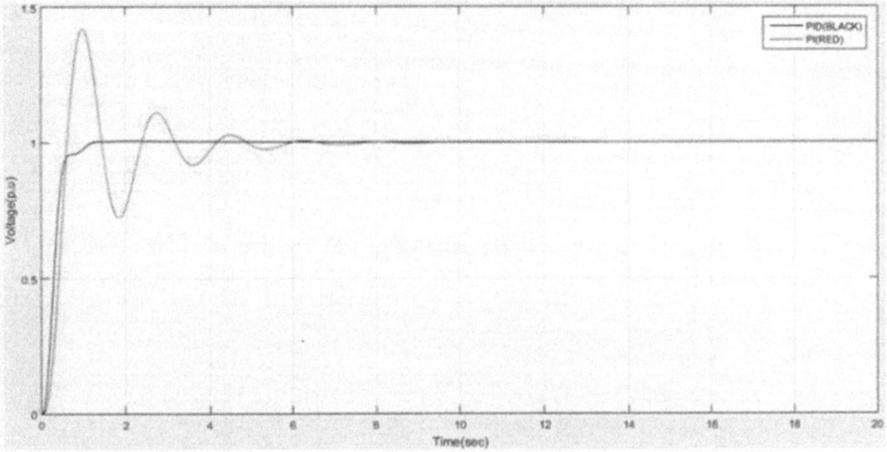

Fig. 7 Terminal voltage performance of PID- and PI-based AVR system (Kg = 0.8, ζg = 2.0)

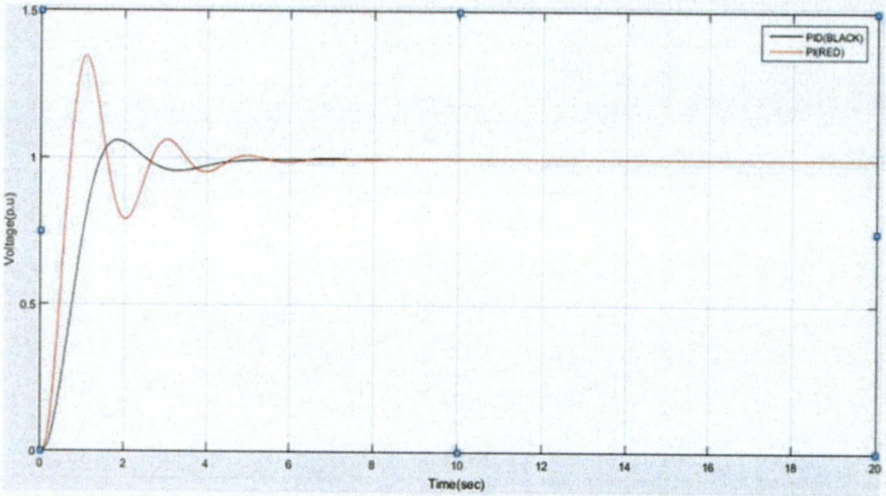

Fig. 8 Terminal voltage performnace of PID- and PI-based AVR system (Kg = 0.9, $\zeta g = 1.0$)

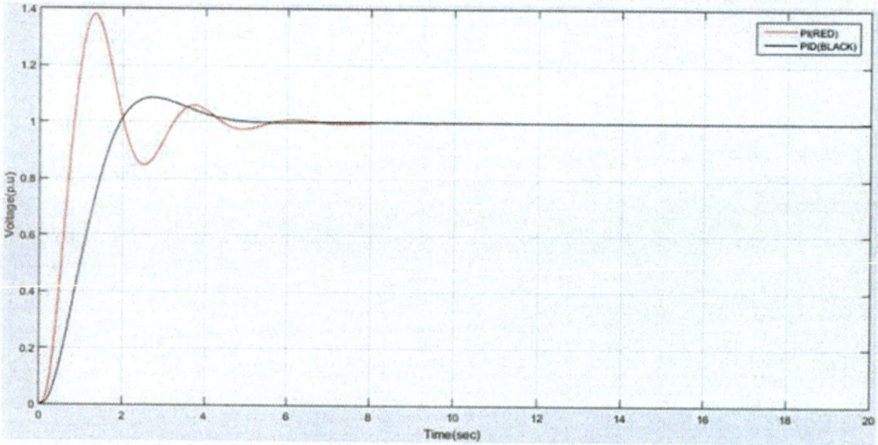

Fig. 9 Terminal voltage performance of PID- and PI-based AVR system (Kg = 0.9, $\zeta g = 2.0$)

Table 3 Optimized PID gains using MRSPSO approach for AVR system with Kg = 0.8 and ζg = 1.0, 2.0

Kg	Controller	ζg	MRSPSO	Obj. Function	KP	KI	KD
0.8	PID	2	1	0.8714	0.5099	0.2496	1.8728
			2	0.4526	1.7579	1.8916	0.2835
			3	0.2452	1.433	0.3121	1.2821
			4	0.3944	1.8248	1.9517	0.3277
			5	0.3553	1.9868	1.2832	0.695
	PI		1	1.0386	0.7835	0.3373	0
			2	0.8568	0.4307	0.2181	0
			3	0.8697	0.3835	0.1648	0
			4	1.1284	0.7956	0.384	0
			5	1.1092	0.8567	0.3609	0
	PID	1	1	0.9539	0.3248	0.1741	1.2366
			2	0.8822	0.52	0.2823	0.2143
			3	0.2388	1.7412	0.4227	0.4395
			4	0.1609	0.9341	0.3832	0.1537
			5	0.1124	1.7147	0.4627	0.3359
	PI		1	1.2059	0.718	0.3559	0
			2	0.8074	0.4472	0.2923	0
			3	0.7259	0.2298	0.2044	0
			4	1.0093	0.3541	0.3433	0
			5	0.724	0.3435	0.2544	0

Table 4 Optimized PID gains using MRSPSO approach for AVR system with Kg = 0.9 and ζg = 1.0, 2.0

Kg	Controller	ζg	MRSPSO	Obj. Function	KP	KI	KD
0.9	PID	1	1	2.723	1.7818	1.8352	1.6127
			2	1.4783	1.8408	0.1022	0.9787
			3	2.2501	0.9393	0.1252	1.8047
			4	1.6328	1.4182	0.1186	1.7021
			5	0.7367	0.202	0.2625	1.3948
	PI		1	1.0027	0.536	0.2625	0
			2	1.1002	0.2299	0.2614	0
			3	0.8882	0.3248	0.1741	0
			4	0.9672	0.52	0.2823	0
			5	1.0102	0.4729	0.3306	0
	PID	2	1	0.8807	0.4307	0.2181	0.8757
			2	0.8443	0.3835	0.1648	1.5863
			3	1.0554	0.2359	0.1298	1.95
			4	0.8615	0.4885	0.2194	1.2825
			5	1.0458	0.7221	0.281	0.9392
	PI		1	0.8638	0.4504	0.1868	0
			2	0.9153	0.5099	0.2496	0
			3	1.1646	0.7189	0.3559	0
			4	0.885	0.4462	0.2243	0
			5	0.9349	0.5222	0.2589	0

References

1. Elgard OI (1982) Electric energy systems theory. Mc Graw-Hill, New York, pp 299–362
2. Madinehi N, Shaloudegi K, Abedi M, Abyaneh HA (2011) Optimum design of PID controller in AVR system using intelligent methods. In: 2011 IEEE trondheim powerTech, Trondheim, pp 1–6
3. Krohling RA, Rey JP (February, 2001) Design of optimal disturbance rejection PID controllers using genetic algorithms. IEEE Trans Evol Comput 5:78–82
4. Krohling RA, Jaschek H, Rey JP (July, 1997) Designing PI/PID controller for a motion control system based on genetic algorithm. In: 12th IEEE conference on symposium on intelligent control, pp 125–130
5. Gaing Z (June, 2004) A particle swarm optimization approach for optimum design of PID controller in AVR system. IEEE Trans. Energy Convers 19(2)
6. Devaraj D, Selvabala B (February, 2009) Real-coded genetic algorithm and fuzzy logic approach for real-time tuning of proportional-integral-derivative controller in automatic voltage regulator system. IET Gener Transm Distrib 3(7):641–649
7. Hasanien HM (December, 2013) Design optimization of PID controller in automatic voltage regulator system using Taguchi combined genetic algorithm method. IEEE Syst J 7(4):825–831
8. Devaraj D, Selvabala B (July, 2009) Real-coded genetic algorithm and fuzzy logic approach for real-time tuning of proportional-integral-derivative controller in automatic voltage regulator system. IET Gen Trans Dist 3(7):641–649

9. Chatterjee S, Mukherjee V (May, 2016) PID controller for automatic voltage regulator using teaching–learning based optimization technique. Int J Elect Power Energy Syst 77:418–429
10. Lakshmi Sowjanya K, Ravi Srinivas L (2015) Tuning of PID controllers using particle swarm optimization. Int J Ind Electron Electric Eng
11. AlRashidi MR, El-Hawary ME (August, 2009) A survey of particle swarm optimization applications in electric power systems. IEEE Trans Evol Comput 13(4):913–918
12. Dash SPM, Baliarsingh AK, Mohapatra SK (2018) Economic load dispatch using moderate random search PSO with ramp rate limit constraints. In: IEEE International Confrence on Technologies for Smart-City Energy Security and Power (ICSESP), pp 1–4

A Wearable Neckband for Creating Personalized Secure Space

Vijay A. Kanade

Abstract The coronavirus disease 2019 (COVID-19 pandemic) has exposed our limitation in tackling respiratory viral diseases. As the world continues to battle the pandemic, we have only been able to comprehend defense mechanisms in the form of face masks, clinically approved drugs, sanitizers, and immune-boosting eatables. However, the challenge to develop full-fledged therapeutics or vaccines against COVID-19 still seems a far-fetched idea. In light of this, the research paper proposes a novel wearable device to safeguard individuals against COVID-19 as well as polluted air. The wearable device creates a personalized secure space for an individual in the fight against airborne viruses and contaminants. The device harnesses the photo-catalytic property of photo-catalysts and air-disinfecting property of non-thermal plasma for keeping the space around the device wearer clean and harmless. The proposed device resembles a neckband and can be worn around the neck for its effect in the wearer's vicinity.

Keywords Photo-catalysis · Non-thermal plasma · Air pollutants · COVID-19 pandemic · Pathogens · Wearable neckband

1 Introduction

The COVID-19 outbreak has caused havoc globally. Following the current COVID-19 infection rate and the number of fatalities, it seems that COVID-19 may cause more tremors with chances of second and third waves of virus looming high [1]. This demands for solutions that are effective and can be adopted by large masses apart from masks, sanitizers, etc.

COVID-19 is transmitted between individuals through air via ejection of droplets during events such as sneezing, coughing, and talking. Apart from airborne transmission, evidences show that the virus persists on surfaces and clothes for a few hours to days. Further, there is also plentiful literature that discloses potential link

V. A. Kanade (✉)
Researcher, Pune, India

S. S. Dash et al. (eds.), *Sixth International Conference on Intelligent Computing and Applications*, Advances in Intelligent Systems and Computing 1369,
https://doi.org/10.1007/978-981-16-1335-7_21

between air quality and COVID-19 transmission [2]. This hints towards the fact that polluted environment could possibly enhance the transmission rate of COVID-19 virus. Making the matter worse are the recent stats for air quality index. According to a recent report, air pollution caused about 67 lakh deaths globally in 2019, thereby making it the fourth highest death risk factor [3].

Looking at the current scenario, there seems to be a long pressing need to utilize current technological advancements for developing alternative solutions to tackle the COVID-19 pandemic along with the air pollution problem.

2 A Wearable Neckband

The research proposes a wearable neckband that can be worn around the neck by a user, similar to the neckband headset or headphones. The structure of the band is as shown in Fig. 1.

The device has two prominent modules: One module handles air quality parameter, and the second module keeps the personalized space free of virus, bacteria, or pathogens. These modules are detailed out in the below sections.

Fig. 1 Neckband

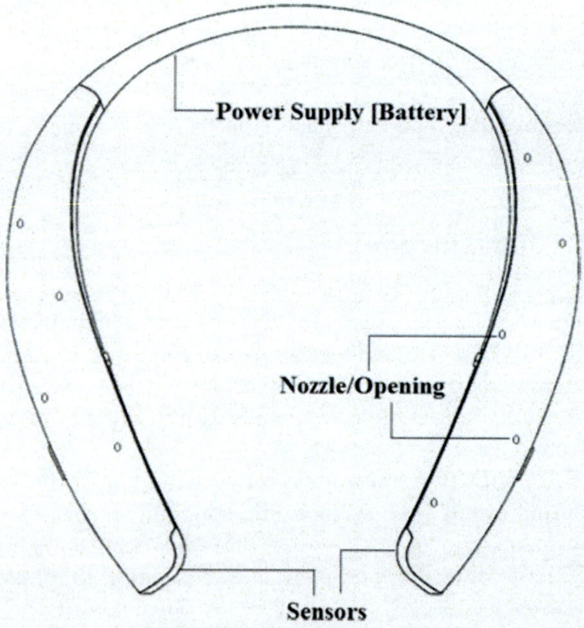

2.1 Photo-Catalysis

The architecture of the Internet-of-Things (IoT) device includes air-pollutant sensors for sensing the air quality (i.e., air pollutant level) in the vicinity of the band wearer. Post-analyzing the air quality, the neckband releases photo-catalysts for neutralizing the pollutant-filled air by converting the pollutants (NO_x and CO) into harmless components such as H_2O, O_2, or nitrates.

The device can employ any visible light-driven photo-catalysts such as tungsten oxide or ZnO, or in some cases, the device may also use UV-light-based photo-catalysts such as TiO_2. In visible light-mediated photo-catalysis, photo-catalyst's ability to absorb low-energy visible light and engage in energy transfer processes with organic substrates is necessary. While in UV-light-based process, the photo-catalysts depend on the UV spectrum of the electromagnetic range to activate the photo-catalysis reactions.

Photo-catalysis in general is seen to unfold through the below process [4]:

1. When light (visible or UV) hits the photo-catalyst (X), electrons (negatively charged particles) are released in the air and/or on the available surfaces (i.e., cloth fabric).

$$X + hv \rightarrow X * \left(h_{vb}^{+} + e_{cb}^{-}\right)$$
$$\left[\text{Charge carrier i.e. electrons } e^{-}\right]$$

2. The electrons react with the water vapors in the air and break them into hydroxyl radicals (OH·). These radicals are highly reactive and are uncharged forms of hydroxide ions (OH).

$$e^{-} + H_2O \rightarrow (OH\cdot)\left[\text{Hydroxyl radical}\right]$$

3. These hydroxyl radicals then react with organic (carbon-based CO) pollutant molecules, breaking their chemical bonds and turning them into harmless substances such as carbon dioxide and water. This is an example of photo-catalytic oxidation (PCO).

$$(OH\cdot) + CO \rightarrow CO_2 + H_2O \text{ (Harmless substances)}$$
$$\text{(Carbon-based pollutant)}$$

4. Similarly, NO particles in the air react with (OH•) to yield harmless nitrates as disclosed below:

(a) Oxidation using (OH•) radicals: (pollutant: NO_x)

$$1.\,NO + 2OH\cdot \rightarrow NO_2 + H_2O$$

$$2.NO_2 + OH\cdot \rightarrow NO_3^{-} + H+ \text{ (Harmless nitrates)}$$

(b) Oxidization using "active oxygen": $O_2{}^-$

1. $O_2 + e^- \rightarrow O_2^-$ (Electron trapping)
(Oxygen in air)

O_2^-

2. $NO_x \rightarrow NO_3^-$ (Harmless nitrates)
(Nitrogen oxides)

The above disclosed are general photo-catalytic reactions that essentially neutralize active pollutants in the air to generate harmless substances.

2.2 *Non-thermal Plasma*

The second module embedded in the wearable neckband includes 'air disinfection' module. In this module, we use the non-thermal plasma, also known as cold plasma. Non-thermal plasma is known to inactivate about 99.9% of all airborne bacteria and viruses within blink of an eye [5–7]. Apart from killing viruses, it acts as a potent alternative for face masks that allow us to breathe in clean air.

It is important to understand what non-thermal plasma really is? Non-thermal plasma is similar to a flame without the heat, where the flame is created by using an electromagnetic arc. Plasma is represented by the ionized or charged air particles formed around the spark of the arc.

In the proposed wearable IoT device, non-thermal plasma is manifested in the form of air plasma spray. The device is incorporated with biosensors for monitoring airborne pathogens present in the wearer's vicinity and in turn triggering the non-thermal plasma spray. The circuit structure of the plasma spray unit within the wearable neckband is as disclosed below.

3 Wearable Device Circuitry

The neckband consists of a pair of central electrodes (each electrode has width = 3 mm) running along the length of the wearable. The electrodes are surrounded by three magnets (i.e., ring magnets), which in combination of the electrodes act as electromagnetic source of the plasma. These three ring-shaped permanent magnets ($12 \times 5.3 \times 2$ mm each in dimension) are inserted through the central electrode and held inside the outer electrode of the torch as seen in Fig. 2 [8].

Such arrangement of electrodes and magnets is done to provide stability in the discharge and also avoid formation of hot spot in the arc discharge. Here, the three

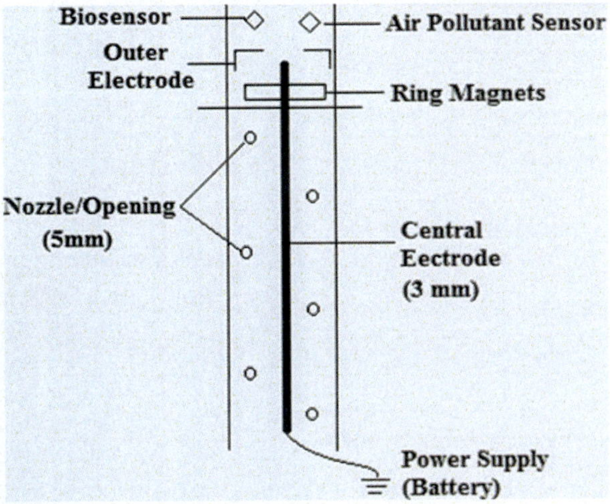

Fig. 2 Cross-sectional view of internal circuitry of the neckband

magnets enable the rotation of the current discharge to minimize unwanted arc constriction. The wearable also has intermediate openings/nozzle (diameter = 5 mm) at various junctures to allow the air flow through the internal circuitry, while taking care of the fact that high-voltage central electrodes are not over exposed. The air passes between the electrodes and then outward through the openings. Such openings allow the plasma spray to be operated in omnidirectional ways.

Further, the band is also facilitated with air pollution sensor and a biosensor for monitoring the air quality near the wearer. These sensors in turn activate their respective modules on detection of foul air quality in the band's vicinity.

Non-thermal Plasma Spray

The proposed wearable neckband generates plasma spray of $\leq$1 cm height and about $\leq$1 cm in diameter in the open region. The plasma discharge gives good reactive atomic oxygen (ROS) that is effective component to kill microorganisms or pathogens. The neckband has low temperature and is therefore safe to be in contact with skin or clothes as like the conventional headsets.

The illustration of the neckband creating a personalized, virtual, secure, and safe environment in the wearer's vicinity is as shown in Fig. 3.

4 Preliminary Results

The feasibility of the research proposal was thoroughly studied by conducting a rigorous literature survey. In the survey, we were able to identify the safety standards of the air quality parameters at public places. We have tabulated our preliminary research findings in Table 1.

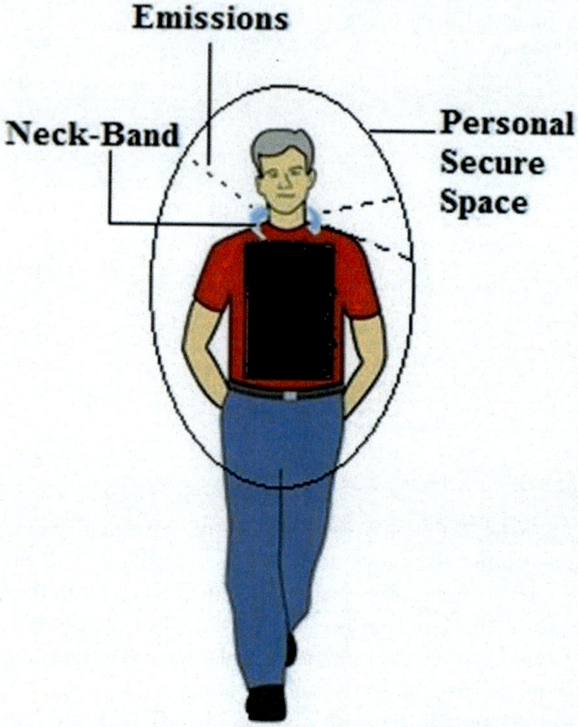

Fig. 3 Illustration of wearable neckband creating personal secure space

Table 1 Air quality standards for offices and public places (for 8 h) [9]

Parameter	Unit	8 hrs average	
		Excellent class	Good class
Room temperature	°C	20 to <25.5	<25.5
Relative humidity	%	40 to <70	<70
Air movement	m/s	<0.2	<0.3
Carbon dioxide (CO_2)	ppmv	<800	<1000
Carbon monoxide (CO)	$\mu g/m^3$	<2000	<10.000
Respirable suspended particulates (PM_{10})	$\mu g/m^3$	<20	<180
Nitrogen dioxide (NO_2)	$\mu g/m^3$	<40	<150
Ozone (O_3)	$\mu g/m^3$	<50	<120
Formaldehyde (HCHO)	$\mu g/m^3$	<30	<100
Volatile organic compounds (VOC)	$\mu g/m^3$	<200	<600
Radon (Rn)	Bq/m^3	<150	<200
Airborne bacteria	cfu/m^3	<500	<1000

Taking the research forward, the proposed wearable device is being built that will keep the air quality in check and maintain the safety standard of the air as disclosed in the table above.

5 Applications

The disclosed wearable neckband has the following applications:

1. The neckband creates a personalized space for a wearer which is free of contaminants (i.e., air pollutants) or pathogens, thereby making environment (i.e., air) in the immediate vicinity of the wearer clean and safe.
2. The neckband can serve as an apt solution for COVID-19 like situations, where virus proliferation through droplets and air seems prominent.

6 Conclusion

The proposed wearable device uses photo-catalysis and air disinfection phenomenon as a weapon against air pollution and the spread of infectious diseases (COVID-19), where pathogens remain suspended in the air for hours. The wearable creates a secure personalized space that maintains the hygienic atmosphere for the band wearer, which stays free of air pollutants, viruses and bacteria. The research proposal discloses first-of-its-kind wearable device which could pave way for future technological developments.

7 Future Work

Considering the scope of the wearable neckband, we intend to expand the application of the device. One such area could be cosmetics. In cosmetics application, we intend to collect user biological data such as skin moisture, sebum content, and melanin index and in turn customize the emission of photo catalysts or plasma in a manner that suits an individual's skin. This will make the virtual surrounding more personalized and biocompatible for any user.

Acknowledgements I would like to extend my sincere gratitude to Dr. A. S. Kanade for his relentless support during my research work.

Conflict of Interest
The authors declare that they have no conflict of interest.

References

1. Talebian S, Wallace GG, Schroeder A, Stellacci F, Conde J (2020) Nanotechnology-based disinfectants and sensors for SARS-CoV-2
2. Manoj MG, Satheesh Kumar MK, Valsaraj KT, Sivan C, Vijayan SK (2020) Potential link between compromised air quality and transmission of the novel corona virus (SARS-CoV-2) in affected areas
3. TWC India Edit Team (2020) Air pollution killed more than 16 lakh people in India last year; poses fourth highest death risk globally: report. TWC India Edit Team, 28 October, 2020, TWC India
4. Kanade VA (2018) A bio-inspired unsupervised algorithm for deploying [BoT]: symbiotic intelligence. In: IOT '18: proceedings of the 8th international conference on the internet of things, October 2018, Article No.: 24, pp 1–5
5. Vaze ND, Arjunan KP, Gallagher MJ, Vasilets VN, Gutsol A, Fridman A, Anandan S (2007) Air and water sterilization using non-thermal plasma. In: 16th IEEE international pulsed power conference, pp 17–22
6. Non-thermal plasma: new technology could kill 99.9% of the deadly germs in the air. April 29, 2019
7. University of Michigan, Cold plasma can kill 99.9% of airborne viruses, April 8, 2019
8. Kuo SP (2008) US7777151B2. Portable plasma sterilizer
9. Nath RK, Zain MFM, Jamil M (2016) An environment-friendly solution for indoor air purification by using renewable photocatalysts in concrete: a review. Renew Sustain Energy Rev 62:1184–1194

Design and Implementation to Find Multi-Issues in Brain Images with Mixture Clustering Techniques

Vaddepalli Shravya, I. Adum Babu, and Srinivas Bachu

Abstract Tumour and edoema regions are also difficult due to the intricate anatomy of brain tumours, blurred lines and environmental influences, such as noise. First, adaptive Wiener filtering is used to denoise, and morphological operations are used for the elimination of nonbrain tissue, thereby essentially minimizing the sensitiveness of the system to noise. The main project contribution is as follows. Second, clustering with K-means++ is coupled with a fuzzy C-means algorithm based on a Gaussian kernel. This clustering not only increases the reliability of the algorithm but also reduces the cluster sensitivity. Finally, the tumour images collected are post-treated with morphological operation and median filtering to ensure that the brain tumours are correctly displayed. The findings reveal that the suggested algorithm is more reliable, sensitive and recalled.

Keywords EECS algorithm · Segmentation · Medical image processing · K-means clustering · Edge detection · Image enhancement

1 Introduction

Digital images are usually used to provide visual viewpoints of details from or from a consumer scene. Thus, imaging methods are critical in restoring the degradation including its information transmitted in the viewer or in the software for further research. There are various claims to visual knowledge from technicians and engineers, for example tomography, medical diagnosis and discovery of distant sensing capabilities on Earth. Imaging technology, for example, is used during service in ferrogram to track the conditions of machinery's health [1]. In comparison, infraround images can be used image recognition techniques. Furthermore, vision is used for target recognition and tracking [1, 2].

The image enhancement problem was aimed at preserving or improving the edges of objects in the figure [3, 4]. The colour saturation work is performed first, and

V. Shravya · I. Adum Babu · S. Bachu (✉)
Department of ECE, Marri Laxman Reddy Institute of Technology and Management Dundigal, Medchal, Hyderabad, Telangana, India

S. S. Dash et al. (eds.), *Sixth International Conference on Intelligent Computing and Applications*, Advances in Intelligent Systems and Computing 1369,
https://doi.org/10.1007/978-981-16-1335-7_22

then, the edge is preserved. A second approach to improve the sharpness of the resulting image is suggested using morphological filters [1]. In addition, a block-based improvement approach has been adopted [2, 3] to resolve the issue of improved comparison. These localized improvement approaches may be more complex for these applications compared to the global method of equalization histogram. Histographic image enhancement algorithms are often classified in the statistical as well as global approach [3, 4]. Essentially, this equation aims to re-map the intensity or other colour of the channel to a given density of chance. Direct implementation of this programme can add some unnecessary artefacts; thus, alternate procedures for applying this programme have been suggested and advanced [1].

There are various approaches that can be used in the case of histograms offsetting. Relevant improvements should be used, for example, empirically. In the other hand, reaching the appropriate optimum parameters is typically a task. An unfurnished approach has been developed for these purposes. This calculates the coefficient of a transition in the power law for the average picture strength. Another approach is designed to boost the input image and retain the same average output light. Since this algorithm is oriented to a flatter mapping density, the complete degree of brightness is not completely used to express the scene data. The researchers then set out to find ways to preserve the initial mean light [3]. In the original work, the pixels are isolated according to their mean that the values are illuminated by a lower group. But even for the picture of system of intensity, a complete preservation of the mean value is not feasible.

An effort was made to reduce the difference between the input and the best brightness of the image media. The image data is first divided by the mean brightness, followed by a median trimmed of the tops of the input image histograms [1]. There was no fixed rationale for using the median value as the clipping limit, although the average brightness error might shrink. Later on, a method of restricting the reach was developed. The limits of brightness are narrow, such that the resulting luminosity is approximate to the original image, instead of aligning it with the entire permitted brightness. A weighted sum approach has been developed to solve the medium brightness problem [2]. Based on the average brightness value, lower and higher pixel groups were formed. Since weighting variables are not always practicable, however, the requirement to reduce the average luminosity error was relaxed so that possible weights were matched. A modification was further introduced to the histogram trimming principle [5]. A comparative assessment of the typical colour space in colour images was carried out [2–4], using the available methods. The journalistic work revealed that a green space is used to level the histogram, where this channel approximates the image's luminosity.

In addition to the above-mentioned division of the image into high and low light, an improvement in the contrast could be rendered by adjusting and defining in the histogram equalization of the target density profile. For example, an intensity-based window width [2] was introduced for the input histogram. You may also expand the policies mentioned there in order to restore the output image brightness that is set to the input frame. The image will be corrected to make sure the mean luminosity is similar to the image input.

2 Fuzzy K-Means and C-Means Clustering

2.1 K-Means Algorithm

The fundamental structure of K-means clustering is shown in Fig. 1, and it is explicitly discussed here.

Let $A = \{ai| \ i = 1,\ldots,f\}$ be attributes of f-dimensional vectors and $X = \{xi \mid i = 1,\ldots,N\}$ be each data of A. K-means clusters which X is SK = {Si | I 1, b…, = k} where M is mi εx = {M = 1, n (SI), the… J |} SI members, where $n(si)$ is number of members for si. Each cluster has cluster centre of $C = \{ci \mid i = 1,\ldots,k\}$. The following steps will be involved in the K-means clustering algorithm [1, 5].

1. Generate the random starting points with centroids C.
2. By utilizing the Euclidean separation, discover the separation d between X to C.

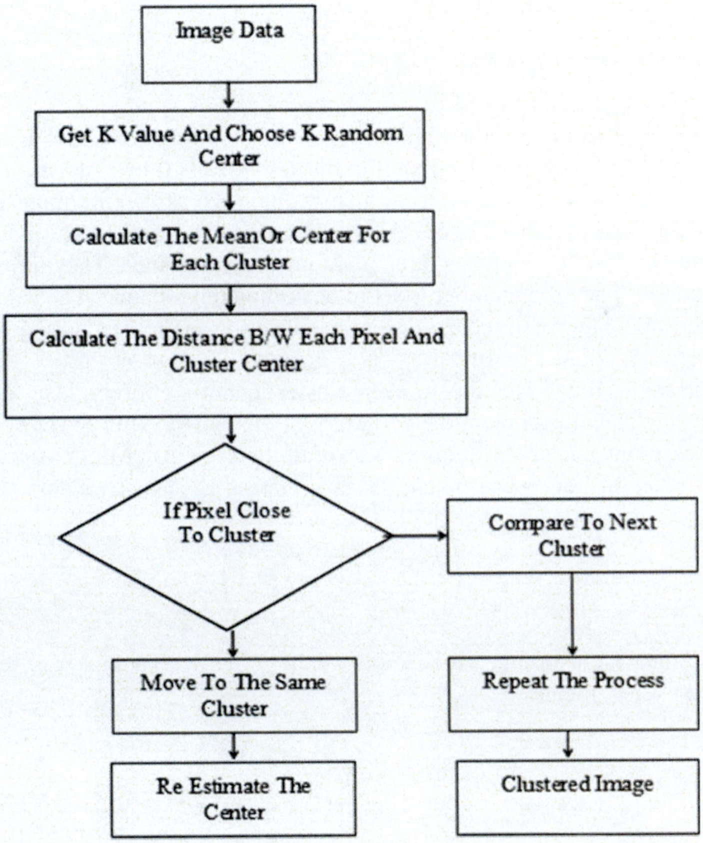

Fig. 1 K-means clustering algorithm

3. Ascertain the base $d(xi, C)$ from the partition of xi for $i = 1 \ldots N$ into.
4. Ascertain the new centre ci for $i = 1 \ldots k$ characterized as:

$$C_i = \frac{1}{n_i} \sum_{j=1}^{n(s_i)} m_{ij} \in s_i$$

5. Rehash the procedure stage 2 until the point that all centroids are concurrent.

The centroids, in case if they do not change their position , will be said as converged in a particular cycle. In addition, it may end with a threshold ε in the t focus if those positions have been refreshed by the distinction below ε:

$$\left| \frac{c^t - c^{t-1}}{c^t} \right| < \varepsilon$$

2.2 *Fuzzy C-Means Clustering*

Fuzzy logic to process data through partial membership in reflection is a method of each pixel value. Fuzzy membership in the set value is 0 to 1 ranges. The fuzzy cluster basically allows a multi-value logical value, such as the intermediate I. E., a member of the same member, can be set in fuzzy sets blurred picture. The membership function that is involved includes three main primary attributes. They have support and restrictions. The core member is set to be completely opaque. The subscription is supported by a non-intermediate or partial subscription and is a border that is set to value between 0 and 1 [4, 6].

Obscure logic, fuzzy clusters, in each cluster location entirely, just one degree from a cluster. The cluster is on the periphery of the cluster, with fewer points than points. Each point x is given status as we are in the kth cluster uk(x) $u_k(x)$ digital head [5]. The contribution coefficient for any given × 1 is usually clear:

$$\forall x \left(\sum_{k=1}^{num.\text{clusters}} u_k(x) = 1 \right)$$

Fuzzy c-means clustering, which kantroad all points with a cluster of his degree of leverage over it, means:

$$\text{centre}_k = \frac{\sum_x u_k(x)^m x}{\sum_x u_k(x)^m}$$

The distance to the cluster centre is related to the inverse state:

$$u_k(x) = \frac{1}{d(\text{centre}_k, x)}$$

Then, coefficients is a true parameter to distribution >1 So, their is 1.

$$u_k(x) = \frac{1}{\sum_j \left(\frac{d(\text{centre}_k, x)}{d(\text{centre}_j, x)}\right)^{\frac{2}{(m-1)}}}$$

The equivalent of m = 2 for coefficients to equal their money to 1 along a linear normalizing. When 1 m is close, and at this point, the cluster nearest to the middle is weighted far more than others, and it is similar to the algorithm of K-means.

The algorithm K-means fuzzy *c*, which is identical to the algorithm:

- Select the number of clusters.
- Clusters assigned to go to each endpoint are adjacent coefficients.
- Repeat algorithm (i.e. the change of the threshold of coefficients sensitivity between two points is from someone else): • Calculate centre for each cluster using the formula above.
- Using the formula above, calculate their coefficients for each location in the clusters.

Intra-cluster analytics K-means are less than the algorithm; however, there are problems, in the same way that there is at least one local minimum depending on weights and the initial selection of results. They know they have given precedence to properties and simple fuzzy C-means.

3 Proposed Segmentation Method and Simulation Results

Here, in this field, we suggested that our hybrid fuzzy K-means acronym (AFKM) cluster will be used, and it is shown in Fig. 2. First, what the average used for

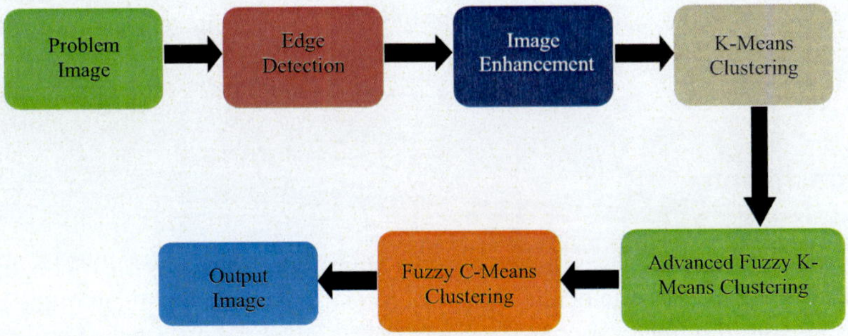

Fig. 2 Proposed system block diagram

Fig. 3 Edges detection

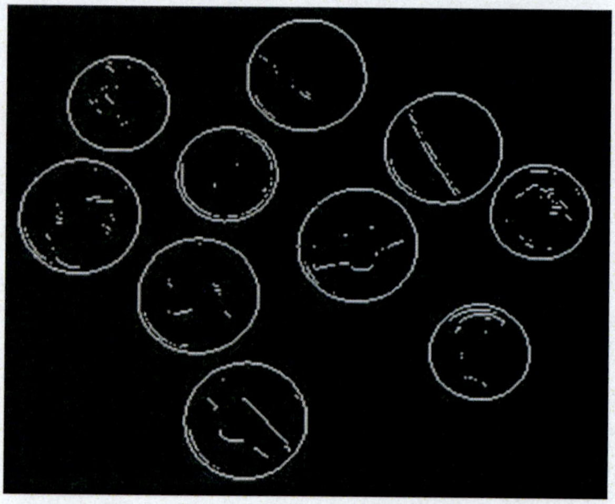

preprocessing will be to remove from digital photographs using filter noise and improve image quality.

The product of first phase will be able to identify the margins of the image, and then it's K-i.e. the Segmented generated mines of the cluster: image. Now, the fuzzy cluster signalling accuracy and precise detection of the cancer of the capsule will be applied to the product of MR images with the improve K-roots. Size of the tumour will be detected. The algorithm that steps up for the proposed system is shown in the diagram of a block.

3.1 Simulation Results

Simulation has been done in MATLAB. Figure 3 represents the edges found in the image, Fig. 4 presents the original image, and Figs. 5, 6 and 7 display the K-means clustering segmented images, fuzzy C-means clustering and the proposed segments.

Table 1 shows the comparison results with K-means clustering and C-means clustering, which have produced the better results in terms of time.

4 Conclusion

In this article, an approach was presented that explicitly defines a profile for histogram equalization-based enhancement of image contrast. A linear adjustment of the target histogram is used in the proposed approach to minimize the difference between the mean brightness between the input and the enhanced image. This approach avoids the need to divide the image into sub-groups and simplifies a single run of the

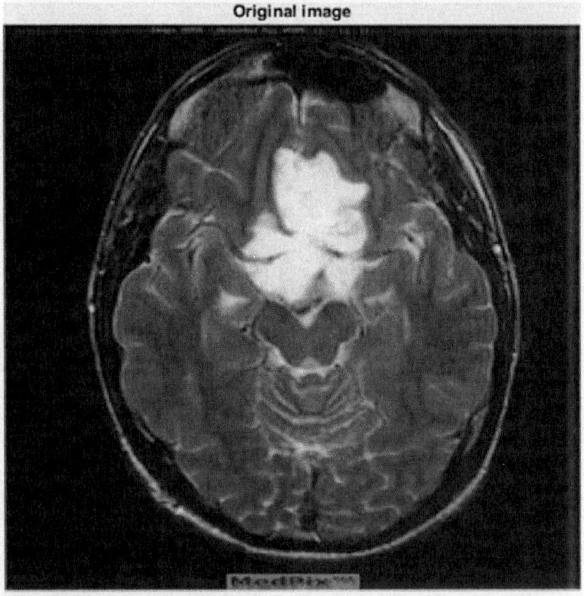

Fig. 4 Original image

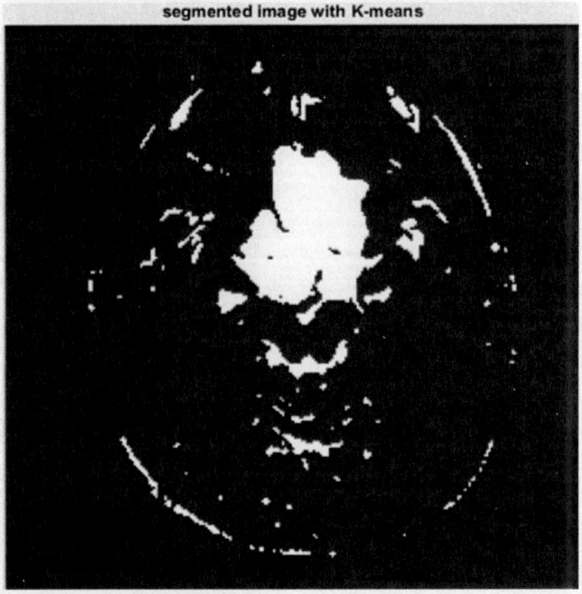

Fig. 5 Segmented image with K-means clustering

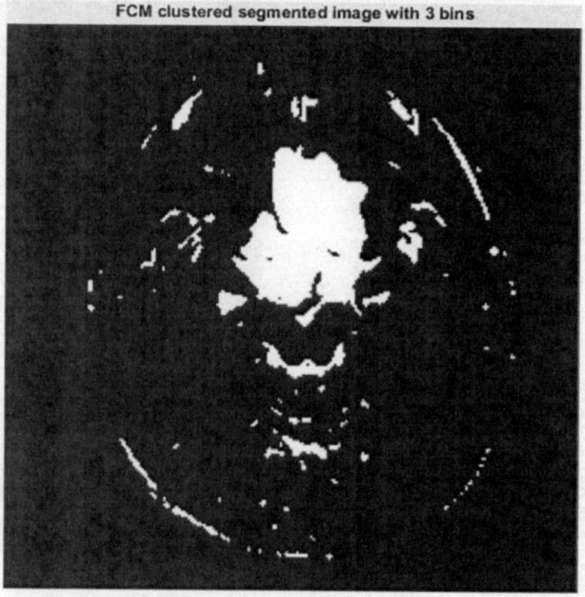

Fig. 6 Segmented image with fuzzy C-means clustering

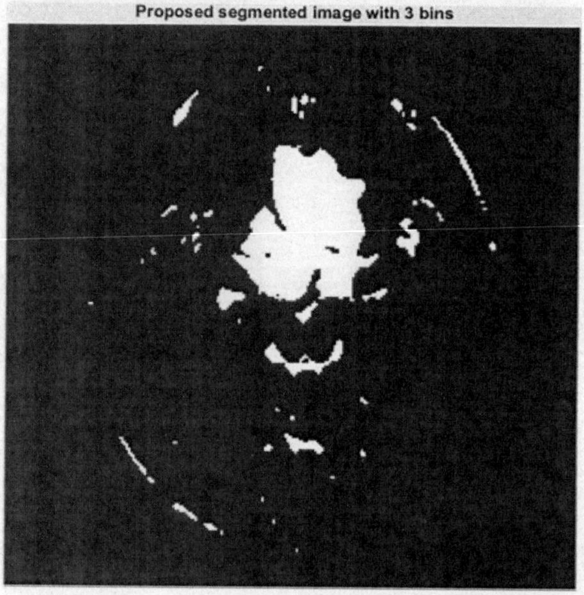

Fig. 7 Segmented image with proposed structure algorithm

Table 1 Comparison with other methods

S. No.	Segmentation method	Time (s)
1	K-means clustering	3.625
2	Fuzzy c-means clustering	4.0625
3	AFKM algorithm	2.8433
4	Proposed algorithm	1.0692

equalization process. Moreover, if a balancing constraint was satisfied, a rationalized option of threshold was devised. Hence, it is called Compliance with the minimum input–output brightness error criterion. Experiments on a broad natural image data collection demonstrate that while there is no single approach that can perform best across all performance metrics, there is no single technique.

References

1. Zheng Y, Wachowiak MP, Elmaghraby AS (2005) Resolution improvement and detail enhancement for computed tomography scout images. J Elect Imaging 14(1)
2. Wu T, Wu H, Du Y, Kwok N, Peng Z (2014) Imaged wear debris separation for online monitoring using gray level and integrated morphological features. Wear 316(1–2):19–29
3. Zhang G, Jia X, Kwok NM (2012) Super pixel based remote sensing image classification with histogram descriptors on spectral and spatial data. In: proceedings 2012 IEEE ınternational geoscience and remote sensing symposium (IGARSS), pp 4335–4338
4. Zhao J, Chen Y, Feng H, Xu Z, Li Q (2014) Fast İmage enhancement using multi-scale saliency extraction in infrared imagery. Optik 125:4039–4042
5. Yu Y-H, Kwok NM, Ha QP (2011) Color tracking for multiple robot control using a system-on-programmable-chip. Autom Constr 20(6):669–676
6. Su F, Fang G, Kwok NM (2012) Adaptive colour feature identification in image for object tracking. Math Probl Eng https://doi.org/10.1155/2012/509597. Article ID 509597, 18

H-infinity Robust Control of DC-DC Converter

Nabin Kumar Patra, Man Mohan Garg, Anup Kumar Panda, and Rakesh Rajan Shukla

Abstract Controller design of DC-DC power converter with pulse width modulation (PWM) technique for application for EV applications plays an important role. The function of any designed controller is to give fixed output voltage despite perturbations in the parameters of the converter, input voltage, and resistance at the load. In this paper, a controller with the H-infinity control method is designed with the help of algebraic Riccati equation and their solution. The H-infinity controller gives almost constant voltage with significantly lesser undershoot, settling time, and peak overshoot in the presence of parameter variations, sensor noise, change in resistance at the load and supply voltage. The results are analyzed with MATLAB/Simulink for different cases to observe the stability and robust performance.

Keywords Buck converter · Controller design · H∞ control · Riccati equation · Modeling

1 Introduction

Renewable energy sources are going to play a very important role in the future as most of the smart grids take power from these resources. The photovoltaic (PV) energy is one of the most popular non-conventional energy resource [1, 2]. The PV-based energy system needs a DC–DC converter for having a regulated voltage demand from unregulated demanded as required by the load or smart grid. The DC-DC power

N. K. Patra · A. K. Panda
Department of Electrical Engineering, National Institute of Technology Rourkela, Rourkela, India
e-mail: akpanda@nitrkl.ac.in

M. M. Garg
Department of Electrical Engineering, Malaviya National Institute of Technology Jaipur, Jaipur, India

R. R. Shukla (✉)
Department of Electrical Engineering, Government College of Engineering Keonjhar, Keonjhar, India
e-mail: rakeshshukla_fee@gcekjr.ac.in

S. S. Dash et al. (eds.), *Sixth International Conference on Intelligent Computing and Applications*, Advances in Intelligent Systems and Computing 1369,
https://doi.org/10.1007/978-981-16-1335-7_23

converter includes boost converter, buck converter, buck–boost converter, and many other converters. Buck converter finds application to step down the supply voltage, whereas a boost converter finds applications to step up the supply voltage. Both the converters have a very significant real-world application like buck converter with MPPT in a PV system battery charging, telecommunication system, hybrid electric vehicle, etc. [3–5]. Buck converter suffers from poor output voltage regulation when operated in open-loop configuration [6]. In an open-loop system, settling time and peak overshoot are very high. Therefore, there is a need to improve the performance of these converters. To get a better transient response, stability, and robustness of buck converter, the design of an efficient controller becomes very important. On the other hand, the real-world systems like buck converter are uncertain. When a system is modeled, there may be many approximations, many unmodeled terms, some environmental disturbances, some sensor noise, some actuator limitations, etc. All these things give rise to uncertainties in the system model. Uncertain systems can be handled in two ways using adaptive control or robust control. Adaptive control tries to estimate the uncertain elements or estimate the control gains depending on the requirements [7]. On the other hand, robust control methods try to develop the controller in such a way that it can handle the maximum level of uncertainty in the system model [8]. There are many robust control methods like sliding mode, H_2, H-infinity, model predictive control, etc. H-infinity control is a type of robust control methodology, where the supremum/maximum bound of the uncertainty is considered and the control law is designed according to it. Algebraic Riccati equation (ARE) is a good example of H-infinity control [9, 10]. The zest of the controller is to consider an optimal cost function which consists of infinity norm of the uncertainty, and then design a closed-loop optimal control law such that the cost function is minimized.

The algebraic Riccati equation (ARE) is a very well organized tool for designing the controller using the H-infinity method. Using the algebraic Riccati equation technique, an H-infinity controller has been implemented in this paper for a pulse width modulated (PWM) buck converter. The prime objective of this work is to control the inductor current, output voltage and also have a good response in the presence of noise. To test the robustness, the designed controller is tested in the presence of noise, disturbance, variation in load, and also parametric uncertainty [11, 12]. It shows the improved transient performance and the controller designed by the H-infinity method traces the reference voltage effectively.

2 H∞ Controller Design Using Riccati Equation

2.1 Optimal H∞ Control Problem

The block diagram for a standard H-infinity configuration is given in Fig. 1. The H-infinity control problem is to find the stabilizing controller(s) $K(s)$ which makes the closed-loop system internally stable and min $\|T_{ZW}(s)\|_{\infty}$. With the help of this

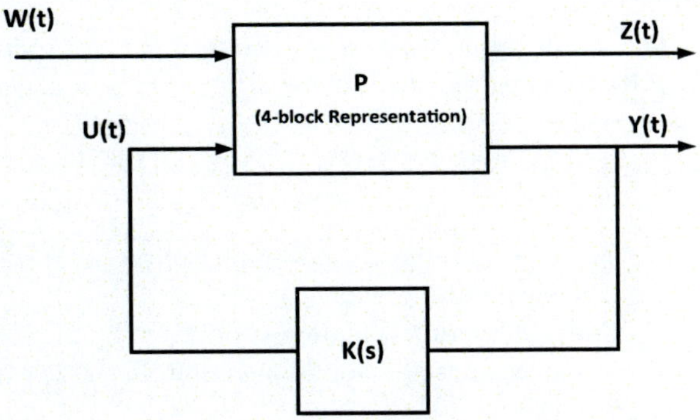

Fig. 1 Standard H-infinity configuration [13]

control signal, we have to minimize the effect of noise and disturbance signal on the regulated output.

where
$w(t)$: noise input signal,
$U(t)$: input control signal,
$z(t)$: regulated output signal,
$y(t)$: output signal.

2.2 Solution of H∞ Control Problem with State-Space Formulation

Plant description

$$\begin{aligned} \dot{x}(t) &= Ax(t) + B_{10}w(t) + B_{20}u(t) \\ z(t) &= C_{10}x(t) + D_{110}w(t) + D_{120}u(t) \\ y(t) &= C_{20}x(t) + D_{210}w(t) + D_{220}u(t) \end{aligned} \tag{1}$$

This can be represented in the packed notation a

$$\begin{aligned} P(s) &= \begin{bmatrix} A & B_1 & B_2 \\ C_1 & D_{11} & D_{12} \\ C_2 & D_{21} & D_{22} \end{bmatrix} \\ &= \begin{bmatrix} C_1 \\ C_2 \end{bmatrix} [sI - A]^{-1} \begin{bmatrix} B_1 & B_2 \end{bmatrix} + \begin{bmatrix} D_{11} & D_{12} \\ D_{21} & D_{22} \end{bmatrix} \end{aligned} \tag{2}$$

$$\begin{bmatrix} z(t) \\ y(t) \end{bmatrix} = \begin{bmatrix} P_{11} & P_{12} \\ P_{21} & P_{22} \end{bmatrix} \begin{bmatrix} w(t) \\ u(t) \end{bmatrix} \tag{3}$$

For stabilization, the aim is to control the closed-loop transfer function $T_{ZW}(s)$ ($u \neq 0$) such that the effect of 'w' on 'z' can be reduced.

$$T_{zw} = P_{11} + P_{12}K(s)[I - P_{22}K(s)]^{-1}P_{21} \tag{4}$$

where T_{zw} represent the closed-loop (from w to z) transfer function with $u = Ky$, where K is the controller gain.

H-infinity suboptimal solution is obtained by the following condition.

1. For the existence of stabilizing controller *K(s)*

($\mathbf{A}$,$\boldsymbol{B}_2$) must be stabilizable.

($\mathbf{A}$,$\mathbf{C}_2$) must be detectable.

2. The rank of $\begin{bmatrix} A - jwI & B_2 \\ C_1 & D_{12} \end{bmatrix}$ should be (N + m_2)

The rank of $\begin{bmatrix} A - jwI & B_1 \\ C_2 & D_{21} \end{bmatrix}$ should be (n + p_2).

This condition ensures the stabilizing solution to algebraic Riccati equations and also to prevent the result from closed-loop instability, the H-infinity controller does not try to cancel zero and pole on the imaginary (jw) axis.

3. The rank of D_{12} must be full column rank, i.e., m_2 the rank of D_{21} must be full row rank, i.e., P_2

As per the fundamental method based on the concept and application of a algebraic Riccati equation, the method of the Riccati equation is stabilizable. Now for analysis assume three matrices **A**, **R**, and **Q** of dimension m-by-m are given, here Q and **R** are symmetric. The algebraic Riccati equation involved with **A**, **R**, and **Q** is [13]:

$$A^{'}X + XA + XRX - Q = 0 \tag{5}$$

Matrix **X** is a solution of the above equation and the obtained solution is stabilizable if and only if:

1. **X** represents the solution of the Riccati equation.
2. **A + RX** is symmetric.
3. **X** is symmetric.

The Hamilton matrix used for algebraic Riccati equation is described by

$$H = \begin{bmatrix} A & R \\ Q & -A^{'} \end{bmatrix} \tag{6}$$

With the help of the above matrix, the algebraic Riccati equation can be obtained completely. The conditions for the solution which is stabilizable is that the Hamiltonian H matrix should not contain any eigenvalue on the imaginary axis and (**A**, **R**) should be controllable [14]. Here, the stable solution obtained is unique and we signify it by Ric(**H**).

Theorem [14]: *Let $\gamma > 0$, for which it includes a controller C for the plant P which is stabilizing such that $\|T_{ZW}(s)\|_\infty < \gamma$ if and only if*:

1. Existence of stable solution $X = \mathrm{Ric}(M)$;
2. Existence of stable solution $Y = \mathrm{Ric}(J)$;
3. $\rho(XY) < \gamma^2$(ρ radius of spectral)

γ_{opt} is the supremum of those $\gamma > 0$ for which $\|T_{ZW}(s)\|_\infty < \gamma$. With the help of iterative search, we can get a controller with $\|T_{ZW}(s)\|_\infty$ as near to γ_{opt}. This search is done by the hinfsyn function present in the MATLAB's robust control toolbox, which will provide a controller that is nearly the most suitable. To get this controller, the only requirement by the function is linearized and averaged state-space representation of the DC-DC converter. The above-discussed buck converter model is presented in the next section.

3 Model of Buck Converter

Figure 2 shows the circuit diagram for the buck converter. It consists of a switch, load resistance, diode, input voltage, capacitor, and inductor. Here, ideal converter is considered and for simplicity, the load is taken as resistive. However, the load may be replaced by a battery of an electrical vehicle.

The state-space equations of the linearized and averaged model of buck converter [15–19] are given by

$$\dot{\hat{x}}_1 = 0 - \frac{1}{L}\hat{x}_1 + \frac{v_o}{DL}\hat{d} \tag{7}$$

$$\dot{\hat{x}}_2 = \frac{1}{L}\hat{x}_1 - \frac{1}{RC}\hat{x}_2 \tag{8}$$

$$\begin{bmatrix} \dot{\hat{x}}_1 \\ \dot{\hat{x}}_2 \end{bmatrix} = \begin{bmatrix} 0 & -1/L \\ 1/C & -1/RC \end{bmatrix}\begin{bmatrix} \hat{x}_1 \\ \hat{x}_2 \end{bmatrix} + \begin{bmatrix} 0 \\ 0 \end{bmatrix}\hat{v}_g + \begin{bmatrix} v_o/DL \\ 0 \end{bmatrix}\hat{d} \tag{9}$$

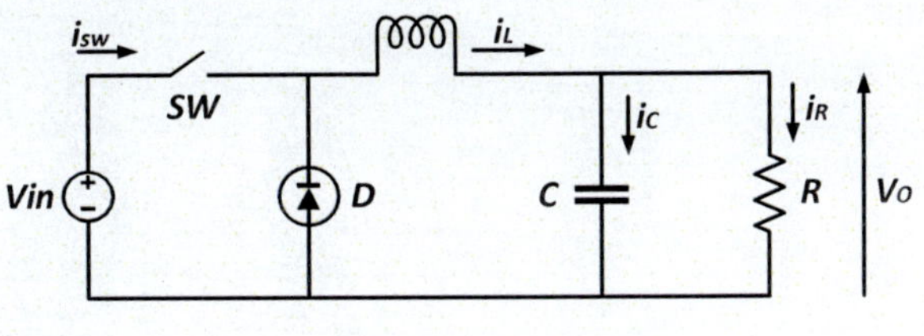

Fig. 2 Structure of buck converter

$$\hat{y} = \begin{bmatrix} 0 & 1 \end{bmatrix} \begin{bmatrix} \hat{x}_1 \\ \hat{x}_2 \end{bmatrix} \tag{10}$$

where $\widehat{x}_1, \widehat{x}_2\widehat{d}$ the state-space variable and duty are cycle, respectively.

Three parameters including ***L***, ***C***, and ***R*** in a buck converter suffer from many uncertainties like inaccurate values of components, change in operating point, etc. This problem can be solved by assuming their values within the interval given by.

$$L_1 = L(1 + \delta_L \rho_L),\ C_1 = C(1 + \delta_C \rho_C),$$
$$R_1 = R(1 + \delta_R \rho_R) \tag{11}$$

In a unified approach, it is noted that the quantity $\frac{1}{L}$ may be represented as a linear fractional transformation (LFT) in terms of perturbation δ_L as follows:

$$\frac{1}{L} = \frac{1}{L(1 + \delta_L \rho_L)} = \frac{1}{L} - \frac{\delta_L \rho_L}{L}(1 + \delta_L \rho_L)^{-1} \tag{12}$$

Similarly, LFT may be obtained for parameter 'C' and 'R'.

With the help of (11) and (12), the relevant equations relating inputs to their corresponding outputs are given by

$$\begin{bmatrix} y_L \\ \dot{x}_1 \end{bmatrix} = \begin{bmatrix} -\rho_L & {}^1/_L \\ -\rho_L & {}^1/_L \end{bmatrix} \begin{bmatrix} u_L \\ {}^{v_o}/_D(u - x_2) \end{bmatrix} \tag{13}$$

$$\begin{bmatrix} y_C \\ \dot{x}_2 \end{bmatrix} = \begin{bmatrix} -\rho_C & {}^1/_C \\ -\rho_C & {}^1/_C \end{bmatrix} \begin{bmatrix} u_C \\ (x_1 - v_R) \end{bmatrix} \tag{14}$$

$$\begin{bmatrix} y_R \\ v_R \end{bmatrix} = \begin{bmatrix} -\rho_R & {}^1/_R \\ -\rho_R & {}^1/_R \end{bmatrix} \begin{bmatrix} u_R \\ x_2 \end{bmatrix} \tag{15}$$

From the above model, the following equation is obtained:

$$v_R = -\rho_R u_R + \frac{x_2}{R} \tag{16}$$

$$\dot{x}_1 = -\frac{1}{L} - \rho_L u_L + \frac{v_o}{DL} u \tag{17}$$

$$\dot{x}_2 = \frac{1}{C} x_1 - \frac{1}{RC} x_2 - \rho_C u_C + \frac{\rho_R}{C} u_R \tag{18}$$

$$y_L = -\frac{x_2}{L} - \rho_L u_L + \frac{v_o}{DL} u \tag{19}$$

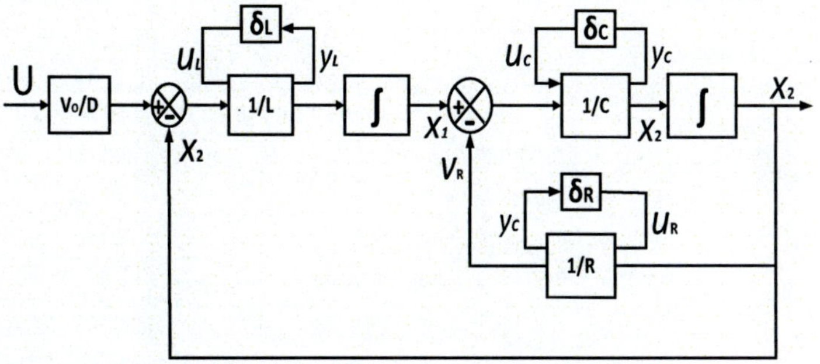

Fig. 3 Schematic diagram of a buck converter with uncertain parameters

$$y_C = \frac{1}{C}x_1 - \frac{1}{RC}x_2 - \rho_C u_C + \frac{\rho_R}{C}u_R \tag{20}$$

$$y_R = -\rho_R u_R + \frac{1}{R}x_2 \tag{21}$$

In matrix (6 * 6) form,

$$\begin{bmatrix} x_1 \\ x_2 \\ y_L \\ y_R \\ y_C \\ y \end{bmatrix} = \begin{bmatrix} 0 & -{}^1\!/_L & -\rho_L & 0 & 0 & {}^{v_o}\!/_{DL} \\ {}^1\!/_C & -{}^1\!/_{RC} & -\rho_L & 0 & {}^{\rho_R}\!/_C & 0 \\ 0 & -{}^1\!/_L & -\rho_L & 0 & 0 & {}^{v_0}\!/_{DL} \\ {}^1\!/_C & -{}^1\!/_{RC} & 0 & -\rho_C & {}^{\rho_R}\!/_C & 0 \\ 0 & {}^1\!/_R & 0 & 0 & -\rho_R & 0 \\ 0 & 1 & 0 & 0 & 0 & 0 \end{bmatrix} \begin{bmatrix} x_1 \\ x_2 \\ u_L \\ u_C \\ u_R \\ u \end{bmatrix}$$

Representing these equations, the entire model can be fitted in a block diagram shown in Fig. 3, where it includes all the uncertainties present for analyzing buck converter which can affect its operation. With the help above diagram, the H-infinity controller is designed with the help of MATLAB's robust control toolbox.

4 H-Infinity Controller Design of Buck Converter

The parameters values for the buck converter that is used in the simulation are given [11] in Table 1.

To make the output waveform smooth and to get continuous inductor current, the switching frequency is made high for the PWM. Also, with the increase in switching frequency, the circuit components size decreases significantly, i.e., capacitor, inductor which makes the system compact. Generally, the switching frequency is ranging from kHz to a MHz for getting a smooth waveform. In this paper, 300 kHz of switching frequency is used for PWM in simulation.

Table 1 Buck converter parameters

Parameter	Value
Input voltage	50 V
Inductance ($\boldsymbol{L_c}$)	22 mH
Capacitance ($\boldsymbol{C_c}$)	75 μf
Load resistance ($\boldsymbol{R_c}$)	20 Ω
Perturbations $\boldsymbol{\delta_L}, \boldsymbol{\delta_R}, \boldsymbol{\delta_C}$	0.3

To design the controller using H-infinity, a robust control toolbox in MATLAB is used. It includes a process of gamma iteration, which solves two algebraic Riccati equations for different values of gamma and stops when there is no solution of two ARE's for any gamma. The equation which is used to get the two algebraic Riccati equation is given by

$$A^T P + PA - PBR^{-1}B^T P + Q = 0 \tag{22}$$

The solution to the well-defined H-infinity control problem cannot be obtained physically as it is a very messy process. Iterative search done by the hinfsyn function present in the robust control toolbox of MATLAB gives a gamma value. With the help of this gamma value, a controller is designed which lessens the supremum of the closed-loop transfer function and reduces the effect of different disturbances on the output. The controller transfer function K(s) and γ with the help of MATLAB programming are given by

$$K(s) = \frac{418.3\ \mathrm{s}^3 + 953.7\ \mathrm{s}^2 + 104.6\mathrm{s} + 238.4}{\mathrm{s}^4 + 1.03 * 10^6\ \mathrm{s}^3 + 7.91 * 10^6\ \mathrm{s}^2 - 3.97 * 10^6\ \mathrm{s} - 4976} \tag{23}$$

and $\gamma = 0.906$.

The controller behavior is analyzed using Simulink model and MATLAB toolbox. In Fig. 4, the simulation results are analyzed with band-limited white noise and without band-limited white noise. Figure 5 shows the output voltage with the change

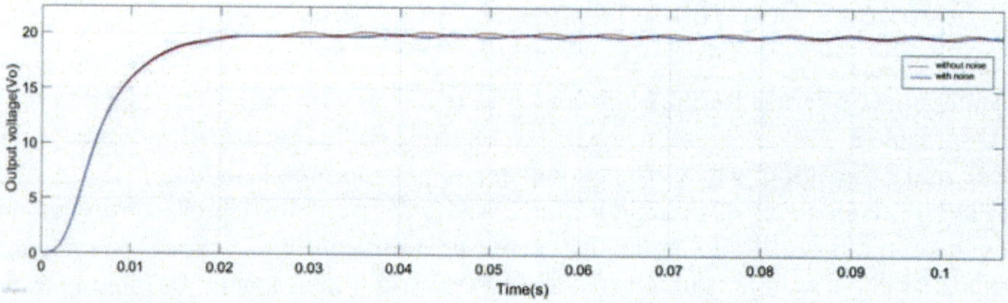

Fig. 4 Transient performance of voltage at the output in the presence of noise

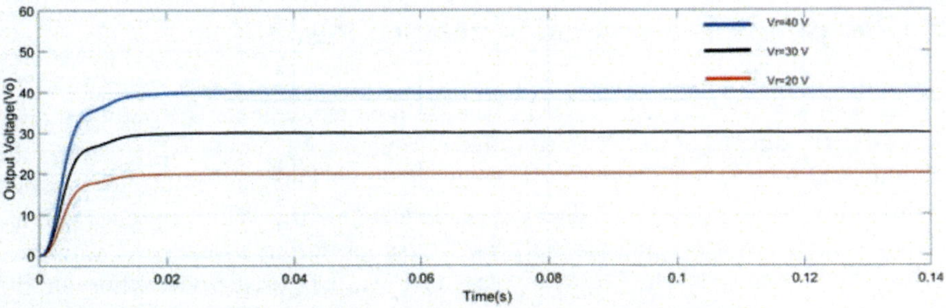

Fig. 5 Transient performance of output voltage at various reference voltages

in reference value. The controller is also tested by varying the input voltage and resistance across load in Figs. 6 and 7, respectively.

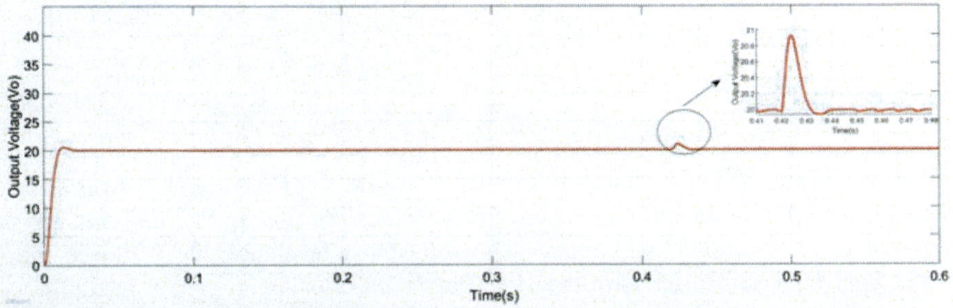

Fig. 6 Buck converter transient performance when load resistance *R* is varied

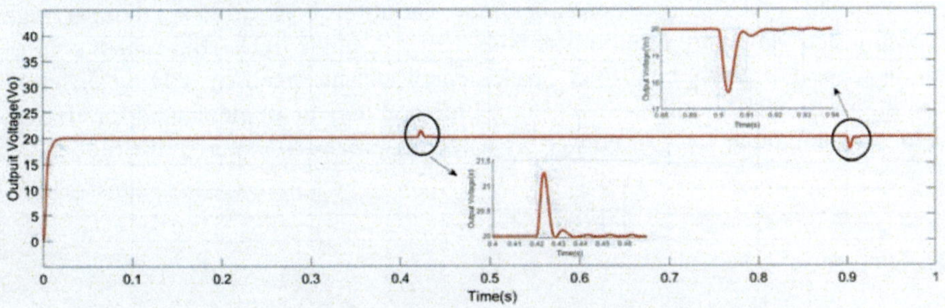

Fig. 7 Buck converter transient performance when V_g is varied

5 Robustness Analysis and Simulation Results

Figure 4 shows that in the presence of band-limited white noise, the output voltage is nearly the same as in the absence of band-limited white noise, which shows the stability and robustness of the controller. In Fig. 5, the designed controller is tested at different reference voltages. The voltage at the output traces the reference voltage effectively. In this case, there is no peak overshoot during a change in reference voltage and it takes very less time, i.e., around 0.0226 s to settle down. Further, the robustness is analyzed by changing the resistance across load and input voltage from their actual value. The transient performance by varying the resistance (R) at the load from 20 Ω to 10 Ω at 0.42 s is shown in Fig. 6, and observation was made that the voltage at the output takes only 0.021 s to settle down to its set value. Similarly from Fig. 7, the robustness is tested by varying the input voltage from 50 to 53 V at 0.43 s and 50 V to 47 V at 0.901 s and it results that the voltage at the output traces the reference in significantly small time with lower peak deviation.

6 Conclusion

Algebraic Riccati equation-based H-infinity controller for PWM buck converter has been presented with the parametric uncertainty, sensor noise, and variation in resistance at the load and input voltage. The robustness and analysis of closed-loop stability of buck converter with the H-infinity controller are achieved in the presence of 30% parametric uncertainty in the inductor and capacitor. With simulation results, it is also observed that while varying the reference voltage, the controller gives no peak overshoot and very less settling time for output voltage. Further, the performance of the designed controller under variations in load and the input voltage is also tested which are found reasonably better. Using MATLAB's robust control toolbox, the H-infinity controller implemented and designed by solving different conditions in ARE's toolbox. This control method may be implemented for electric vehicles using other DC-DC converters with suitable modification in the model.

References

1. Rahim APN, Pagano DJ, Lenz E, Stramosk V (2015) Modeling and stability analysis of Islanded DC microgrids under droop control. IEEE Trans Power Electron 30(8):4597–4607
2. Zabalawi SA, Mandic G, Nasiri A (2008) Utilizing energy storage with PV for residential and commercial use. In: 34th Annual Conference of IEEE Industrial electronics, 2008. IECON 2008, pp 1045–1050
3. Khaligh A, Cao J, Lee Y (2009)A multiple-input DC-DC converter topology. IEEE Trans Power Electron 24(3):862–868
4. Chan CY (2007) A nonlinear control for DC–DC power converters. IEEE Trans Power Electron 22(I):216–222

5. Green TC, Weiss G, Zhong Q-C, Liang J (2004) H∞ repetitive control of DC-AC converter in microgrids. IEEE Trans Power Electron 19(1)
6. Alejo D, Maussion P, Faucher J (2003) Multiple model control of a Buck dc/dc converter. Math Comput Simulat 63(3–5):249–260
7. Xu M, Zhao Z, Chen Q, Lai JS, Cho Y (2012) Derivation, analysis, and implementation of buck-boost converter based high-efficiency PV inverter. IEEE Trans Power Electron 27(3):1304–1313
8. Hochschulverlag AG, Kokame H, Kobayashi H, Mori T (1998) Robust performance for linear delay differential systems with time-varying uncertainties. IEEE Trans Automat Control 43(2):223–226
9. Lam HK, Tan SC (2009) Stability analysis of fuzzy-model-based control systems: application on the regulation of switching DC-DC converter. IET Control Theo Appl 3(8):1093–1106
10. Cao G, Fan S, Xu G (2004) The characteristics analysis of magnetic bearing based on H-infinity controller. In: Proceedings of the 51st world congress on intelligent control and automation, June 15–19, 2004, Hangzhou, P.R. China.
11. Vidal-Idiarte E, Martinez-Salamero L, Valderrama-Blavi H, Guinjoan F, Maixe J (2003) Analysis and design of H control of nonminimum phase-switching converters. IEEE Trans Circuits and Systems Part I 50(10):1316–1323
12. Vinicius FM, Pedro LDP (2003) H control with pole location for a DCDC converter with a switched load. In: IEEE International Symposium on Industrial Electronics, Brasil, pp 550–555
13. Naim R, Weiss G, Shmuel BY (1997) H- infinity control applied to boost power converters. IEEE Trans Power Electron 12(4):677–683
14. Safonov MG, Limebeer DJN, Chiang RY (1989) Simplifying the H1 theory via loop shifting, matrix pencil and descriptor concepts. Int J Control 50(6):2467–2488
15. Erickson RW, Maksimovic D (2004) Fundamentals of power electronics: Kluwer Academic Publishers (2004)
16. Verghese GC, Bruzos CA, Mahabir KN (1989) Averaged and sampled-data models for current-mode control: a reexamination. In: IEEE Power Electronics Specialists Conference. Record, pp 484–491.
17. Garg MM, Hote YV (2014) Leverrier algorithm based reduced-order modeling of dc-dc converters. In: India international conference on power electronics, IICPE, pp 1–6
18. Garg MM, Hote YV, Pathak MK (2015) Design and performance analysis of a PWM DC-DC buck converter using PI–Lead compensator. Arab J Sci Eng 40(12)
19. Garg MM, Hote YV, Pathak MK, Behera L (2018) An approach for buck converter PI controller design using stability boundary locus. In: 2018 IEEE/PES transmission and distribution conference and exposition (T&D), pp 1–5

Gravitational Search Algorithm-Based PID Controller in AGC for Stability Improvement in the Interconnected Power System

Sangram Keshori Mohapatra, P. M. Dash, Subhransu Sekhar Dash, and A. K. Baliar Singh

Abstract In this work, analysis of LFC in two-area interconnected non-reheat thermal power system with PID controllers optimized gravitational search algorithm is highlighted. The disturbance of 0.01 pu is injected in control area-1, area-2 separately and also simultaneously in both areas. System dynamic performances in terms of settling time, peak overshoot, and undershoot with PID controller are compared with those of PI controller to depict the supremacy of the present work. Different system parameters in the range of ($\pm 25\%$ to $\pm 50\%$) have been considered to analyze the effectiveness.

Keywords Automatic generation control · PID controller · Gravitational search algorithm · LFC

1 Introduction

Electrical power system is the largest complex network performing the task of generation, transmission, and distribution which are interconnected in large scale. In the control area of this system, power and system frequency of the tie line should be closed to nominal value [1–5] with respect to the load frequency control. For this purpose, researchers are trying to use different types of controllers and optimization techniques in automatic generation control of such systems as well as to make a balance between load variation and generation [1–5]. Some researchers also analyzed

S. K. Mohapatra (✉) · S. S. Dash
Department of Electrical Engineering, Goverment College of Engineering, Keonjhar, Odisha, India

S. S. Dash
e-mail: subhransudash_fee@gcekjr.ac.in

P. M. Dash
EE Department, Bhubaneswar Engineering College, Bhubaneswar, Odisha, India

A. K. Baliar Singh
GCE, Kalahandi, Odisha, India

S. S. Dash et al. (eds.), *Sixth International Conference on Intelligent Computing and Applications*, Advances in Intelligent Systems and Computing 1369,
https://doi.org/10.1007/978-981-16-1335-7_24

the performances in single-area hydro and thermal, and two-area hydro and thermal interconnected power system [6–8] applying several types of controllers and techniques for optimization. From the observation, it has been found that the conventional PI controller attempts in the LFC-control shows poor dynamic performances in terms of settling time and frequency overshoot in single and two-area power system [9]. In the literature, fuzzy PI-based tabu search algorithm, generalized neural network, fuzzy logic control, hybrid neuro-fuzzy approach have been implemented for automatic generation control of multi-area interconnected power system [10–13]. In the purposed analysis, two-area automatic generation control of interconnected power system using GSA optimization in the presence of PID controller is considered. The LFC of the system is compared with PI-based controller in terms of settling time, frequency over and under shoot, and variation in tie line power flow of the two-area interconnected power system. Performances of the system under consideration with both PI and PID controller are studied under different loadings conditions (±25% and ±50%). Comparison of performances is also done with GA- and GSA-based PI controllers for the same system [14].

2 Model of the System

2.1 The Power System Under Study

In this work, a two-area non-reheat thermal power system is considered toward automatic generation control (AGC) using PID controller for analysis as shown in Fig. 1. In the simulation model, Δf_1 and Δf_2 are taken as deviations from nominal system frequency, Δptie represents the tie line power mismatch, AEC_1 and ACE_1 are area control errors, ΔPG_1 and ΔPG_2 considered to be changes in governor's valve positions, ΔPC_1 and ΔPC_2 are representing governors speed changer positions, ΔPT_1 and ΔPT_2 are change in turbine outputs, ΔPD_1 and ΔPD_2 are load demand changes, T_{PS1} and T_{PS2} are taken as power system time constants, synchronizing coefficient T_{12}, control outputs are represented by U_1 and U_2 and speed regulation parameters of the governors as R_1 and R_2, power system gains are represented by K_{PS1} and K_{PS2},B_1 and B_2 represent frequency bias parameters. Speed governor time constants are represented by TG_1 and TG_2, whereas T_{T1} and T_{T2} are representing turbine time constants. Different parameters of the purposed system are taken as per values given in the appendix. Controller's structures (PI and PID) are shown in Eqs. (1) and (2) in terms of proportional gain (K_P), integral gain (K_I), and derivative gain (K_D). Same areas are considered in the purposed anlaysis for both controllers. Inputs of PI/PID controller are taken as ACE_1 and ACE_2, and the controller outputs in area-1 and area-2 are taken as U_1 in area-1 and U_2 in area-2.

$$K(S) = K_P + \frac{K_I}{S} \tag{1}$$

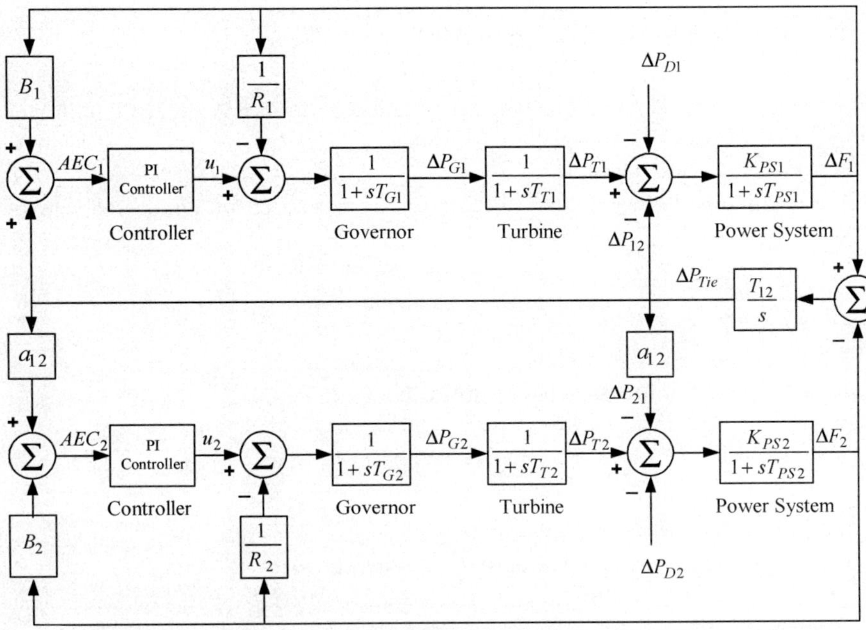

Fig. 1 Power system with area-1 and area-2

$$K(S) = K_{\mathrm{P}} + \frac{K_I}{S} + SK_{\mathrm{D}} \tag{2}$$

2.2 *Objective Function*

The objective function (*J*) of integral of time multiplied absolute error (ITAE) has been taken in this work. The mathematical expression of J is shown in Eq. (3)

$$J = \int_{0}^{t_{\mathrm{sim}}} (|\Delta f_1| + |\Delta f_2| + |\Delta P_{\mathrm{Tie}}|) \cdot t \cdot \mathrm{d}t \tag{3}$$

where Δf_1, Δf_2 are the frequency changes in area-1 and area-2 of the system and $\Delta P_{\mathrm{tie}}$is the change of power flow in tie line.

3 Gravitational Search Algorithm

The purposed heuristic technique is the inspiration of Newton's laws of gravity and motion [15, 16] which is highlighted as follows

Step 1: Agents Initialization:

Initialization of ith agent, dth dimension with n-space of dimension can be expressed as

$$X_i = \left(x_i^1, \ldots, x_i^d, \ldots x_i^n\right)$$

Step 2: Evolution and computation of best fitness for each agent:

The best and worst fitness value of fitness computation is

$$\text{Best}(t) = \min_{j\in\{1\ldots n\}} \text{fit}_j(t),\ \text{Worst}(t) = \max_{j\in(1\ldots n)} \text{fit}_j(t),$$

Step 3: Computation of gravitational constant G:

The gravitational constant G with time 't' is expressed

$$G(t) = G0\, e\left(-\acute{\alpha}\, t/T\right)$$

Step 4: Calculation of mass of agents:

The calculation of mass of agents is updated

$$M_a i = M_p i = M_i i = M_i \text{ Where } i = 1, 2, 3 \ldots n$$

$$m_i(t) = \frac{(\text{fit}_i(t) - \text{worst}(t))}{(\text{best}(t) - \text{worst}(t))},\ M_i(t) = \frac{(m_i(t))}{\left(\sum_{(j=1)}^{N} m_j(t)\right)}$$

Step 5: Calculation of accelerations of the agents:

$a_i^d(t) = \frac{F_i^d(t)}{M_{ii}(t)}$ where the total force of ith agent $F_i^d(t) = \sum_{j=1, j=i}^{N} .rand_j F_{ij}^d(t)$

Step 6: Updating the velocities & positions of masses.

Velocity and position of ith agent are updated as per the equation

$$v_i^d(t+1) = \text{rand}(i * v_i^d) + a_i^d(t),\ x_i^d(t+1) = x_i^d(t) + v_i^d(t+1)$$

Step 7: Repetition of above steps is to be continued till iterations attain their maximum limit.

4 Results and Discussion

In this work, GSA is applied to automatic generation control of the proposed power systems (interconnected) for optimizing outputs of PI and PID controllers. Performances of simulation results obtained through MATLAB/SIMULINK are checked in three cases (1) increasing the demand in area-1 stepwise (2) increasing the demand in area-2 stepwise (3) increasing the demand in both the area simultaneously. Tables 1 and 2 show the preferred values of the controller parameters and settling time with frequency deviation of the system in the optimization process.

Case-1:*Increasing the demand of the area-1 in step load.*

Case-2:*Increasing the demand of the area-2 in step load.*

10% step rises in demand applied for both area is considered in case-1 and case-2 at time '0' sec with the same operating conditions. It is observed from Figs. 2, 3, 4, 5, and 6 that change in frequency in both cases along with the tie line power is highly oscillatory with respect to the conventional controller. However, the system stability is maintained upto some extent with PI controller. But the system shows better response in terms of settling time & frequency overshoots with PID controller.

Case-3. *Increasing the demand of the area-1 & area-2 in step* simultaneously.

In this case, the step increase in demand is considered for both areas simultaneously to verify the supremacy and effectiveness of the system under consideration. The frequency mismatches (Δf_1)& (Δf_2) in areas-1&2, respectively, in the closed-loop system are shown in Figs.7 and 8. From Figs. 2, 3, 4, 5, 6, 7, and 8, it is observed

Table 1 Optimization parameters of controllers

Controllers	Parameters	J(ITAE)
PID controller employed with GSA	K_P	0.8932
	K_I	0.9211
	K_D	0.9376
PI controller employed in GSA [15]	K_P	−0.3100
	K_I	0.2670
PI controller employed in GA [14]	K_P	−0.2346
	K_I	0.2662
Conventional PI controller	K_P	0.7005
	K_I	0.3802

Table 2 Different value of settling time of the system (Ts Sec.)

Parameters		Conv	GA Optimized PI controller	GSA Optimized PI controller	GSA Optimized PID controller
	Δf_1	45	10.2	8.7	7.67
	Δf_2	45	10.6	8.45	6.13
ΔP_{tie}		29.8	9.2	8.87	7.54

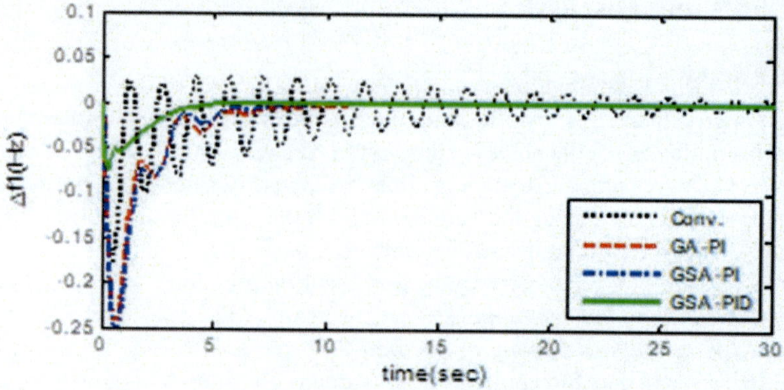

Fig. 2 Frequency response the area-1 with by disturbances taken in area-1

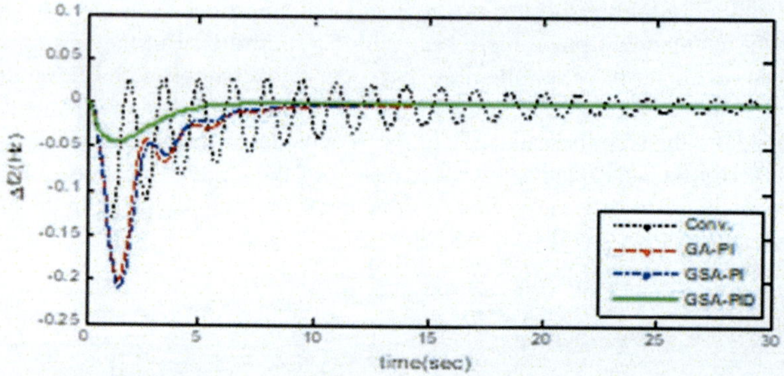

Fig. 3 Frequency response area-2 by disturbances taken in area-2

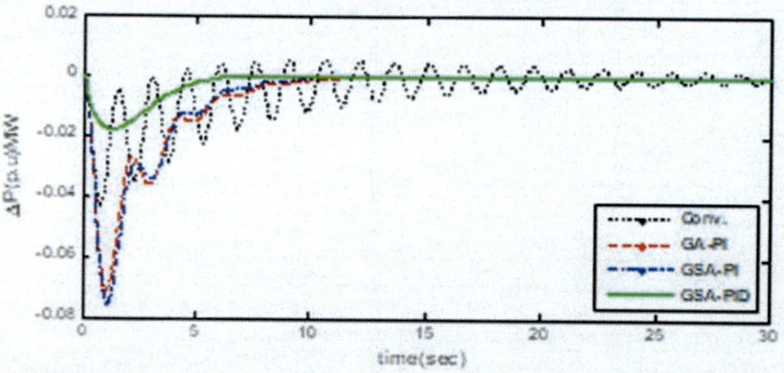

Fig. 4 Response of ΔPtie in area-1 by disturbances taken in area-1

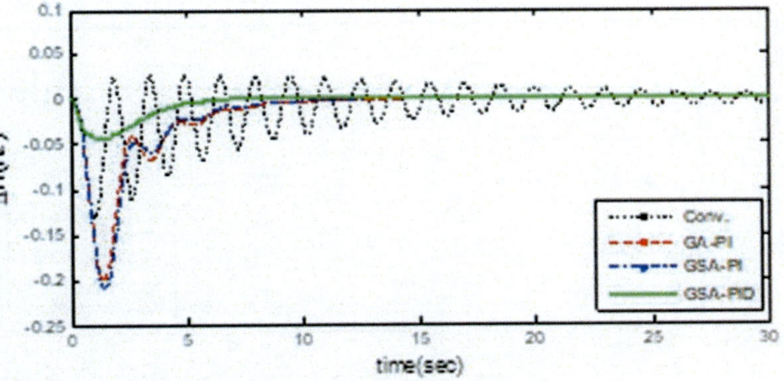

Fig. 5 Frequency deviation area-1 by disturbance taken in area-1

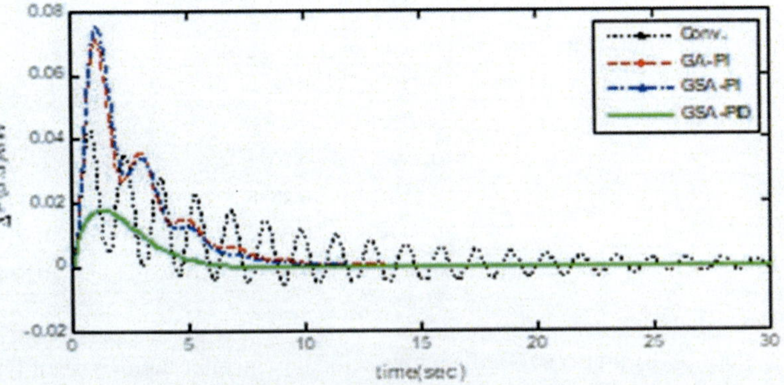

Fig. 6 ΔPtie response in area-2 with disturbances taken in area-2

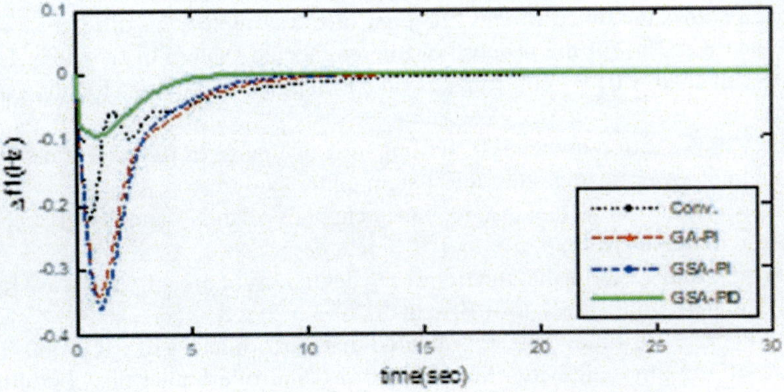

Fig. 7 Frequency response of area-1 due to disturbances in area-1 and area-2

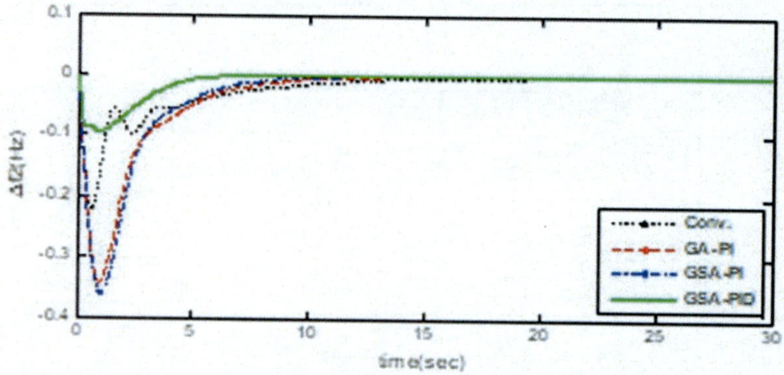

Fig. 8 Frequency deviation of the area-2 due to disturbances in area-1 and area-2

Table 3 Tuned controller parameters for different loadings

Controller	Controller parameters	+25%	+50%	−25%	−50%
PI	K_P	−0.4989	−0.5342	−0.5034	−0.5183
	K_I	0.2395	0.2082	0.2345	0.2486
PID	K_P	0.1165	0.5578	0.1789	0.3411
	K_I	0.8001	0.9569	0.8879	0.9345
	K_D	0.4456	0.6772	0.5122	0.6342

that very poor performances found in conventional controller but in case of GA- and GSA-based PI controllers, the damping characteristic is improved. However, PID controller gives much better response than PI controller of the same system as system response quickly reaches steady state.

Table 3 shows the tuned PI and PID controller parameters for different loadings (±25% and ± 50%) for the propose of interconnected system of two area. For the tuning of all load, PI/PID controller is taken range from [−1, 1] for the purpose analysis.

Table 4 shows different values of settling time of change of frequency and change of the tie line power mismatch of the system under study.

Figures 9 and 10 show that the frequency changes of area-1 and 2 with respect to rise in load demands from +25% to +50% in area-1.

Figures 11 and 12 show that the frequency deviations of area-1 and area-2 for rise in load demands from −25% to −50% in area-1.

From Figs. 9, 10, 11, and 12, it is cleared that performances of PID controller in terms of settling times are better than those of PI controller under this condition.

Table 4 Settling time at different loading

Ts	GSA optimized PI controller		GSA optimized PID controller	
	Parameters	Value	Parameters	Value
+25%	$\Delta f1$	7.63	$\Delta f1$	6.45
	$\Delta f2$	6.93	$\Delta f2$	5.15
+50%	$\Delta f1$	8.24	$\Delta f1$	4.95
	$\Delta f2$	7.45	$\Delta f2$	6.38
−25%	$\Delta f1$	7.23	$\Delta f1$	6.76
	$\Delta f2$	8.21	$\Delta f2$	5.36
−50%	$\Delta f1$	7.86	$\Delta f1$	4.89
	$\Delta f2$	7.63	$\Delta f1$	6.45

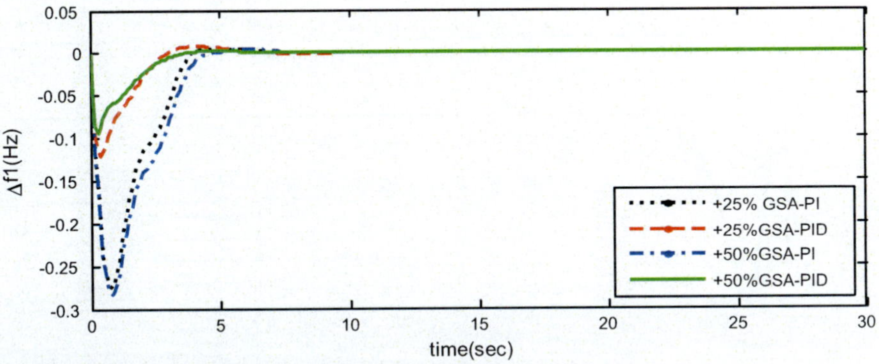

Fig. 9 Frequency response of area-1 by rise in load demands in area-1

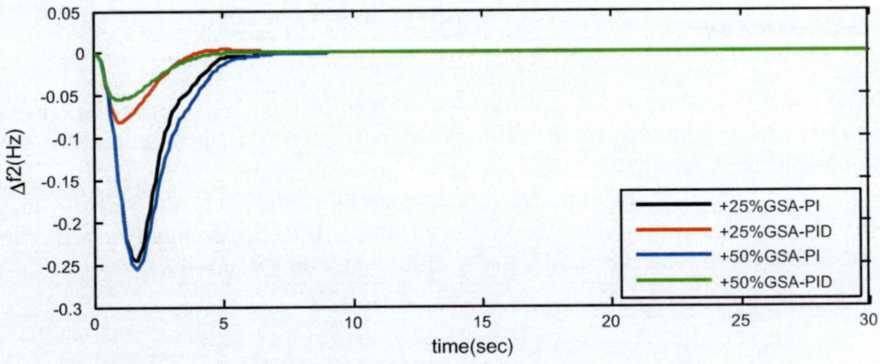

Fig. 10 Frequency response of area-2 by rise in load demands in area-1

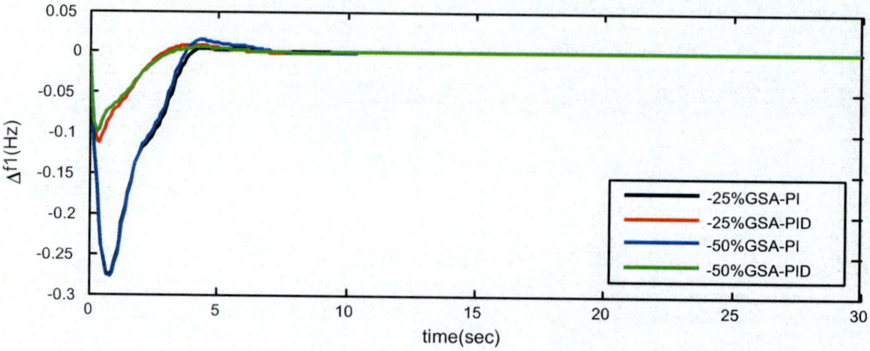

Fig. 11 Frequency response of area-1 by fall in load demands in area-1

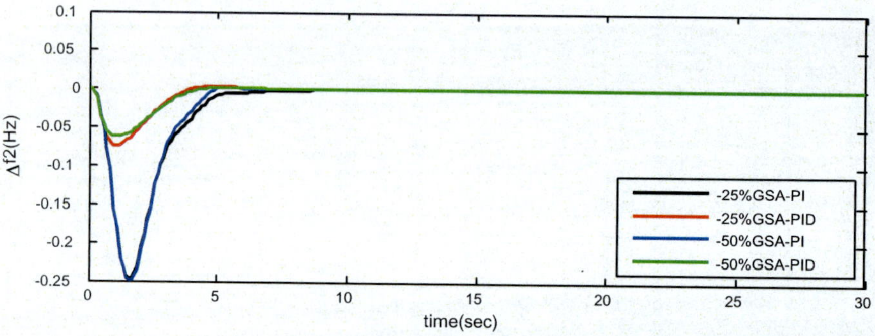

Fig. 12 Frequency response of area-1 by fall in load demands in area-2

5 Conclusion

In this work, GSA-based PID controller with ITAE objective function has been used for automatic generation control of proposed power systems. It observed that the tuning controller parameters give reduction of overshoot and converging to steady state without much oscillation and very less settling time. This study shows that performances found by tuning the PID controller is better in comparison with the same action taken for PI controller under different loading conditions.

Appendix

System parameters under study as follows.

$TPS_1 = TPS_2 = 20s$, $B_1 = B_2 = 0.045$ p.uMW/Hz, $R_1 = R_2 = 2.4$ Hz/p.u,

$TG_1 = TG_2 = 0.08s, TT_1 = TT_2 = 0.3s,\ KPS_1 = KPS_2 = 120$ Hz/pu MW, $T_{12} = 0.545$ p.u, $a_{12} = -1$

References

1. Shayeghi H (2007) Robust decentralized neural networks based LFC in a deregulated power system. Elect Power Syst Res 03
2. Mohanty B, Panda S, Hota PK (2014) Controller parameters tuning of differential evolution algorithm and its application to load frequency control of multisource power system. Int J Electric Power Energy Syst
3. Nikmanesh E, Hariri O, Shams H, Fasihozaman M (2016) Pareto design of load frequency control for interconnected power systems based on multi-objective uniform diversity genetic algorithm (MUGA). Int J Electr Power Energy Syst
4. Sambariya DK, Shrangi S (2017) Optimal design of PID controller for load frequency control using harmony search algorithm. Indonesian J Electr Eng Comput Sci
5. Panda S, Yegireddy NK (2013) Automatic generation control of multi-area power system using multi-objective non-dominated sorting genetic algorithm-II. Int J Electr Power Energy Syst
6. Sahu BK, Pati S, Mohanty PK, Panda S (2015) Teaching–learning based optimization algorithm based fuzzy-PID controller for automatic generation control of multi-area power system. Appl Soft Comput
7. Pandey SK, Mohanty SR, Kishor N (2013) A literature survey on load–frequency control for conventional and distribution generation power systems. Renew Sustain Energy Rev
8. Nikmanesh E, Hariri O, Shams H, Fasihozaman M (2016) Pareto design of Load Frequency Control for interconnected power systems based on multi-objective uniform diversity genetic algorithm (MUGA). Int J Electr Power Energy Syst
9. Parmar KPS, Majhi S, Kothari DP (2012) Load frequency control of a realistic power system with multi-source power generation. Int J Electr Power Energy Syst
10. Hassan LH, Moghavvemi M, Almurib HAF, Steinmayer O (2013) Current state of neural networks applications in power system monitoring and control. Int J Electr Power Energy Syst
11. Sonker B, Kumar D, Samuel P (2018) Design of two degree of freedom-internal model control configuration for load frequency control using model approximation. Int J Model Simul
12. Sahu RK, Panda S, Padhan S (2015) A novel hybrid gravitational search and pattern search algorithm for load frequency control of nonlinear power system. Appl Soft Comput
13. Mohapatra SK et al (2016) Stability enhancement with SSSC based controller design in presence of non-linaer voltage dependent load. Int J Intell Syst Technol Appl 15(2):163–186
14. Kumar Sahu R, Panda S, Biswal A, Chandra Sekhar GT (2016) Design and analysis of tilt integral derivative controller with filter for load frequency control of multi-area interconnected power systems. ISA Trans (2016)
15. Rashedi E, Nezamabadi-pour H, Saryazdi S (2009) GSA: a gravitational search algorithm. Inf Sci 179:2232–2248
16. Mohapatra SK, Panda S (2014) Coordinated design of power system stabilizer with FACTS based damping control by using gravitational search algorithm. J Bioinf Intell Control, Am Sci Publ 3:13

Coordinated Design of STATCOM Based Power System Stabilizer Using BFOA

Sangram Keshori Mohapatra, M. Madhumita Pal, and Duriya Murali

Abstract In this article, Static Compensator (STATCOM) based power system stabilizer (PSS) coordinate control design is scrutinized. The purposed analysis PID types STATCOM based PSS coordinate controller (STATCOMPSSPID) is compared with LL types STATCOM based PSS coordinate controller (STATCOMPSSLL) in single machine infinite bus and multi machine power system. The optimal coordinately controller parameter are obtained by using BFOA algorithm in both the system. The effectiveness analysis of the systems different loading and different fault condition is applied to scrutinize to shows the power system stability improvement.

Keywords STATCOM · BFOA · LL and PID controller

1 Introduction

The interconnection of power system main advantages to increases security, economy and power system stability but few tie lines operate maximum capacity with increasing the demand of electrical energy [1]. The power system oscillation is increases of the weakly of the interconnected system may occurs the inter area mode of oscillation and low frequency oscillation observed the power system. The effective enhancement of power system stability, the power system stabilizer (PSS) it is widely used in the single and multi-machine power system [2]. The different types of conventional PSS were analyzed the performance of power system stability by changes the parameters of PSS and it increases the transient stability under large changes the loading conditions [3, 4]. The increases the damping of oscillation the Static Synchronous Compensator (STATCOM) one of the supplementary controllers can be used to enhance the power system stability.[5]. The coordinated control of parameters tunning PSS and STATCOM was analyzed a nonlinear programming model [6]. The particle swarm optimization (PSO) based algorithm was based controller design of STATCOM and Optimal location of the transmission line has been formatted as [7],

S. K. Mohapatra · M. Madhumita Pal (✉) · D. Murali
Department of Electrical Engineering, G.C.E., Keonjhar, Odisha, India
e-mail: madhumitapal@gcekjr.ac.in

S. S. Dash et al. (eds.), *Sixth International Conference on Intelligent Computing and Applications*, Advances in Intelligent Systems and Computing 1369,
https://doi.org/10.1007/978-981-16-1335-7_25

the coordinate control of neuro fuzzy controller [8], Genetical algorithm [9], Differential evolution algorithm [10] etc. was presented in the literature. In the present study both lead-lag (LL) and proportional-integral-derivative (PID) used in the coordinated control of STATCOM based controller as it is very simple design and low cost with industry accepted controller. The different type of input signal with coordinated control of STATCOM based damping controller using heuristic algorithm employed to tunning the controller parameters for the improving the power stability [11]. In the present analysis of this paper is the coordinated design of input signal of $\Delta\omega$ based PSS with STATCOM based $\Delta\omega$-based lead lag damping controller is considered in SMIB power system. In the same system coordinated control of PID controller has been attempt for the transient analysis. The optimal controller parameters for both the cases can be obtained by applying the BFOA optimization techniques. The effectiveness and robustness analysis can be check by taking multi machine power system in the same way as per demonstrated in the SMIB power system.

2 The Proposed Approach

2.1 System Modelling with SMIB

The SMIB power system with PSS is installed in the generator excitation system and shunt FACTS device (STATCOM) is connected at bus-2as shown in Fig. 1 of single line diagram in the proposed study [11]. The simulation modeling can be done by considering parameters are given in Appendix.

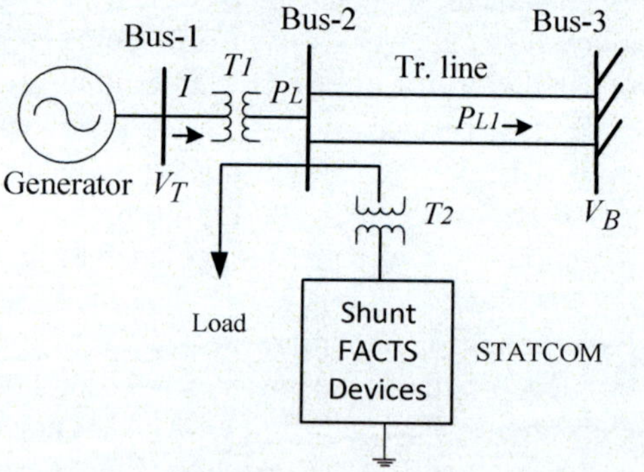

Fig. 1 System modelling in SMIB and STATCOM

3 Design of STATCOM-Based Controller

3.1 Structure of the STATCOM Controller

Figures 2 and 3 shows the STATCOM based PID and STATCOM based LL controller respectively. The STATCOM-PID control comprise of gain blocks of gains K_p, K_i and K_d. In STATCOM-LL controller structure comprises of a gain block with gain K_S, a signal filter and two-stage phase compensation block. The signal filter act as a high-pass filter, with the time constant T_W in the present analysis taken wash out time constant $T_W = 10$second in the proposed controller.

3.2 Power System Stabilizer

Figure 4. shows the block representation of power system stabilizer (PSS) which comprises gain block with gain K_{PS}, washout block act as high pass filter and two stage lead-lag block which compensate phase lag among synchronous generator output torque and input excitation. The input to the PSS is taken remote signal (speed deviation) and output of PSS is the additional voltage of input excitation.

Fig. 2 STATCOM based PID controller

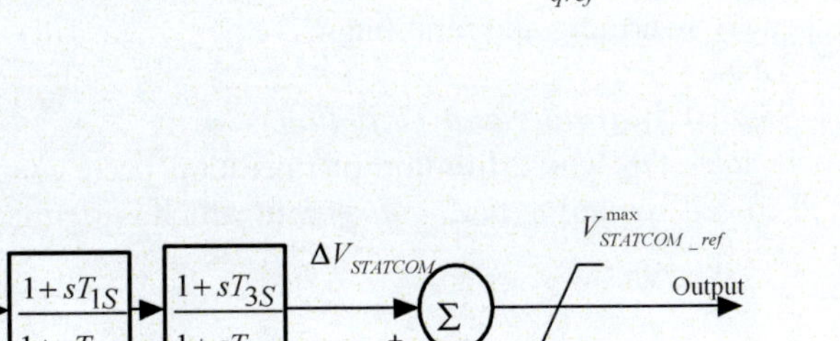

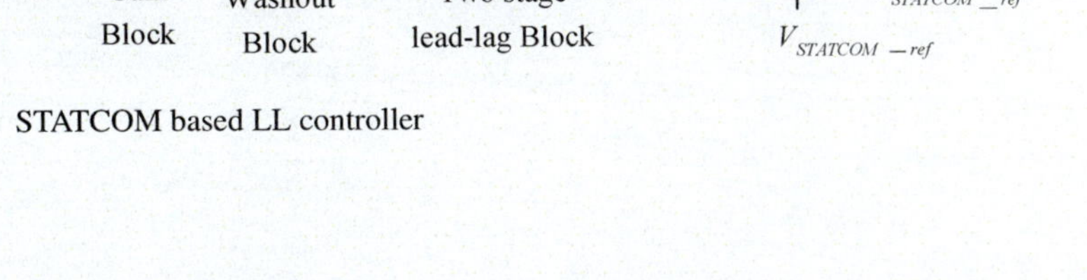

Fig. 3 STATCOM based LL controller

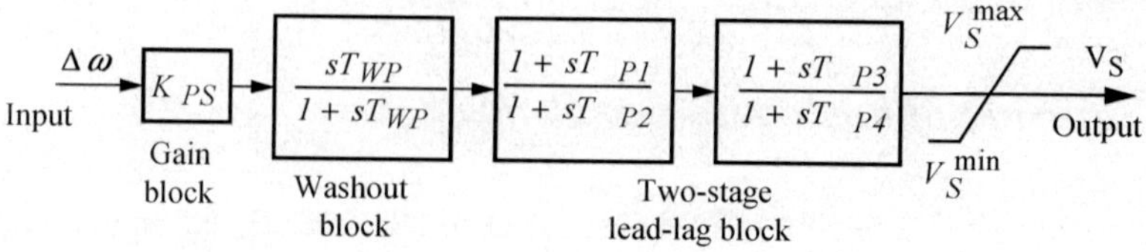

Fig. 4 Power system stabilizer

3.3 *Objective Problem*

The objective function (J) for SMIB power system with speed deviation ($\Delta\omega$) can be expressed as in Eq. (1)

$$J = \int_{t=0}^{t=t_{\text{sim}}} \mid \Delta\omega \mid \cdot t \cdot \mathrm{d}t \tag{1}$$

where t_{sim} is the time range of simulation.

Equation (1) is the objective which is minimize the value for both LL and PID controller of STATCOM controller subjected to.

$K_p^{\min} \le K_p \le K_p^{\max}$, $K_i^{\min} \le K_i \le K_i^{\max}$ (For STATCOM-PID Controller)

$K_S^{\min} \le K_S \le K_S^{\max}$, $T_{1S}^{\min} \le T_{1S} \le T_{1S}^{\max}$, $T_{2S}^{\min} \le T_{2S} \le T_{2S}^{\max}$, $T_{3S}^{\min} \le T_{3S} \le T_{3S}^{\max}$, $T_{4S}^{\min} \le T_{4S} \le T_{4S}^{\max}$ (For STATCOM-LL Controller)

$K_{PS}^{\min} \le K_{PS} \le K_{PS}^{\max}$, $T_{1P}^{\min} \le T_{1P} \le T_{1P}^{\max}$, $T_{2P}^{\min} \le T_{2P} \le T_{2P}^{\max}$, $T_{3P}^{\min} \le T_{3P} \le T_{3P}^{\max}$, $T_{4P}^{\min} \le T_{4P} \le T_{4P}^{\max}$(For PSS)

4 Bacteria Foraging Optimization

The Bacteria Foraging Optimization algorithm (BFOA) is a nature inspired optimization algorithm as anticipated by passion[xx]. This algorithm consists of four steps as taken (i) Swarming and tumbling (ii) Chemotaxis (iii) reproduction (iv) elimination (v) Dispersal.

Step-(i) Swarming and Tumbling.

In this step whole lifetime of operation bacterium moves two modes and the cell-to-cell signalling in *E. coli* swarm which is signified by:

$$J_C(\alpha, P(j,k,l)) = \sum_{i=1}^{N} J_C(\alpha, \alpha^i(j,k,l))$$

$$= \sum_{i=1}^{N}\left[-d_{\text{at}}\exp\left(-w_{\text{at}}\sum_{m=1}^{P}(\alpha_m - \alpha_m^i)^2\right)\right] + \sum_{i=1}^{N}\left[h_{\text{re}}\exp\left(-w_{\text{re}}\sum_{m=1}^{P}(\alpha_m -)\alpha_m^i)^2\right)\right] \quad (2)$$

where, $J_C(\alpha, P(j,k,l))$ is the objective function value, N is the total number of bacteria, P is the number of variables to be optimized, $\alpha = [\alpha_1, \alpha_2, \ldots \alpha_P]^T$ is a point in the P-dimensional search domain and $d_{\text{at}}, w_{\text{at}}, h_{\text{re}}, w_{\text{re}}$ are different attractant and repellent coefficients.

Step –(ii): Chemotaxis (N_C).

In the chemotaxis, computational chemotaxis the drive the step, suppose specified by the tumble. It is characterised by:

$$\alpha^i(j+1,k,l) = \alpha^i(j,k,l) + S(i)\frac{\delta(i)}{\sqrt{\delta^T(i)\delta(i)}} \quad (3)$$

where δ designates a vector varies between -1 to 1 in the random direction And $\alpha^i(j,k)$ characterises *i*-th bacterium at *j*-th chemotactic, *k*-th reproductive and *l*-th elimination-dispersal step. *S*(*i*)is represented the size of the step.

Step-(iii): Reproduction (N_{re}).

In step-(iii), the healthier bacteria give the best value of objective function where as the less healthy bacteria ultimately die and the new one is sited the identical location subjected to size of swarm constant.

Step-(iv): Elimination and dispersal (N_{ed}).

In step-(iv) elimination and dispersal action may abolish chemotactic process and to simulate this process in BFOA few bacteria are liquidated at random with a less probability while the new proxies are randomly initialized over the search space.

5 Simulation Result and Analysis

5.1 Comparison of Δω-Based Input PSS with LL and PID with STATCOM Controller by Taking SMIB Power System

The coordinated control of $\Delta\omega$ input to the PSS with STATCOM based $\Delta\omega$-based LL controller is compared with $\Delta\omega$ based PSS with STATCOM based PID controller has been analysed in SMIB power system. The tunning controller parameters both the cases can be find out by applying the BFOA techniques Table 1 is showing the tunning the coordinated controller parameters of LL/PID based STATCOM based damping controller

To check the performance of transient analysis in the coordinated PSS with STATCOM damping controller by applying the 3-phase disturbance between the intermediate of one transmission line joining between bus 2 and 3 at time ($t = 1.0$ s), the disturbance is detached by opening the defective lines are reclosed after 3 cycles in three different loading condition (nominal loading Pe = 0.8 pu, $\delta_0 = 51.50^\circ$, light loading Pe = 0.5 pu, $\delta_0 = 28.56^\circ$, heavy loading Pe = 1 pu, $\delta_0 = 59.8^\circ$). In all the Figures, the responses without control are shown with dotted line with legend NC; and responses for $\Delta\omega$ based input signal PSS with STATCOM based PID controllers is shown case-1 and $\Delta\omega$ based PSS with STATCOM LL controllers is shown case-2 respectively. Figures 5, 6 and 7 show the transient response of coordinated PSS with LL and PID STATCOM based damping controller under three loading condition respectively. This is shows that coordinated control of PSS with STATCOM under nominal loading the case-2 is better than case-1 in term of transient response.

Table 1 Coordinated tunning parameters of LL/PID controller

Signal/parameters	K_S	T_{1S}	T_{2S}	T_{3S}	T_{4S}
$\Delta\omega$-based STATCOM-LL	0.121	1.001	1.934	1.003	0.3003
Signal/parameters	K_{PS}	T_{1P}	T_{2P}	T_{3P}	T_{4P}
$\Delta\omega$-based PSS	16.201	0.410	1.981	0.104	2.002
Signal/parameters	K_P		K_i		K_d
$\Delta\omega$-based STATCOM-PID	84.616		92.782		14.345
Signal/parameters	K_{PS}	T_{1P}	T_{2P}	T_{3P}	T_{4P}
$\Delta\omega$-based PSS	9.713	0.429	0.674	0.341	0.992

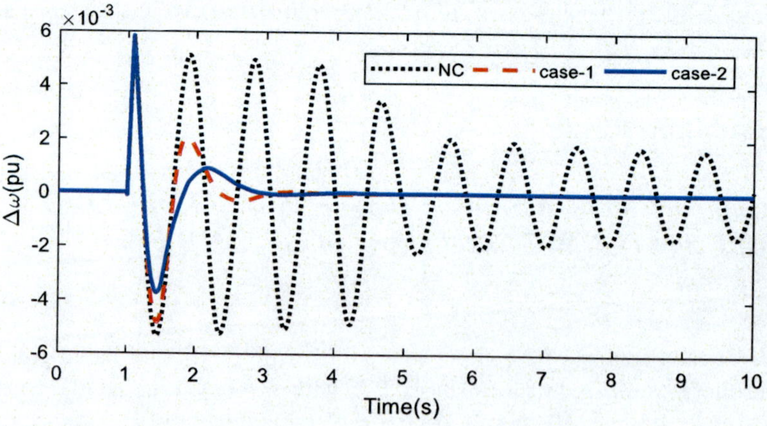

Fig. 5 Speed response for nominal loading case

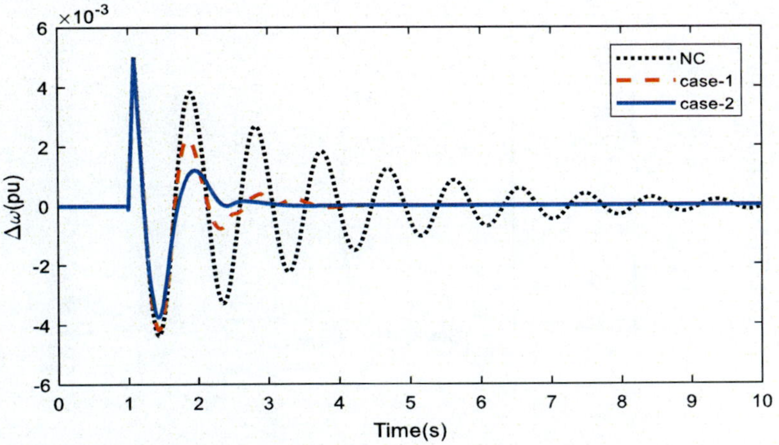

Fig. 6 Speed response under light loading case

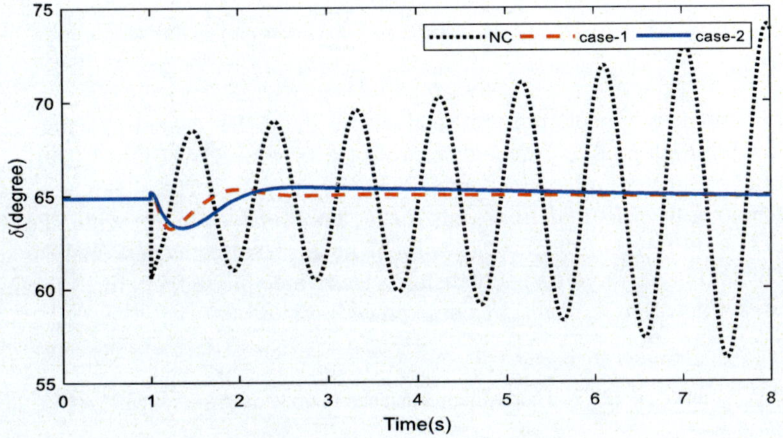

Fig. 7 Speed response for heavy loading case

5.2 Extension of Three Machine Power System

Now for the verification and usefulness of anticipated work can be further demonstrate in three machine power system. The shunt connected STATCOM is connected at bus-5 to inject the current in the transmission line. In this system three PSS is installed in each synchronous generator to enhance the excitation system and transient analysis of the coordinated control of the three generators with STSTCOM based LL/PID control can be verified in three machine power system (Fig. 8).

The fitness function (J) with inter area ($\Delta\omega_I$) and local mode ($\Delta\omega_L$) of speed deviation is defined as:

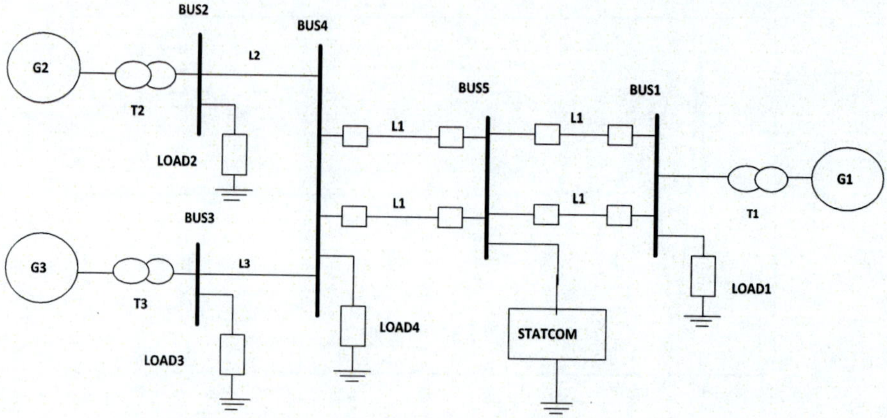

Fig. 8 Three machine power system with STATCOM

$$J = \int_{t=0}^{t=t_{\text{sim}}} \left(\sum |\Delta\omega_L| + \sum |\Delta\omega_I|\right) \cdot t \cdot dt \tag{4}$$

The demonstration of transient analysis as in SMIB system as same way to optimize the tuning of coordinated control of PSS with STATCOM LL/PID based controller parameters value in three machine power system. The global best value of fitness among the 30 simulations runs for the coordinated control with STATCOM based LL and PID controller are shown in Table 2. The effectiveness analysis can be check by taking the three-transient disturbances (three phase fault, line tripping and small disturbance) in the multi machine power system.

Table 2 Optimized coordinated controller parameters in three machine system

Signal/parameters	K_S	T_{1S}	T_{2S}	T_{3S}	T_{4S}
$\Delta\omega$-based STATCOM-LL	68.145	0.592	1.764	0.673	1.842
Signal/parameters	K_{PS}	T_{1P}	T_{2P}	T_{3P}	T_{4P}
$\Delta\omega$-based PSS1	4.962	0.782	0.679	0.9921	1.092
$\Delta\omega$-based PSS2	3.367	0.612	0.914	0.882	0.993
$\Delta\omega$-based PSS3	2.781	0.347	0.891	0.773	1.301
Signal/Parameters	K_p		K_i		K_d
$\Delta\omega$-based STATCOM-PID	66.871		0.444		0.321
Signal/parameters	K_{PS}	T_{1P}	T_{2P}	T_{3P}	T_{4P}
$\Delta\omega$-based PSS1	5.901	0.460	0.881	0.992	1.123
$\Delta\omega$-based PSS2	5.222	1.280	0.762	0.988	0.884
$\Delta\omega$-based PSS3	4.891	0.985	0.553	0.782	0.782

The 3-phase disturbance is used near at bus-1 at time $t = 1$ s and it is removed after 3 cycle, under this condition system speed deviation response between the generator G1 and G2 as shown in Fig. 9. Similarly, the transmission line between bus 1and 6 is tripped at time ($t = 1.0$ s) and after 3 cycles came to original position. The performance of transient response as shown in Fig. 10 and the speed deviation response between the generator G2 and G3 under small disturbance to the load at bus 4 is disconnected for 100 ms. It is clear that variation of speed among the generator under these three cases shows that under no control system is highly oscillating but in case of coordinated control with STATCOM-LL controller is better performance as compared with PID controller of the same system in term of settling time and overshoot (Fig. 11).

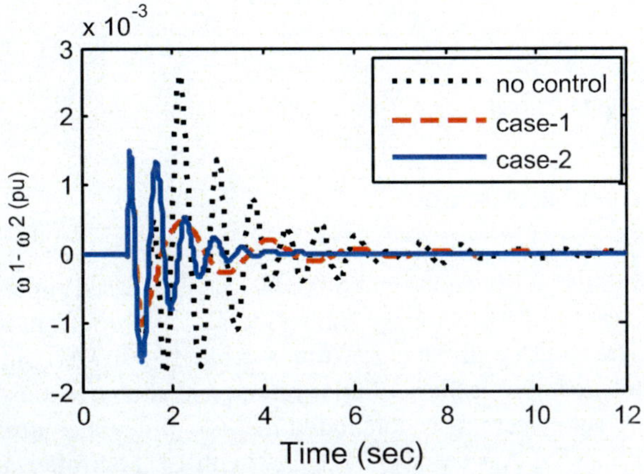

Fig. 9 Speed variation (G1 and G2) for 3-phase fault

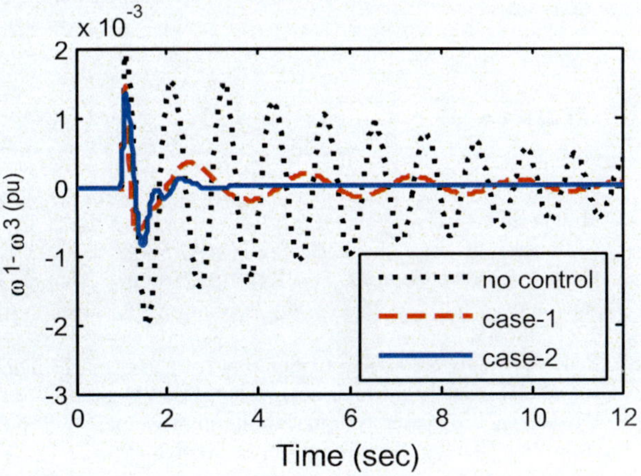

Fig. 10 Speed variation (G1 and G3) response under line trip

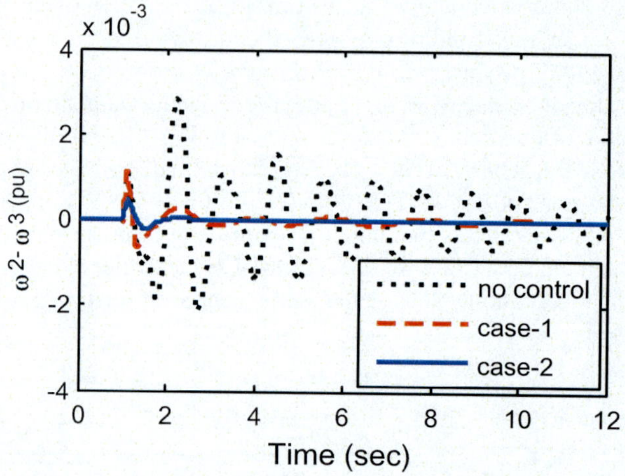

Fig. 11 Speed variation (G2 and G3) for small disturbance

6 Conclusion

In this study, coordinated control of $\Delta\omega$ based input signal PSS with STATCOM based LL/PID damping controller is analysed for improvement of transient stability. The BFOA optimal algorithm is applied to find out the coordinated tunning parameters to improve the settling time and overshoot responses in SMIB and three machine power system. It is concluded that coordinated control of $\Delta\omega$ based input signal to PSS with $\Delta\omega$ based STATCOM with LL controller gives is better performance as compared with $\Delta\omega$ based PSS with $\Delta\omega$ based STATCOM with PID controller in SMIB and multi machine under different fault of the proposed systems.

References

1. Kundur P (1994) Power system stability and control. Mc-Grall Hill, New York
2. Hingorani NG, Gyugyi L (2000) Understanding FACTS: concepts and technology of flexible ac transmission systems. IEEE Press, New York
3. Gyugyi L, Schauder CD, Sen KK (1997) Static synchronous series compensator: a solid-state approach to the series compensation of transmission lines. IEEE Trans Power Delivery 12(1):406–417
4. Panda S, Padhy NP (2008) Comparison of particle swarm optimization and genetic algorithm for FACTS based controller design. Appl Soft Comput 8:1418–1427
5. Panda S (2008) Power system stability improvement by PSO optimized SSSC-based damping controller. Electr Power Comput Syst 36:468–490
6. Abido MA (2005) Analysis and assessment of STATCOM-based damping stabilizers for power system stability enhancement. Int J Electr Power Energy Syst 73:177–185
7. Panda S (2009) Multi-objective evolutionary algorithm for SSSC-based controller design. Electr Power Syst Res 79:937–944

8. Panda S, Padhy NP (2008) Optimal location and controller design of STATCOM using particle swarm optimization. J Franklin Inst 345:166–181
9. Panda S, Narendra KY, Mohapatra SK (2013) Hybrid BFOA-PSO approach for coordinated design of PSS and SSSC-based controller considering time delays. Int J Electr Power Energy Syst 49:221–233
10. Panda S (2009) Multi-objective evolutionary algorithm for SSSC-based controller design. Electr Power Syst Res 79:937–44
11. Mohapatra SK, Panda S, Satpathy PK (2013) Power system stability improvement by simultaneous tuning of PSS and SVC based damping controllers employing differential evolution algorithm. Int J Electr Comput Eng 6(9):1081–1089
12. Mohapatra SK, Panada S (2014) Coordinated design of power system stabilizer with facts based damping control by using gravitational search algorithm. J. Bioinf Intell Cont 3(4):1–13

Design and Implementation of Fuzzy-Based MPPT Technique for PV Panel

A. Geetha, S. Usha, T. M. Thamizh Thentral, and C. Subramani

Abstract In recent year, solar system with battery charging technology takes a lead in renewable energy developments. This article highlights the effect of charging battery via bidirectional power converter and also to the load. A simulation is developed for a solar tracker for estimating the system parameters like current and voltage under maximum power point tracking model of the photovoltaic system. The maximum power available from the sun is tracked using fuzzy-based control system. Under critical loading condition, when the sun's energy is not sufficient to supply the power, the stored energy in the storage system (battery) can supply to the load. The results obtained in the simulation are also validated by a hardware setup.

Keywords MPPT · Photovoltaic · FLC · CUK converter · Bidirectional converter

1 Introduction

With an increased utilization coal, petrol, diesel, etc., the level of pollution has also been increasing which is gradually resulting in global warming [1]. Moreover, these energy resourced are available in limited quantity on the earth, so one utilized, they take thousands of years for their production. As a result, mostly every country in the world is now focusing on finding the alternate of these energy resources. Solar energy and wind energy are some of such sources [2]. The advantage of using renewable resources is that using them causes almost zero pollution and they can be harnessed every time without damaging the earth and they are available in abundant quantity.

A. Geetha (✉) · S. Usha · T. M. Thamizh Thentral · C. Subramani
Department of Electrical and Electronics Engineering, SRM Institute of Science and Technology, Kattankulathur, Tamilnadu 603203, India

S. Usha
e-mail: ushas@srmist.edu.in

T. M. Thamizh Thentral
e-mail: thamizht@srmist.edu.in

S. S. Dash et al. (eds.), *Sixth International Conference on Intelligent Computing and Applications*, Advances in Intelligent Systems and Computing 1369,
https://doi.org/10.1007/978-981-16-1335-7_26

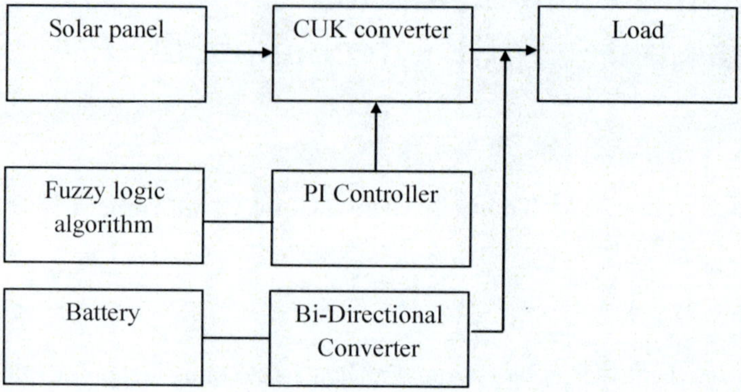

Fig. 1 Schematic diagram of PV charge control system

Among other resources available, the solar energy has been proven to be the most reliable and cheapest form of energy as compared to others [3, 4].

There are various conditions that affect the solar power generated by the panel. Factors such as the weather conditions affect the power generated [5, 6]. The sunlight incident in the panel should have photons having energy greater than the threshold energy, then only the incident photons would be capable of generating electricity from the sunlight. Not only has the weather conditions but also the type of materials used for making the solar panel also affect the electrical powered generated [7]. There are various types of silicon sheets that are used for solar panel constructions depending upon their use and cost. Similarly, they give the output and have definite lifetime [8–10]. Also, for enhancing the solar panel output of solar panel, MPPT which can tracks the maximum power for energy conversion is utilized. A typical system with solar panel, MPPT, and battery used for power generation is known as standalone PV system [11, 12]. In the proposed model as in Fig. 1, the battery is connected in parallel to the solar panel via a bidirectional DC/DC power converter. Hence, the battery current can easily be controlled.

In this paper, battery connected in parallel to the conventional system. The main wok of battery here is to store charge when sufficient amount of sunlight is available. Under normal conditions, when the power generates by the solar panel is enough to drive the load, the panel will feed the supply to load and surplus power generated will be used to charge the battery. This charged battery can further be used to drive the load under abnormal weather conditions. The output of fuzzy-based controller is later fed to PI controller to generate gate pulses of IGBT of CUK converter. Depending upon the gate pulse only, the CUK will work. Also, it is easy to extract power from it and requires less investment when compared to other renewable sources of energy. There are two physical inputs considered for the solar panel, namely irradiance (W/m^2) and temperature (°C).

From the equivalent circuit shown above Fig. 2, output current is,

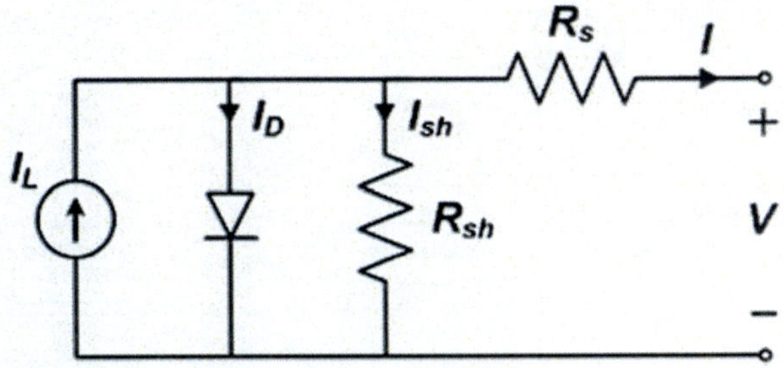

Fig. 2 Equivalent circuit of solar panel

$$I = I_L - I_o\left(e\frac{q(V + IR_s)}{kT} - 1\right) - \frac{V + IR_s}{R_{sh}}$$

q—Charge of electron, k—Boltzman constant, T—Temperature of cell, I_D—Diode current, R_{sh}—Shunt resistance, I_{sh}—Current through shunt resistance, R_s—Series resistance, I_L—Light generated current, I—Output current.

2 Fuzzy Logic-Based Maximum Power Point

A specific arrangement of some physical components such designed so that it alters the behavior of another physical system and exhibits desired resulting characteristic is called a control system. The use of fuzzy controller is for providing the operating point to the PV system for its peak to obtain maximum power output. Advantages are that the response to sudden changes in the atmosphere is best in the case of fuzzy logic controller (FLC) along with it being fast and robust in characteristic.

FLC in this article consists of two inputs and one output: Temperature and irradiance are the two inputs while the output acts as an input for a PI controller that gives a duty cycle pulse basically used to control the switching characteristic of the MOSFET in CUK converter. Fuzzy logic consists of the following components: Fuzzification, fuzzy Knowledge and rule base, fuzzy inference system and de-fuzzification (Fig. 3).

2.1 *Fuzzification*

The two inputs are fuzzified to form three linguistic variables namely: Low, medium, and high. The Gaussian membership functions for the two fuzzy variables are shown in Figs. 4 and 5. Figure 6 denotes the output membership function.

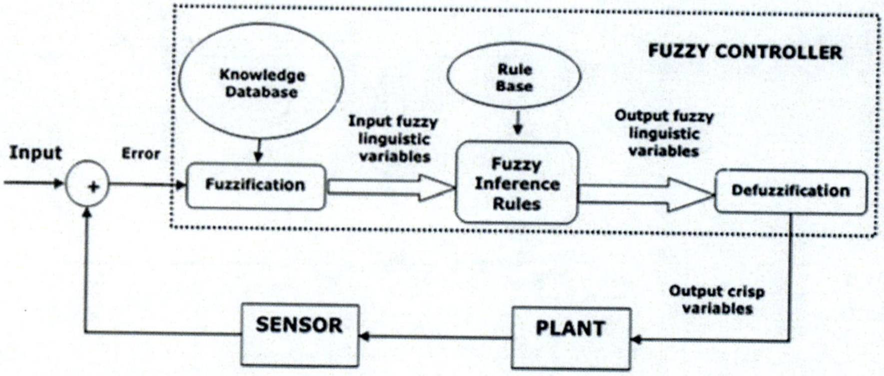

Fig. 3 Fuzzy logic components

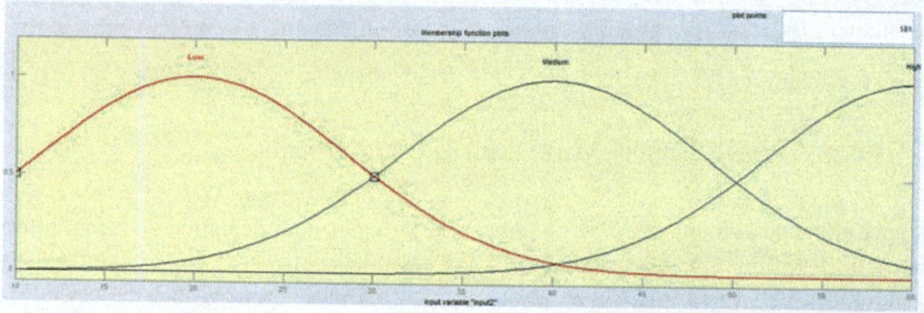

Fig. 4 Membership function plot of temperature

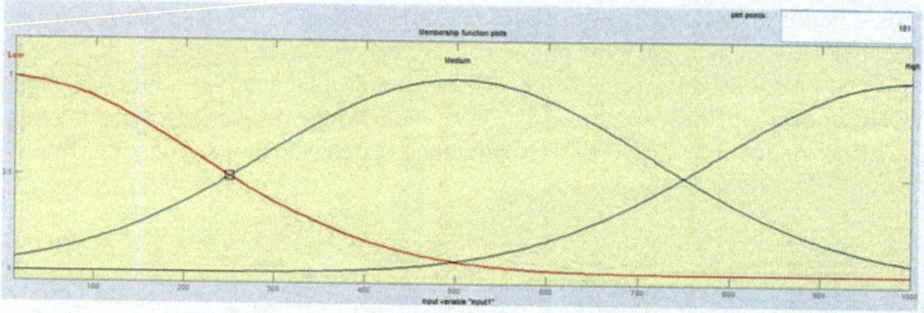

Fig. 5 Membership function plot of temperature error

2.2 *Fuzzy Rule Base and Inference Method*

Based on the three fuzzy variables for both the inputs, a set of If-Then rules are formed to dictate the output of the fuzzy logic for each input. It is shown in the following Table 1. The rule base follows Mamdani implication rules.

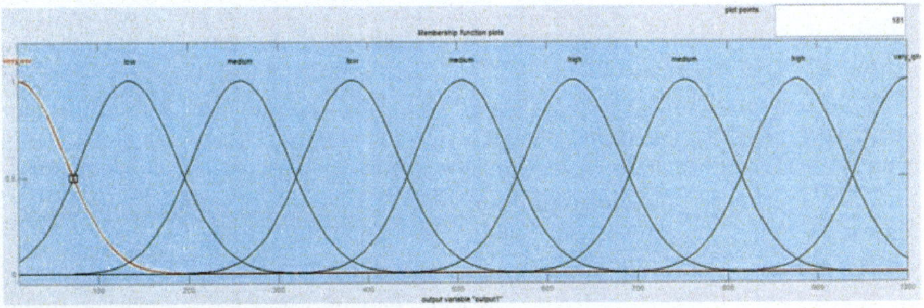

Fig. 6 Membership function plot of output

Table 1 Rule base of FLC

Temperature	Irradiance		
	Low	Medium	High
Low	Very low	Low	Medium
Medium	Low	Medium	High
High	Medium	High	Very high

2.3 De-Fuzzification

The output from the If-Then rule base of the fuzzy logic controller is crisp in nature. This is called de-fuzzification. The error signal obtained from the difference of this crisp output from PV power output is supplied to the PI controller. This controller basically improves the output of the fuzzy logic controller. The output from PI controller is utilized for generating the firing pulses that control the switching of the MOSFET in the CUK converter.

2.4 Power Converter

A controller which has output voltage magnitude greater than or less the input voltage with inverted output is known as CUK converter which is shown in Fig. 7. It is essentially a boost converter followed by a buck converter coupled by a capacitor to

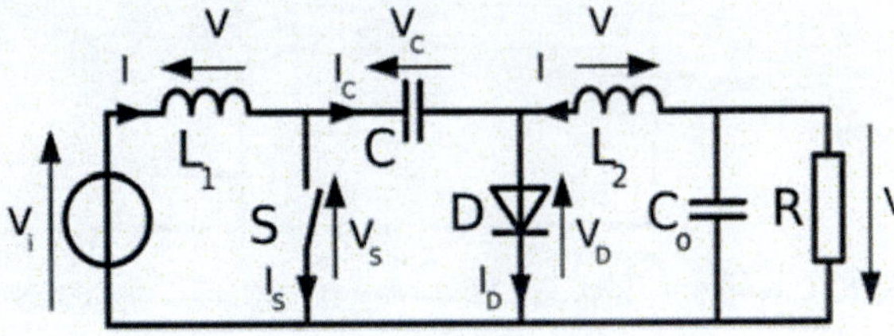

Fig. 7 Circuit diagram of CUK converter

store energy. It has an inverted output, i.e., the output of the converter has opposite polarity as compared to the input signal. The capacitor is the main energy storage device unlike the other types of converters which use an inductor as their primary source of energy storage. The CUK converter has minimum EMI radiations and is subject to less ripple as there is continuous flow of power through the capacitor. The CUK converter allows the electricity to flow in two directions (bidirection) using a diode and a switch.

Since the converter is an inverting one, the output has the negative polarity as compared to supply. The capacitor "C" is meant for energy transfer which is internally connected. The two inductors (L1, L2) are utilized for converting the input and output voltage into the current sources respectively.

2.4.1 Bidirectional DC–DC Converter

An ordinary DC–DC converter has been known for its operation in only one direction. But nowadays, with advancement in electrical field and applications of converters, a new type of converter has been introduced. These are called bidirectional DC–DC converters. These are responsible for flow of power in both the directions. They enable the flow of power from both source to load as well as load to source. In places where solar panels and wind turbines have been installed to generate power, various fluctuations are there at the output side.

2.4.2 Bidirectional Buck-Boost Converter

It is a non-isolated bbidirectional converter. The bbidirectional buck-boost converter given in Fig. 8 is an anti-parallel combining both buck and boost operation. During

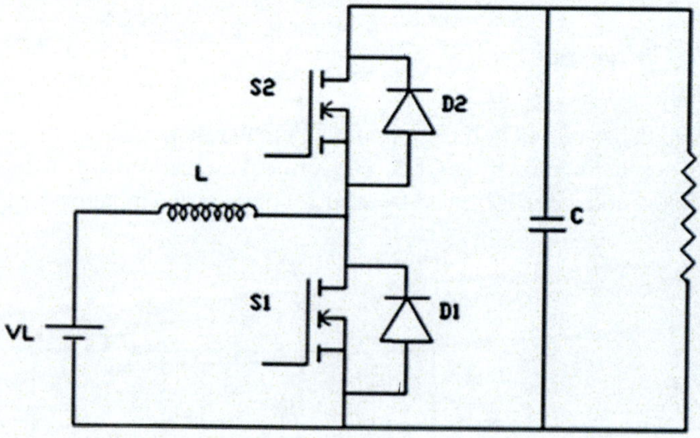

Fig. 8 Circuit of bidirectional buck-boost converter

step-up operation, S2 is in conduction mode, whereas S1 remains in cut off mode. During the step-down mode, S1 conducts, whereas S2 remains switched off. Both S1 and S2 operate according to the duty cycle. A small dead time is provided between the operation of S1 and S2.

3 Simulink Model

The model as shown in Fig. 9 comprises of PV array connected to MPP tracker (fuzzy logic), Bidirectional buck-boost converter for battery control and CUK converter. The boosted voltage of is required in order to facilitate charging of battery. The duty cycle is applied for driving the switch of the CUK converter. Switching converters are more efficient in terms of power. The circuit diagram of CUK converter is implemented in the Simulink model.

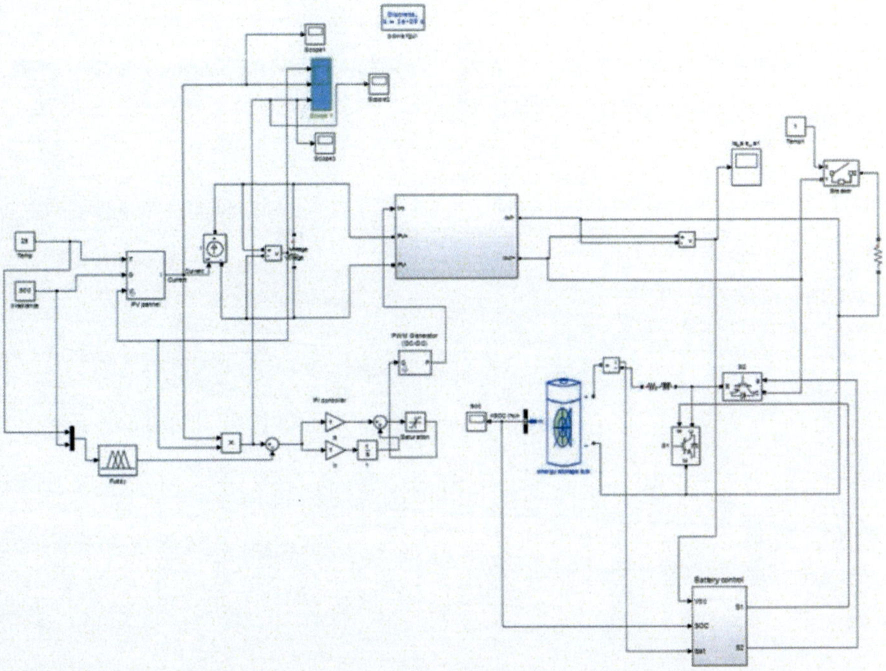

Fig. 9 Simulation of solar charge controller model

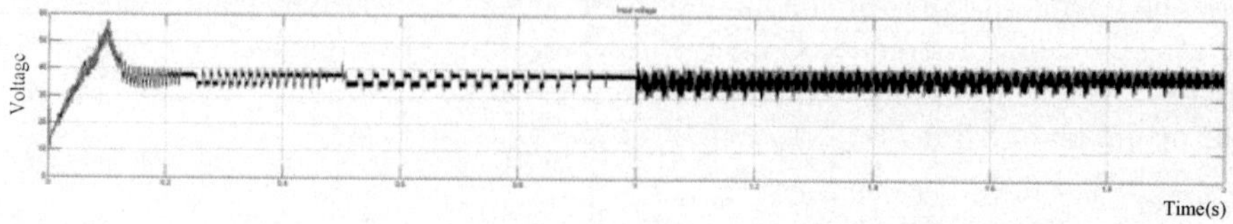

Fig. 10 *V-t* curve obtained for the solar panel at 500 W/m^2 irradiance and 25 °C temperature

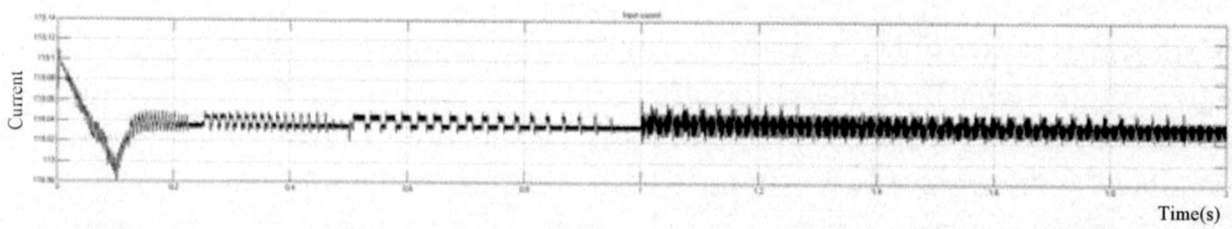

Fig. 11 *I-t* curve obtained for the solar panel at 500 W/m^2 irradiance and 25 °C temperature

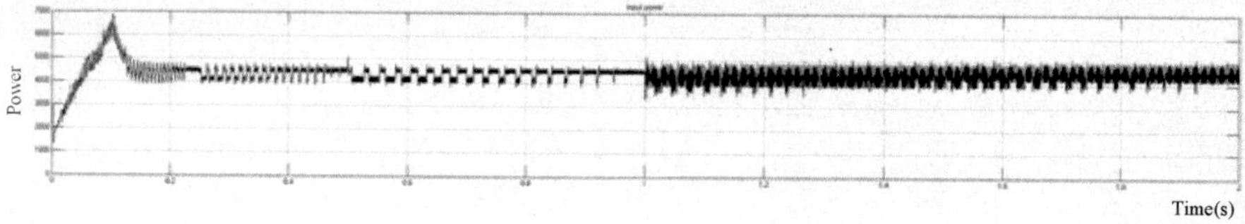

Fig. 12 *P-t* curve obtained for the solar panel at 500 W/m^2 irradiance and 25 °C temperature

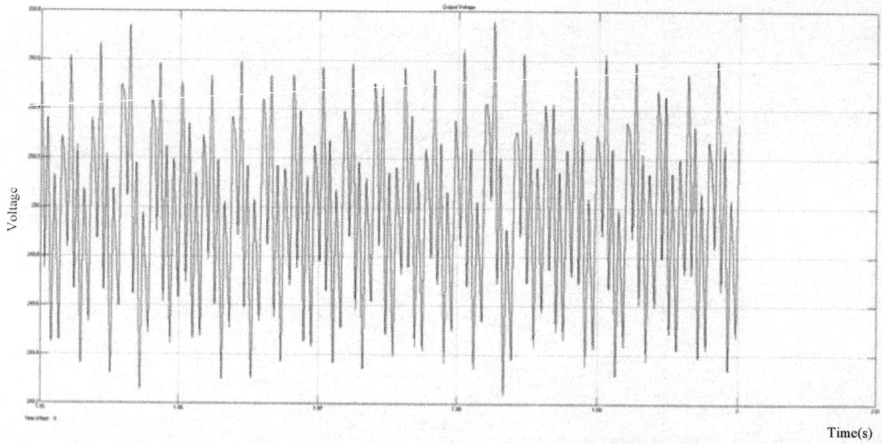

Fig. 13 Output voltage curve obtained at the load side

3.1 Results and Discussions

Given Figs. 10, 11, 12 and 13 are the voltage-time, current-time and power-time waveforms respectively, of the solar panel designed using MATLAB simulation obtained at initial SOC = 60 and Irradiance = 500 W/m^2 and Temperature = 25 °C.

The distortion that occurs in the output voltage curve is due to the charging and discharging cycle of the battery that takes place simultaneously during the supply of power from source to load side.

The hardware setup and the corresponding outputs are shown in Figs. 14 and 15. The 12 V two tap transformers are utilized for converting 231 V supply into

Fig. 14 Solar output to CUK converter

Fig. 15 CUK converter output

12 V output. One of the transformers is used to run the PIC microcontroller circuit which is connected with other primary components that convert ac to dc. The other transformer is used to run the driver circuit. The driver circuit is used to convert 5 V pulses of micro controller to 15 V of the gate pulses for triggering the gate of MOSFETs of CUK as well as bidirectional converter. These optocouplers also act as isolation between microcontroller and gate pulse.

The solar panel converts the light energy into usable electrical energy. The PV panel is connected to the CUK converter which boosts the output of the solar panel and gives the input to the load. A battery is connected in parallel to the solar panel using a bidirectional converter. The bidirectional converter acts as path between the CUK-Load-Battery. Also, MPPT utilizing FLC is utilized for tracking the maximum point available. The codes of design for fuzzy logic are done in the PIC microcontroller. The battery is used to drive the load when sufficient amount of sunlight is not available to drive the load. Also, the battery is charged simultaneously using solar panel when there is excess of energy available from the sun. The load used here is a 1 KΩ resistor.

4 Conclusion

By using the MPPT algorithm, we have ensured that we are able to extract the maximum usable power from the PV system. The factors on which the output of a PV system relies on, like ambient cell temperature and irradiance, are variable, and hence, the operating point is rarely at its peak output power. The CUK converter is used to transform the impedance of the source circuit to the load circuit. By altering the duty ratio (T_{on}/T), the change in impedance is adopted by the PV system. At some specific impedance, the operating point reaches the peak power transfer point. But because of the dynamically changing conditions, it is not possible to fix the duty cycle of panel. It is so observed that at lesser temperatures and higher values of irradiances, maximum output power can be obtained from the PV system. So, the solar panels can give a maximum output in winters during the hours when the sun is at its peak. Using the fuzzy algorithm, it is managed to utilize the maximum power from the panel successfully under different weather conditions. Hence, the maximum power was successfully transferred from the source circuit to the load circuit.

References

1. Wang C, Chen W, Shao S, Chen Z, Zhu B, Li H (2011) Energy management of stand-alone hybrid PV system. Energy Procedia 12:471–479
2. Rokonuzzaman M, Mishu M, Hossam-E-Haider M, Islam MS (2017) Design of a MPPT solar charge-controller in Matlab-Simulink GUI environment. In: International conference on mechanical engineering and applied science (ICMEAS), Dhaka, Bangladesh, Feb 22–23
3. Koutroulis E, Kalaitzakis K (2004) Novel battery charging regulation system for photovoltaic applications. IEE Proc Electr Power Appl 151(2):191–197

4. Nedumgatt JJ, Jayakrishnan KB, Umashankar S, Vijayakumar D (2011) Perturb and observe MPPT algorithm for solar PV systems-modeling and simulation. In: IEEE December annual India conference, Hyderabad, India, Dec 16–18
5. Khaehintung N, Sirisuk P (2004) Implementation of maximum power point tracking using fuzzy logic controller for solar-powered lightflasher applications. In: Proceedings of the 47th IEEE international midwest symposium on circuits and systems, Japan, pp III-171–174
6. Won C-Y, Kim D-H, Kim S-C, Kim W-S, Kim H-S (1994) A new maximum power point tracker of photovoltaic arrays using fuzzy controller. In: Proceedings of 1994 power electronics specialist conference—PESC'94, Taipei, Taiwan, pp 396–403
7. Khera N et al (2015) Design of charge controller for solar PV systems. In: 2015 International conference on control, instrumentation, communication and computational technologies (ICCICCT), Kumaracoil, pp 149–153
8. Lin B (1993) Analysis of fuzzy control method applied to DC–DC converter control. In: Proceedings eighth annual applied power electronics conference and exposition, San Diego, CA, USA, pp 22–28
9. Takum P, Kaitwanidvilai S, Jettasen C (2011) Maximum power point tracking using fuzzy logic control for photovoltaic systems. In: Proceedings of international multiconference of engineers and computer scientists, Hongkong, March 16–18
10. Cheik MS, Larbes C, Kebir GFT, Zerguerras A (2007) Maximum power point tracking using a fuzzy logic control scheme. Rev Energies Renouvelables 387–395
11. Vasantharaj S, Vinodhkumar G, Sasikumar M (2012) Development of a fuzzy logic based photovoltaic maximum power point tracking control system using boost converter. In: IET Chennai 3rd international conference on sustainable energy and intelligent systems, Tiruchengode, India, Dec 27–29
12. Ramakumar R, Bigger JE (1993) Photovoltaic systems. Proc IEEE 81:365–377

Design and Implementation of Smart Classroom Using Microcontrollers and Advanced Sensors

T. M. Thamizh Thentral, K. Vijayakumar, A. Geetha, and S. Usha

Abstract This paper focuses on smart class rooms in college by installing small Internet-connected devices and various sensors in each classroom. Arduino is used as the brain of the whole controlling system. The input of the Arduino is connected to the multiple sensors which are fitted across the classroom. Relay, an electromechanical switch, is connected to the appliances of the classroom. This relay module receives signal from the Arduino and operates the physical appliances. Sensory units will sense the students and automatically control the switching on and off of fans and lights, opening and closing of doors using RFID scanners and automatically update attendance. This project aims to simplify life and conserve electricity, which is one of the most demanded resources in scarcity, and make a secure and safe environment.

Keywords RFID · Piezoelectric · Smart classroom · Image processing · Node microcontroller unit

1 Introduction

In this century, giving and grasping knowledge in the most efficient manner is necessary. Also, conserving energy is important. But the reality is that this is not achieved and a lot of energy is wasted during teaching in classrooms. The main objective of our project is to bring a transformation in the conventional classroom, making it more secure and effective. This is done with the help of a central controlling unit which is branched into sub-units. Each sub-unit consists of sensory units, communication unit, and controlling unit. The primary goal of the project is to provide a system that offers detection of phone usage, smart doors and power saving through both conventional and non-conventional methods. The implementation of the system in educational institutions will have a great impact on students and also will enhance the teaching and learning process. The system strives in minimizing the physical

T. M. Thamizh Thentral (✉) · K. Vijayakumar · A. Geetha · S. Usha
Department of Electrical and Electronics Engineering, Faculty of Engineering and Technology, SRM Institute of Science and Technology, Kattankulathur 603203, India

S. S. Dash et al. (eds.), *Sixth International Conference on Intelligent Computing and Applications*, Advances in Intelligent Systems and Computing 1369,
https://doi.org/10.1007/978-981-16-1335-7_27

efforts with the help of digital technology. Making classrooms smart would be great step in the process of making India digitalized.

2 Components Used

2.1 *Piezoelectric Sensor*

Piezoelectric sensors work on the piezoelectric effect. This effect involves a source of mechanical stress upon the sensor due to which electric charge accumulates inside it. The accumulated charge then acts as a voltage source. This piezoelectric effect is a result of the interaction of the mechanical and electrical states. They have a wide variety of applications ranging from medical, nuclear, and aerospace industries to quality assurance and process control in various industries [1]. Piezoelectric sensors have sufficiently high modulus of elasticity ranging up to 106 N/m^2 making them withstand high amounts of stress. They also show negligible deflection and do not react to compression. Also, they are insensitive to electromagnetic radiations enabling operations under harsh conditions. Some of the major disadvantages of piezoelectric are, it has high temperature sensitivity and is used for dynamic measurement only. Piezoelectric crystals are water soluble and get dissolve in high humid environment.

2.2 *Node Microcontroller Unit (Node MCU)*

Node MCU, an open-source platform module which runs on ESP8266 WiFi system on a chip (SOC), is also based on ESP-12 module. It refers firmware rather than referring development kits. Node MCU is a single board microcontroller-based type having XTOS as an operating system. Some specifications of Node MCU includes it memory of 128 kB, storage capacity of 4 MB, and power source through USB. Node MCU is much powerful and compact comparing Arduino allowing inbuilt WiFi function. Comparing with arduino ATmega328, it has more powerful processor, 4 MB of memory, and more RAM. It is much cheaper than Raspberry Pi and also used for tracking device interactions. This Node MCU does not compile the application rather interprets it while running. This reduces the free space available. When every time there is a need to create free space, the system has to be reset. When the space gets full, it might stop responding to commands. Besides having less space, sometimes the restart system also ceases to start.

2.3 Servomotor

Here, the servomotor used is a rotary actuator. It consists of four components, namely a DC motor, control, gearing set, and a potentiometer which acts as a position sensor. The small DC motor is coupled with a sensory feedback. The feedback is the major difference between a servomotor and a stepper motor. To operate the feedback, a suitable controller is used in this closed-loop mechanism. Servomotor has a unique design due to which they have high torque and are fast. They show very accurate rotation in a small range of angle. They are a perfect alternative to stepper motors with higher performance but a little more complicated setup due to the use of PWM tuning. Major disadvantage of a servomotor is its tuning. Tuning a servomotor is a very difficult process as it requires tuning to stabilize the feedback loop. It also requires safety circuits as it will become unpredictable when something breaks. Its peak torque is limited to 1% duty cycle and can be damaged by sustained overload. Feedback components that are necessarily to be used with servomotor make it more costly than stepper motor.

2.4 Arduino

Arduino UNO is an open-source microcontroller board which is based on microchip ATmega328 microcontroller having number of digital and analog input and output pins. The board consists of 14 digital and 6 analog pins labeled from A0 to A5 and uses one 'type B' USB cable. Each of the digital and analog pins uses functions like pinMode(), digitalWrite(), and digitalRead() to operate at the voltage of 5 V. Each pin can send or receive signals at a maximum rate of 20 mA. One major disadvantage is that if the rate of transmission exceeds 40 mA, then permanent damage may be caused to the microcontroller. The ATmega328 provides 5 V for serial communication with laptop, Arduino board or any other microcontroller. The transmission of any signal is communicated through digital pins 0 and 1, i.e., RX and TX. The LEDs of pins RX and TX will flash when any data is being transmitted through USB to serial chip which is connected to any computer. This serial communication to any pin is provided through a software serial library. Arduino is very versatile as it is compatible with different interfaces, i.e., it can run on different operating systems like Linux, Macintosh OSX, Windows operating systems. It is also user-friendly and is flexible enough for advanced programming. It is an open source for both extensible software and hardware. Arduino uses C or C++ with a few small extensions and is not computational efficiency. A lot of the libraries of Arduino are not enhanced well and a lot of improvement is needed. Language of Arduino has been ported to a lot of microcontroller but still mere knowledge of Arduino would never leverage proper hardware-level expertise which is required very much as embedded engineer/enthusiast. And finally, Arduino strips your hardware-level access barring you from many functions.

2.5 *Radio Frequency Identification (RFID)*

As the name suggests, radio frequency identification (RFID) uses radio signals to identify the tags attached to objects. The tags contain digital data in them. There are two types of tags. Passive tags are the ones which receive energy from any nearby RFID reader which at that time is searching for radio waves in its surroundings. Active tags are the ones in which there exists a local power source such as a battery and because of which they are able to operate hundreds of meters away from the RFID reader, unlike a barcode which needs to be in the line of sight of the reader. RFID is a method for automatic identification and data capture (AIDC) in the line of sight of the reader. Radio frequency identification (RFID) being an automatic ID system like a barcode or magnetic strip provides a unique identification code for scanning devices. It uses radio waves for communication, and these waves get converted to digital data for the identification of the object that contains the tag. Scanning range of RFID is limited within its frequency range. It can also scan multiple items simultaneously within a few milliseconds. RFID systems are more expensive than barcode systems and have technology harder to understand. It is less reliable than barcodes and has usually larger RFID tags than barcode labels. All tags are specific, i.e., no one tag fits all. It has possibility of unauthorized reading of passport, and also many tags can respond at the similar time.

3 Calculation for Energy Savings

The total power consumed per day and working days in month in a classroom is calculated to design prototype. In general case, the classroom is assumed with nine fluorescent lamps and seven fans. Total power consumed by all the fans and lights is calculated and the same is tabulated in Table 1. The total working hours in a day are nine, in which six hours are allocated as lecture hours in a classroom. Two hours are assumed as buffer hours. The reason for the buffer hours is someone may used to sits in the classroom during their free time and switche 'ON' all the lights and fans. Remaining 1 h is assumed as all equipment are in 'off' condition.

Table 1 Power consumed by all equipment

Equipment	Rating
Load-fluorescent lamp	40 W
Load-fan	70 W
Power consumed by all fans in one hour	$70 \times 7 = 490$ Wh
Power consumed by all fluorescent lamps in one hour	$40 \times 9 = 360$ Wh
Total power consumed by all equipment in one hour	$490 + 360 = 850$ Wh

Table 2 Power consumed in a month for 9 h per day

Total power consumed by all equipment in one day	850 × 8 = 6800 Watt-day
Total power consumed in a month (22 working days)	6800 × 22 = 149.6 kW
Cost of electricity per kW for educational institutions in Tamil Nadu	Rs. 6.35
Total cost of electricity consumption in one month per classroom	149.6 kW * 6.35 = Rs. 950

Table 3 Power consumption in buffer hours

Power consumed by 2 lights in one hour	40 × 2 = 80 Wh
Power consumed by 2 fans in one hour	70 × 2 = 140 Wh
Power consumed in one hour	80 + 140 = 220 Wh
Total power consumed in two buffer hours	220 × 2 = 440 W
Total power consumed in a month (22 days)	440 × 22 = 9.68 kW

The total cost based on the power consumed by all the equipment is evaluated and the same is shown in Table 2 by considering total number of working days in a month as twenty-two.

In the second case to monitor the power consumption for the smart classroom in a buffer hour, two fans and two lights are taken by assuming minimal number of students in a classroom. Total power consumption for two hours is listed in Table 3. As per the Tamilnadu, cost of electricity per kilo watt for educational institutions is Rs. 6.35. Total cost of electricity consumption in one month according to two buffer hours is 9.68 kW multiplied by Rs. 6.35 is Rs. 61.

The six hours lecture period is taken as two cases. In the first case, out of six hours, all the equipment is working for three hours with hundred percent student strength. In the second case, for the remaining three hours with based on the student strength, the number fans will be controlled automatically. So that four fans and six lights are considered. With the smart classroom design, wherever the student is not present, the equipment is going to off condition, so that the power can be saved. For the two cases A and B, total power consumed and the cost for the electricity consumption are shown in Table 4.

The power consumption and the electricity bill for the same are calculated for one classroom. Then the energy consumed and the total cost will increase when number of classrooms and number of months increased. The same is calculated for nine months with twenty classrooms. It shows that with the smart classroom system, energy consumption and total electricity bill cost are reduced. It is tabulated in Table 5.

Table 4 Power consumption per six hours

Equipment	Case A	Case B
Power consumed in one hour	850 W	240 + 280 = 520 Wh
Number of hours in this scenario	3	3
Total power consumed in a day	850 × 3 = 2.55 kW	520 × 3 = 1.56 kW
Total number of working days in a month	22 days	22 days
Total power consumed in a month	2.55 × 22 = 56.1 kW	1.56 × 22 = 34.3 kW
Cost of electricity per kW for educational institutions in Tamil Nadu	Rs. 6.35/kW	Rs. 6.35/kW
Total cost of electricity consumption in one month	56.1 kW × 6.35 = Rs. 356	34.3 kW × 6.35 = Rs. 218

Table 5 Power consumption in nine months for twenty classrooms

Total saving in one month for one classroom	Rs. 950 − 635 = Rs. 315
Number of working months per calendar year	9
Total saving in 9 months for one classroom	Rs. 315 × 9 = Rs. 2835
Assumed number of classrooms in one building	20
Total saving in 9 months for 20 classrooms	Rs. 2835 × 20 = Rs. 56,700/year

4 Methodology

4.1 Energy Conservation Piezoelectric Sensor

The lighting is controlled based on the number of students present in the classroom to save energy costs. The lights and fans do not work when the class is empty. And if students are sitting in a particular section of the classroom, then the appliances only above that particular section work. Rest all appliances remain switched off and thus result in energy conservation throughout the day. This process is carried out in following manner. Here, piezoelectric sensors are used which are sensitive pressure sensors. The chairs in the room are equipped with these sensors which respond to the occupancy. The presence of a student is detected by the piezoelectric sensor, and then,it sends the signal to Arduino and the corresponding appliances gets switched on. The circuit diagram for the same is shown in Fig. 1 along with its hardware circuit (Figs. 2 and 3).

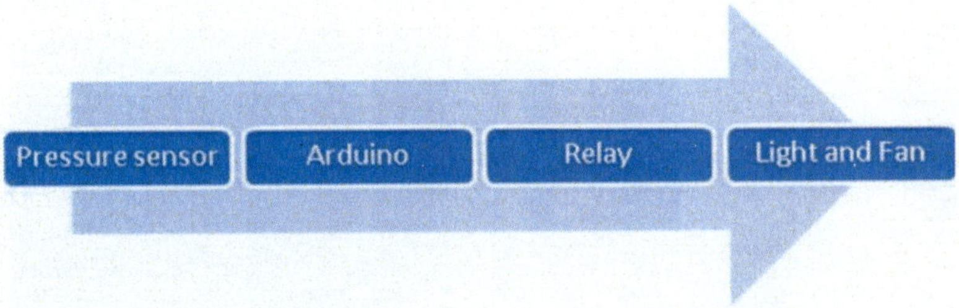

Fig. 1 Flow diagram

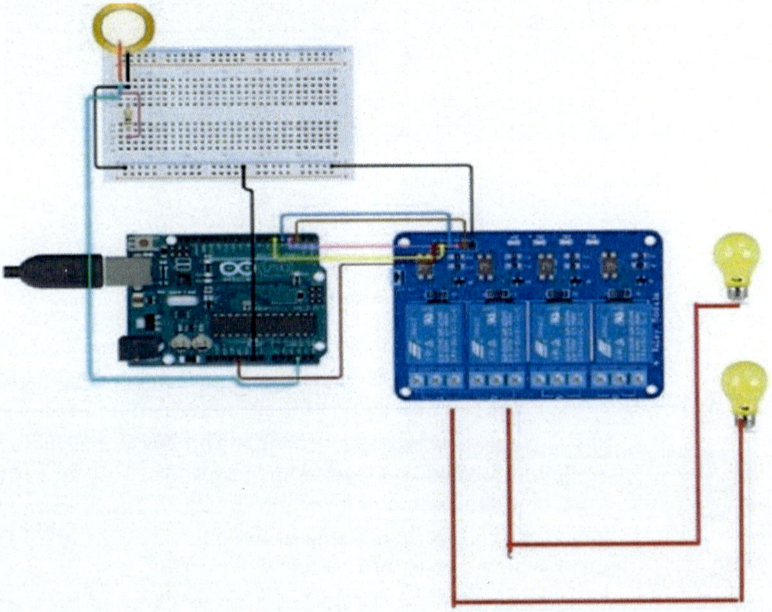

Fig. 2 Circuit diagram

4.2 Energy Conservation Image Processing

Image processing is a process of capturing the data, processing it, compress it, store it, print it, and accordingly displaying that image. Images should be taken using digital camera. Camera should be perfectly placed covering the entire classroom. In this, the camera captured real-time images with some intervals of time, and reference image is taken only once in a time for setting. The images after being captured are run in the main program using certain algorithms [2]. The real-time image which is captured is a colored one. The image is then split into many cells, and each cell is simply the area covered.

The image processing code consists of two parts:

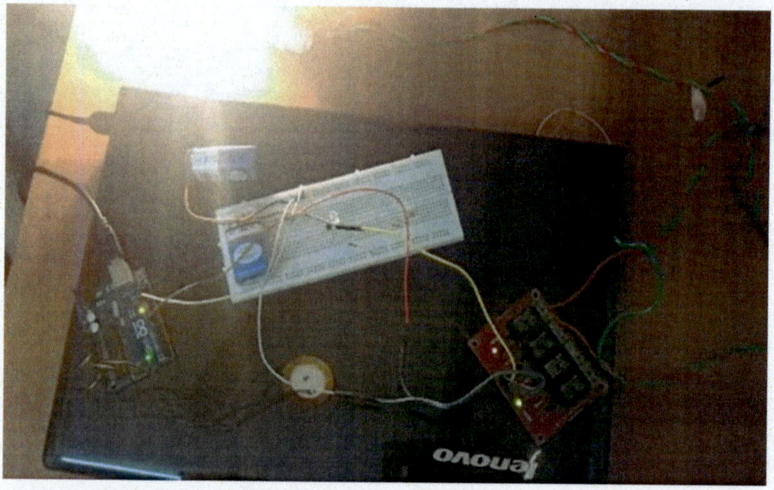

Fig. 3 Hardware circuit

- Face recognition
- Sending the coordinates of detected face to Arduino.

The first face recognition part consists of taking the image from camera input which may or may not consist face. The detected image or camera image checked for face using HAARCASCADE file which gives the coordinates of upper left corner of face in (x, y, w, h) where x, y are the top left corner coordinates and w, h are width and height to make a rectangle for the detected face. Now, the x, y of face is tested to check the region were face is situated. The whole image input is divided into four planes 1, 2, 3, 4 and face detected in any one of the plane changes the value of coordNumber[i] array to 1 which consist only 0 initially [3] (Fig. 4).

For sending the coordinates, the coordNumber[i] array is checked for 0 or 1. If coordNumber[i] has value 1, then using serial writing method '1' is sent to Arduino, and if coordNumber[i] is 0, then '0' is sent to Arduino. In Arduino, the receiving serial input is stored in an array a[i] of size 4. Each array index is linked with Arduino pin 13, 12, 7, and 8. If array input a[0] is '1,' then 13 pin light switches on, and if it is '0,' then 13 pin light switches off. Similarly, the switching for 12, 8, and 7 pin with a[1], a[2] and a[3] is done. Counter variable is also used to put some delay in output so that there will not be any instant switching (Figs. 5 and 6).

On the human face, the parts which are detected are located at a particular relative distance from each other. These are referred to as nodes. There are a total of five nodes on a human face. When all these nodes are detected, it counts as one face. The image is then converted from RGB form to gray form [2]. The circuit diagram for image processing has been depicted below along with its physica demonstration using the laptop camera which also includes its hardware circuit.

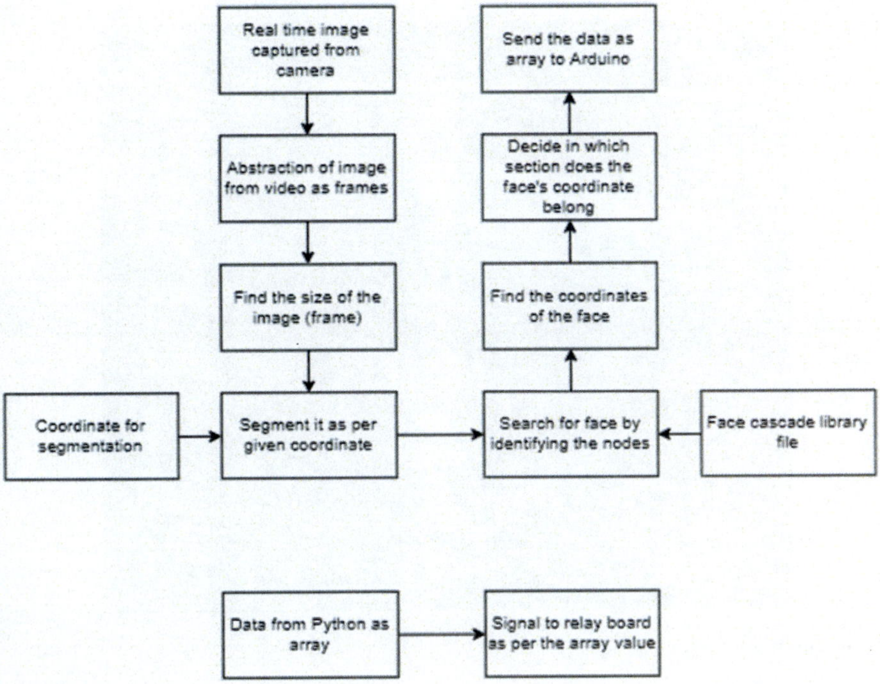

Fig. 4 Flow diagram

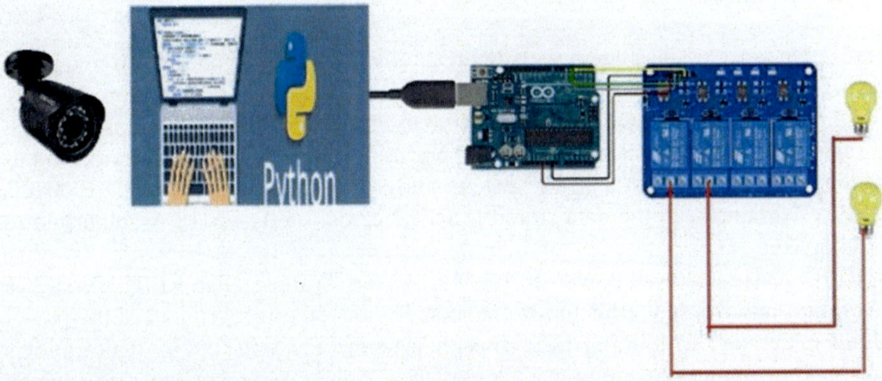

Fig. 5 Circuit diagram

Fig. 6 Hardware circuit

4.3 Auto Update of Attendance Using RFID and Implementation of Smart Door Using Servo Meter

Here, attendance of the student is captured automatically through the radio frequency identification and electromagnetic fields. Every RFID tag has a data holding capacity of 2 kB. This 2 kB of data in RFID can be used to store various information such as registration number, name, post (as in our case can be used to differ a person as student or staff), level of staff or year of study in case of student. RFID-EM18 is used to scan and read the data stored in RFID cards. RFID-EM18 is connected to Arduino (Fig. 7).

RFID-EM18 receives power of 3.3 V from the Arduino. The RFID-EM18 has nine pins, namely VCC (this pin is connected to the positive terminal of the power source in our case Arduino), BUZZ (this pin is connected to the buzzer which allows user to know about the action has taken place), 2 NC pins (no connection is given to this pin), SEL (this is an input pin which is used as a selector for output if SEL = 1 the output is at Rs. 232 and if SEL = 0 then the output is WEIGAND), TX (data pin of Rs. 232), GND, (ground pin is used to power up the IC and generate the electromagnetic wave required to read data from RFID card by completing the circuit).

Once the data is sent to Arduino through SPI interface, it stores the data in an assigned cloud location and sends signal to the servomotor which is used to open or close the door after authentication of the particulars. It is quite often found that

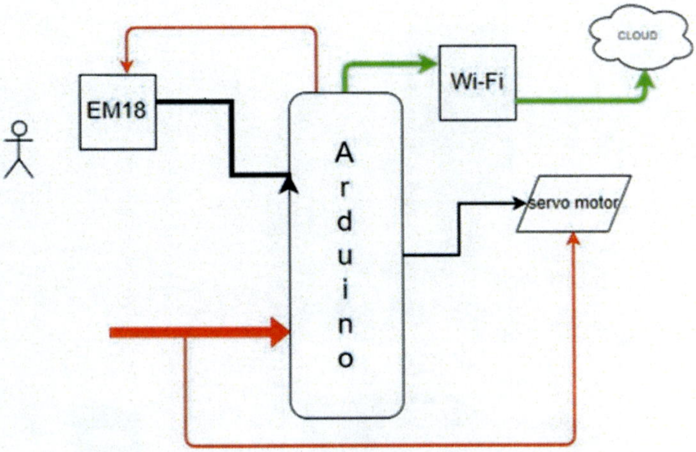

Fig. 7 Flow diagram

students enter class or move out of the class without permission. To avoid this, smart door has been designed which employs opening and closing of the door at stipulated time on scanning the RFID card. Thus, entering and leaving the class have been accounted for by RFID which also helps us to auto-update the attendance making the whole process more systematic. Once the teacher enters the class and scans the RFID card, the reader finds the presence of the teacher and disables the system after a delay of 5 min, so that no one interrupts the class once teaching process has been started [4–7].

Figure 8 shows the circuit diagram involving the servomotor and the RFID card reader. Below the circuit diagram is an image of the hardware circuit involved (Fig. 9).

5 Conclusion

The smart classroom project which includes a number of parts has been proposed. Energy conservation using piezoelectric sensors and image processing, smart doors and auto-update attendance using RFID tags have been demonstrated by implementing the hardware circuit. Also, a rough estimate of cost savings is also calculated if such smart classrooms are adapted in a college building. The huge amount of saved money despite little high initial setting up costs depicts that upon real life implementation, it can be highly effective.

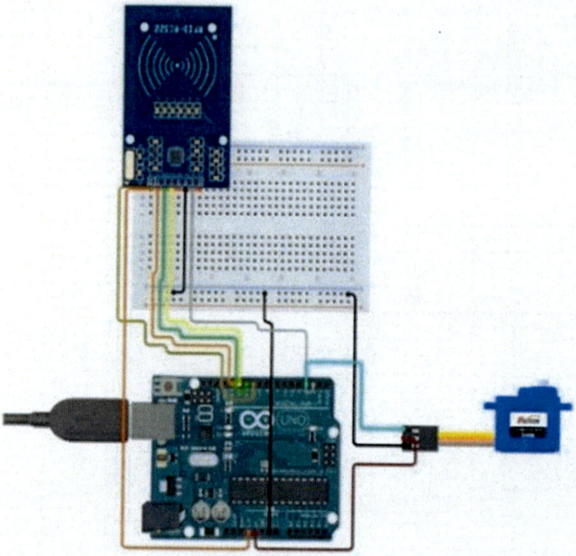

Fig. 8 Circuit diagram

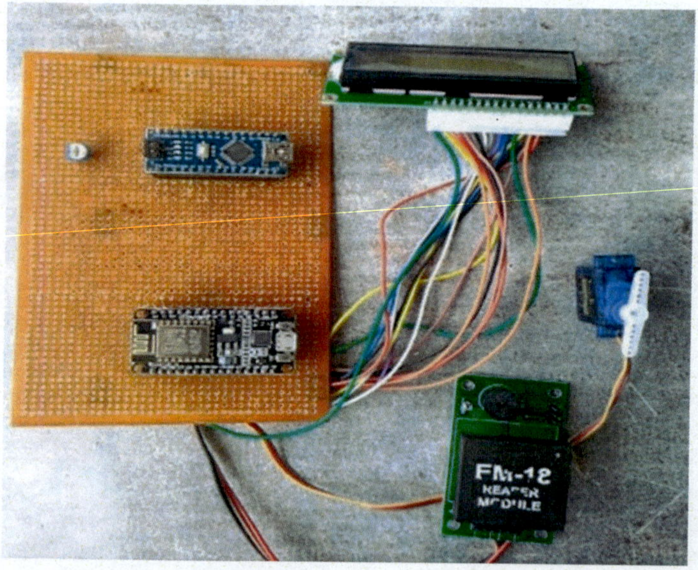

Fig. 9 Hardware circuit

References

1. Tianze L, Xia Z, Chuan J, Luan H Analysis of the characteristics of piezoelectric sensor and research of its application
2. Marques F, Marcotegui B, Zanoguera F, Correia P, Mech R, Wollborn M Partition—based image representation as basis for user-assisted segmentation. IEEE-7803-6297/7/2000
3. Talari S, Shafie-Khah M, Siano P, Loia V, Tommasetti A, Catalão JPS A review of smart cities based on the internet of things concept
4. Lee E-K, Chu P, Gadh R (2013) Fine-grained access to smart building energy resources. IEEE Internet Comput
5. Song S, Zhong X, Li H, Du J, Nie F (2014) Smart classroom: from conceptualization to construction. In: 2014 International conference on intelligent environments. https://doi.org/10.1109/ie.2014.56
6. YueNan Z Based on the Matlab image enhancing technology and implementation. IEEE Smart City—7383995
7. Shah VS, Shah MSA Power saving by using image processing

Performance Analysis of BLDC Motor Using Different Controller

S. Usha, T. M. Thamizh Thentral, A. Geetha, C. Subramani, and Akanksha Lal

Abstract Mainly, the DC motors are employed in most of the application. The main objective is to regulate the DC motor system. A motor which displays the appearances of a DC motor but there is no commutator and brushes is called as brushless DC motor. These motors are widespread to their compensations than other motors in relationships of dependability, sound, efficiency, preliminary torque and longevity. To achieve the operation more reliable and less noisy, brushless dc motors are employed. In the proposed work, dissimilar methods of speed control are analysed. In real-time submission of speed control of BLDC motor, numerous strategies are executed for the speed control singularity. The modified approaches are the employment of PI controller and use of current controller.

Keywords BLDC motor · PI controller · Firing signal generator · Frequency analysis (FFT) · Speed control

1 Introduction

BLDC motors have an outsized number of applications in industries. The only difference between a BLDC motor and a permanent synchronous motor is the back EMF. The back EMF in a permanent synchronous motor is of sinusoidal form, and in a BLDC, machine is of trapezoidal form. Commutation is the process of interchanging the current through the coil which make the motor to spin. The commutation of the currents is done with the help of automatically changes DC to AC which is basically supplied by a DC source. The currents which induced in the rotating part can be failed due to very high resistivity of mutually stainless steel and magnet. Damping winding should be of very less value because due to this, only magnetic flux got decreases [1, 2].

The BLDC motor is run on DC supply rather than AC supply. The DC voltage is converted into three-phase AC voltage, which is then supplied to the motor. AC source

S. Usha (✉) · T. M. Thamizh Thentral · A. Geetha · C. Subramani · A. Lal
Department of Electrical and Electronics Engineering, SRM Institute of Science and Technology, Kattankulathur, India

S. S. Dash et al. (eds.), *Sixth International Conference on Intelligent Computing and Applications*, Advances in Intelligent Systems and Computing 1369,
https://doi.org/10.1007/978-981-16-1335-7_28

can also be used, and then, it can be converted into DC but the main disadvantage of converting AC to DC is that it will cause new problems like THD, poor power factor and poor quality, and that is why BLDC motors are running on DC source and not on AC [3].

The DC source can be directly connected, or it can be given by connecting it across any of the converters like DC–DC converter, SEPIC converter, etc. BLDC fed with diode bridge rectifier has poor power factor. By using different converter techniques, voltage can be controlled and power factor can be improved [4].

Different simulations are performed by researchers to analyse the performance and process of the procedure to get the better speed regulation and higher starting torque of the motor. The researchers are working on tuning PID controllers by performing iterations, trying PSO algorithms, experimenting using Fuzzy logic controllers and also trying the conventional methods [5].

Speed control theatres are necessary role in the current world. The speed control offers many advantages such as higher reliability, higher efficiency, high starting torque and reduced electrical noise. The speed control is of two types: open and closed loop. In this first system, there is no response path, and the desired speed may/may not be obtained. The machine parameters and gate pulses sequences are considered, which needed to be chosen more carefully while designing the system. The closed-loop system is the one in which a feedback loop is always present and the speed control is easy. The desired speed can be obtained using this mechanism [6]. The closed-loop system employed here is the dual closed-loop system.

The comparative study is performed depicting the performance of the machine having standard parameters. This paper focuses to show which of the proposed methods is more reliable, efficient and better in comparison to others [7, 8].

2 Proposed Terminology

2.1 *Reference Speed*

The input to the motor is given by the reference speed. It is the motor speed that is to be obtained.

2.2 *Actual Speed*

It is the speed at which the motor rotates. It is obtained from the machine after running the motor with initial variables. Initially, this speed is obtained on the basis of the machine parameters and the torque given to the motor. The output of the motor in the MATLAB Simulink software is expressed in rps. It is converted into RPM by multiplying it by gain ($k = 30/\text{pi}$).

2.3 Load Torque

The BLDC motor needs an external initial load torque so that the motor can run if the current is not sufficient enough to start the motor. In these simulations, the load torque is given as a constant value.

2.4 Rotor Position

This block gives the information about the location of the rotor. This uses the sensors which send signals corresponding to the position of rotor with signal for each phase. This plays a very important role in triggering of gate pulses of three-phase inverter.

2.5 Controlled Voltage Source

The input to the circuitry is a DC supply. A controlled voltage source is used which has a fixed amplitude but keeps changing with the change in the error speed.

2.6 Three Phase Inverter

It is used to convert the DC supply into the three-phase AC supply, which serves as input to the motor. MOSFET's are used as switching devices. The gate pulses for these switches are obtained from a closed-loop system for the operation of the inverter.

2.7 Decoder

The function of this block is to convert the hall signals into EMF signals corresponding to each phase. The Hall sensors send the signals corresponding to each phase in the motor. The Hall effect signals are given by Ha, Hb and Hc. These signals are converted into EMF signals represented by emf_a, emf_b and emf_c. The decoder can be designed based on this truth table by either using a MATLAB function or by traditional methods of using gates.

2.8 *Firing Signal Generator*

The function of this block is that it provides the gate pulses for the switching of the MOSFET's in the three-phase inverter. This receives the input from the decoder and gives the output as pulses. The number of outputs from this block depends on the number of MOSFET switches present in the inverter. The signals then operate the inverter in either 120° conduction mode or 180° conduction mode, depending on the requirement.

3 Operation Principle

The BLDC motor works on the principle based on the feedback from internal shaft or rotor. Unlike brushed DC motors, where feedback is maintained using commutator and brushes, the brushless DC motor uses certain feedback sensors to attain the position of the rotor [6].

For the three phase BLDC motor, the back EMF is of trapezoidal form, and there is a phase difference of 120°. The model of BLDC motor is illustrated in Fig. 1.

The circuit showing the model of BLDC motor is shown below, and the output waveforms for the back EMF and the output currents are shown Fig. 2.

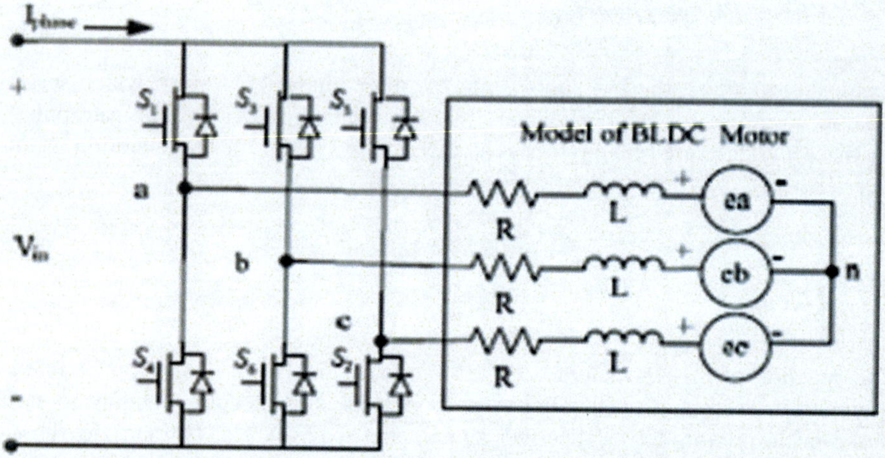

Fig. 1 Model of BLDC motor

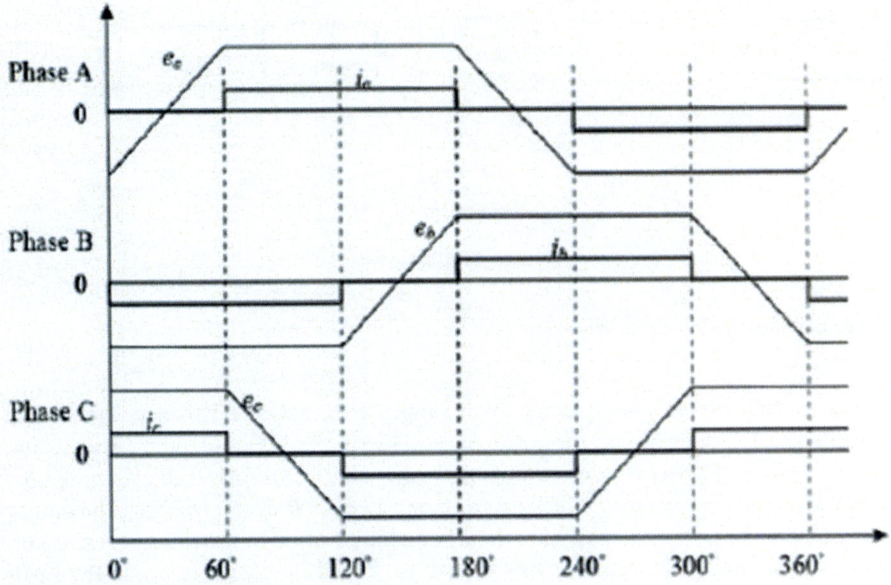

Fig. 2 Output waveforms of back EMF and currents

4 Methodology

The methods secondhand for the speed control of the BLDC motor use the double closed-loop speed control in which the one loop is need to limit the current or govern the current control or for the control of the gate pulses. The other loop is for speed control which is used to get the desired speed as per the requirement. The required speed can be either set as an integer value or can be given as step input. The reference speed is given as a constant value which can be changed based on the requirement, and accordingly, the desired speed can be obtained.

The methods used in this paper are namely speed control using PI controller, using current controller and a CUK converter-based dual closed-loop speed control. These methods are compared based on the values of current total harmonic distortion (THD), output current (RMS value) and on the values of rise time and settling time of the speed response of various methods.

4.1 *PI Control*

This method is a closed loop control in which the PI controller is used for controlling the speed of BLDC motor. In this method, the feedback speed from the motor is compared with the reference speed [7]. The error speed is then passed through the PI controller which then is given directly to the controlled voltage source. The voltage

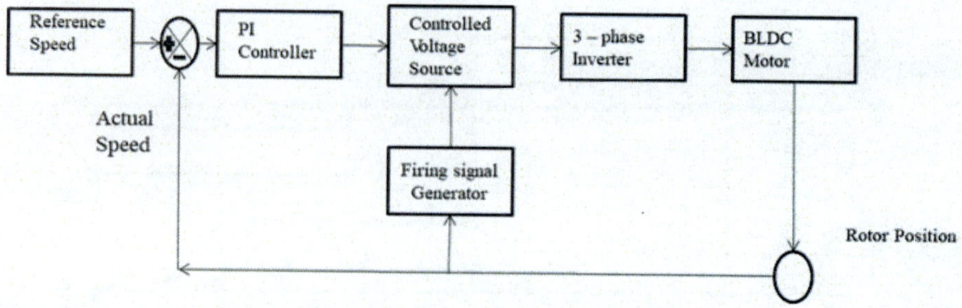

Fig. 3 PI controller implementation

source is a DC voltage source which gives supply to the MOSFET bridge. The bridge converts the DC supply into the three-phase AC supply which is then given to the motor. The motor gives various outputs including stator currents, Hall effect signals, back EMF voltages, electromagnetic torque and speed of the rotor (in rps). The output from the Hall sensors is then used to determine the position of the rotor of the motor. These signals form a closed loop in which these signals are converted into the EMF signals and again are converted to the gate signals using different topologies and are converted to the gate signals based on the number of switches connected in the MOSFET bridge. These signals serve as the gate signals for controlling the voltage which in turn control the input voltage to the motor which in turn helps in controlling the performance of the motor.

The PI controller is need as it helps in eliminating the steady state error. The main disadvantage of the PI controller is that the oscillations cannot be removed, and also, rise time cannot be decreased by use of this controller.

The PI controller gives the output based on the sum of proportional gain and integral gain. Mathematically,

$$\text{Output} = Kpe(t) + Ki \int_{0}^{t} e(t)\mathrm{d}t \tag{1}$$

The block diagram depicting the circuitry of the system is depicted in Fig. 3.

The three-phase inverter operates in 120° conduction mode, which is why at a time only two of the switches conducts.

4.2 Current Control

This method uses a current controller which is used to control the switching states of the MOSFET bridge. This method takes the phasor current from the input supply to the motor and is compared with a reference value the signals are converted in the form of 0's and 1's. The Hall effect signals from the hall sensors which determine

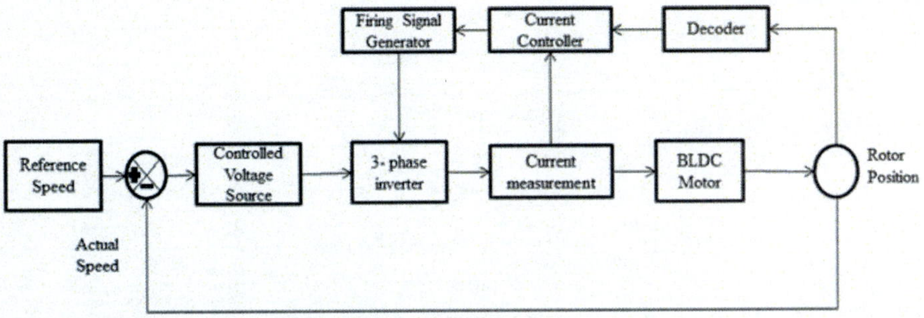

Fig. 4 Block diagram of current controller algorithm

the position of the rotor give the position in the form of 0's and 1's. These signals are converted into EMF signals of the form −1, 0, +1. These signals are obtained based on the decoder truth table. The signals are then compared with the current signals which generate 6 gate pulses for the MOSFET bridge. This method helps in controlling the input supply to the motor, indeed controlling the speed of the motor.

This also uses the dual closed-loop system, one loop for controlling the switching of the MOSFET and the other loop for controlling the speed output. The reference speed and actual speed are compared, and the output signal is given directly to the controlled DC voltage source.

The block diagram depicting this circuitry of this method is as under Fig. 4.

5 Observations and Results

The simulations are carried on a model of BLDC motor. The parameters of the machine are mentioned below with their standard values as shown in Table 1.

The simulations are performed in MATLAB/Simulink software, and the outputs are obtained for different models and are given below.

Frequency analysis is also performed for all the models. It is done in order to know which model is better and which shows less distortion while operating at a certain frequency. This also allows to know which method is advantageous over other in terms of both efficiency and economically.

The hardware implementation of model which is much better than the others is also done. The results are shown below.

5.1 Speed Control Using PI Controller

The speed control is achieved by proportional and integral controller. The stator current, back EMF, speed and electromagnetic torque performance are analysed.

Table 1 Values of machine parameters

S. No.	Parameters	Value
1	Stator phasor resistance (Rs)	0.7 Ω
2	Armature inductance	2.72e−3 H
3	Flux linkage established by magnets	0.105 Wb
4	Voltage constant	87.9646 (V/krpm)
5	Torque constant	0.84 (N/A)
6	Inertia	08e−3 J
7	Viscous damping	1e−3
8	Pole pairs	4
9	Static friction	0
10	Number of phases	3
11	Back EMF	Trapezoidal
12	Mechanical input	Torque
13	Back EMF flat area (degrees)	120°

5.2 *Speed Control Using Current Controller*

The speed control is achieved by current controller. The stator current, back EMF, speed and electromagnetic torque performance are analysed.

The output waveforms for different characteristics are shown above from Figs. 5, 6, 7, 8, 9, 10, 11, 12, 13, 14, 15, 16, 17, 18, 19, 20 and 21. These characteristics of the BLDC motor can help in determining which method is best among these three. The comparison among these three methods is depicted below in a tabular form.

For a reference speed of 1000 RPM, the actual speed is obtained among different control methods. The 5% error is allowed for the error in the value of the actual speed, as the error may occur due to the switching states, friction force in the machine or due to the operation under variable frequency (Table 2).

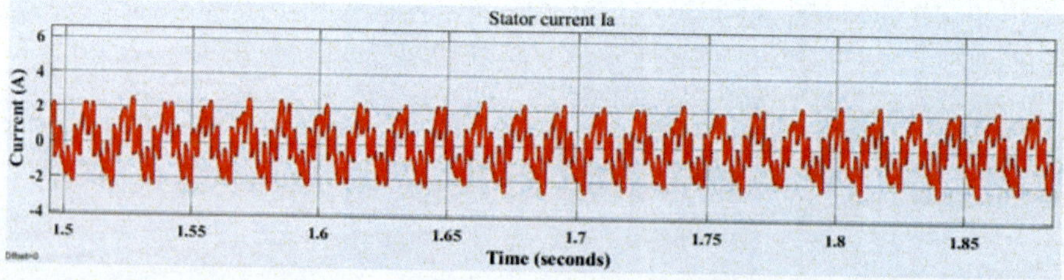

Fig. 5 Output stator current (Ia)

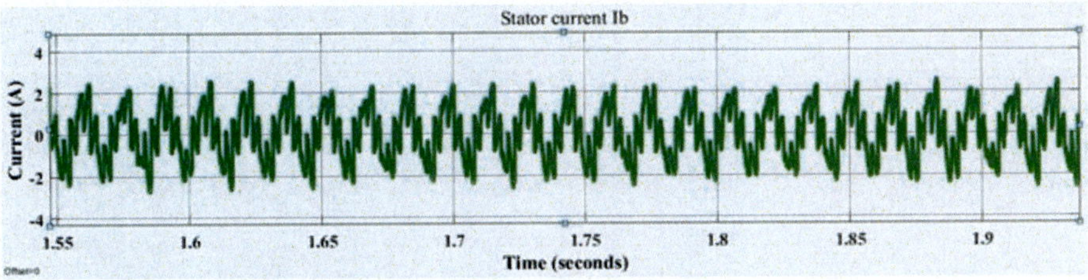

Fig. 6 Output stator current (Ib)

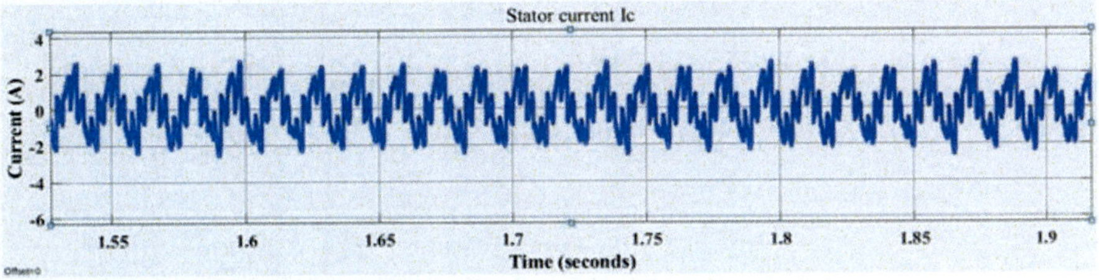

Fig. 7 Output stator current (Ic)

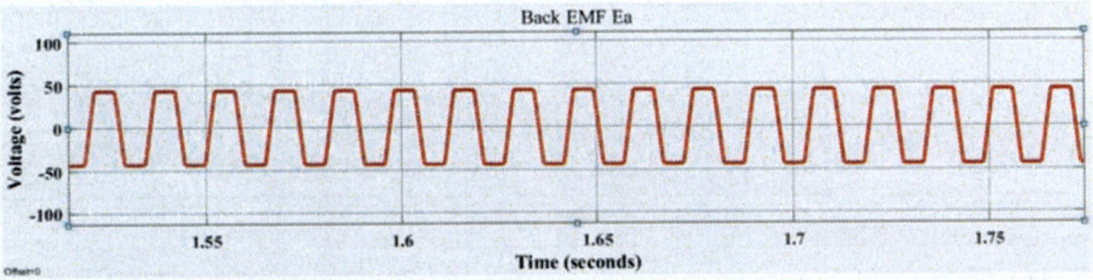

Fig. 8 Output back EMF (Ea)

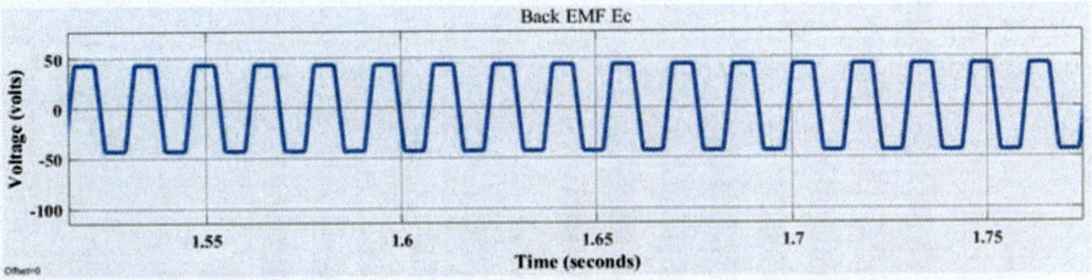

Fig. 9 Output back EMF (Eb)

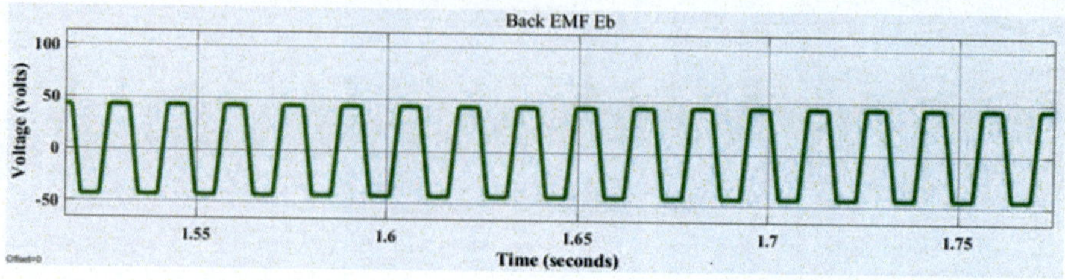

Fig. 10 Output back EMF (Ec)

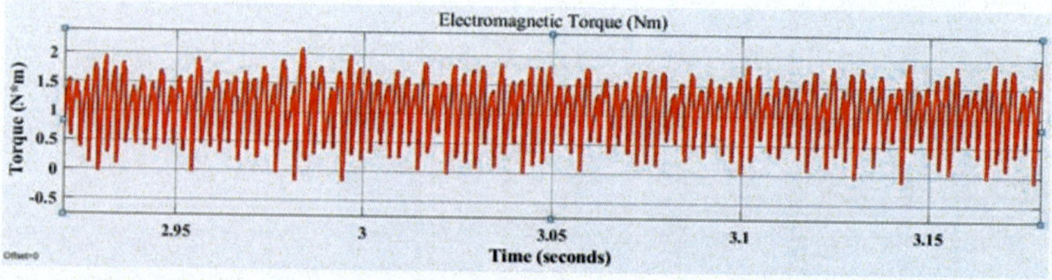

Fig. 11 Output electromagnetic torque

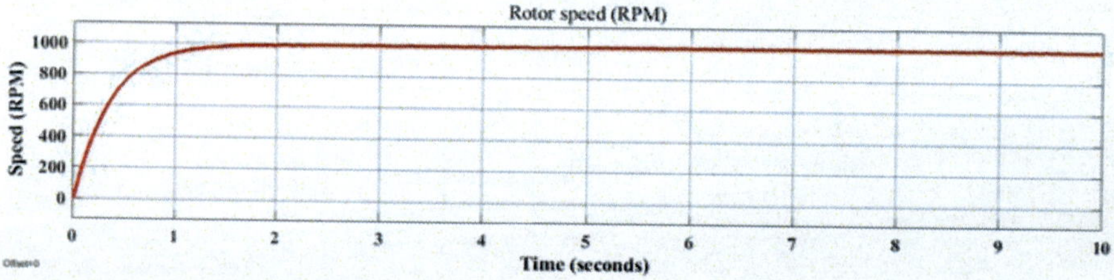

Fig. 12 Output speed

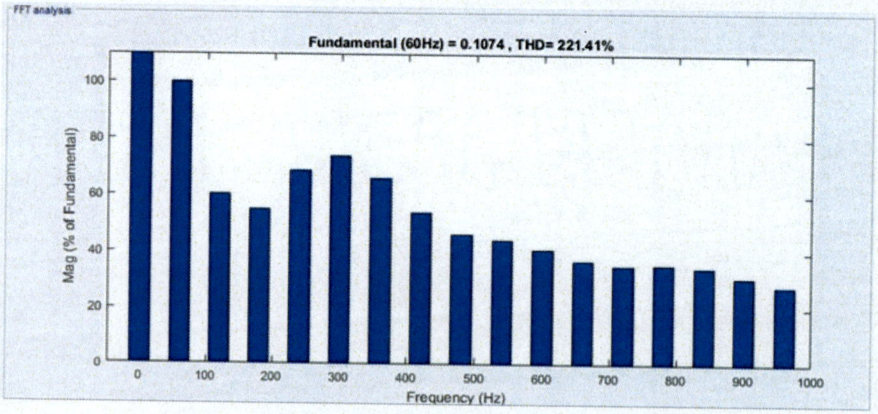

Fig. 13 Frequency analysis at fundamental frequency of 60 Hz

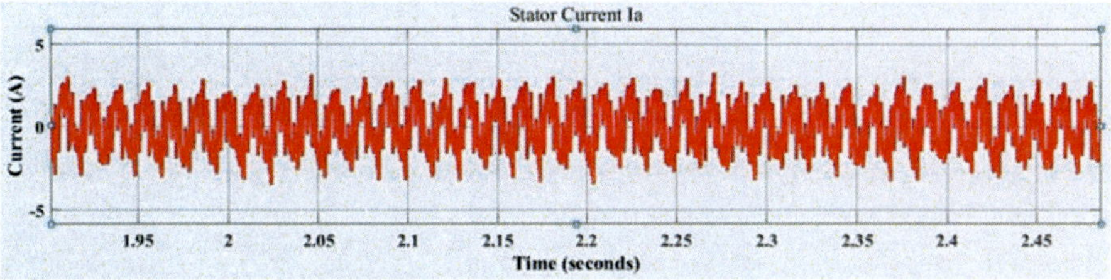

Fig. 14 Output stator current (Ia)

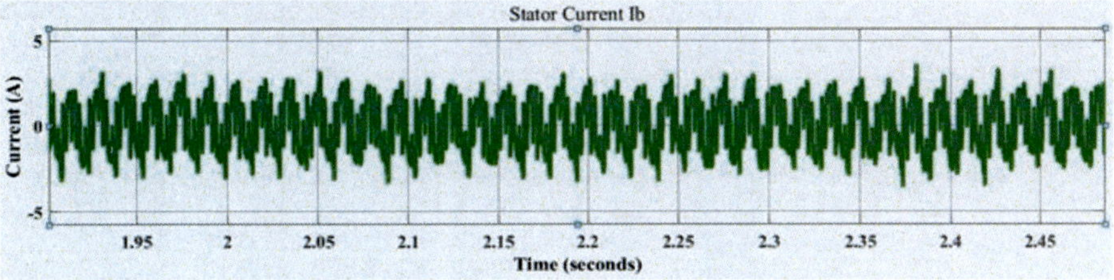

Fig. 15 Output stator current (Ib)

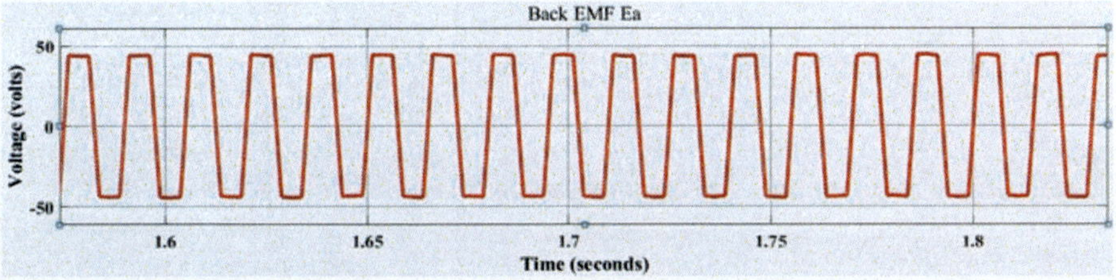

Fig. 16 Back EMF (Ea)

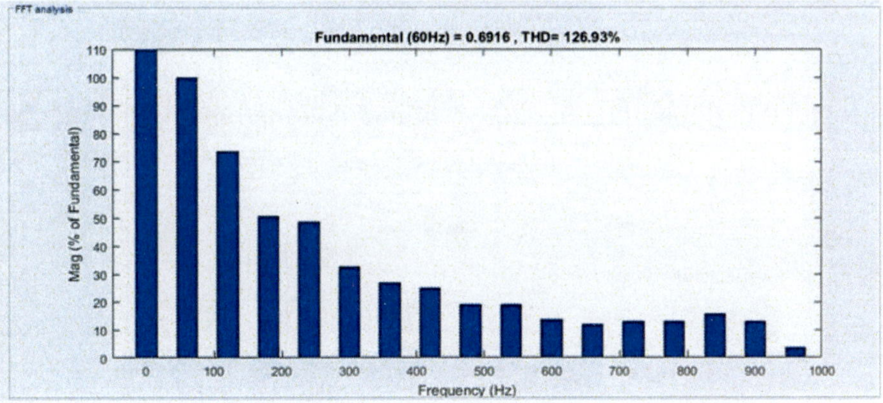

Fig. 17 Total harmonic distortion

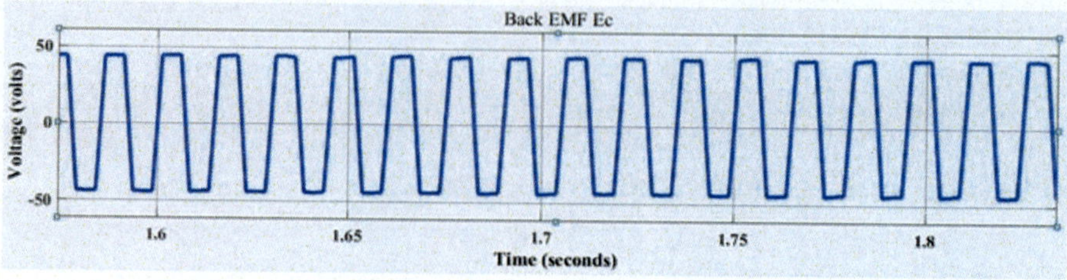

Fig. 18 Output back EMF (Ec)

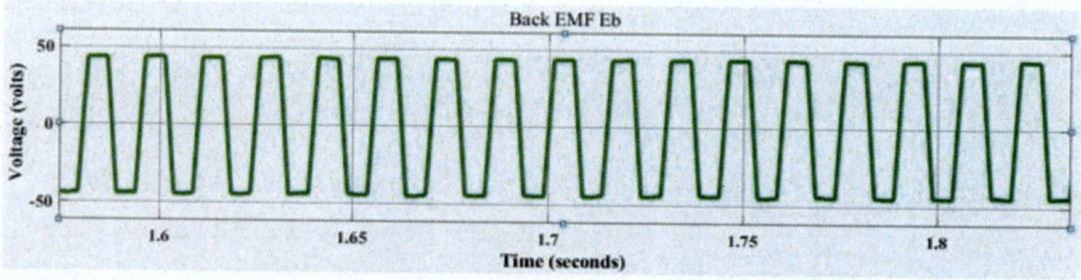

Fig. 19 Output back EMF (Eb)

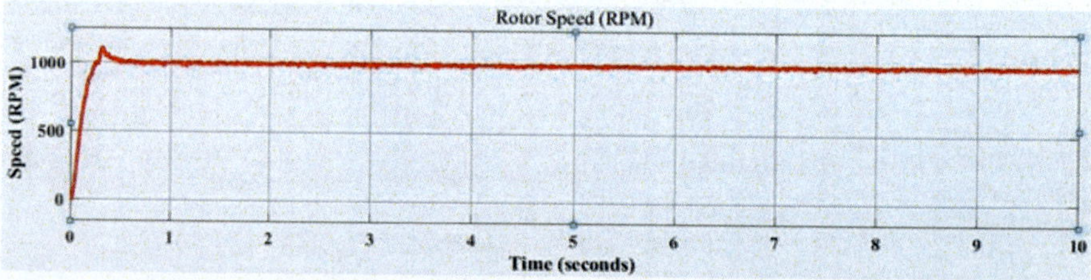

Fig. 20 Output speed

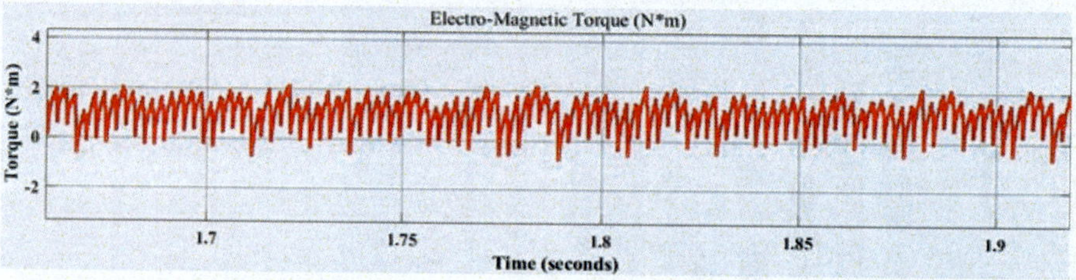

Fig. 21 Electromagnetic torque

Table 2 Performance analysis

Performance parameter	PI controller	Current controller
Output speed (RPM)	1002	1002
Current THD (%)	221.41	126.93
Signal ripple	Very high	Medium
Settling time (seconds)	0.75	0.2

6 Conclusion

Different methods of speed control are discussed. These methods implement the use of PI controller and use of current controller. These models were modelled in MATLAB Simulink, and the outputs were obtained for different parameters like stator phasor current, back EMF voltages, output speed, electromagnetic torque and FFT analysis. Different parameters were compared so as to conclude the method more efficient among the three.

Frequency analysis was done with the fundamental frequency of 50 Hz on the stator current (Ia) which showed the total harmonic distortion (THD) in the machine. FFT analysis is done on all of the controller techniques, and the results were obtained. The PI controller at fundamental frequency of 50 Hz showed THD of 221.41%, whereas the current controller at fundamental frequency of 50 Hz showed 126.93% THD. This showed that current controller is the most suitable option among these control techniques.

The machine was run for a total of 10 s, and the corresponding waveforms were obtained. Rise time and settling time of the output speed of the machine were also calculated.

The simulation results showed that the use of current controller for the speed control of BLDC motor proves to be more efficient.

References

1. Deepa AB, Pawar M (2014) Implementation of solar powered PMBLDC motor drive. Int J Sci Res Dev 2(8):3642–3658
2. Jayachandran A, Lakshmi GRP (2014) Analysis and implementation of efficient BLDC motor drive with different converter systems. Int J Eng Serv Res Technol 2:2034–2040
3. Obed AA, Kadhim AK (2018) Speed and current limiting control strategies for BLDC motor drive systems—a comparative study. Int J Adv Eng Res Sci 5(2):10–14
4. Dudhe SS, Thosar AG (2014) Mathematical modeling and simulation of three phase BLDC motor using MATLAB/Simulink. Int J Adv Eng Technol 7(5):1426–1433
5. Agarwal LK, Chauhan BK, Banerjee GK (2018) Speed control of brushless DC motor using conventional controllers. Int J Pure Appl Math 119(16):3955–3961
6. Sameera K, Srinivas A. Rao MVR (2014) Improved power factor and reduction of harmonics by using dual boost converter for PMBLDC motor drive. Int J Electr Electron Eng Res 4(5):43–51
7. Sathiyamoorthy S, Gopinath M (2016) Simulation of TI-SEPIC converters for BLDC motor drives. Int J Adv Eng Manag Sci 2(9):1607–1610
8. Pattathurani L, Dwibedi RK (2015) A BL-CSC converter fed BLDC motor drive with power factor correction. IOSR J Electr Electron Eng 10(6):89–97

A Bidirectional DC–DC Resonant Converter for Electric Vehicle Application

B. Vinoth kumar and P. Avirajamanjula

Abstract The bidirectional DC–DC resonant converter is proposed for a wide variety of applications such as electric vehicle, aerospace, electric charger etc., This Dual active bridge converter will perform both buck and boost operation based on the condition. The transformer isolated converter is providing better isolation on various load condition. This converter is can achieve better voltage gain, high-power density in wide range of load condition. The converter topologies and control techniques was verified through MATLAB Simulation.

Keywords Electric vehicle · Battery · Solar · Energy management · DC–DC bidirectional converter

1 Introduction

To improve the converter efficiency bidirectional converter is used. This converter preferred microgrid, electric vehicle application, uninterrupted power supply. Most converter have problem of circulating current because of the switching [1, 2] frequency is not near to the resonant frequency of the resonant tank.

Based energy storage application the design of the bidirectional converter will fulfil the following criteria High voltage gain, High conversion efficiency, wider voltage regulation range, linear forward and backward mode operation.

In power system the power converter is used for wide range of application. The unidirectional power converter was introduced to fulfil the requirement of load. But the energy storage system application is suitable because BDC have features of the wide range of voltage control, bidirectional current flow, a smaller number of components, high efficiency. Mostly BDC is preferred for electric vehicle (EV), photovoltaic (PV), smart grid, aerospace etc., The block diagram of Bidirectional converter as shown in Fig. 1.

B. Vinoth kumar (✉) · P. Avirajamanjula
Department of Electrical and Electronics Engineering, PRIST University, Thanjavur, India

S. S. Dash et al. (eds.), *Sixth International Conference on Intelligent Computing and Applications*, Advances in Intelligent Systems and Computing 1369,
https://doi.org/10.1007/978-981-16-1335-7_29

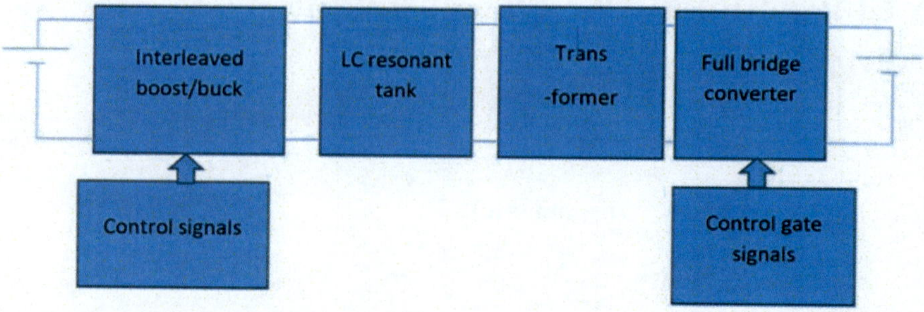

Fig. 1 Block diagram of BDC converter

2 Bidirectional Converter

LLC resonant converter is having feature of soft computing through this, and it is easy to achieve the better efficiency, reduce reverse energy and high-power density. In this paper, primary side interleaved bidirectional converter is designed. The advantages of interleaved boost converter compared to the classical boost converter are low input current ripple, faster transient response, reduced electromagnetic emission improved reliability, and high efficiency, in wide power range. In the secondary side, full bridge circuit is connected to load [3–5]. Based on the current rating, the control algorithm was designed. When the primary current is lower than the secondary current, the current sensor based the controller will decide the current flow, i.e., the current flow will happen from high voltage side to low voltage side. When the primary current is higher than the secondary current, the current flow will happen from low voltage side to high voltage side [1, 6–9]. For most of the electric vehicle application, the current direction control using controller strategy will be decided. Especially, for the regenerative case, the current flow will happen from motor to battery. It is like for many applications, the bidirectional control will improve the efficiency of the system.

Rule-based algorithm is most useful for controlling the current flow direction and gets the high efficiency [10, 11]. For regenerative control, the rule-based algorithm is mostly suitable compare to optimize method because of the familiarized the travel route will have nighty percentage of same date to control parameter for find the optimize solution of the system [12–15]. So fuzzy type of control will be used as supervisory control system for energy management system.

2.1 Design Consideration

A Duty cycle (D)

Formula for the voltage conversion ratio is as below.

$$V_o/V_d = 1/(1 - D)$$
$$D = (V_o - V_d)/V_o$$
$$D = (95-30)/95 = 0.684 \tag{1}$$

B Current ripple (ΔIL)

$$\text{Load current } I_o = \text{Output power/Output voltage}, \; I_o = 95/95 = 1$$
$$\text{IRMS} = 1.414 \text{ IPeak}$$
$$\Delta I_o = 40\% \text{ of load current}(1.414\text{ A}) = 0.565\text{ A}$$

Relation between input and output currents is as below.

$$\text{IIN} = I_o/(1 - D) = (1)/(1 - 0.684) = 3.16\text{ A} \tag{2}$$

$$2\text{IL} = I_o/(1 - D)$$
$$\text{IL} = (0.5) * I_o/(1 - D))$$
$$\text{IL} = (0.5) * 3.16\text{ A} = 1.58\text{ A} \tag{3}$$

$$\text{IRMS} = 1.414 * 1.58 \text{ IPEAK} = 2.23 \text{ IPEAK}$$
$$\Delta\text{IL} = 20\,\% \text{ of IL}(2.23\text{ A})$$
$$\Delta\text{IL} = 0.4\text{ A}$$

C. Inductor value
For CCM, output current (IO) should be greater than the boundary current (IOB)

$$\text{IO} > \text{IOB}$$
$$V_o/R > T_s V_o D(1 - D)^2/2L$$
$$L_{\min} > RT_s D(1 - D)^2/2 \tag{4}$$

$$\text{Load resistor } R = V_o/I_o = 95/1 = 95\,\Omega$$
$$T_s(\text{Time period}) = 1/f_s = 1/100\,\text{kHz} = 10\,\mu\text{s}$$
$$V_d = 30 \text{ VRMS} = 42.42 \text{ Vpeak}$$
$$V_o = 95 \text{ VRMS} = 134.35 \text{ Vpeak}$$

The minimum value of inductance for CCM is given by the below formula:

$$L_{\min} > RT_s D(1 - D)2/2$$
$$L_{\min} > (95) * (10 * 10 - 6) * (0.684) * (1 - 0.684)2/2$$

$$L_{\min} > 32.44\,\mu\text{H}$$

If the inductance value is 32.44 μH, it will result in boundary condition.

$$\begin{aligned} L &= V_d DT_s/\Delta\text{IL} \\ L &= (42.42)(0.684)(10 * 10 - 6)/0.5 \\ L &= 580.305\,\mu\text{H} \end{aligned} \tag{5}$$

As we can see, this is much greater than the boundary value resulting in CCM.

D. Capacitance value

$$\begin{aligned} C &= I_o D(T_s/2)/\Delta V \\ C &= (1)(0.684)(10 * 10 - 6/2)/0.2 \\ C &= 17.1\,\mu\text{F} \end{aligned} \tag{6}$$

Controller Techniques:

The bidirectional converter is basically a buck-boost converter and is also sometimes called a step-up/down power stage. In this converter, the output voltage is inverted, and its magnitude can be varied to be higher or lower than the magnitude of the input voltage. This converter consists of two different isolated topologies forward and backward mode control. To ensure that the voltages and current on both the lower voltage and higher voltage side are within safety limits, the initial condition of the compensator as well as its gains must be tuned accurately. SVPWM is used to control the switching operation of the secondary side of power converter. The resonant frequency of the series resonant tank based, the primary transformer side swiches are operated. So the magentic component design is simple. The better voltage gain can be obtaining through the fixed switching control, and wide range of voltage regulation is achieved.

3 MATLAB Simulation

See Figs. 2, 3, 4, 5 and 6.

4 Conclusion

The bidirectional resonant converter is preferable for high power density application. Using soft switching method, wide range of voltage regulation and high voltage gain can be obtained. The significant difficulties of the bidirectional converter,

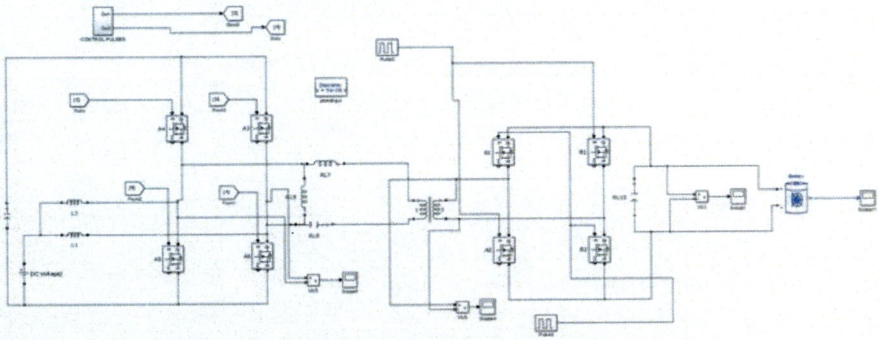

Fig. 2 MATLAB simulation of BDC converter

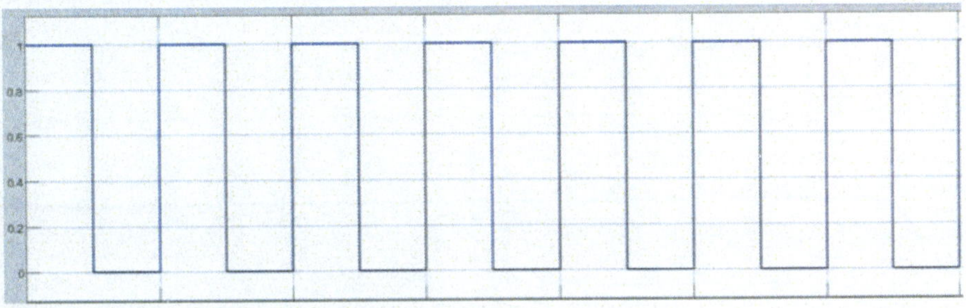

Fig. 3 Gate pulse for buck operation

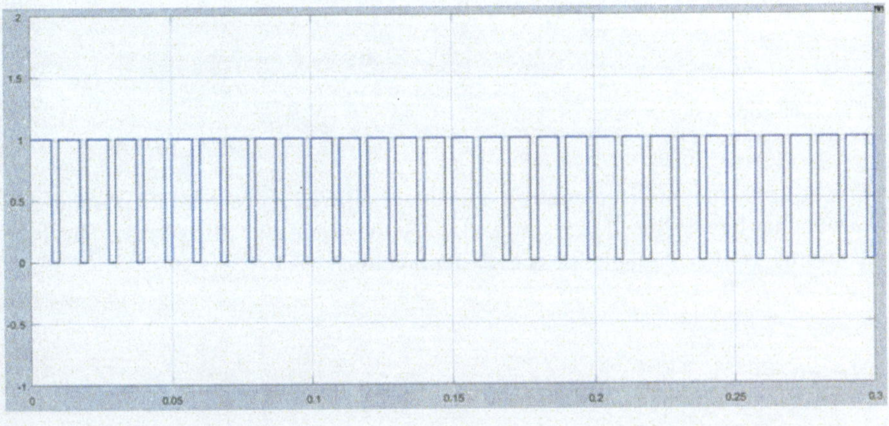

Fig. 4 Gate pulse for boost mode

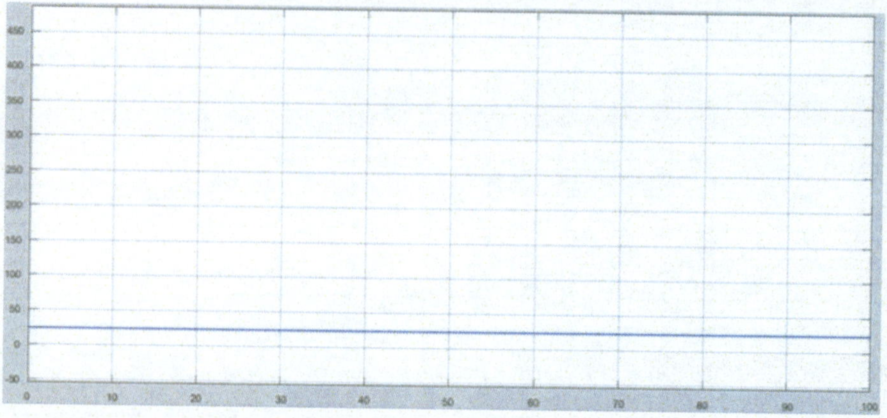

Fig. 5 Input voltage waveform

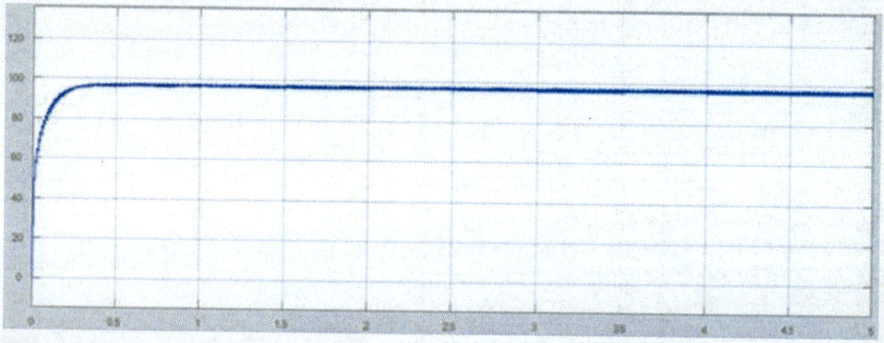

Fig. 6 Simulation result of BDC

which utilizes computerized control, are settled. The LLC resonant tank frequency controller has been reduced the complexity in control and switching stress. This type of converter is more suitable for energy application.

References

1. Wu T-F, Yang J-G, Kuo C-L, Wu Y-C (2014) Soft-switching bidirectional isolated full-bridge converter with active and passive snubbers. IEEE Trans Ind Electron 61(3):1368–1376
2. Oggier GG, Ordonez M (2016) High-efficiency DAB converter using switching sequences and burst mode. IEEE Trans Power Electron 31(3):2069–2082
3. Do H-L (2011) Non-isolated bidirectional zero-voltage-switching DC–DC converter. IEEE Trans Power Electron 26:2563–2569
4. Gnanavel C, Albert Alexander S (2018) Experimental validation of an eleven level symmetrical inverter using genetic algorithm and queen bee assisted genetic algorithm for solar photovoltaic

applications. J Circ. Syst. Comput 27(13):1850212-1–1850212-23. https://doi.org/10.1142/s0218126618502122
5. Liang T-J, Lee J-H (2015) Novel high conversion ratio high efficiency isolated bidirectional DC–DC converter. IEEE Trans Ind Electron 62(7)
6. Wu H, Sun K, Chen L, Zhu L, Xing Y (2016) High step-up/step-down soft-switching bidirectional DC–DC converter with coupled-inductor and voltage matching control for energy storage systems. IEEE Trans Ind Electron 63(5):2892–2903
7. Ramachandran R, Nymand M (2016) A 98.8% efficient bidirectional full-bridge isolated DC–DC GaN converter. IEEE APEC 609–614
8. Zhao B, Song Q, Liu W, Sun Y (2014) Overview of dual-active-bridge isolated bidirectional DC–DC converter for high-frequency-link power-conversion system. IEEE Trans Power Electron 29(8):4091–4105
9. Bai H, Mi C (2008) Eliminate reactive power and increase system efficiency of isolated bidirectional dual-active-bridge DC–DC converters using novel dual-phase-shift control. IEEE Trans Power Electron 23(6):2905–2914
10. Huang W, Qahouq JAA (2016) Energy sharing control scheme for state-of-charge balancing of distributed battery energy storage system. IEEE Trans Ind Electron 62(5):2764–2776
11. Xuewei P, Rathore AK (2014) Novel bidirectional snubberless naturally commutated soft-switching current-fed full-bridge isolated DC/DC converter for fuel cell vehicles. IEEE Trans Ind Electron 61(5):2307–2318
12. Tummuru NR, Mishra MK, Srinivas S (2016) Dynamic energy management of renewable grid integrated hybrid energy storage system. IEEE Trans Ind Electron 62(12):7728–7737
13. Choi W, Rho K-M, Cho B-H (2016) Fundamental duty modulation of dual-active-bridge converter for wide-range operation. IEEE Trans Power Electron 31(6):4048–4064
14. Daneshpajooh H, Bakhshai A, Jain P (2012) Modified dual active bridge bidirectional dc–dc converter with optimal efficiency. IEEE APEC, pp 1348–1354
15. Xuewei P, Rathore AK (2015) Naturally clamped zero-current commutated soft-switching current-fed push-pull DC/DC converter: analysis, design, and experimental results. IEEE Trans Power Electron 30(3):1318–1327

Semantic Relation from Biomedical Text Documents Using Machine Learning Algorithm

R. Srinivasan and C. N. Subalalitha

Abstract Semantic relations are underlying relation between the concepts present in sentences. Semantic relation plays an important role for building applications namely, information retrieval (IR), information extraction (IE), question answering (QA) and chatbots. This paper proposed a model to extract entities, and identifying the semantic relations exists between the entities. This paper is used to extract entities using novel feature extraction techniques with Naïve Bayes approach. Furthermore, it is used to identify the semantic relations using various machine learning approaches. This paper addresses eight semantic relations namely, prevent, no cure, disease only, side effect, vague, treatment only, cure and none. The model is evaluated by standard metrics and produced the result of 68.42% F-score for this data set.

Keywords Semantic relation · Entity extraction · Machine learning

1 Introduction

Biomedical Natural Language Processing (BioNLP) is a method of applying data extraction techniques into biomedical domain. It helps to extract the useful information or identifying entities from the biomedical document. BioNLP expands its research towards the wide variety of biomedical text. This helps out to identify the named entity recognition, protein name extraction, genre analysis, document classification and information extraction [1]. This paper is mainly focusing on information extraction from the biomedical documents. Numerous techniques also implemented to extract information from different datatypes such as image, text, video and audio [2]. This paper addresses on extracting information in the form of biomedical text document. This paper tries to extract the information using biological entities such as "disease" and "treatment". Based on the entities, the informations are extracted and then classified into the eight semantic relations namely, prevent, no cure, disease only,

R. Srinivasan (✉) · C. N. Subalalitha
Department of Computer Science and Engineering, SRM Institute of Science and Technology, Kattankulathur 603203, India

S. S. Dash et al. (eds.), *Sixth International Conference on Intelligent Computing and Applications*, Advances in Intelligent Systems and Computing 1369,
https://doi.org/10.1007/978-981-16-1335-7_30

side effect, vague, treatment only, cure and none. The major challenge of extracting above-mentioned entities are:

1. Different forms: The disease name and treatment entities can be written in different forms [3], e.g. "pre-eclampsia" is disease entity. But the same disease can be written in "preeclampsia".
2. Abbreviations and Acronyms: The data set is extracted from the medline abstracts. Due to that some sentences, the entities are not given directly. It is represented in the form of acronyms, e.g. "omphalocele-exstrophy-imperforate anus-spinal defects" is a disease entity. It is represented as "OEIS" in the data set.
3. Descriptive modifiers: Some disease entities are descriptive modifiers. These modifiers refer to the parts of the human body [4]. These modifiers cause difficulties to identify the other type of medical entities, e.g. lung cancer and lung transplantation. Lung is the descriptive modifier but based on the next word, it may conform that the word refers to disease or treatment entity.
4. Word affixes: Most of the disease entities are derived from Latin and Greek root suffixes, e.g. haemophilia. To extract entities in morphological aspect is a difficult task.

Once the entities are extracted, the next task is to classify the sentence based on the semantics. Based on the entities, we can classify the sentences into two groups: informative sentence and non-informative sentence. The informative sentences consist of either disease entities or treatment entities or both. But non-informative sentences do not carry any of the entities. This helps to extract the semantics relation from the biomedical documents easily. Out of eight relations, "None" category does not need any entities so it falls into non-informative sentences. The remaining seven relations fall into the informative sentences because it carries either any of the entities or both. The main contribution of our works is:

1. To extract entities named as "disease" and "treatment" using feature extraction techniques along with Naive Bayes algorithm. It helps to classify the informative and non-informative sentences
2. To extract the semantic relation from the informative sentences using different ML approaches.

2 Literature Survey

There are lot of approaches to solve the Biomedical Named Entity Recognition (BNER) system. Approaches can be broadly classified into rule-based approaches, dictionary-based approaches, machine learning (ML) approaches and deep learning (DL) approaches. Rule-based approaches mainly depend on existing vocabularies to identify the biomedical entities.

Asmaa et al. used a rule-based approach with embedding vector to classify the text document. The author compared the rule-based approach with other algorithms, namely ZeroR, OneR, JRip to the Reuters data set [5]. The author also proved that the rule-based approach performs good compared to the other algorithms. Mykowiecka et al. extract the information from patient's clinical data using rule-based approach [6]. The author has designed an ontology-based approach to extract the information from the polish medical text and produces 80% of precision and recall value. Apart from this, the author also addressed the rule-based approaches to extract the information from multidomain text [7, 8]. Due to the information explosion, it is challenging and time consuming to extract features from the rule-based approaches.

Mostly modern methods use the machine learning approach to extract the information from the text data. ML approaches are categorized as unsupervised learning, supervised learning and semi-supervised learning. From 1960 to 2010, statistical methods played a dominant role in the text classification model [9]. Those models are turned into supervised and unsupervised approaches. Khreisat has proposed a machine learning approach to classify the Arabic text using N-gram frequency [10]. The text was extracted from the Jordanian Arabic newspaper and apply the Manhattan distance metric to find the dissimilarity measure.

He et al. recognize drug name using conditional random field combined with dictionary look-up method [11]. Alimova et al. extract the clinical relation extraction on benchmark data set namely MADE 2018 and n2c2 2018 [12]. The author compares the state-of-the-art techniques with proposed model and improves the F-measure of 3.5% on the MADE corpus. A lot of work focused on information extraction using machine learning approaches on multiple languages such as Bangla [13], Arabics [14], Italian [15] and so on. The author also addressed the ensemble methods to extract the relation from the entities [16].

Deep learning models are introduced to reduce the feature selection or improving the feature extraction process. Human experts do not design features in the deep learning approaches [17]. Empirical studies proved that classification accuracy increases in the deep learning approaches compared to the machine learning approaches in natural language and bioinformatics application [18]. Suárez-Paniagua et al. introduced a hybrid model to extract the information from the biomedical text [19]. This hybrid model includes BI-LSTM, and CNN helps to detect, classify and identify the relation between entities.

Sarrouti and Said developed fully automatic biomedical question answering (QA) system and achieves good performance compared to the state-of-the-art techniques [20]. Although deep learning or machine learning approaches for identifying the semantic relation produce a satisfactory result, still there is a huge gap for improvement. On the other hand, identifying eight semantic relations from the biomedical entities are very difficult and results are quite low. The proposed methodology introduces a two-stage ML approach to extract entities and identify the semantic relation exist between entities.

3 Proposed Work

Figure 1 describes the overall architecture of our work. The proposed work is divided into two tasks: (1) Identification of entities named as "disease" and "treatment". It helps to classify the informative and non-informative sentence. The informative sentence carries any one of the entities or both, whereas non-informative sentence does not contain any of these entities. (2) Identification of semantic relation exists in informative sentence. Machine learning technique is applied to classify the sentence based on the semantic relation.

3.1 Data Set Details

The data set used in the proposed approach is created by rosario and hearst [21]. The data set consists of 3580 sentences and eight semantic relations, namely prevent, no cure, disease only, side effect, vague, treatment only, cure and none sentences. After performing the data cleaning process, duplicate sentences have been removed from

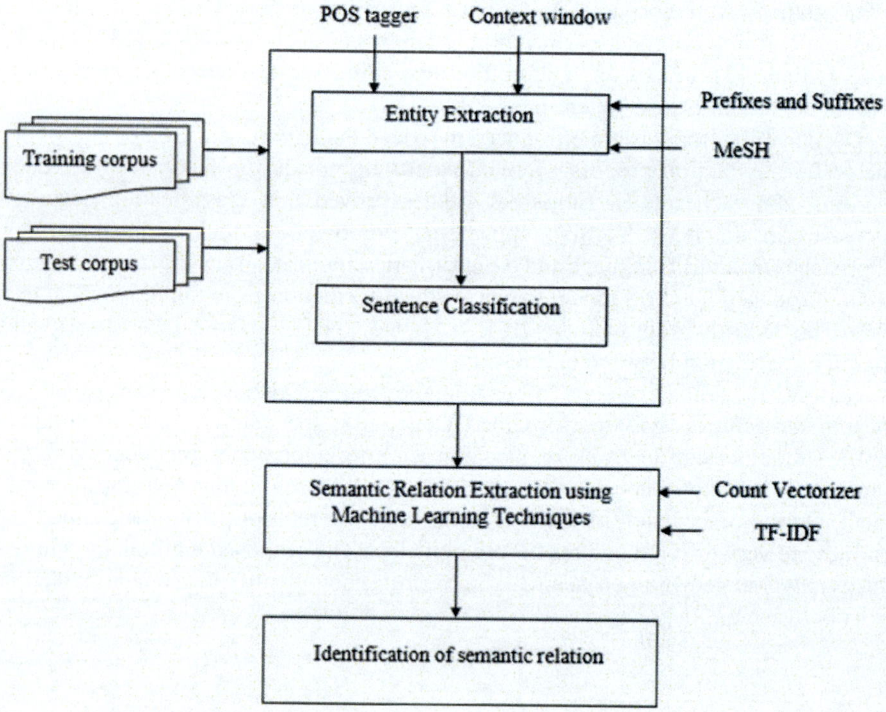

Fig. 1 Architecture of proposed model

Table 1 Data set description

Semantic relations	Entities present	Definition
Cure 810 [648, 162]	Disease and treatment	Treatment cures the disease
Prevent 63 [50, 13]	Disease and treatment	Treatment prevents the disease
Side effect 29 [24, 5]	Disease and treatment	Disease is result of the treatment
Disease only 616 [492, 124]	Disease	Disease entity alone present and not mentioned about treatment
Treatment only 166 [132, 34]	Treatment	Treatment entity present and not discussed about disease
Vague 36 [28, 8]	Disease and treatment	Not a clear relationship between disease and treatment
No cure 4 [1, 3]	Disease and treatment	Treatment does not cure the disease
None 1771 [1416, 355]	–	Both the entities are not present

the data set. Table 1 gives the description of data set used in our work, e.g. side effect consists of 29 sentences (24 training data, 5 test data).

3.2 Entity Extraction

The first step is to extract disease, treatment entities from the given sentences. The entities are extracted by various feature extraction techniques, namely parts of speech, context window, prefixes and suffixes and Medical Subjective Headings (MeSH). POS tagger helps to identify the parts of speech of the text such as noun, adjective and verb. Context window technique helps to identify the previous word and next word from the current word. The context window size is fixed as 1. Prefixes and suffixes of the word help to identify the beginning and ending of characters of the word. MeSH is medical thesaurus consist of entities and are classified into A–Z categories. These four techniques form the pattern of each word and help to identify the word is disease entity or treatment entity. Once the pattern is formed, Naïve Bayes algorithm is applied to identify the entities of each and every word.

Let $S = \{w_1, w_2, \ldots, w_n\}$, "S" be sentence and "w" means word. The sentence is break down into words. Each word is checked with two different clusters, φ and χ. If any one of word in the sentences matches with any of the word in the clusters, then the entire sentence (S) will be treated as informative sentence (α), otherwise the sentence (s) is considered to be non-informative sentence (β).

Algorithm for classifying the sentence based on thesaurus

Input: dataset
Output: α, β
Step 1: Read each sentence{ Si } i ∈ ℕ
Step 2: for each sentences Si ={ w1, w2,……., wn }
Step 3: for each word w
        If w ⊃ φ ^ χ then
                Label α, goto step 2
        end for
        Label β, goto step 2
    end for
Step 4: Exit

The algorithm describes the detailed procedure, to classify the sentence based on entity extraction. The informative sentence describes the seven sematic relations except "None". Because the "None" category falls into the non-informative sentence.

3.3 *Semantic Relation Using Machine Learning Techniques*

The identification of semantic relation is the next task in the proposed system. Various ML approaches are experimented in the proposed system to extract the semantic relations. Two feature extractions are applied in the various machine learning approaches. Count vectorizer helps to calculate the word frequency and build a vocabulary of known words. This helps to convert the text into integer values based on the number of times word occurs in the data set.

Count vectorizer consists of a lot of parameters that help to pre-process the data, setting up n-gram range, stop word removal, etc. [22]. Term frequency and inverse document frequency (TF-IDF) is alternative feature extraction technique that converts the text into vector form. TF-IDF helps to take the important word in the larger proportion. TF helps to build the vocabulary, whereas IDF helps to encode the documents. IDF assigns the lowest frequency to the frequently occurred words in the document. Once the text is converted into vectors, the next step is to apply the different machine algorithms. This paper targets three ML algorithms, namely random forest multinomial Naïve Bayes (MNB) and support vector machine (SVM).

4 Results and Discussion

Recall, precision, *F*-measure and accuracy have been used as a performance metrics to evaluate the data set. A confusion matrix is built to evaluate the classification model [23].

True positive (TP) denotes predicted sentence matches the actual sentence, whereas false positive (FP) means predicted sentences are misclassified as "positive". False negative (FN) means predicted sentence do not matches the actual sentence "negative". True negative (TN) defines the sentences are correctly predicted as "negative". Equation 1 is used to calculate the accuracy,

$$\text{Accuracy} = \frac{\text{TP} + \text{TN}}{\text{TP} + \text{FP} + \text{FN} + \text{TN}} \tag{1}$$

Equations 2 and 3 determine the precision and recall value

$$\text{Precision} = \frac{\text{TP}}{\text{TP} + \text{FP}} \tag{2}$$

$$\text{Recall} = \frac{\text{TP}}{\text{TP} + \text{FN}} \tag{3}$$

F-score combines the precision and recall value describes in Eq. 4

$$F\text{-score} = \frac{2 * \text{recall} * \text{precision}}{\text{recall} + \text{precision}} \tag{4}$$

Sections 4.1 and 4.2 briefly describe the result and error analysis of this data set. Section 4.1 describes the entity extraction and classifying the sentence using SCE. Section 4.2 explains the relation extraction using various machine learning approaches.

4.1 Sentence Selection

This section describes the number of sentences used to evaluate the model. Out of 3495 sentences, 2793 sentences are considered as training data and remaining 702 sentences are considered to be test data. Table 2 clearly depicts the sentence used for first task.

Applying the feature extraction techniques mentioned in Sect. 3.3 helps to identify the disease and treatment entities. These combined feature extraction techniques along with Naive Bayes techniques help to extract approximately 80% of disease and treatment entities in the given sentences. Out of 702 sentences used in the test data, 610 sentences are correctly classified and remaining sentences are misclassified. Out

Table 2 Sentences used for first task

Sentence	Informative (α)	Non-informative (β)
Training	1377	1416
Testing	347	355

Table 3 Output of sentence classifier

Approach	Recall	Precision	*F*-score	Accuracy
SCE	93.52	78.96	85.63	86.89

Table 4 Output of semantic relation

Method	Count vectorizer			TF-IDF		
	Recall	Precision	*F*-score	Recall	Precision	*F*-score
MNB	68.14	64.42	65.63	68.29	60.89	62.37
SVM	70.55	68.32	67.89	71.7	68.3	68.42
Random forest	58.47	49.1	47.54	56.04	47.63	44.08

of 610 sentences, 274 sentences are classified as "α" and 336 sentences are classified as "β". The overall performance metrics are defined in Table 3. As a result, some of the disease and treatment entities are not able to identify using the combined feature extraction technique mentioned in Sect. 3.2.

4.2 Relation Identification

The second task is to identify the semantic relation exist in an informative sentence. The second task is depending on the result of the first task. Table 4 describes the results obtained from the machine learning algorithm. SVM outperforms best *F*-score value compared to other machine learning algorithms. Out of 702 sentences, 629 sentences are used as test data for identifying the semantic relation, because the rest of the sentences are misclassified as non-informative sentence. The reason for obtaining 68.42% *F*-score value is class imbalance problem. Class imbalancing means number of sentences in one relation is higher than the other class, e.g. "No cure" consist of only 4 sentences, 1 is used as test data. Table 1 clearly describes the sentences used in the semantic relation.

5 Conclusion

This paper addressed a two-stage machine learning approach to perform semantic relations from the biomedical text documents. The first step is to extract entities by combining different feature extraction technique with Naïve Bayes algorithm. The second step is to produce output for various classification algorithms using count vectorizer and TF-IDF. Evaluation results clearly shows that SVM produces the good result compared to other classification algorithms. The future direction is to increase the F-score value in the second task and to avoid imbalancing problem in this data set.

References

1. Muhammad A, Parwez MA, Jahiruddin (2019) DiseaSE: a biomedical text analytics system for disease symptom extraction and characterization. J Biomed Inform 100
2. Adnan K, Akbar R (2019) An analytical study of information extraction from unstructured and multidimensional big data. J Big Data 6:91
3. Gridach M (2017) Character-level neural network for biomedical named entity recognition. J Biomed Inform 70:85–91
4. Xu K, Zhou Z, Gong T, Hao T, Liu W (2018) SBLC: a hybrid model for disease named entity recognition based on semantic bidirectional LSTMs and conditional random fields. BMC Med Inform Decis Mak 18(Suppl 5):114
5. Aubaid AM, Mishra A (2020) A rule-based approach to embedding techniques for text document classification. Appl Sci 10(11):4009
6. Mykowiecka A, Marciniak M, Kupść A (2009) Rule-based information extraction from patients' clinical data. J Biomed Inform 42(5):923–936
7. Boufrida A, Boufaida Z (2020) Rule extraction from scientific texts: evaluation in the specialty of gynecology. J King Saud Univ Comput Inf Sci
8. Tubishat M, Idris N, Abushariah M (2021) Explicit aspects extraction in sentiment analysis using optimal rules combination. Future Gener Comput Syst 114:448–480
9. Li Q, Peng H, Li J, Xia C, Yang R, Sun L, Yu P, He L (2020) A text classification survey: from shallow to deep learning
10. Khreisat L (2009) A machine learning approach for Arabic text classification using N-gram frequency statistics. J Informetrics 3:72–77
11. He L, Yang Z, Lin H, Li Y (2013) Drug name recognition in biomedical texts: a machine-learning-based method. Drug Discovery Today 19. https://doi.org/10.1016/j.drudis.2013.10.006
12. Alimova I, Tutubalina E (2020) Multiple features for clinical relation extraction: a machine learning approach. J Biomed Inform 103:103382
13. Kowsher M, Sanjid MZI, Das A, Ahmed M, Sarker MMH (2020) Machine learning and deep learning based information extraction from Bangla names. Procedia Comput Sci 178:224–233
14. Albukhitan S, Alnazer A, Helmy T (2020) Framework of semantic annotation of Arabic document using deep learning. Procedia Comput Sci 170:989–994
15. Guarasci R, Damiano E, Minutolo A, Esposito M, De Pietro G (2019) Lexicon-grammar based open information extraction from natural language sentences in Italian. Expert Syst Appl 143:112954
16. Thomas P, Neves M, Solt I, Tikk D, Leser U (2011) Relation extraction for drug-drug interactions using ensemble learning. In: CEUR workshop proceedings 761
17. Altınel B, Ganiz M (2018) Semantic text classification: a survey of past and recent advances. Inf Process Manage 54:1129–1153
18. Larochelle H, Bengio Y, Louradour J, Lamblin P (2009) Exploring strategies for training deep neural networks. J Mach Learn Res. 1
19. Suárez-Paniagua V, Zavala RMR, Segura-Bedmar I, Martínez P (2019) A two-stage deep learning approach for extracting entities and relationships from medical texts. J Biomed Inform 99:
20. Sarrouti M, El Alaoui SO (2020) SemBioNLQA: a semantic biomedical question answering system for retrieving exact and ideal answers to natural language questions. Artif Intell Med 102:101767

21. Rosario B, Hearst MA (2004) Classifying semantic relations in bioscience texts. In: Proceedings of the 42nd annual meeting on association for computational linguistics (ACL '04). Association for computational linguistics, USA, 430–es
22. Yang JS, Zhao C, Yu HT, Chen HY (2020) Use GBDT to predict the stock market. Procedia Comput Sci 174:161–171
23. Zhang W, Gao F (2011) An improvement to naive bayes for text classification. Procedia Eng 15:2160–2164

Distributed Updation and Authentication of Aadhaar Leveraging Blockchain Technology

B. Sowmiya, Iyer Harini Umamaheshwaran, Chhetri Priyanka, and B. Ida Seraphim

Abstract Blockchain is a distributed, decentralized public ledger. It allows digital information to be distributed and is immutable, creating a new backbone for the Internet. It is a series of immutable time stamped records with no central authority. Data or transactions are stored in the form of blocks that are cryptographically sealed and chained to one other. We seek to exploit blockchain's exemplary advantages of security and decentralized distribution in a government application. Currently when a person wants to create or change his identification details, he has to navigate individually to different offices of the government. The government information is stored and managed by a centralized administrator. In order to eliminate the centralized dependency and to save the time and energy of the applicant, data acquired while creating a government identification card or number can be stored in a blockchain. Any changes made by the applicant to his data can be distributed to all the government departments that are the members of the blockchain network. It also uses a self-executing contract with a set terms of agreement called a smart contract. We create a consortium blockchain to optimize the parameters of time and security.

Keywords Aadhaar · Applications · Authentication · Bitcoin · Blockchain · Distributed · JavaScript · Proof of work

1 Introduction

Blockchain is defined as a distributed, decentralized public ledger. Blockchain technology has enabled the decentralized, trustworthy transactions of digital assets that enable the elimination of an intermediary using a peer-to-peer framework. Before the age of blockchain, transactions took place in a centralized manner where if one

B. Sowmiya (✉) · I. Harini Umamaheshwaran · C. Priyanka · B. Ida Seraphim
SRM Institute of Science and Technology, Kattankulathur, India
e-mail: sowmiyab@srmist.edu.in

B. Ida Seraphim
e-mail: idaserab@srmist.edu.in

S. S. Dash et al. (eds.), *Sixth International Conference on Intelligent Computing and Applications*, Advances in Intelligent Systems and Computing 1369,
https://doi.org/10.1007/978-981-16-1335-7_31

wants to sell their asset, say a music album, they need to trust that the seller has enough assets to buy the said music album, and this trust is established by relying on an intermediary like PayPal or a bank that verifies if the buyer has enough credit in their bank account to make this purchase. The asymmetry of trust in this transaction is called the Byzantine General's problem. Say there are four generals who attack an army protecting a city, the four men have to have a consensus on what commands to follow, even if one of the generals is a traitorous one and gives out false commands the entire mission will fail. The only way to ensure the command issuing general is a trusted one is by checking the history of all his previous commands and measuring its correctness. In the music album buying scenario, we cannot see the buyer's history of all transactions and hence cannot detect if the buyer is traitorous and have to rely on a centralized authority like PayPal that adds cost to the transaction. To overcome problems such as these, in November 2008, a pseudonymous individual or group called Satoshi Nakamoto published a whitepaper called "Bitcoin: A Peer-to-Peer Electronic Cash System."

In the revolutionary bitcoin framework, a set of transaction is verified and authorized by a miner, who in order to add a block to the blockchain, solves a complicated, intentionally high computational hashing puzzle. The bitcoin, the first blockchain, had the characteristics of redundancy as data are replicated and stored in multiple computers, decentralization as the involvement of a central intermediary is primarily eliminated, immutability as the blockchain is a append only ledger and hence cannot be censored making it anti-fraudulent, permissionless framework which is open to all and does not need the authorization from a bank or such parties.

1.1 Blockchain

Blockchain consists of blocks that are linked to one another. The details of the transactions that occur in the network get stored in each discrete block. The blockchain characteristic of decentralization is supported by digital signature, cryptographic hash as well as the consensus algorithms. Its decentralized and distributed characteristics are used to eliminate the participation of third parties for transfer of digital assets. In other words, blockchain allows digital information to be transferred from peer to peer. It represents a distributed system that eliminates the involvement of central authority who may control the content of the block in the blockchain. Transactions once committed is immutable and hence cannot be altered. Each transaction or activity performed by the different members or nodes taking part in the blockchain network gets recorded and hence security and reliability are achieved. The blockchain is comprised of many blocks which has a header and a body in it. As shown in Fig. 1, genesis block is the first block present in the blockchain.

The transactions in the blockchain network are usually a process of transferring the digital assets. Sender nodes initiate a transaction by supplying its own private key and digital signature. These transactions are forwarded to a pool of transactions where they are stored. The miners execute the task of verifying and validating the

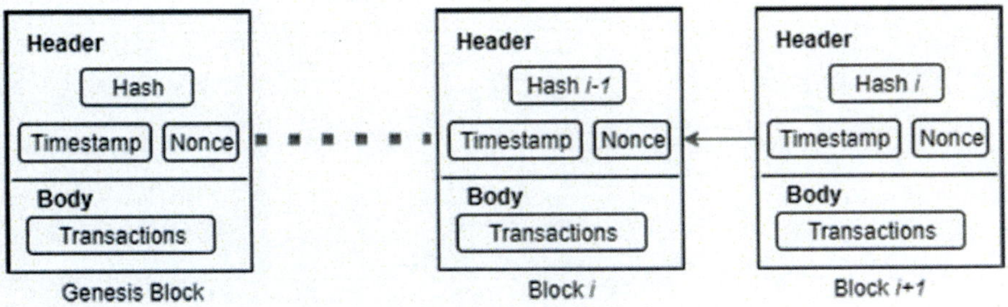

Fig. 1 Block structure

transactions and attaches them to the block. The miners compete against one another to solve the complex computational puzzle by utilizing their own power of computation. The first miner who succeeds in solving the puzzle using his computation becomes the winner and is provided with a reward for his effort. The winning miner gets a chance to mine a new block. The peers use the consensus mechanism to verify the newly formed block, where they agree upon the blockchain network status.

1.2 Existing System

There are two main processes, enrollment and authentication. Biometric data such as two irises, ten fingerprints, and a picture are collected for registration, and demographic information includes address, age, place of birth, phone number, and other profile information (Fig. 2).

Main processing pipeline. Quality check—Manual process verifies the gathered user data with their photographs and biometric data are checked for validity against

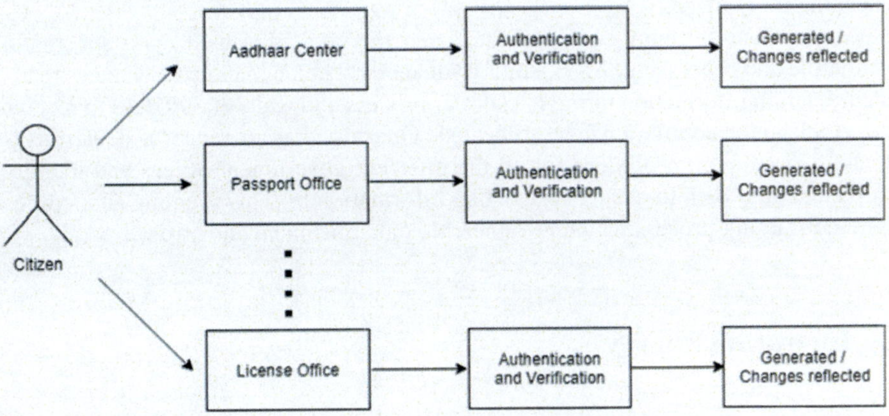

Fig. 2 Existing system

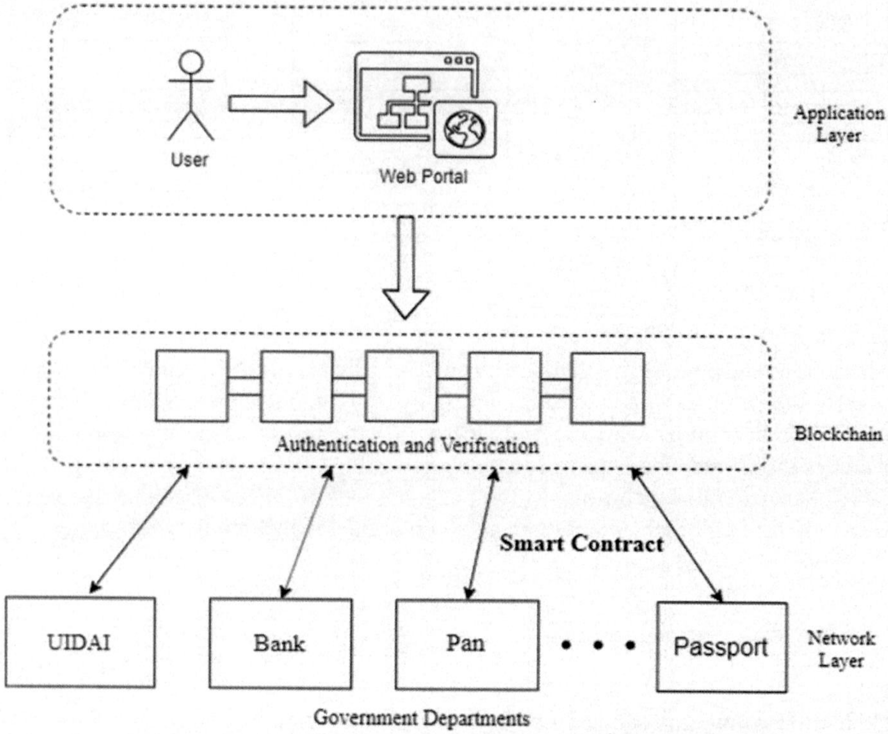

Fig. 3 Proposed architecture

parameters like gender. For instance, if a female has submitted a male's photograph, etc. Sender validation—Collected data at the Aadhaar data collection centers by verified UIDAI employees or volunteers are sent to central office where more intricate checking takes place.

An Aadhaar ID number is generated and the card is printed upon successful validation and processing. An existing Aadhaar card holder can make changes to his identification information through a similar process and waits 30–90 days to receive the Aadhaar or see the changes reflected. One also has to repeat a department-specific enrollment procedure for all the different government offices and identity cards that they wish to change an existing information in. This is a long and tedious time-consuming process which is vulnerable data mismatch and errors.

2 Literature Survey

The problem addressed by Wu and Tsai [1] is the lack of any authentication technique in the IoT networks. The intelligent agriculture system contains essential data required to manage and keep a check on the growth of different crops according to the environmental conditions of the agricultural field area. The primary reasons for the global crisis of food are aging of labor force working in agricultural fields and also important role played by climatic conditions that are at its extreme that results in damaging and destroying the crops.

The problem addressed by Nguyen et al. [2] is the issues faced in the user authentication process in the various e-learning systems where the users retrieve the resources of the course and take online exams. Earlier solutions to authentication include biometric-based, image-based, and profile-based authentication (PBA). These solutions do not provide security and integrity and hence faced few drawbacks. The profile-based authentication (PBA) has been improved that solves the issues related to user's identity misuse and also maintains the integrity of their academic results using Merkle tree and blockchain technology.

Mahmoud et al. [3] address the problem of vulnerability of the customer and employee-related information to various attacks in the water distribution system (WDS). The prevention of attacks in the information technology (IT) and operational technology (OT) can be done by controlling and monitoring the water distribution system (WDS). In order to prevent cyberattacks, an intelligent WDS can be developed. A customer data protection is used to achieve anonymity of the user identity, and smart water network in combination with blockchain technology is used to solve the problem of trust and security.

The problem addressed by Jamal et al. [4] is identity theft and misuse of it which can arise in systems that stores users' personal information. The security of the system that stores personal records is ensured using blockchain technology that makes use of its decentralization feature. The system allows personal records to be accessed by the third parties while making it immutable. Smart contract allows social networks to view the authentication records for registration and the verification purposes.

Monrat et al. [5] describe the blockchain distributed ledger technology as blocks chained or linked together where each block uses hash code to link to the preceding block. This technology characteristics include decentralization, immutability, and transparency of block data. The new blocks keep on getting attached to the end of the blockchain and keeps on growing. It enables peer-to-peer exchange of the knowledge between the nodes that are involved in the network of blockchain. It removes the presence of the various middlemen who perform tasks of verification and validation.

Hasan and Salah [6] use smart contracts to ensure proof of delivery in a commerce between buyers and sellers. Previously, it involved a trusted third-party system varying from bank to transport intermediaries. It also depended on handsigned papers or device-based authentication relying heavily on the honesty of the transporter for proof of delivery.

Tang et al. [7] shed light on the issue of information sharing that takes place between hospitals and various organization of the health records. Transcending from physical storage of health records, one of breakthrough technologies included the electronic health records (EHR) which ensured easy storing and collection of information of health management and medical records that were stored in cloud-based infrastructures. The sharing of records from one hospital to another organization was a non-trustable task in a sever storage infrastructure with independent databases managed by centralized authorities creating a lack of authority of patients over their own information that could be easily tampered.

Bosri et al. [8] propose a blockchain-based system using hashing to make the voting process unbiased and independent ensuring the much-needed transparency and security. Before the invention of electronic voting, ballot-based voting faced issues of booth capturing, ballot paper stealing, unfair counting, and other such frauds. This paper uses a public Ethereum blockchain-based voting system that ensures two main factors of information security in voting systems—anonymity and privacy of storing votes. This proposes faster, legal, and trusted way of storing e-votes.

Feng et al. [9] handle the issue of secure authentication of messages transmitted in vehicular ad hoc networks (VANETs) and maintaining conditional anonymity to keep certain information in the vehicular infrastructure private, hence addressing the problems in VANETs' accurate and credible message transmission. VANETs' main purpose involves real-time traffic data exchanged between vehicles to monitor activities and to provide assistance with smart routing, traffic lighting and avoiding collisions and accidents, etc. All previous PKI-based blockchain implementation to store transmitted messages suffers from problems with the use of a trusted third party that can be easily compromised and also involves high computational costs, and KGC uses the key Escrow issue and variable public key authentication is affected by scalability and membership management, providing tight transparency yet not preventing malicious behaviors.

Cai et al. [10] conduct an analytical survey on the growth of the technologically advanced blockchain systems to expose the significance and potential value of decentralized applications (dApps). Some of the problems in the decentralized systems included synchronization, which was combatted with consensus algorithms and double spending caused by sybil attacks. Turning-complete language like solidity is used in public platforms like Ethereum for creating immutable self-executing contracts. DApps are distinguished by open source, internal crypto-currency support for the calculation of all credits and transactions, global consensus, and no central failure point.

3 Proposed Work

In the Aadhaar system, we propose using a consortium blockchain (Fig. 3). The blockchain's numerous nodes, which may be privately permitted to access and download the data, are those of government departments such as the tax division, the license department, approved banks, which serve as the blockchain's private members. Such private members can access a copy of the entire blockchain; hence, all the information of Aadhaar including the changes a person has made. Aadhaar's applicants or those who want to make improvements to their current Aadhaar can access a Web portal (Fig. 4). The Web site allows the department of Aadhaar to introduce

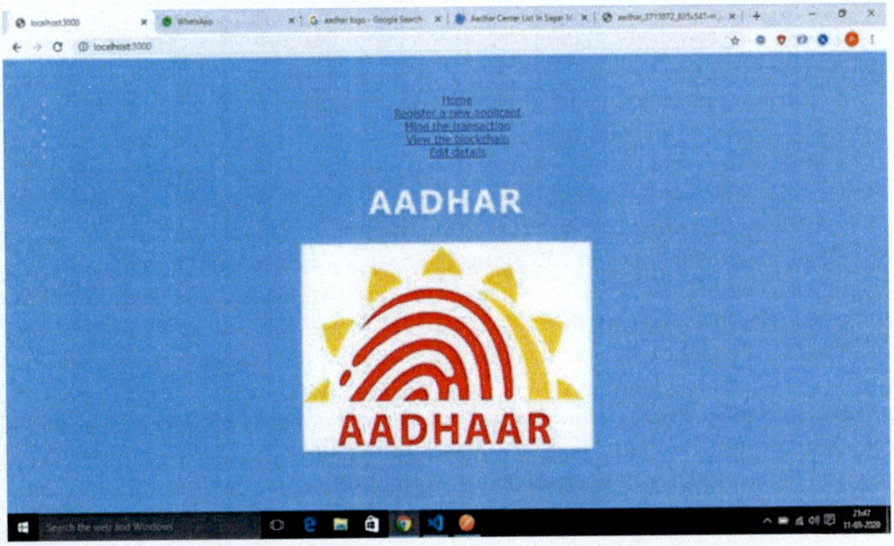

Fig. 4 Main page portraying all the options given to a user, including registration form and edit details form

new departments after authentication. All registration document details and modifications are maintained in an immutable manner. We use a proof-of-work algorithm reach a consensus that validates and verifies the applicant and their biometric and demographic data. It makes uploading data to all government agencies an simpler process and prevents central third-party security risks, data loss, and fraud. It also helps to save the user's time and energy.

4 Implementation

4.1 Modules

Creating the blockchain. In the project, Node.js is used to create a blockchain and the output is retrieved onto the Postman API or user interface (UI). The user-defined classes used are register, block, blockchain, transaction, Aadhaar (and) Smart Contract. Aadhaar Number for a new applicant gets generated when an object for the class register gets created. The app.js file uses HTTP methods like get and post. The applicant details are recorded as a transaction when the set terms mentioned in smart contract gets satisfied. The blockchain class contains functions such as getNextBlock() and addBlock() which are used to mine the transactions into a new block. It also contains a function named generateHash() to generate a hash value by continuously incrementing the nonce value until the constraint mentioned for the hash is satisfied.

The Aadhaar number retrieved is hashed for security and is only visible to the Aadhaar department and can be provided upon request and approval. All these values are printed in the form a block by as a JSON instance to display the previous hash, current hash, block information content, and timestamp. A nonce value should be set based upon the difficulty criteria in order to attach a block to the chain. When the block gets mined, one can authenticate, verify duplicates, and cryptographically secure information for further use. The validity of the block is checked during the mining process by verifying if the hash value of previous block has been altered or tampered. The changes made in the blockchain are notified to every participant node involved in the blockchain.

Peer creation and synchronization. A block once mined in a blockchain becomes visible to every node taking part in the network of blockchain. The multiple peers involved in the blockchain network should be registered first. In the project, we used various ports that behave as members or peers in the network. The registration of multiple ports is done using HTTP methods. Each port is acquainted with every other participant nodes. The nodes known to a specific node can also be displayed by making use of a get method.

A resolve function is used to synchronize blockchain data contained in a node with every other participant nodes of the network. The longest chain logic is implemented for the synchronization. The blockchain of the node having the longest chain gets copied to all other peers synchronizing them.

Creating UI and integrating with the blockchain.

Integrating UI with the blockchain is the final step where the well-crafted, user-friendly user interface with all the options to enroll for a new Aadhaar allows the user to fill up a form. The option to make changes to the existing Aadhaar is available which also leads to filling a form and providing the Aadhaar details to verify and authenticate the user. This information and data are to be dynamically uploaded into the blockchain after it has been verified by the smart contract. Here, the blockchain will check and store this data in a distributed network. An Aadhaar number gets displayed to the applicant interface after mining is done. In case of changes, the old card is deleted and the new one is stored in its place. This information is again updated as status updates to the user and the user can either print it or avail the shipping option. This entire process is highly secure and ensures there are no duplicates. It cannot be easily tampered and significant amount of data can be modified fast, within a few minutes.

4.2 Screenshots

See Figs. 5 and 6.

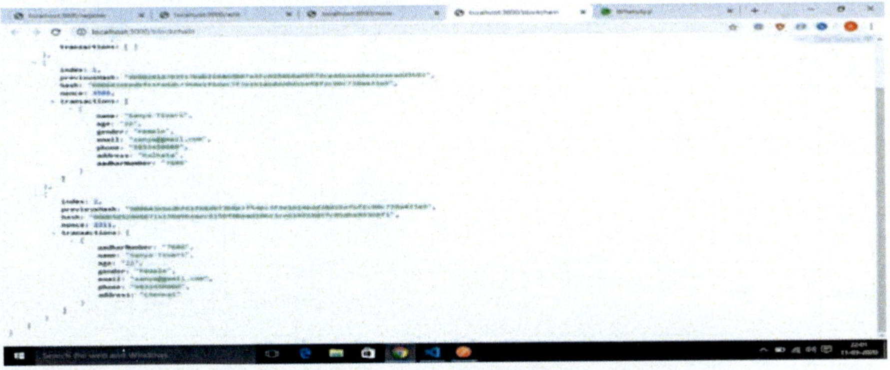

Fig. 5 Blockchain immutably displaying the edited details (change in address)

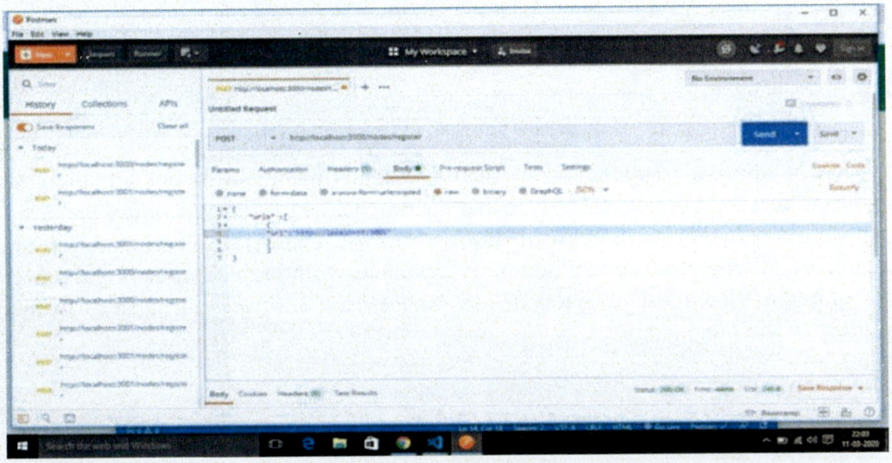

Fig. 6 Peer addition to whom a copy of the blockchain is deployed upon resolving

4.3 Analysis

A nonce (number only used once) value in a blockchain is a 32-bit number that is solved for upon which a block is awarded to a miner. The correct nonce value in a blockchain is usually preceded with a pre-set number of zeros. The miners compete to solve the cryptographic puzzle that requires heavy computation and skill to find this set number within a short period of time and ahead of others. Once found, the block is then awarded to the miner where he can mine the transactions of Aadhaar.

The leading number of zeros determine the difficulty in solving this puzzle. The greater number of zeros makes it harder to find the correct combination for the nonce value. This number is unique thus the more number of zeros eliminate the chances of duplication (Fig. 7).

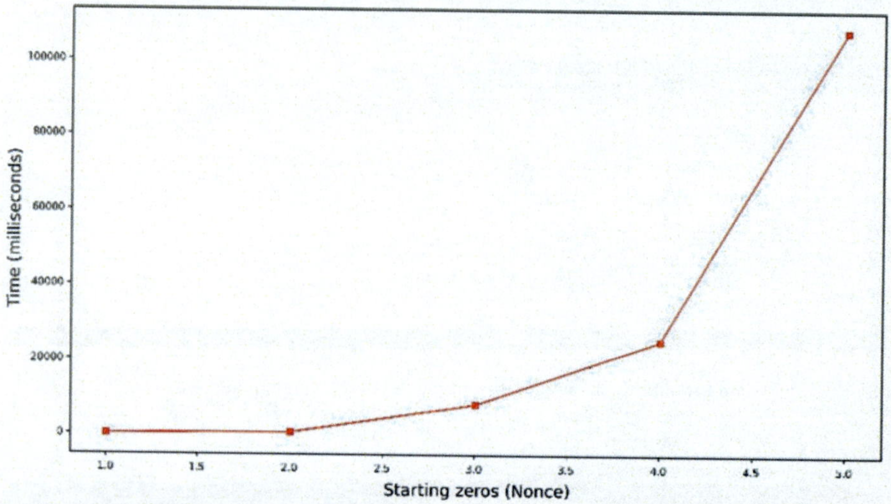

Fig. 7 Leading zeroes in nonce versus time

Here, we have a graph that shows the increase in time in our model with increase in the number of zeros. It is observed that the time taken to crack the nonce increases exponentially with the increase in the number of leading zeros. Hence, for the system of Aadhaar to function optimally, computer with high computational power is required to mine these transactions in a small period of time ensuring security and validity of the blockchain.

5 Conclusion

In the Aadhaar application, we propose to use a consortium blockchain that has the government departments like the tax division, the license department, authorized banks, etc., as the private members of the blockchain. These private members can download a copy of the entire blockchain using their private key. The citizen applicants of Aadhaar or the people who want to make changes to their existing Aadhaar are the public members of the blockchain. They are given a public key the first time they come to generate an Aadhaar and this public key is used to authenticate them the next time they want to make changes. We use a proof-of-work consensus algorithm to reach a consensus validating and verifying the applicant and his biometric and demographic data. This makes the updation of data to all the departments of the government an easier task and avoids the security threats of a central third party, data mismatch, and fraud.

References

1. Wu H-T, Tsai C-W (2019) An intelligent agriculture network security system based on private blockchains. J Commun Netw 21(5):503–508
2. Nguyen D-M et al (2019) MB-PBA: leveraging Merkle tree and blockchain to enhance user profile-based authentication in E-learning systems. In: 2019 19th International symposium on communications and information technologies (ISCIT). IEEE, 2019
3. Mahmoud HHM, Wu W, Wang Y (2019) Secure data aggregation mechanism for water distribution system using blockchain. In: 2019 25th International conference on automation and computing (ICAC). IEEE, 2019
4. Jamal A et al (2019) Blockchain-based identity verification system. In: 2019 IEEE 9th international conference on system engineering and technology (ICSET). IEEE, 2019
5. Monrat AA, Schelén O, Andersson K (2019) A survey of blockchain from the perspectives of applications, challenges, and opportunities. IEEE Access 7:117134–117151
6. Hasan HR, Salah K (2018) Blockchain-based proof of delivery of physical assets with single and multiple transporters. IEEE Access 6:46781–46793
7. Tang F et al (2019) An efficient authentication scheme for blockchain-based electronic health records. IEEE Access 7:41678–41689
8. Bosri R et al (2019) Towards a privacy-preserving voting system through blockchain technologies. In: 2019 IEEE international conference on dependable, autonomic and secure computing, international conference on pervasive intelligence and computing, international conference on cloud and big data computing, international conference on cyber science and technology congress (DASC/PiCom/CBDCom/CyberSciTech). IEEE, 2019
9. Feng Q et al (2019) BPAS: blockchain-assisted privacy-preserving authentication system for vehicular Ad-Hoc networks. IEEE Trans Industr Inf
10. Cai W et al (2018) Decentralized applications: the blockchain-empowered software system. IEEE Access 6:53019–53033

Convolutional Neural Network-Based Automatic Object Detection on Aerial Images

C. Vimala, T. Thenmozhi, M. Jagadeesh Kumar, and C. Subramani

Abstract Object identification is the one of the most developing area in image processing. In this paper proposed a neural network based object identification for aerial images. The computer is able to understand high level image are video for processing. Convolution neural network is detected the single object first and then detects the multiple objects. The input to an R-CNN is an image consisting of various instances, and output will be bounding boxes with the class label of the object. The input images are passed through discrete wavelet transform for pre processing. In this paper, it is mainly focused on object detection with segmentation.

Keywords Aerial images · Object detection · CNN · Neural network

1 Introduction

But the position of the object is identified in the image. This role is referred to as object detection, which usually consists of various subtasks that admire face detection, pedestrian detection and skeleton detection, mutually of the elemental PC vision issues, and object detection is ready to generate useful data for linguistic interpretation of images and videos and is applied to a range of applications, including image classification, human behavior analysis, facial recognition and autonomous driving [1, 2]. Meanwhile, inheritable from neural networks and linked learning systems, developments in these fields can establish neural network algorithms and can even

C. Vimala (✉)
Department of Electronics and Communication Engineering, SRM Institute of Science and Technology, Kattankulathur 603203, India

T. Thenmozhi · M. Jagadeesh Kumar
Department of Electrical and Electronics Engineering, SRM Institute of Science and Technology, Kattankulathur, India
e-mail: thenmozhi.eee@sairamit.edu.in

C. Subramani
Department of Electrical and Electronics Engineering, SRM Institute of Science and Technology, Kattankulathur 603203, India

S. S. Dash et al. (eds.), *Sixth International Conference on Intelligent Computing and Applications*, Advances in Intelligent Systems and Computing 1369,
https://doi.org/10.1007/978-981-16-1335-7_32

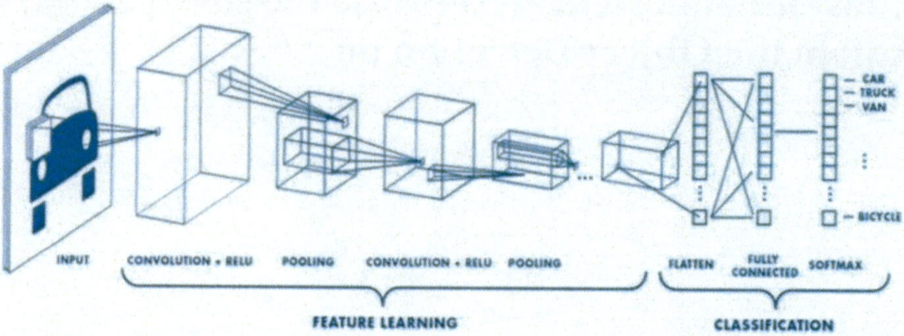

Fig. 1 CNN architecture

have pleasant effects on object detection techniques that could be known as learning systems. Because of immense differences in views, poses, occlusions and lighting conditions, however, it is difficult to fully Perform object detection with another function of localization of objects. Most attention has been centered in recent [3, 4].

The object detection is basically involved with the choice of informative area, extraction and classification features. Informative choice of region. Since completely different objects can appear in any location of the image and have different facet ratios or sizes, scanning the full image with a multi-scale window is a normal choice [5, 6]. While this comprehensive strategy will determine all the objects' doable positions, its shortcomings are evident. It is computationally high ticket due to an outsized number of candidate windows and generates too many redundant windows. However, if only a hard and fast set of window templates is implemented, failed regions may be generated as well [7]. The structure of CNN is shown in Fig. 1.

2 Methodology

The flowchart of the proposed model is shown in Fig. 2.

2.1 Image Processing

In image processing, Gaussian blurring plays a key role in processing, and Gaussian blurring is generally referred as a process used to blur an image using Gaussian function. Gaussian function is a mathematical function used to create Gaussian blur to remove noise and reduce details. The performance of the Gaussian filter is shown in Fig. 3. The Gaussian function in one dimension is

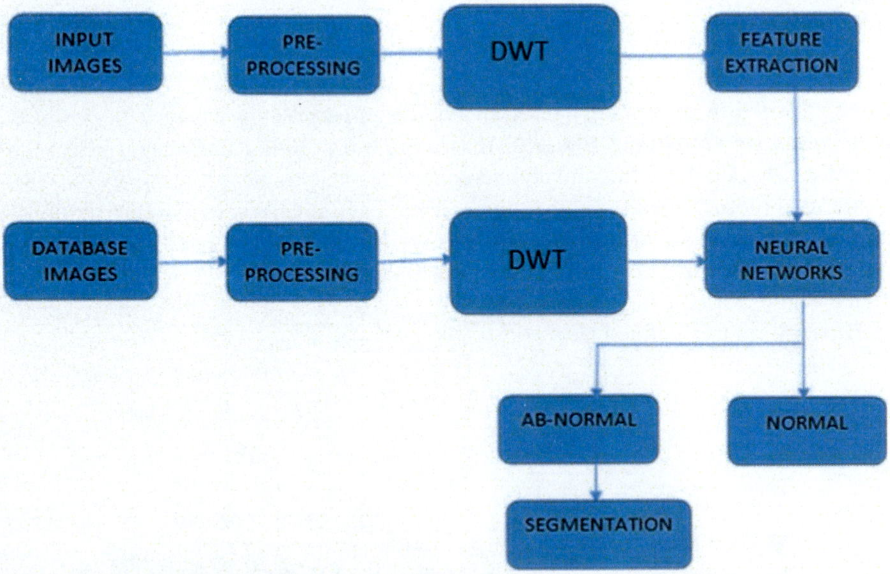

Fig. 2 Flowchart of the model

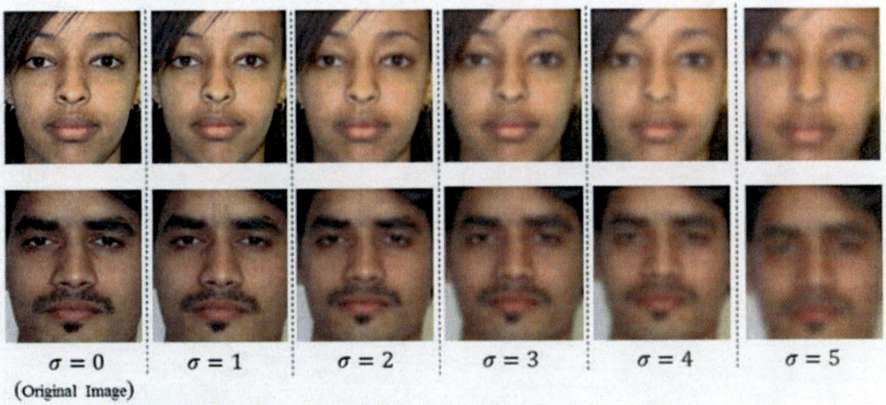

Fig. 3 Gaussian filtering

$$G(x) = \frac{1}{\sqrt{2\pi\sigma^2}} e^{-\frac{x^2}{2\sigma^2}}$$

The Gaussian function in two dimension

$$G(x, y) = \frac{1}{2\pi\sigma^2} e^{-\frac{x^2+y^2}{2\sigma^2}}$$

2.2 Bilateral Filtering

Bilateral filtering is generally referred for image processing where it performs as noise reducing smoothing filtering. It is a edge preserving the image processing technique. It calculates the ranked average of intensity values from pixels closer to it and uses these values to replace intensity of each pixel. Gaussian distribution is used to calculate weight average. But this weight is mainly depends on Euclidean distance between the pixels. The bilateral filter output is shown in Fig. 4 and the image segmentation example in Fig. 5.

Bilateral filter is defined as

$$I^{\text{filtered}}(x) = \frac{1}{W_p} \sum_{x_i \in \Omega} I(x_i) f_r(\|I(x_i) - I(x)\|) g_s(\|x_i - x\|),$$

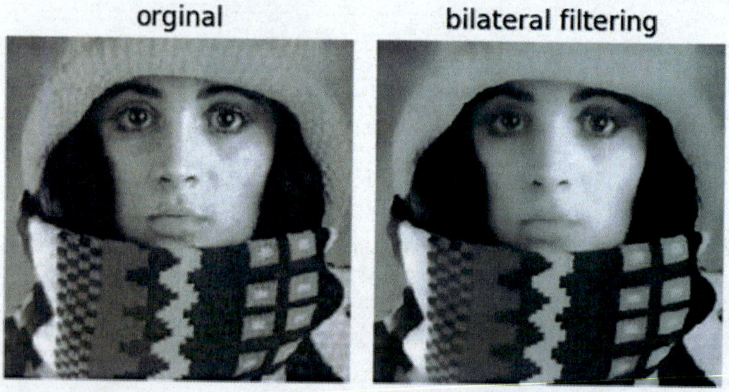

Fig. 4 Bilateral filtering

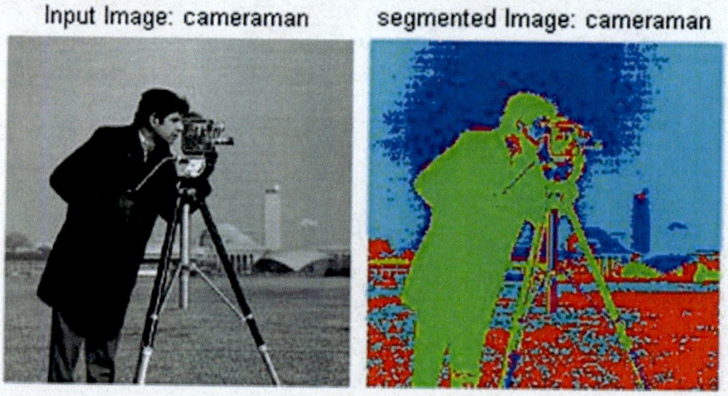

Fig. 5 Image segmentation

Normalization term W_p is defined as

$$W_p = \sum_{x_i \in \Omega} f_r(\|I(x_i) - I(x)\|) g_s(\|x_i - x\|)$$

2.3 *Image Segmentation*

It is a clustering algorithm used for image segmentation. This clustering is used for segmenting an image into further different categories on few basis without loss of any data and values. This K-means is an unsupervised learning used for segmentation of an image. Segmentation of images is defined as division of an image into smaller sections, without losing any pixels. This K-means algorithm is applicable when we have any uncategorized data or ungrouped data, and this is carried out by finding the most similar value denoted by K.

Steps involved in K-means segmentation

1. Choose the clusters number K.
2. Select the random points at K centroids.
3. Assign each data point to nearest centroid.
4. Valuate and place the new centroid of cluster.
5. Continue step 4 till each data is segmented.

2.4 *Image Classification*

The classification system aims to categorize all pixels during a digital picture into one of several groups of land cowls or "themes." Then, this hidden knowledge could be used to carry out the land cowl gift in a image thematic maps. Multispectral knowledge is typically used to conduct classification, and thus, the gift of spectral pattern between the information for each component is used because of the numerical basis for categorization. The aim of the image classification is to detect and portray, as a novel gray level (or color), the possibilities that occur in the picture with regard to the article or shape of the land cowl that these features are really reflected at the bottom. Probably, the most important aspect of digital image processing is photo classification. Getting a "beautiful picture" or image is wonderfully fun, showing a magnitude of colors that demonstrate various choices of the underlying piece of land, but it is rather useless unless you understand the meaning of the colors PCI (2007). 2 The main methods of classification are supervised classification and unattended classification.

2.4.1 Supervised Classification

We tend to create samples of the data categories (i.e., land cowl type) of interest within the picture with supervised classification. These are known to be testing grounds. The image process software package is then used for each data form to create an applied math characterization of the reflection feature. This stage is generally referred to as "signature analysis" and may entail developing a characterization as straightforward because of the mean or craze dimension of reflection on each band, or as advanced as comprehensive analyzes of mean, variance and variance across all bands. If a characterization of applied math has been achieved for each data group, the picture is then identified by analyzing the reflection factor for each variable and making a call with respect to the signatures it most resembles (Eastman, September 1995).

2.5 Feature Extraction

Extraction of features means reducing the amount of resources needed to describe an over dimensional collection of details. When an advanced intelligence analysis is played, one of the most critical issues comes from the quantity of variables involved. Study with an excessive number of variables typically involves a significant amount of memory and processing resources, as well as allowing a classification rule to overfit samples for coaching and generalizing poorly into new samples. Extraction of features can be a general term for strategies to create combos of variables to induce around these issues while also representing the information with adequate accuracy. If the code file to AN rule is just too large to be processed and is assumed to be redundant (e.g., a similar measure in each feet and meters, or the verboseness of pictures given as pixels), then it is also re-modeled into a reduced set of options (also called a vector functions). This method is called option of function. The selected options should contain the data from the file, so the appropriate function is often performed through mistreatment of this example rather than all initial knowledge.

3 Results and Discussion

Using MATLAB software and CNN, we take the aerial images taken from the UAV's and use it for the detection of objects. Image processing, classification and segmentation are done to filter the noise and define the borders and boundaries of the images, and later, using CNN and tensor flow, the detection of objects is done. The simulation input and corresponding output are shown in Figs. 6 and 7, respectively.

Fig. 6 Input image

Fig. 7 Output image

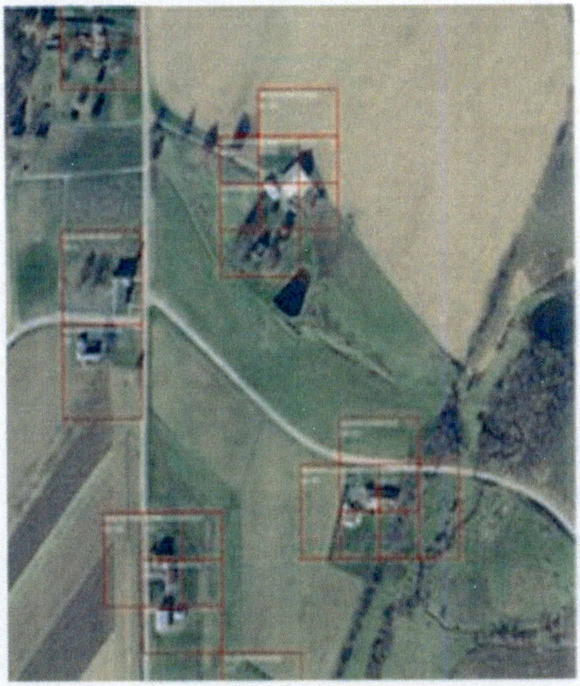

4 Conclusion

Mask R-CNN outperforms all existing single-model entries on every task. This method is one of the suitable object detection for aerial images. Feature map is extracted using CNN, which is of pixel size. Which is used to identify the objects during flood time. For these types of factors, we can enhance data, improve the edge enhancement and make the masking of each image more accurate.

References

1. Ortiz A, Bonnin-Pascual F, Garcia-Fidalgo E (2014) Vessel inspection: a micro-aerial vehicle-based approach. J Intell Robot Syst 76:151–167
2. Snavely N, Seitz SM, Szeliski R (2006) Photo tourism: exploring photo collections in 3D. ACM Trans Graph (TOG) 25:835–846
3. Vimala C, Aruna Priya P (2019) Artificial neural network based wavelet transform technique for image quality enhancement. Comput Electr Eng 76:258–267
4. Qu T, Zhang Q, Sun S (2016) Vehicle detection from high-resolution aerial images using spatial pyramid pooling-based deep convolutional neural networks. Multimed Tools Appl 1–13
5. Huerzeler C, Naldi R, Lippiello V, Carloni R, Nikolic J, Alexis K, Siegwart R (2013) AI robots: innovative aerial service robots for remote inspection by contact. In: Proceedings of the IEEE/RSJ international conference on intelligent robots and systems (IROS), Tokyo, Japan, 3–7 Nov 2013, p 2080
6. Wang Q, Rasmussen C, Song C (2016) Fast, deep detection and tracking of birds and nests. In: Proceedings of the international symposium on visual computing, Las Vegas, NV, USA, 12–14 Dec 2016, pp 146–155
7. Vimala C, Aruna Priya P (2014) Noise reduction based on double density discrete wavelet transform. IEEE international conference on smart structures and systems, IEEE digital library, pp 15–18, Jan 2015

Determining the Close Price of Bitcoin Using Regression Based on Blockchain Information

B. Ida Seraphim, Shivangi Dash, Khushi Ambastha, and B. Sowmiya

Abstract Bitcoin is a worldwide virtual currency. It is globally used as a financial asset and as a currency factor for buying and selling products and services in exchange of fractions or numbers of Bitcoins. Bitcoin is not owned by a lone authority or organization but rather is decentralized. Bitcoins can be sold, bought, or traded on platforms called "Bitcoin exchanges." Exchanges permit individuals to buy/trade/sell Bitcoins using an array of currencies using a P2P (i.e., peer-to-peer) network. This paper presents regression models for determining close price of Bitcoin in Bitstamp exchange based on characteristics of the cryptocurrency and blockchain information. Two live datasets have been acquired that provide real-time data consisting of various attributes of Bitcoin price and blockchain information and the two datasets have been combined to prepare final dataset for the investigation. Times-series model of the live data has been developed. The motive of this research is to provide an insight on prediction of Bitcoin close price. The underlying catalyst for implementing machine learning techniques is to meticulously forecast time-series data. Machine learning techniques have better demonstrated to outplay nonlinear techniques including neural network-based algorithms. The ultimate goal is to contribute an observation into the applications of various predictive models that can be used to predict Bitcoin prices.

Keywords Bitcoin · Machine learning techniques

1 Introduction

Bitcoin is a digitally available currency which uses cryptographic methods for security. New Bitcoins are originated by a competitive process known as mining. Miners

---

B. I. Seraphim (✉) · S. Dash · K. Ambastha · B. Sowmiya
Department of Computer Science and Engineering, SRM Institute of Science and Technology, Chennai, India
e-mail: idaserab@srmist.edu.in

B. Sowmiya
e-mail: sowmiyab@srmist.edu.in

S. S. Dash et al. (eds.), *Sixth International Conference on Intelligent Computing and Applications*, Advances in Intelligent Systems and Computing 1369,
https://doi.org/10.1007/978-981-16-1335-7_33

are paid by the network for their services. Miners gain Bitcoins by processing transactions and by procuring the network using specialized hardware. It was invented by an unknown person(s) using the name Satoshi Nakamoto in the year 2008 and was started in 2009. There exists a total of 17,982,562.5 BTC globally out of the 21 million there ever can be. It is a form of money which is convenient to use and is commonly accepted by people since it is accessible. Bitcoin's worth comes from its code, framework, security, and its acceptance. Bitcoin has the monetary attributes of durability, portability, fungibility, scarcity, divisibility, and recognizability. Bitcoin is now an easy digitally available financial asset. There are two important platforms when it comes to dealing with Bitcoin. One of them is the network which uses peer-to-peer framework that is used to maintain the blocks of the blockchain. The second is a Bitcoin exchange that is used to transact Bitcoins using various fiat currencies. The basis of prediction models for Bitcoin pricing generally constitute of a dataset and various algorithms used to statistically determine one or more of the pricing attributes. This paper provides a general survey of these algorithms and techniques. The trends in the Bitcoin price can be analyzed and also how it is affected by different parameters. There are few studies in existence which were performed in order to evaluate and predict the Bitcoin price. The construction of the Bitcoin price [8] is grouped into factors such as market forces, attractiveness for investors, and global macro-financial factors.

Prediction of Bitcoin's price has been researched at length in this paper. Various methods have already been used to predict the price or difference on the price of Bitcoin in the reference papers: BNN [1], RNN, and LSTM [2]. Since Bitcoin is significantly popular these days, its prediction shows in what way the Bitcoin features and other information affects the cost of Bitcoin. The price trend is shown using time-series modeling which is done with the help of candlestick graph.

The purpose of this paper is to predict the Bitcoin's price using different machine learning techniques and gets the accuracy for the same. Forecasting is also done for next few days using two regressor techniques. To facilitate the prediction, feature extraction is carried out using f-test which helped in elimination of nine attributes which were not affecting the Bitcoin prices significantly and the correlation of the remaining attributes was plotted using heatmap which helped in further obtaining the relevant features. The dataset is collected from combining two different sets of data obtained from Quandl and Blockchain.info. The dataset taken from Quandl include features such as Open, High, Low, Close price among others, and Blockchain.info dataset includes attributes such as total block size, Average block size, transaction count, mempool size, and others.

2 Literature Survey

Jang and Lee [1] conducted an experimental observational study based on facts and evidences to compare effective performance of a Bayesian neural network (BNN) with that of linear and nonlinear prediction models. A time-series analysis of the

Bitcoin information was done by developing a Bayesian neural network (BNN). From this study, we can credit a better performance ability to the BNN in comparison with other benchmark methodologies. The neural network model accomplished to better predict the price and explain the volatility in trend of Bitcoin price.

McNally et al. [2] aim at finding out the certainty with which Bitcoin price can be predicted. The close price is taken as the independent variable from CoinDesk Bitcoin Price. Random forest classification was used to extract features. After feature extraction, RNN and LSTM models were used for implementation. The accuracy from both the models is compared and presented.

Velankar et al. [3] The aim was to predict Bitcoin price. Dataset is fetched from Quandl and CoinmarketCap selecting specific attributes that are truly effective. It is normalized thoroughly using five techniques with regard to smoothening. Prediction model techniques used were bayesian regression, generalized linear model (GLM), and support vector machine (SVM) techniques.

Yogeshwaran et al. [4] introduce a trained machine methodology for accurately predicting Bitcoin price given appropriate data and sufficient computational power. To assist the framework of the model, support vector machine algorithm was used followed by a model generated employing convolutional neural network and recurrent neural network. Accuracy of both models was compared checking the number of layers. This paper concluded that the three-layered approach in both the models defeated the other approaches. However, models can be improved with the help of resources having better computational power.

Wu et al. [5] This research proposes a framework with model which is equipped with long short-term memory networks. The performance for the models was checked, and the research flow with different stages of the prediction process is demonstrated. Different tests are conducted to assess the performance. The dataset is analyzed and visualized to show the connection between different characteristics. The predicted values of the two models are close to the actual values. The trend direction of the deviations is also highly consistent. The models' predictions are the error values for prediction to show the comparison. The utility can be seen in various predictive works.

Madan et al. [6] This work aims to estimate the sign change in Bitcoin price by employing machine learning algorithms to conclude sign of priced change with an accuracy of 98.7%. Out of the 26 features, 16 were employed for initial model and seventy percent was used for the purpose of training. This paper employed random forest, support vector machine, and binomial logistic regression algorithms for the predictive model setup. It was concluded that the random forest approach gave the best result in contrast to the other proposed models.

Ciaian et al. [7] This paper puts forward the analytical study as to how the Bitcoin rate is impacted by the basics of the Bitcoin market. The Bitcoin's rise, framework, the interaction between supply and demand, and the financial developments are explained. The data comprises Wikipedia views and variables are derived from Bitcoin.org. Bitcoin price formation is impacted by supply and demand driving forces which are among the key factors in determining the stability of the price. Econometric approach is used for the analytical work where the explanatory variables are considered, and the Bitcoin Price is mutually interdependent. A series of specification tests are implemented to ensure the adequacy of estimated models.

Almasri and Arslan [8] This study uses neural network for forecasting cryptocurrency prices. A feedforward neural network was integrated, and data was normalized. Resilient backpropagation (Rprop) was applied to train the neural network. Tests were performed on various cryptocurrencies at a close price of hour and day. Encog machine learning framework was used for tests to be conducted on Dash, Bitcoin Cash, and Bitcoin. Two trials, one for 31 days and another for 150 h, were employed to conclude that hour-based predictions were more stable in nature.

Dyhrberg [9] presented the hedging capabilities of Bitcoin. Bitcoin has been put in a place where comparisons are made to gold since they have similarities. Different models are incorporated to see how Bitcoin used is correlated with the FTSE index asset. Threshold-GARCH is estimated to study the correlation in detail. The data used is collected from datastream, and the Bitcoin price is sourced from Coindesk Bitcoin Price Index. The data is modeled against different assets and portfolios. In this paper, the models used helped in showing the restricting ability of Bitcoin and also how the high regularity trading of Bitcoins creates fitting conditions for the conducted hedging. Hence, Bitcoin has now joined the list of instruments along with gold and other assets in order to minimize risks.

Vujičić et al. [10] provide a synopsis of Blockchain with cipher currencies. Firstly, reminiscence of older solutions for cipher currencies that prevailed before Bitcoin is provided. Then an insight of its kernel function of different cryptocurrencies like Bitcoin and Ethereum is provided. Following it an overview of the history of Bitcoin origin is bestowed, and following an addition of how a Bitcoin is mined and transacted along with scalability issues. Then there is a brief synopsis of Ethereum origin and history along with transactions, blockchain, and tokens are discussed. This paper furnishes a progressive array of primordial cipher currencies and its up-and-coming employment of Bitcoin and Ethereum.

3 Proposed System Overview

The approach being proposed in this paper for the prediction of Bitcoin price is as follows.

The first step in the price prediction of Bitcoin is data collection. The dataset that we have contains various attributes of Bitcoin and Blockchain Information. The Bitcoin data is acquired using an API key created in Python. From the Quandl website which provides datasets for financial data, we have collected one part of the dataset which is real-time data. The Bitcoin dataset consists of various attributes like open price and close price. From Blockchain.info, we have collected the Blockchain information which consists of features like total block size, average block size, transaction count, and others. The final dataset to be considered for implementation is obtained by combining the Bitcoin and Blockchain datasets together. Next step consists of data visualization which is performed using time-series modeling. A time series is a sequence of data points arranged (or listed or graphed) in time order. The time series visualization has been implemented with the help of a candlestick graph. Various features of Bitcoin are plotted in a regular timestamp to visualize the data. Then feature extraction is done using two methods which are F-test and heatmap. For the next step of implementation, Bitcoin's price is predicted by training the data for each algorithm first and performing prediction. The accuracy is checked for both the algorithms to show comparison. The final step is to project Bitcoin's close price for the particular exchange. Forecasting the Bitcoin's close price for the next five days, we have used two algorithms: decision tree regression and linear regression technique.

4 Data Specification

Bitcoin features are gathered from https://www.quandl.com/, from 1st April, 2014, till date (9th March, 2020). There are seven attributes which are selected—Open, High, Low, Volume BTC, Volume Currency, Close Price, and Weighted Price. Blockchain attributes are accumulated from https://www.blockchain.com/, starting from the same date as that for Bitcoin features to aid combining the two datasets into one and prevent incorrect mapping of data on identical dates for the dual datasets. The blockchain variables considered are: Total block size, Average block size, Number of Transactions, Median Transaction Confirmation Time, Hash rate, Difficulty, Miners Revenue, Transaction Fees, Cost % of Transaction Volume, Cost per Transaction, Unique Addresses, Total number of Transactions per day, Output Volume. The dual datasets after being obtained are combined into one final dataset which contains 20 attributes out of which close price is our target variable. It must be remembered that at any time, the Python implementation code is run, daily data from April 2014 is collected at run time. The dataset is stored in csv format. The blockchain information helps in identifying the appropriate features which affects the Bitcoin price. The combined dataset has 2100 + rows and 20 attributes in columns.

5 Time Series Visualization

After acquiring the daily data for six years, we have visualized it using a candlestick chart. A candlestick chart is a way to figure out the price change based on previous patterns. In this paper, Bitcoin price is visualized using this technique. This helps in showing the trend in the Bitcoin price. Candlestick charts are similar to bar charts. It represents all four important pieces of information for that day: open, close, high, low. This helps us in understanding the changes over time for the cryptocurrency's price. In Fig. 1, we have represented rise of price with green color and fall of price with red color. A red candle represents a price motion where the closing price is lower than the prior candle's close and a green candle represents a higher closing

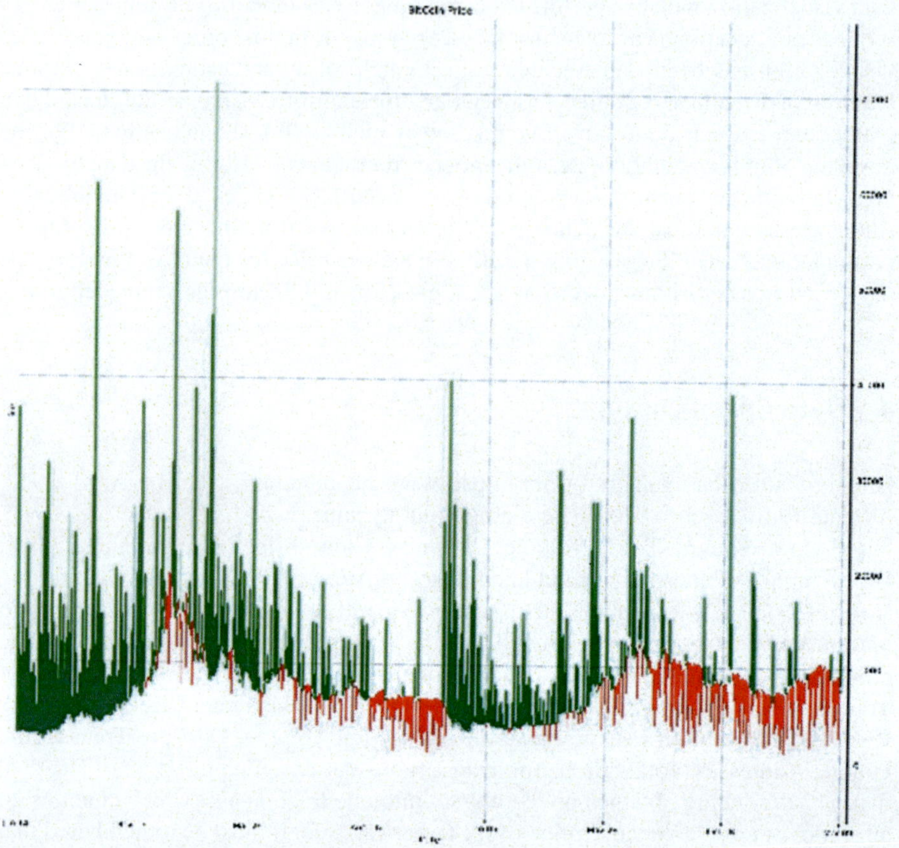

Fig. 1 Candlestick graph for Bitcoin price since Apr 2011

price than the prior candle's close. We can assign any color to rising or falling price candle bars.

6 Feature Extraction

Post-data collection, we cleaned the data record by replacing non-available values as zero. The next stage is to consider parameters or features which are relevant enough to be fed to the predictive network.

This is done with the help of two techniques: F-test and heatmap. The initial data contains a total of 20 parameters.

We used F-test to curtail the number of parameters by identifying the more significant features among them. After applying f-test, the number of attributes is reduced to ten. The ten parameters are: Open, High, Low, Volume of currency, Weighted price,

Difficulty, Miner's Revenue, Cost per Transaction, Number of transactions, output volume. Considering the quantity of data, to strengthen the model and accomplish the further contraction of features in the dataset based on correlation, a heatmap is generated. The heatmap gives a correlation between the selected ten features after the F-test, to further reduce the final number of attributes to be chosen for the model. It is often detected that correlated features add noise and inaccuracy to a model which can make it difficult to reach an outcome with desired accuracy.

So, heatmap gives an effective way to identify the correlated variables and allow us to streamline the feature selection process. Using the heatmap, we are mapping the ten parameters with themselves to understand intercorrelation of the parameters. It is obvious that all diagonal values of the heatmap will show correlation as 1. Using heatmap, we were able to diminish the number of parameters from ten to six. All parameters showing correlation of 0.9 and above were analyzed and eliminated (Fig. 2).

The final six parameters will be used for the prediction process (Table 1).

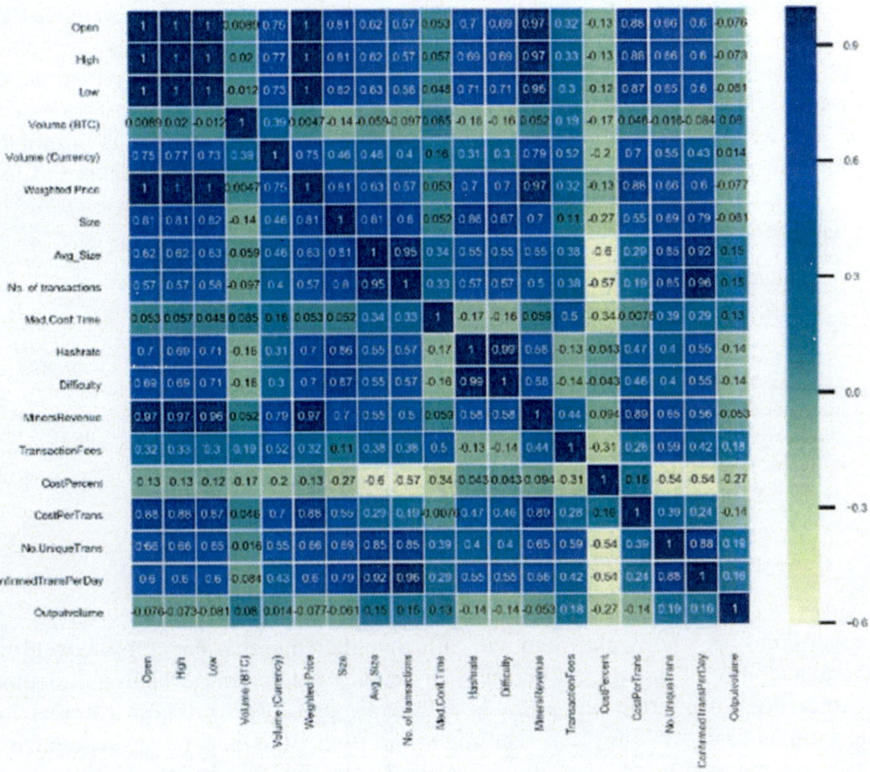

Fig. 2 Heatmap showing correlation between parameters after F-test

Table 1 Final list of parameters to be fed to training model

Features	Definition
Open	Opening price
Volume currency	Total price volume of day in USD
Difficulty	Relative measure of how difficult it is to fins new block
Cost per transaction	Miner's revenue divided by the number of transactions
No. of unique transaction	Unique addresses used
Output volume	Total value of all transaction outputs per day

7 Prediction and Analysis

In the prediction phase, to model the data we have divided the dataset into train and test data. The data from the csv file format is read into the data frame. We have then used 65% of the dataset as training data and 35% of the dataset as test data. The trends learned from the trained set of data will be used to assess the close price estimation of the test dataset. The trained data is fed to the machine learning algorithm model, and then accuracy is checked for results produced when the train data is fit into the model. Accuracy of the predicted values will be tested against the test dataset to give an account of how the technique applied is able to perform or not.

This procedure is followed for the two techniques used.

A. Linear Regression.
B. Lasso Regression (Fig. 3).

Further, we incorporated different techniques to foretell the Bitcoin close price for the coming five days. Decision tree regressor and linear regression method are used. Various error values are determined for the two techniques such as mean-squared error, mean absolute error [MSE], root mean squared error [RMSE], and R-squared error [R2] to check the efficiency of the models.

8 Conclusion and Future Scope

Cipher coins are a growing trend and influence the emerging economy. Given the unpredictability of this digital currency, it becomes important to enable a method to foretell an estimation of its price. This paper presents a predictive model for the estimation of Bitcoin price. The main purpose of this paper is to recognize a suitable technique to forecast the value of Bitcoin for a significant time period. The prediction and forecast of the Bitcoin close price have been achieved through the implementation which has helped in understanding the basic trend in the price instability of Bitcoin and also how different criteria affects the price. The accuracy check shows how their accuracy values differ. Both the prediction frameworks have

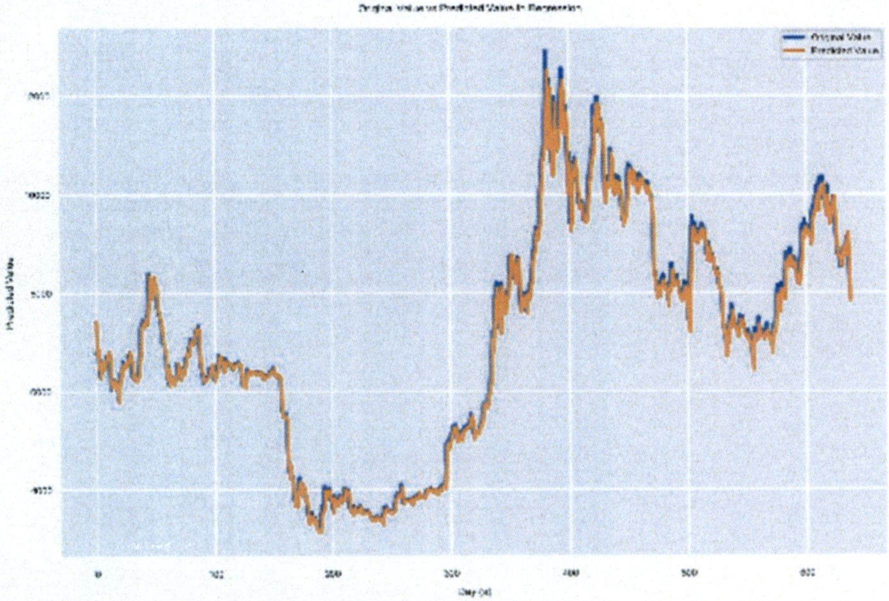

Fig. 3 Plot of original and predicted values for linear regression model

performed well by showing a precision of more than 95%. For further study, how the price of Bitcoin is influenced by market mechanism, investor's interest, and other global macro-financial factors can be analyzed. Different modelling techniques can be exploited for prediction, and forecasting can be done for a greater number of days. Neural network models can be of worthy use.

References

1. Jang H, Lee J (2018) An empirical study on modeling and prediction of bitcoin prices with bayesian neural networks based on blockchain information. IEEE Access
2. McNallyS, Roche J, Caton S (2018) Predicting the price of bitcoin using machine learning. In: IEEE 26th Euromicro international conference on parallel, distributed and network-based processing (PDP)
3. Velankar S, Valecha S, Maji S (2018) Bitcoin price prediction using Machine Learning. In: International conference on advanced communications technology (ICACT), 11–14 February 2018
4. Yogeshwaran S, Kaur MJ, Maheshwari P (2019) Predicting bitcoin prices using Deep Learning. In: IEEE global engineering education conference (EDUCON), 9–11 April 2019
5. Wu C-H, Ma Y-F, Lu C-C, Lu R-S (2018) A new forecasting framework for Bitcoin price with LSTM. In: IEEE international conference on data mining workshops (ICDMW), 17–20 November 2018
6. Madan I, Saluja S, Zhao A (2015) Automated Bitcoin trading via machine learning algorithms. Tech Rep

7. Ciaian P, Rajcaniova M, Kancs D (2016) The Economics of Bitcoin price formation. Appl Econ 48(19):1799–1815
8. Almasri E, Arslan E (2018) Predicting cryptocurrencies prices with neural networks. In: 6th international conference on control engineering and information technology (CEIT), 25–27 October 2018
9. Dyhrberg AH (2016) Hedging capabilities of Bitcoin. Is it the virtual Gold? Finance Res Lett
10. Vujičić D, Jagodić D, Ranđi S (2018) Blockchain technology, Bitcoin, and Ethereum: A brief overview. In: 17th international symposium INFOTEH-JAHORINA, 21–23 March 2018

Automatic Number Plate Recognition Using Convolution Neural Network

B. Rajasekar, B. M. S. Krishna Roshan, B. Chandrababu Naidu, and V. Vijaya Kumar

Abstract Automatic number plate recognition system (ANPR) finds the vehicle that violates traffic signal by detecting from the number plate of the vehicle. After finding the details of the vehicle, it sends a violation message to the owner. This ANPR will take the digital image as input. It uses segmentation techniques to detect the number plate region. The segmented image is compared to the original number plate image by correlation technique. The obtained data is checked with the database. We used MATLAB software which extracts the number plate from the input image, and it further uses GSM module (SIM900) to send a violation message to the person after violating traffic signal. Here the input image is further processed using convolution neural network algorithm for more accuracy purpose. After segmentation, each character made to compare with the thousands of characters for more accuracy. After obtaining the number plate number, it checks the data base for details of the owner. Later the process sends a violation message to the owner of the vehicle. Foreign ANPR systems are not able to connect with the Indian systems due to less compatibility. So this paper decreases the difficulty of the system.

Keywords OCR · CNN · GSM module · MATLAB

1 Introduction

World without vehicles like cars and bikes is very difficult in present generation. Each and every vehicle has their identity which is known as number plate displayed at the front. So by using that many vehicle-based applications are developed. Due to increase in population, the traffic violations also getting increased.

For controlling these issues, an ANPR system is the best thing. Automatic number plate recognition is a useful technique for detecting vehicles. Mainly, an ANPR system contains three core steps: (1) preprocessing, (2) breakdown of characters, and (3) convolutional neural network.

B. Rajasekar (✉) · B. M. S. K. Roshan · B. C. Naidu · V. V. Kumar
Sathyabama Institute of Science and Technology, Chennai, Tamil Nadu, India

S. S. Dash et al. (eds.), *Sixth International Conference on Intelligent Computing and Applications*, Advances in Intelligent Systems and Computing 1369,
https://doi.org/10.1007/978-981-16-1335-7_34

Many papers have been researched for getting the exact information about ANPR system. The methods followed in those papers are using OCR technique and some machine learning techniques. In machine learning techniques, there are different types like convolutional neural network, support vector machine, and neural networks using sine functions. All these techniques will not give the accuracy for number plate. In this paper, we used optical character recognition using convolutional neural networks. The process used here gives the utmost accuracy for extracting the number plate. After getting the information about the number plate, a traffic violation notification will be sent to the owner. The violation message is sent to the owner using GSM module.

2 Existing System

In India, there are many papers published based on automatic number plate detection. All those papers have their own pros and cons. We have the tried the best method for obtaining high accuracy. Some of the algorithms are mentioned as below. In [1 5], the proposed algorithm is based on feature extraction which is used to extract the character from the given input. In this, they had also given training for the characters to get the high accuracy level.

In [2 6], they proposed the segmentation process which brings a lot of accuracy level. Not only had that it also used to get correct extraction of the number plate. By using [3–6], we can be able to train the images for the correct recognition of the characters. This can be trained based on the artificial neural networks.

3 Proposed System

The main aim of this part is to give well-described information regarding the number plate detection and extraction. Generally, the image of the number plate is taken by a cc camera present at the traffic signals. After that the ANPR system undergoes the following process. In the last stage, after getting information of the violated vehicle, a violated notification is sent through GSM module (Fig. 1).

4 Methodology

As we mentioned earlier, the methodology is as shown below.

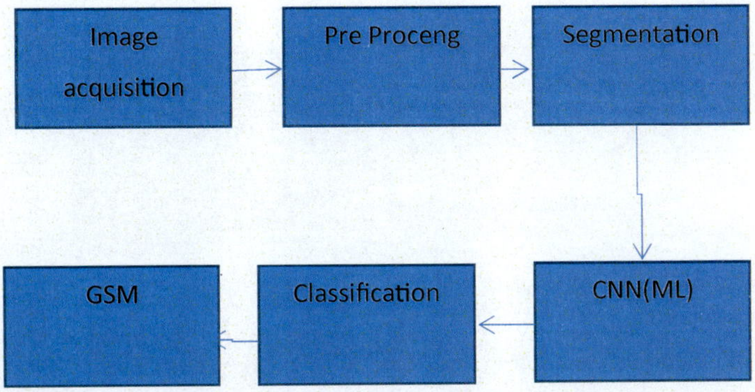

Fig. 1 Block diagram for proposed system

Fig. 2 Image of a car number plate

4.1 *Image Acquisition*

In general, it is a process of obtaining the image from the hardware components like camera. These cameras are present at the traffic signals which captures the image of a vehicle when violates the traffic signal. Here we are placing an image which we taken from the camera in Fig. 2.

4.2 *Pre Processing*

The input image after getting from the server may consist in different colors. So the image will get converted into the standard background, and then it will be further processed to another process. Here in this paper, it will be converted into black background and processed to next process as shown in Fig. 3.

4.3 *Segmentation*

Segmentation is a process of dividing the image into different segments. It is mainly used to find the boundaries that are present in an image. We can divide image segmentation into different methods. For example, methods based on compression techniques

Fig. 3 Preprocessing of number plate

Fig. 4 Loss in CNN system

propose the best method of segmentation. It is the one which minimizes data's coding length and the general probable segmentations.

4.4 Character Recognition Using CNN

After the segmentation process, we used neural networks downloaded from [5] to train our process. Even after training, we used the same neural networks for recognizing the characters. Firstly, we use neural networks with two layers at the beginning and two layers at the end. The dataset consists of 1000 sample images for each of the 36 characters. Out of 36,000 samples, we used the first 30,000 samples as training data and remaining 6000 samples as test data (Figs. 4 and 5).

4.5 GSM Module

Basically, GSM module is a hardware component which is used to send the message using Sim (900). By [7], we got the maximum information about the GSM module and so that we can be able to send the message for alert notification as shown in Fig. 6.

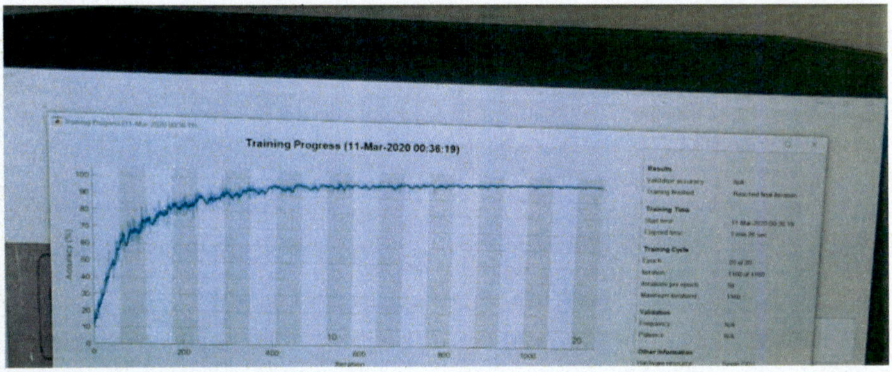

Fig. 5 Accuracy in CNN system

Fig. 6 GSM module

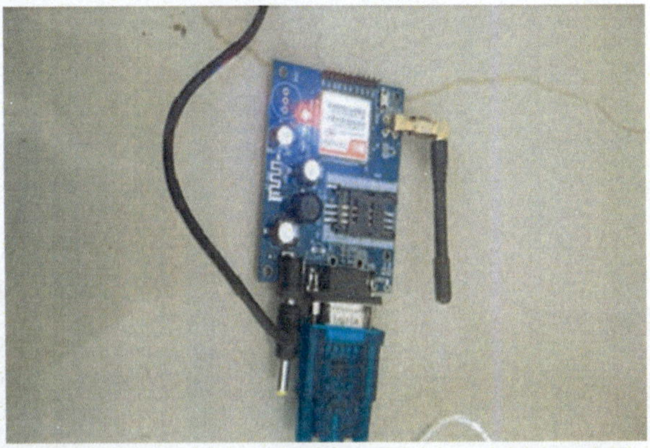

5 Result

The training which we gave using the convolutional neural network has worked very well bringing the accuracy of the number plate. It even shows the loss decrease and accuracy increase in the form of graph as shown in Fig. 7.

Fig. 7 Output of the number plate

6 Conclusion and Future Scope

Automatic number plate recognition is a very big field where it can be implemented and can be developed by many number of techniques. Here we used optical techniques for obtaining the number plate and implemented convolution neural network or artificial neural network techniques for recognizing the characters. After that a traffic violated message is sent to the owner if he violates the traffic rules.

As the population is growing very vastly, it is very difficult to control people on traffic roads manually. This is done based on the simple font's styles which are taken from the reference. By our technique, we are able to bring the accuracy of recognizing the number plate.

In future, we can even develop for different font styles of number plates. Besides that, we can even bring the accuracy of 100% by using different techniques. In our project, we got accuracy up to 95% and above. In this growing world, we definitely need an accuracy of 100%.

References

1. Jain AS, Kundargi JM (2015) Automatic license plate recognition using artificial neural network. IRJET 2(4). e-ISSN 2395-0056; ISSN 2395-0072
2. Bhat R, Mehandia JM (2016) Recognition of vehicle number plate using matlab. https://ijireeice.com
3. Russell SJ, Norvig P Artificial intelligence a modern approach, 2nd ed. Pearson Education
4. Rajasekaran S, Vijayalakshmi Pai GA Neural networks, fuzzy logic and genetic algorithms. PHI Learning
5. Sivanandam SN, Deepa SN Principle of soft computing, 2nd ed. Wiley Publication
6. Indian Central Motor Vehicle Act, Bear Act, Rule no. 49, 50
7. Chaithra B, Karthik KV, Ramkishore D, Sandeep R (2017) Monitoring traffic signal violations using ANPR and GSM. In: 2017 international conference on current trends in computer, electrical, electronics and communication (CTCEEC)

Coordinated Design of PSS and AVR for Stability Enhancement Using Differential Evolution Algorithm

Sangram Keshori Mohapatra, Monoj Kumar Senapati, and Srikanta Kumar Dash

Abstract In a coordination design of a power system stabilizer (PSS) and automatic voltage regulator (AVR) for power system, reliability enhancement is introduced in this paper. The design problem is that formula is constructed as an optimization problem. Optimization problem and differential evolution algorithm (DE) is used as an approach for optimal controller parameter search. Various objective functions like ITAE, IAE, ITSE, and ISE are employed, and a comparison has been made. The performance of the proposed DE optimized coordinated PSS and AVR is evaluated under different loading condition, type of fault, and location of fault in the studied power system. Moreover, the proposed algorithm is more effective than traditional algorithm which is demonstrated in this work by simulation.

Keywords Power system stability · Power system stabilizer (PSS) · Automatic voltage regulator (AVR) · Differential evolution algorithm (DE)

1 Introduction

The power system is routinely subjected to a variety of disturbances. The power system stability can be well defined as the capability of the power system to maintain the transition following a disturbance and achieve an acceptable operating condition [1, 2]. With the increasing deregulation practices of the electric sector presently, the practical application of power system stability not only shapes it as one of the most important but also at the same time makes it the most difficult issue to asses and control [3]. There are many different techniques used to damp out the oscillation in the rotor angle. The terminal voltage magnitude of synchronous generator is maintained to a certain level by the AVR. It also plays a key role in its reactive power controlled and stability. Then PSS is used to add damping to the rotor angle oscillation and

S. K. Mohapatra · M. K. Senapati (✉)
Department of Electrical Engineering, G.C.E, Keonjhar, India

S. K. Dash
Department of Electrical Engineering, GIFT, Bhubaneswar, Odisha, India

S. S. Dash et al. (eds.), *Sixth International Conference on Intelligent Computing and Applications*, Advances in Intelligent Systems and Computing 1369,
https://doi.org/10.1007/978-981-16-1335-7_35

compensate the negative damping of the AVR. In this work, a coordinated design of PSS and AVR for stability enhancement is presented [4–10].

Over the last few years, several inventive computation techniques have been implemented by different researchers to interpret various optimization problems. One such optimization algorithm is differential evolution technique which is very easy to understand and computationally technically advantageous. With this advancement, DE comes to existence which is flexible and well-balanced mechanism to improve the local and global exploration abilities. Hence, DE algorithm is used to enhance the stability and damped the oscillation in power system [4].

2 Systems Under Study

2.1 *Single Machine Infinite Bus*

A single machine infinite bus power system is taken as system under study as shown in Fig. 1. The system configuration of the power system in Fig. 1 consists of a 2100 MVA, 13.8 kV, 60 Hz synchronous generator which is connected to a 300 km long double-circuit transmission line through a 3-phase 13.8/500 kV step-up transformer. The synchronous generator is provided with hydraulic turbine and governor (HTG), an automatic voltage regulator (AVR), and a power system stabilizer (PSS). All of the other relevant parameters are given in Appendix. The Sim Power Systems (SPS) toolbox is carried out for all simulations as well as SSSC-based damping controller design. SPS is a MATLAB-based advanced build tool that enables scientists and researchers to easily and quickly build designs to simulate power systems with help of Simulink. Other components of SPS include powerlib, model of typical power equipment such as machines, governors, excitation systems, transformers, and transmission lines. The load flow and machine initialization option of the Powergui block performs their respective functions.

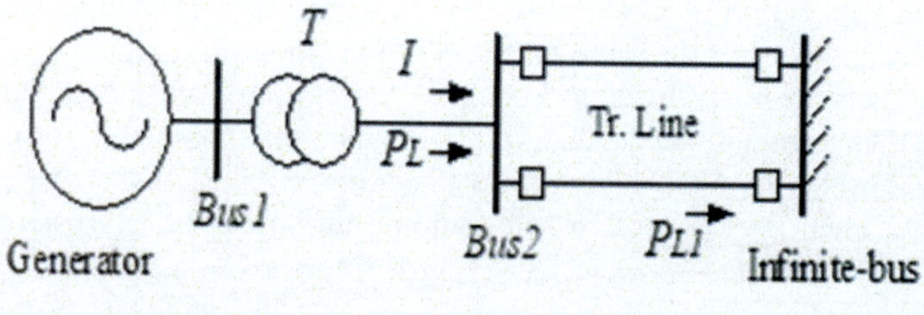

Fig. 1 Single machine infinite bus power system

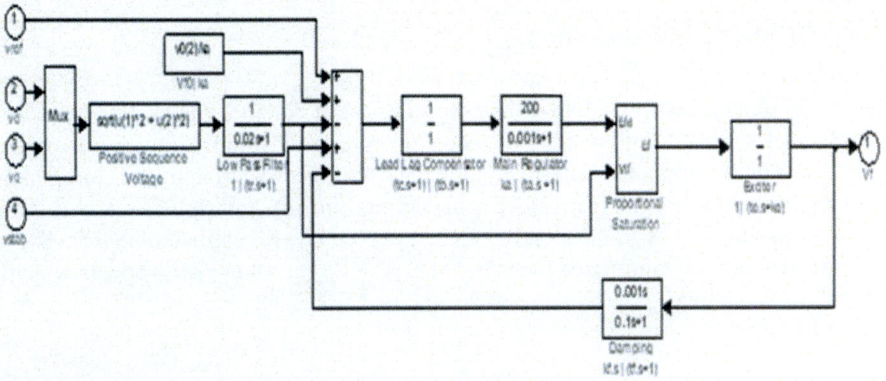

Fig. 2 MATLAB/Simullink model of excitation system

2.2 Modeling the Excitation System

The excitation system provides excitation voltage for synchronous machine and regulates its terminal voltage. The MATLAB/Simulink block for excitation system is shown in Fig. 2. It consists of (i) a low-pass filter, (ii) main regulator with gain Ka and time constant Ta, (iii) exciter, (iv) transient gain reduction,, (v) damping filter, and (vi) output limiters.

2.3 Modeling the Power System Stabilizer

The functions of PSS mentioned here as follows. PSS is used to enhance damping of the synchronous generator by regulating its excitation. The input signal of PSS can either be the speed deviation of synchronous generator or acceleration of power which is the difference between the mechanical input power and the electrical power ($P_a = P_m - P_{e0}$), and the output signal acts as an additional input voltage (V_{stab}) to the excitation system block. The PSS model for MATLAB/Simulink block comprises of (i) a sensor, (ii) a gain block with gain K, (iii) a washout block, (iv) multistage lead lag blocks with appropriate time constants, (v) damping filter, and (vi) output limiters as shown in Fig. 3.

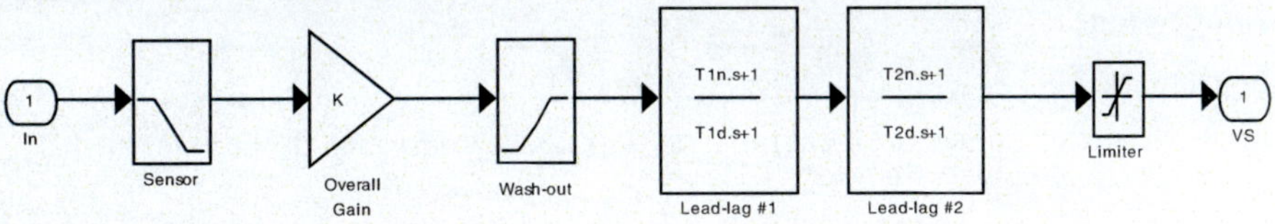

Fig. 3 MATLAB/Simulink model of PSS

2.4 Problem Formulation

The main objective of PSS and AVR controllers is to enhance the system response in terms of settling time, overshoot, and undershoot by minimizing the objective function. The performance criteria include the four-objective function: (i) integral of time multiplied absolute error (ITAE), (ii) integral of time multiplied squared error (ITSE), (iii) integral of squared error (ISE), and (iv) integral of absolute error (IAE). The speed deviation minimized the value of Eqs. 1, 2, 3 and 4 in SMIB power system. These are given by:

$$\mathrm{ITAE} = \int_{0}^{t_{\mathrm{sim}}} (|\Delta\omega|) \cdot t \cdot \mathrm{d}t \tag{1}$$

$$\mathrm{ISE} = \int_{0}^{t_{\mathrm{sim}}} (\Delta\omega)^2 \cdot \mathrm{d}t \tag{2}$$

$$\mathrm{ITSE} = \int_{0}^{t_{\mathrm{sim}}} \left[(\Delta\omega)^2\right] \cdot t \cdot \mathrm{d}t \tag{3}$$

$$\mathrm{IAE} = \int_{0}^{t_{\mathrm{sim}}} (|\Delta\omega|) \cdot \mathrm{d}t \tag{4}$$

where $\Delta\omega$ is the speed deviation in and t_{sim} the time range of the simulation.

3 Differential Evolution Algorithm

DE algorithm is newly introduced advanced algorithm base population optimization technique that operates with two sets of population. However, the population size of the algorithm is controlled by the parameter NP and the population comprises of real valued vectors with dimension D that is equal to the number of control variables. The evolutionary operators are mentioned below [4];

Initialization:

For each parameter j with upper and lower bound taken as X_j^U, X_j^L, respectively. The initial parameter values are generally randomly set uniformly in the interval [X_j^L, X_j^U].

Mutation:

For a given parameter vector $X_{i,G}$, three vectors are selected randomly such as $(X_{r1,G} X_{r2,G} X_{r3,G})$ and $r1$, $r2$ and $r3$ are random numbers generated. Then a donor vector is found by summing the weighted difference between the two vectors to the third vector in Eq. (5) where mutation factor F varies between 0 and 2.

$$V_{i,G+1} = X_{r1,G} + F.(X_{r2,G} - X_{r3,G}) \tag{5}$$

Crossover:

This is the final step where all the three parents are considered for recombination, and the resultant child is a perturbation of one of them. The trial vector is developed from the elements of the target vector as well as the donor vector, and the elements of the donor vector enter the trial vector with recombination probability *Cp* as in given Eq. (6)

$$U_{j,i,G+1} = \begin{cases} V_{j,i,G+1} & \text{if } \operatorname{rand}_{j,i} \le Cp \text{ or } j = I_{\text{rand}} \\ X_{j,i,G+1} & \text{if } \operatorname{rand}_{j,i} > Cp \text{ or } j \ne I_{\text{rand}} \end{cases} \tag{6}$$

where $i = (1, 2, ..., N; j = 1,2, ..., D)$ and

With $\operatorname{rand}_{j,i} \sim U(0, 1)$, I_{rand} is a random integer from $(1, 2, \ldots, D)$

Selection:

In this step, the target vector is compared with trial vector which gives the better fitness value that is admitted to the next successive generation. The selection operation can be denoted by Eq. (7)

$$X_{i,G+1} = \begin{cases} U_{i,G+1} & \text{if } f(U_{i,G+1}) < f(X_{i,G}) \\ X_{i,G} & \text{otherwise.} \end{cases} \tag{7}$$

where $i \in [1, N_P]$.

4 Results and Discussions

The PSS and AVR controllers are used for the transient response of the proposed system in term of settling time, peak time, and overshoot by minimizing the fitness function.

The four type of objective function (ITAE, IAE, ISTE, and ISE) has attempt in the proposed single machine infinite power system. The controller parameters are obtained using DE algorithm and in each case of objective function take run the system minimum 20 time and best value of objective taken as in the given Table 1.

Table 1 Controller parameter of SMIB by using DE algorithm

Objective function	K_p	$T1$	$T2$	$T3$	$T4$	K_a	T_a	Best fitness
ITAE	7.2835	0.2392	0.1487	0.4537	0.1377	286.0081	0.0041	0.0122
IAE	6.9667	0.2740	0.1366	0.3459	0.1044	214.6521	0.0079	0.0067
ISTE	8.2875	0.1957	0.1117	0.3809	0.1031	222.1841	0.0041	4.4181×10^{-5}
ISE	7.2777	0.3932	0.1159	0.4206	0.2626	266.0023	0.0097	3.0154×10^{-5}

A 3-Φ fault occurs at the middle of the transmission line which causes transient instability in the power system. DE-based coordinated AVR-PSS controller is used to damp out the oscillation to make the system stable. This is shown by the simulation results of sending active power, rotor angle, and speed deviation at different loading conditions. Three types or loading conditions are considered in the proposed system which is reflected as given in Table 2. Table 3 shows the best value of objective functions (ITAE, ITSE, ISE, IAE) of SMIB power system and the peak time, maximum overshoot, and settling time under these objective functions.

Case-I: Nominal Loading

In this case, the loading is 80%. The fault is triggered at 1 s, then be cleared after 0.083 s. Figs. 4, 5, 6 and 7 represent the response of the system during severe disturbances. Power system oscillations have been improved significantly by the DE-based coordinated AVR-PSS controller as compared to non-coordination of these devices.

Table 2 Different loading conditions

Loading conditions	Active power (pu)	Rotor angle (°)	Fault clearing time (s)
Small	0.30	16.25	0.083
Nominal	0.80	48.47	0.083
Heavy	0.95	57.67	0.083

Table 3 Comparison between different fitness functions

ITAE parameters	ITAE	IAE	ITSE	ISE
IAE	0.0140	0.0072	5.6749e−005	3.3106e−005
ITSE	0.0128	0.0066	4.6530e−005	2.8892e−005
ISE	0.0125	0.0065	4.5105e−005	2.7922e−005
Peak time	0.0121	0.0065	4.9386e−005	3.0208e−005
Max. overshoot	0.085	0.085	0.085	0.085
Settling time	0.00541	0.005413	0.005414	0.005419

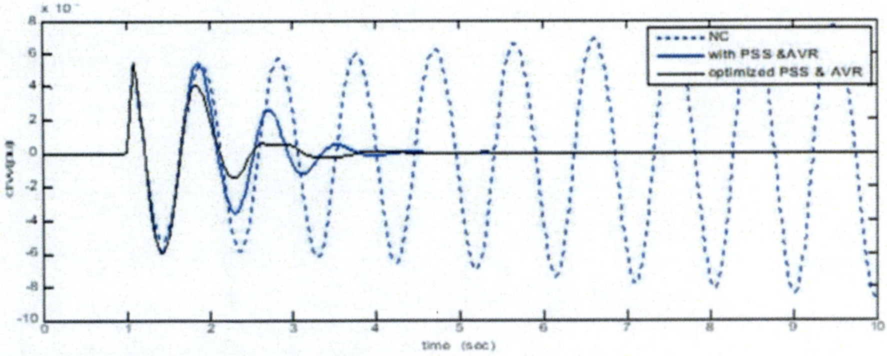

Fig. 4 Speed deviation response under case-I

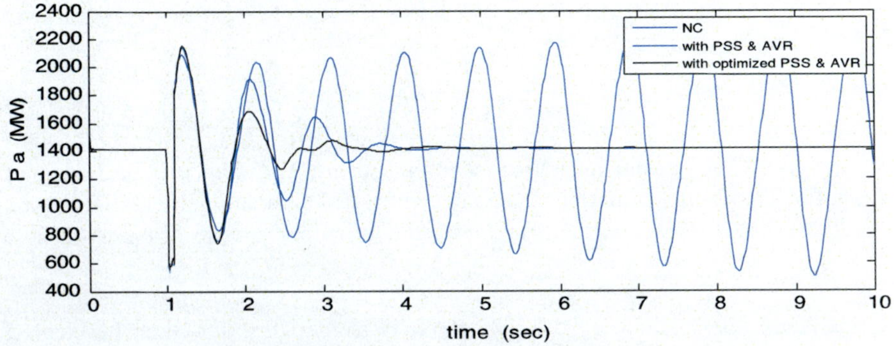

Fig. 5 Active power response for case-I

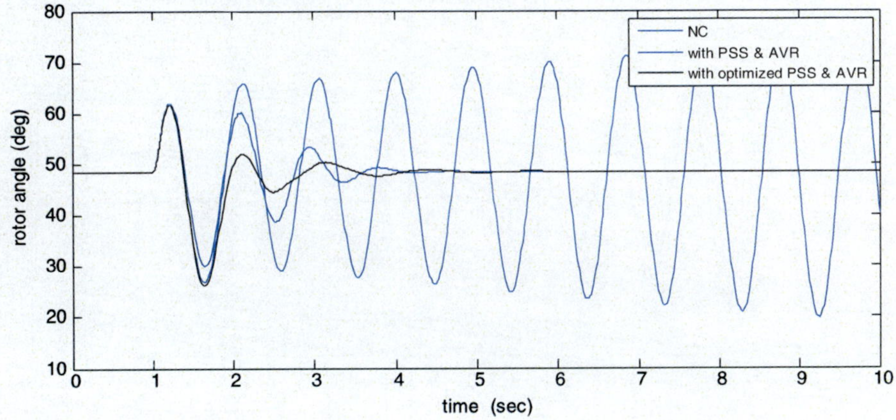

Fig. 6 Rotor angle response for case-I

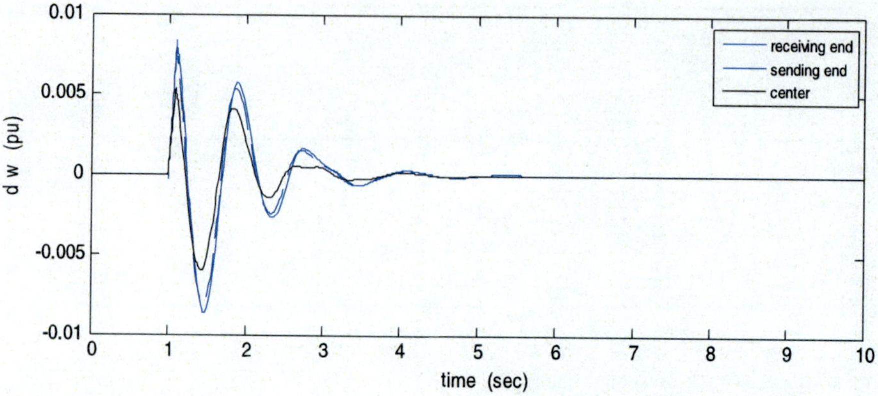

Fig. 7 Speed deviation response for case-I with different location of the transmission line

Case-II: Small Loading:

In this work, the loading is 30%. The fault is triggered at 1 s, then be cleared after 0.083 s. The system response under this severe disturbance is presented in Fig. 8. Power system oscillations have been enhanced compelling by the DE-based coordinated AVR-PSS controller as compared to non-coordination of these devices.

Case-III: Heavy Loading

In this study, the loading is 95%. The fault is triggered at 1 s, then be cleared after 0.083 s. The system response under this severe disturbance is presented in Fig. 9. Power system oscillations have been enhanced compelling by the DE-based coordinated AVR-PSS controller as compared to non-coordination of these devices.

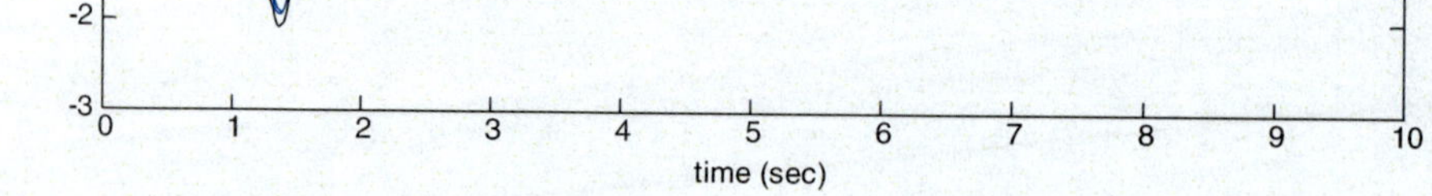

Fig. 8 Speed deviation response for case-II

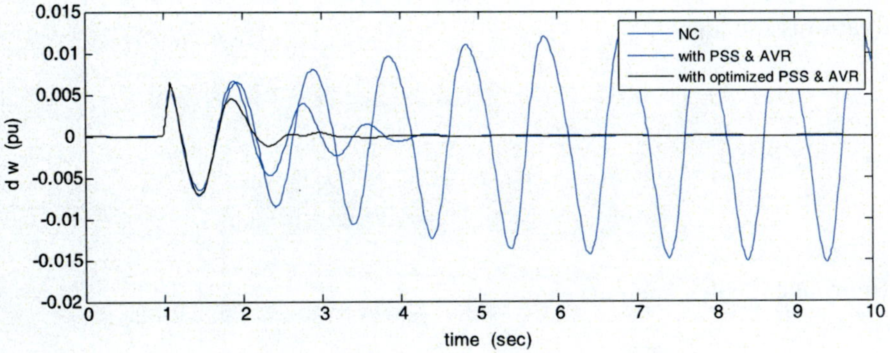

Fig. 9 Speed deviation response for case-III

5 Conclusions

In this paper, transient stability improvement using coordinated PSS and AVR is studied in the single machine infinite bus power system. Differential evolution optimization technique is used to optimize the controller parameters of the PSS and AVR. Time domain simulations are carried out to demonstrate comparison between presented method and conventional method.

Appendix

Single machine infinite bus power system (SMIB):

Generator:

$$\begin{aligned} S_B &= 2100\,\text{MVA},\ H = 3.7\,\text{s},\ V_B = 13.8\,\text{kv}, \\ f &= 60\,\text{Hz},\ R_s = 2.8544e-3,\ X_d = 1.305\,\text{pu}, \\ X'_d &= 0.296\,\text{pu},\ X''_d = 0.252\,\text{pu},\ X_q = 0.474\,\text{pu}, \\ X'_q &= 0.243\,\text{pu},\ X''_q = 0.18\,\text{pu},\ T_d = 1.01\,\text{s} \\ T'_d &= 0.053\,\text{s},\ T''_{q0} = 0.1\,\text{s} \end{aligned}$$

Transformer:

$$\begin{aligned} &2100\,\text{MVA},\ 13.8/500\,\text{kv},\ 60\,\text{Hz},\ R_1 = R_2 = 0.002\,\text{pu}, \\ &L_1 = 0,\ L_2 = 0.12\,\text{pu},\ D_1/Y_g\text{connection}, \\ &R_m = 500\,\text{pu},\ L_m = 500\,\text{pu} \end{aligned}$$

Transmission line:

$$
\begin{aligned}
&3-\mathrm{ph},\ 60\,\mathrm{Hz},\ L_1 = 300\,\mathrm{km},\ L_{2-1} = L_{2-2} = 150\,\mathrm{km},\\
&R_1 = 0.02546\,\Omega/\mathrm{km},\ R_0 = 0.3864\,\Omega/\mathrm{km},\\
&L_1 = 0.9337\mathrm{e}-3\,\mathrm{H/km},\ L_0 = 4.1264\,\mathrm{H/km},\\
&C_1 = 12.74\mathrm{e}-9\,\mathrm{F/km},\ C_0 = 7.751\mathrm{e}-9\,\mathrm{F/km},
\end{aligned}
$$

Hydraulic turbine and governor:

$$
\begin{aligned}
&K_a = 3.33,\ T_a = 0.07,\ G_{\min} = 0.01,\ G_{\max} = 0.97518,\\
&V_{g\min} = -0.1\,\mathrm{pu/s},\ V_{g\max} = 0.1\,\mathrm{pu/s},\\
&R_p = 0.05,\ K_p = 1.163,\ K_i = 0.105,\ K_d = 0,\\
&T_d = 0.01\,\mathrm{s},\ \beta = 0,\ T_w = 2.67\,\mathrm{s}
\end{aligned}
$$

Power system stabilizer (PSS): $T_s = 30\mathrm{e}-3\,\mathrm{s}$, $V_s^{\max} = 0.15\,\mathrm{pu}$, $V_s^{\min} = -0.15\,\mathrm{pu}$.

References

1. Kundur P (1994) Power system stability and control. McGrawHill, New York
2. Mohapatra SK, Panda S (2014) Coordinated design of power system stabilizer with FACTS based damping control by using gravitational search algorithm. J Bioinform Intell Control Am Sci Publ 3:1–13
3. Panda S, Swain SC, Rautray PK, Malik RK, Panda G (2010) Design and analysis of SSSC-based supple-mentary damping controller. In: Simulation modelling practice and theory, vol 18(9)
4. Mohapatra SK, Panda S, Satpathy PK (2012) Power system stability improvement by simultaneous tuning of PSS and SVC based damping controllers employing differential evolution algorithm. Int J Electr Sci Eng (WASET) 6(9):1244–1252
5. Panda S, Padhy NP (2008) Comparison of particle swarm optimization and genetic algorithm for FACTS-based controller design. Appl Soft Comput 8(4):1418–1427
6. Gozde H, Taplamacioglu MC (2011) Comparative performance analysis of artificial bee colony algorithm for automatic voltage regulator (AVR) system. J Franklin Inst 348:1927–1946
7. Quinot H, Bourles H, Margotin T (1999) Robust coordinated AVR + PSS for damping large scale power systems. IEEE Trans Power Syst 14(4):1446–1451
8. Rao PS, Sen I (2000) Robust pole placement stabilizer design using linear matrix inequalities. IEEE Trans Power Syst 15(1):3035–3046
9. Kundur P, Lee DC, Zein El-Din HM (1981) Power system stabilizers for thermal units: analytical techniques and on-site validation. IEEE Trans Power App Syst PAS-100(1):81–95
10. Cao Y, Jiang L, Cheng S, Chen D, Malik OP, Hope GS (1994) A nonlinear variable structure stabilizer for power system stability. IEEE Trans Energ Convers

Retrieval Services Using Mobile Computing and Blockchain

J. Albert Mayan, Tharangini Venna, and S. Vignaya

Abstract In present world, everyone is busy and has no concern on other individual person. This application is totally for this technological world where no one has time to think about others. Is anyone there to spend time for the welfare of other individual? Definitely not! This proposed application concentrates on the abandoned child, injured animal or any crime activity that takes place in their vicinity. Unforeseen circumstances can happen at any time, and since everyone is running for their livelihood, they do not find time for other individuals. These types of circumstances can be solved using this "Retrieval services using mobile computing and block chain" application. The concerned application has many advantages and very secured. Since smartphones are getting trending and famous, this proposed activity can be achieved using any Android or iOS devices. Blockchain is the major concept used in this application which provides an effective database. This application helps the other person who is in danger situation who seeks the help of another person. The process of this application is very fast, and there will not be any delay. The pride of saving the life of other individual can be entitled to the person who uses this application from their mobile phone during unforeseen situations. The procedure for using this proposed application is very simple, and the person who installs the application from app store or play store has to enter their details—name, email id and phone number. If the user notices any unfair situation, the user can immediately capture the moment and send to the authorized person who is appointed for the particular location which is done through blockchain. It also intimates the user about the final status of the victim.

Keywords Android · Blockchain · Abandoned · Longitude and latitude · Rescue

---

J. A. Mayan (✉)
School of Computing, Sathyabama Institute of Science and Technology, Chennai, India

T. Venna · S. Vignaya
Department of CSE, Sathyabama Institute of Science and Technology, Chennai, India

S. S. Dash et al. (eds.), *Sixth International Conference on Intelligent Computing and Applications*, Advances in Intelligent Systems and Computing 1369,
https://doi.org/10.1007/978-981-16-1335-7_36

1 Introduction

Android operating system is an essential operating system because of user friendly, low memory consumption, platform independence and portability, and its scope never gets down in the technology. Because of using Android operating system in many laptops, mobile phones and tablets, it has increased its demand on high rate. Now, everyone is using smart phones. Since the Android is playing an important part in everyone's life, lot of people started emerging many applications day by day, and it is free of cost. Smart phones made life secured in various aspects. For better future, the existing condition made both the parents to work. The safety of the present younger generation is very important in such scenarios. The Android and iOS contain global positioning system (GPS) which helps to find the location of anyone who is away from home. The proposed system that is using the Android technology builds a secure application for injured animal, abandoned children and any additional crime activity. To extract the data at any time, the proposed system uses the blockchain which is an important database. The data which is in the block chain cannot be distorted. The exact location of the target can be traced by latitude and longitude positions. The main benefit of proposed application is that the application can be worked in both iOS and Android. The code which is written is very transportable because the code which is implemented is transcribed in React Native. The main advantage of React Native is that the written code is same enough for both the iOS and Android and no separate code is needed. The proposed system has a benefit that the prominence of the happening is updated to the person who takes the picture. The reorganized information directly goes to the nearby police station if the person who authorized is unable to get the injured animal or abandoned children.

2 Review of Literature Survey

Lih-jenkau et al. [1] proposed an algorithm "pocket fall accident detection positioning and rescue system". In this scheme, fall mishap spotting and correlate saving system is suggested to cell phone using 3G Web. With the help of viewpoints generated by the waveform sequences and electronic compass, the mobile phone could be used as arrangement inputs. The contributions that are acquired are used for recollection purposes. After characteristics are confirmed on gong state, the situation can begin for the following state. On other hand, the arrangement will be amended for the baseline condition that can be hold back for the further sequences. If any plummet occurrence takes place, the manipulator's spot can be procured by the GPS, the position is dispatched to saving organization over 3G network, hence the component is once substantiated at the current circumstance national, and they can spring to the future state. And the shortcoming of this is that the activity should be compelled and trailed at differing circumstances.

UpalMahbub et al. [2] proposed a technique where he had explored about how an handset can be used for computerized end-user attestation. The use of non-identical data files and the Computer-Assisted Structure Elucidation (CASE) programs help in user attestation identification. Location-based subsequent place divination and biometric identification are prearranged that designate the additional vigorous practice. The fine-tuned for the mobile party line is necessary to realize adequate authentication correctness. The data obtained is available to the research center for assisting supplementary research.

Michael cheffena [3] proposed an "fall detection system builds on mobile phone audio landscapes" are developed. Linear predictive coding, matching pursuit features and mel frequency cepstral coefficients of various machine learning classifiers—minimum square method, support vector device, k-nearest neighbor classifiers and unnatural standard network—are examined for the difference between no-fall events and fall, and for each individual audio feature, the concert in terms of responsiveness, conscientiousness, computational complexity and exactness is obtained. And the demand for training and testing allotment also has a suitable calculus.

Nehabhati [4] proposed a system for the "fall related injuries that cause severe health damage, psychological problem, immobility and physical damage for senior citizens". Frequently, the treatment for fall-related injuries is hindered due to long days without treatment after falling because of delayed or absence acceptance of the fall incident since there will not be anyone around to inform the concerned person. The proposed system explains the design model for fall detection. The proposed system that was designed comprises the main dispensation and sensing units which have a temperature sensor, heartbeat sensors and accelerometer and pressure sensor. This architype makes use of smart phones which are widely used, on reach over a short period of time.

The rhythmic sensor values are transmitted to intended mobile application and further to the movable sensor for uninterrupted keeping track of the senior person. This arrangement has the important features of giving an alarm and telling the health camps, friends, emergency contacts, relatives and the health care camps during much needed situations.

Allen et al. [5] proposed an established approach for the people those are suffering from the post-stroke by the sensor-based assignment. The ultimate aim of proposed system is to use their initial activity capacity component to systematize fall risk experiments. By the paper, the proposed work has shown present state of system.

Together with the proposed happening, scoring techniques for spotting the events such as their action, movement and the steps involved throughout the test. Signal processing tools which are used for determining the patient's balance are also with the system. And the proposed system is delivered as running applications on the smart phones.

3 Methodology

The person who has noticed about the wounded animal and the abandoned children in their locality will take the picture by the application that was created for the mobile phones. The details of the person who uploaded the picture in the developed application will be stored. When the picture sent by the person is received by the authorized person, then the person will form a team for the variety of cause, and the team will be sent to the locality of the shattered picture. As soon as they have seen the incident, then they decide whether they need to send the injured animal to the hospital or the abandoned children to the orphanage. After that, the details are sent back to the person who has clicked the image.

Algorithm: For Identifying the Location

Step 1: Initializing the user Id as zero, current name is stored in app, and application program interface of current name stored in API, cross origin resource sharing of application (cors(api)).

Step 2: Requesting the HTTP Protocol by request parsing ().

Step 3: Adding the arguments latitude, longitude, effDate in string format.
parser.add_argument('latitude',type=str).
parser.add_argument('longitude',type=str) parser.add_argument('effDate',type=str).

Step 4: Getting the user Id, user name, Email Id.

Step 5: Importing the OS.

Step 6: In class Register.

printing "processResource1".

latitude=str(args['latitude']).

longitude=str(args['longitude']).

effDate=str(args['effDate']).

printing latitude, longitude, effDate.

Step 7: Calculating file length for validation.

if len(request.files).

file=request.files['filetoupload'].

if file.filename!='':data_file_namefile.filename.

printing"processResource2".

Step 8: Parsing arguments and storing in args printing"processResource3".

Step 9: Connecting to server by pyodbc package storing in conn1 printing "processResource4".

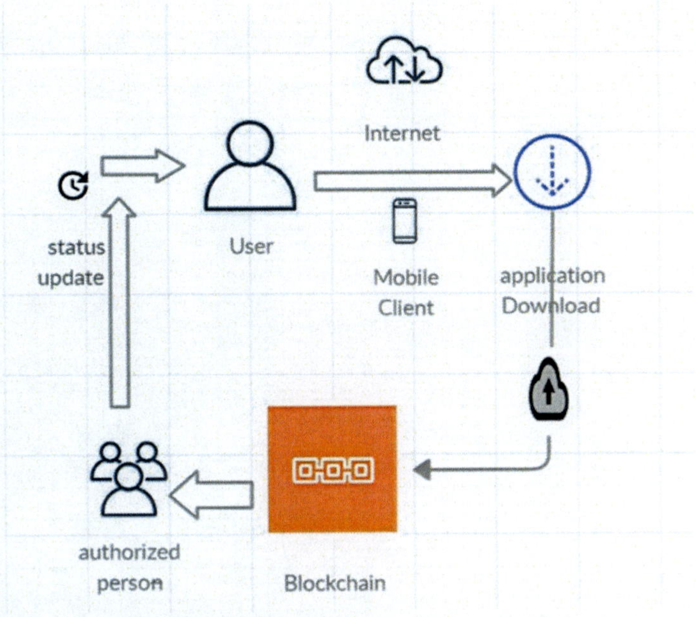

Fig. 1 Design architecture of retrieval service

Step 10: ifname=='main': app.run (host='192.168.1.17′, port=5009,debug=True).

Step 11: Stop the process.

4 System Architecture

The above-mentioned algorithm portrays the entire process of proposed application. This depicts step-by-step flow. The login details of the user are acquired to avoid misusing the application. Once the information is obtained, the user does not give his or her details each time while logging. The details stored and retrieved whenever needed (Fig. 1).

5 Results and Discussion

Figure 2 is the framework for the proposed application. The proposed application is available in iOS store and the store in Android, and the application can be downloaded from it.

Figure 2 is the page for uploading the picture of the injured animal, abandoned system and any other criminal doings that take place. And the application that was

Fig. 2 Proposed framework

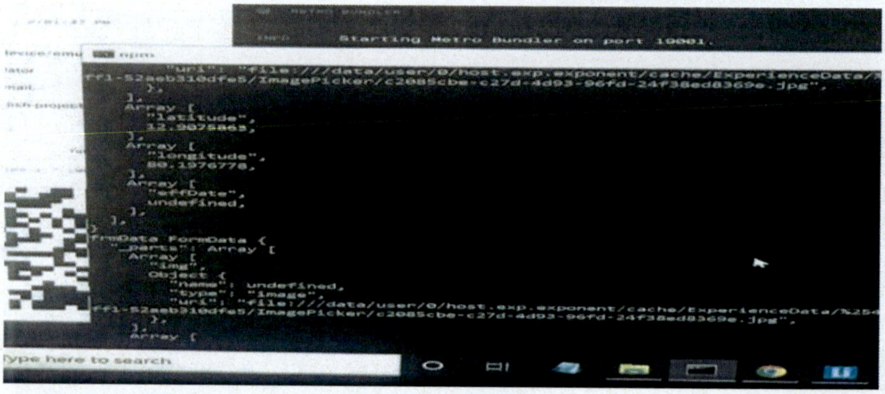

Fig. 3 Getting details of the location

proposed contains two options "pick image—in which the person is going to select the pictures from the gallery". And the option "send to authorize person—is the option for forwarding the image to the person who is authorized", and the person can retrieve the picture from the blockchain database. And the location details will be sent to the authorized person.

Figure 3 shows that the details of the victim are getting by the authorized person. The details are sending in the form of the latitude and longitude, and with these exact details of the location, they send the setup team to the location. With this exact location, the authorized person can easily find the location details of target.

As the proposed system uses the blockchain, the data is blunder free, and no replica data is received. And it is the main benefit of blockchain, and this data cannot be misused by the third party.

Figure 4 illustrates the login page for application in which it requests the user to log in. By the way of asking the login details, the proposed system can save the data from the third-party people. If anyone has entered the wrong details, then that person will not be logged in.

Figure 5 shows the blockchain in which the picture details, date and address are taken. It will have the option "send mail" by which the person who is authorized will send the mail to the person after rescuing by the team in persons locality. The response is directed by the user to person for knowing whether the problem is solved in their location. Blockchain is a distributed database technically known as neck-to-neck ledger.

All the data is saved in the form of blocks, each block has a unique code termed as "HASH", and it gives a unique identification to the blocks that are created. If any data misusing activity occurs, it would be stopped to process at that instant of time itself making it safe and secured.

It is a mobile application, and when we find any abandoned child while moving on road, we will just click the picture of them, capture in gallery and upload that picture in the application. These pictures are moved into the blockchain for Security purpose so that authorized person may not misuse the information. When the information is sent to the authorized person, he immediately takes care of data received and sends the updates of the child to the person who took the picture (Fig. 6).

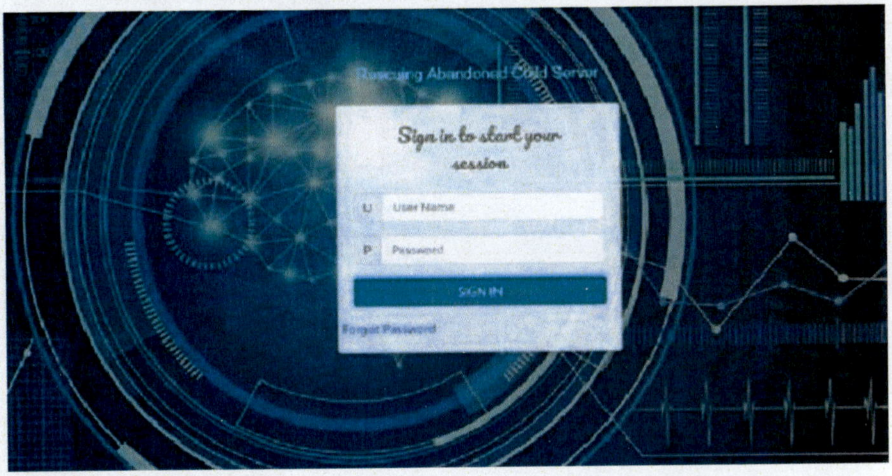

Fig. 4 Blockchain login page

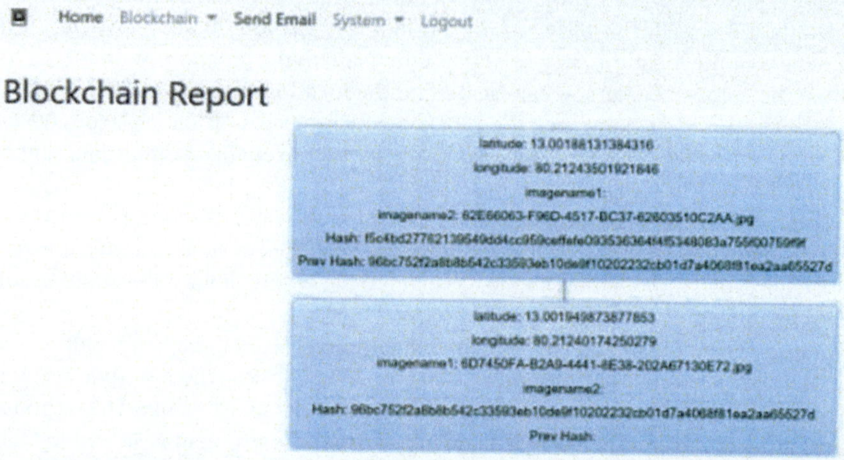

Fig. 5 Blockchain generated

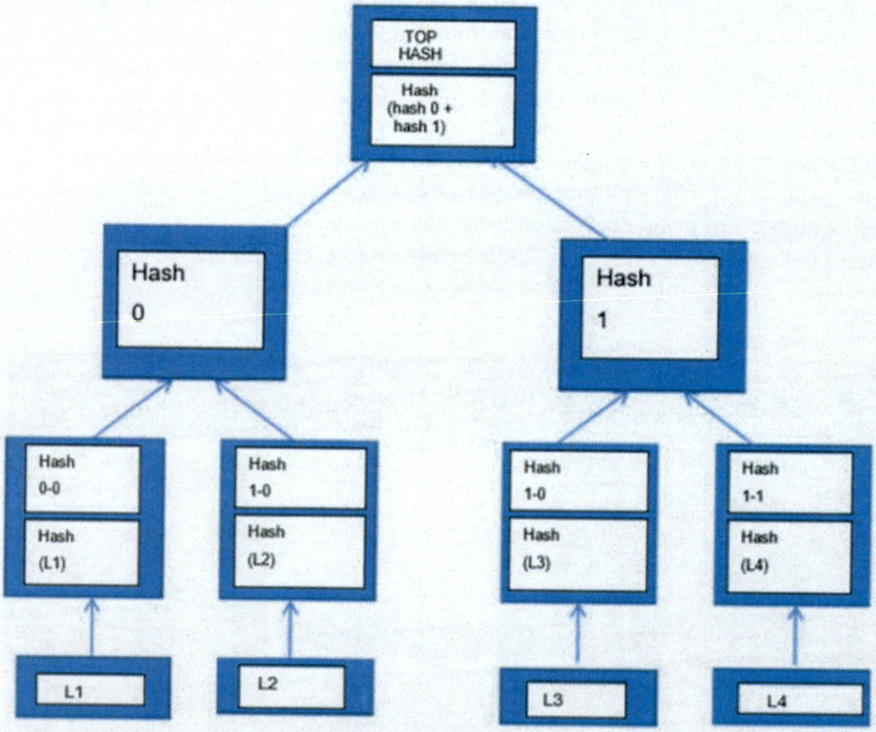

Fig. 6 Demonstrating hash in blockchain

This application is mainly for saving the abandoned child and also for saving time of user by just clicking the picture.The Application Developed are Tested [6, 7] and can be developed to suite both Android User and IOS Users.

6 Conclusion

The proposed has an advantage of rescuing the abandoned children and wounded animal occurred in the locality. The proposed system has a lot of advantage, and the proposed system is also having a disadvantage like if the abandoned children move from the one place to another place, it becomes tougher to find the location of the children. The Limitation noted in this application can be taken as part of the future work. This Application gives access to track the children who are missing with Android and iOS using this application "retrieval services using mobile computing and block chain". Because of the software used by the mobile application, we can be easily track the location of child.

References

1. Dhamodaran S, Albert Mayan J, Saibharath N, Nagendra N, Sundarrajan M (2020) Spatial interpolation of meteorological data and forecasting rainfall using ensemble techniques. AIP Conf Proc 2207:050005
2. Muthuselvan S, Srividhya E, Miruthula SR, Abdul Samad S (2018) Location based orphanage finder application for Google android phones. Int J Pure Appl Math 119(16)
3. Ankayarkanni B, Albert Mayan J, Aruna J Support vector machine for effective robust visual tracking. J Comput Theor Nanosci 18(8):3571–3575
4. Wang C, Lu W, Narayanan MR, Chang DCW, Lord SR, Redmond SJ, Lovell NH (2016) Low-power fall detector using triaxial accelerometry and barometric pressure sensing. IEEE Trans Industr Inf 12(6):2302–2311
5. Eswar VOS, Vinil B, Ankayarkanni B, Albert (2018) Integrated collective node behavior analysis with onion protocol for best and secured data transmission. In: 2018 international conference on communication and signal processing (ICCSP), Chennai, pp 0918–0921
6. Bandaru R, Albert Mayan J (2016) Novel approach for whole test suite generation using metamorphic relations. Indian J Sci Technol 9(10):1–7
7. Mayan JA, Priya KL (2015) Novel approach to reuse unused test cases in a GUI based application. In: IEEE international conference on circuit, power and computing technologies, ICCPCT 2015, pp 1–7
8. Pandian A, Bharathi B, Albert Mayan J, Jacob P, Pravin A A comprehensive view of scheduling algorithms for mapreduce framework in Hadoop. J Comput Theor Nanosci 16(8):3582–3586
9. World Health Organization (WHO) Global report on falls prevention in older age. Accessed on 15 January 2016

Closed-Loop High-Gain DC-DC Converters for Renewable Energy Source Application

D. Ravi Kiran and D. Godwin Immanuel

Abstract The reliability on renewable energy sources has increased in the current period due to the world environmental conditions. Resolution toward green environment by the lead countries of the world has led development and usage of renewable energy sources as alternative source for conventional methods. Emission (majorly) from power plants and vehicles pollutes the environment and is a major concern and technology pertaining to power generation from renewable sources, and electric vehicles are trending these days. In the view of increasing the voltage level of renewable source, high-gain DC-DC converters play a vital role. This paper presents a closed-loop analysis of three high-gain DC-DC converters which steps-up the 40 V output of renewable source to 400 V. The three converters are non-isolated class with only single operating power switch. Simulation results are presented, and the models are developed using MATLAB/SIMULINK software.

Keywords Renewable sources · Single switch · High voltage gain · DC-DC · Closed loop

1 Introduction

Its time to move on leaving behind the fossil fuel age. Conventional fossil fuels used in power generation companies and vehicles pollute the environment [1, 2] and are a concern expressed by almost all countries. The air quality in New Delhi, the capital city of India has reached all time worst in November 2019, and government is insisting for alternative methods to improve the quality of air. The present scenario of the world is not able to meet international environmental standards where in most countries still providing subsidized fossil fuels have reached 11% in 2017, according

D. R. Kiran (✉) · D. G. Immanuel
Department of EEE, Sathyabama Institute of Science and Technology, Chennai, India

D. G. Immanuel
e-mail: godwinimmanuel.eee@sathyabama.ac.in

S. S. Dash et al. (eds.), *Sixth International Conference on Intelligent Computing and Applications*, Advances in Intelligent Systems and Computing 1369,
https://doi.org/10.1007/978-981-16-1335-7_37

to the reports. Meeting global environmental standards immediately is not possible, but if the usage of non-conventional sources increases can yield better results.

Though the technology to use renewable source has progressed, the countries failed to utilize the renewable energy effectively. Flaming the conventional fossil fuels for the generation of energy in large amount emits flue gases that are responsible for global pollution. Majority of the renewable sources in the process of energy conversion emits very less or no dangerous gases. Energy conversion while emitting no gases improves the quality of air and public health as well. Majority of the air is polluted by emissions from road transport, fossil fuel-based power generation, and industrial activities.

Renewable sources like photovoltaic (PV) system, fuel cells can generate electrical energy with very less or no pollution [3–5]. The main constraint with renewable sources is its cost with respect to the power density. Cost of installing renewable sources increases with increase in power density requirement. Government is also providing subsidized loans for promoting renewable-based power generation.

Another important aspect of renewable energy sources like PV system and fuel cells is they generate DC output power which insists for DC-DC converter to gain the voltage level required to drive the load. Figure 1 illustrates the basic flow arrangement of renewable source-based DC-DC converter. The DC output from the renewable source is fed to DC-DC converter which steps-up the voltage level to the load requirement [6]. The output of the DC-DC converter can be fed to DC micro-grid, battery set or can feed the grid through inverter. DC-DC converters [7] play a vital role in renewable-based operations. Isolated and non-isolated DC-DC converter types are used more. Output voltage controlled by resonant converter topology [8], LLC series resonant converter with high-frequency converter explained [9]. Interleaved converter is explained with high voltage gain [10]. Fuzzy-based technology is implemented for power factor correction in [11]. High-frequency isolated converter with half bridge resonant converter is explained [12].

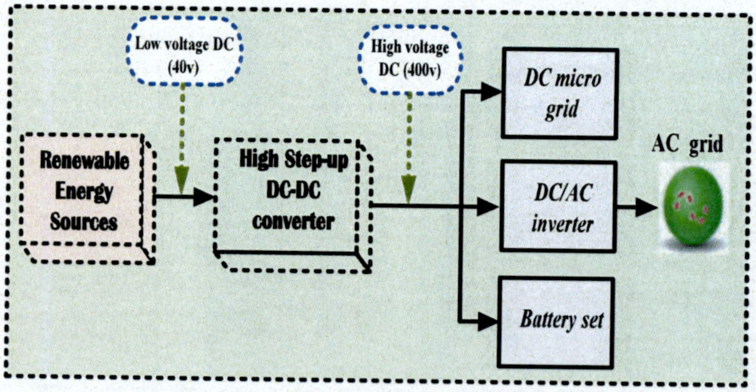

Fig. 1 Schematic arrangement of renewable source-based DC-DC converter

This paper presents a closed-loop analysis of three high-gain DC-DC converters which steps-up the 40 V output of renewable source to 400 V. The three converters are non-isolated class with only single operating power switch.

2 Closed-Loop Converter Topologies

2.1 Topology-I

Figure 2 illustrates the circuit representation of DC-DC converter topology-I employ switched capacitor and voltage doubler circuit for high-gain operation. This circuit is capable of getting high-gain by coupled inductor turns ratio. The performance of the converter relies on loss-less clamping, and the stress across the switch is reduced and thus facilitating to use low $R_{DS(ON)}$ switch.

Capacitor voltages across C_3 and C_b are illustrates as:

$$V_{C_3} = V_{N_3} = n_3.V_{in} \tag{1}$$

$$V_{C_b} = V_{in} + V_{N_2} = (n_2 + 1).V_{in} \tag{2}$$

The voltage output is given as:

$$V_0 = V_{C_1} + V_{C_2} + V_{C_3} \tag{3}$$

The voltage gain is expressed as:

$$\frac{V_0}{V_{in}} = n_2 + \frac{2 - D + n_3}{1 - D} \tag{4}$$

Fig. 2 High-gain step-up converter (open loop) configuration-I

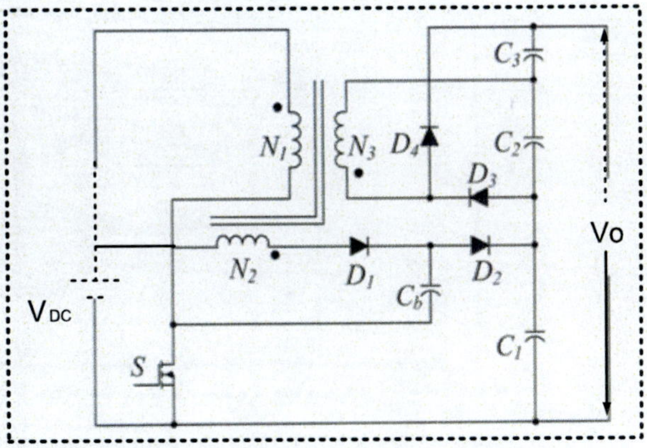

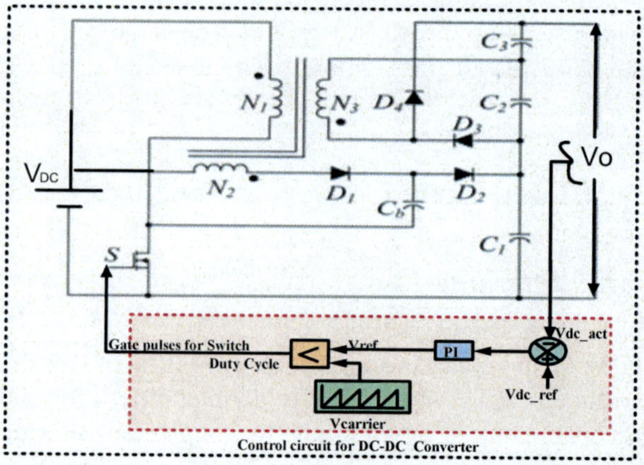

Fig. 3 High-gain step-up converter (closed loop) configuration-I

Figure 3 illustrates the closed-loop converter topology-I where the output voltage feedback is controlled to give precise output. The output actual voltage is compared to reference value, and the error signal is controlled using simple PI controller which gives out reference signal. The reference signal is overlapped on a carrier signal to generate gate pulses to switch of converter.

2.2 Topology-II

Figure 4 illustrates the DC-DC converter structure II with multiplier units and coupled inductors. Coupled inductor discharges its energy to charge capacitor during switch OFF time to obtain high voltage gain. Capacitor C_1 and diode D_1 form a clamping

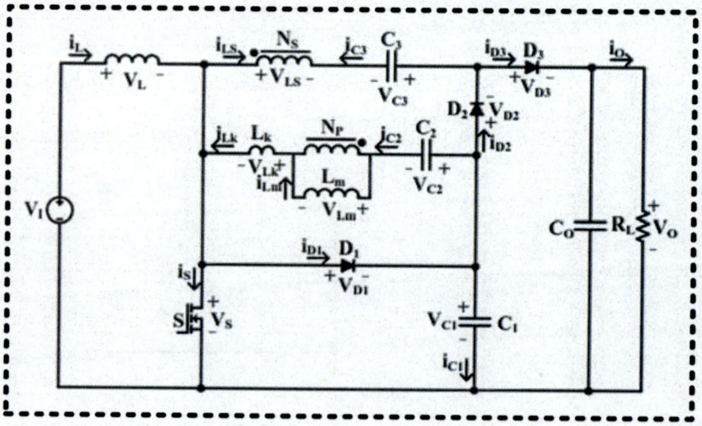

Fig. 4 High-gain step-up converter (open loop) configuration-II

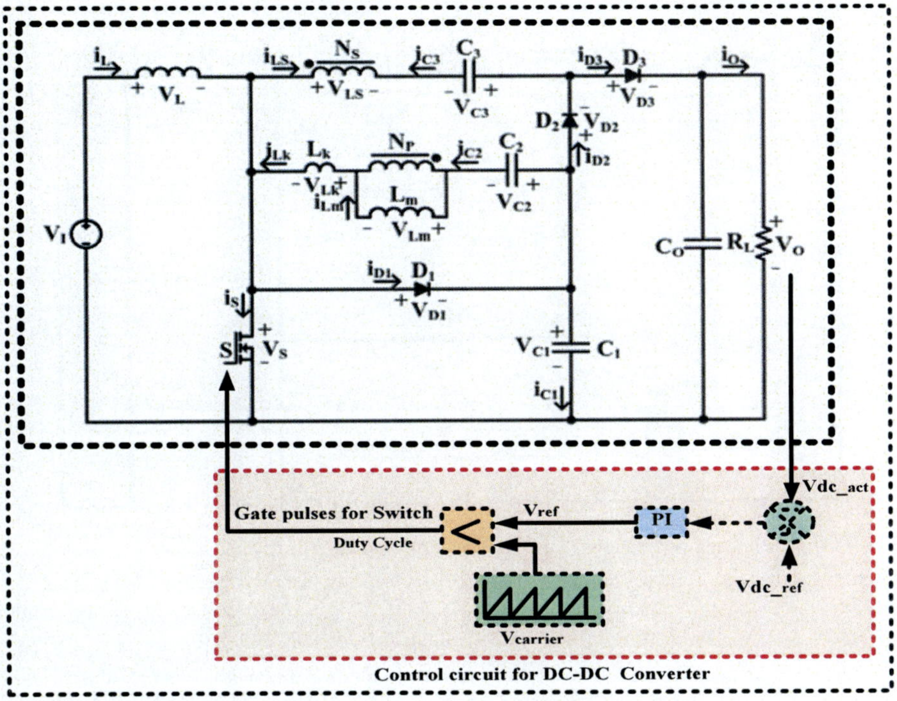

Fig. 5 High-gain step-up converter (Closed loop) configuration-II

circuit, capacitor C_2 and diode D_2 form a multiplier unit, and capacitor C_3 is extended multiplier unit.

The gain in voltage is expressed as:

$$\frac{V_0}{V_{\text{in}}} = \frac{2 + n + nD}{1 - D}.V_{\text{in}} \tag{5}$$

Figure 5 shows the high-gain step-up converter (Closed loop) configuration-II. The output actual voltage is compared to reference value, and the error signal is controlled using simple PI controller which gives out reference signal. The reference signal is overlapped on a carrier signal to generate gate pulses to switch of converter.

2.3 *Topology-III*

Figure 6 shows the converter topology-III where input consists of series inductor to reduce input current ripple. To make conduction losses less, diode D_1 and capacitor C_1 are clamped. Switch is operated in ZCS mode.

The gain in voltage is expressed as:

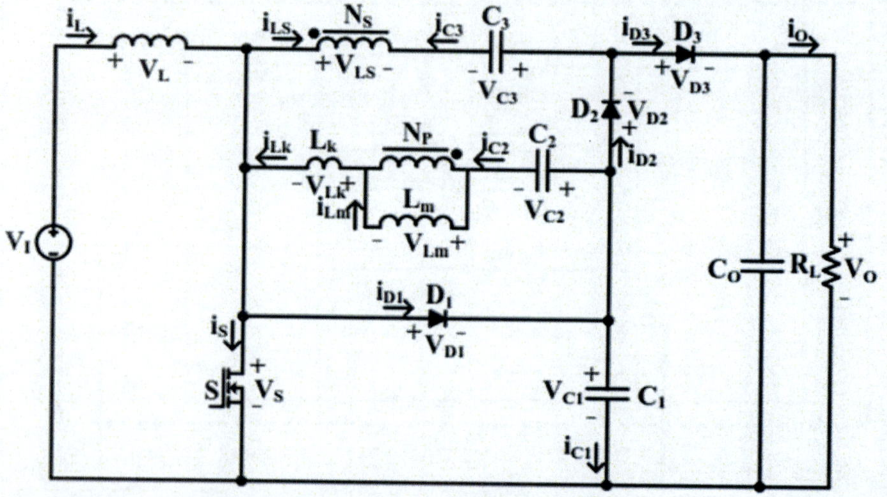

Fig. 6 High-gain step-up converter (open loop) configuration-III

$$\frac{V_O}{V_{in}} = \frac{n+2}{1-D} \tag{6}$$

Figure 7 shows the high-gain step-up converter (Closed loop) configuration-III. The output actual voltage is compared to reference value, and the error signal is controlled using simple PI controller which gives out reference signal. The reference signal is overlapped on a carrier signal to generate gate pulses to switch of converter.

3 Result Analysis of DC-DC Converters

3.1 *Topology-I*

Figure 8 shows voltage fed to converter. Constant 40 V input is given to the DC-DC converter. The output of the converter in open-loop mode of operation is shown in Fig. 9. Open-loop mode of operating converter gives output slightly more than 400 V. The voltage ripple at output is depicted in Fig. 10, and the waveform shows voltage ripple above 400 V.

Figure 11 shows feed voltage to the converter. Constant 40 V input is given to the converter. The output of the converter in closed-loop mode of operation is shown in Fig. 12. Closed-loop mode of operating converter gives output exactly 400 V. The voltage ripple at output is depicted in Fig. 13, and the waveform shows voltage ripple at 400 V.

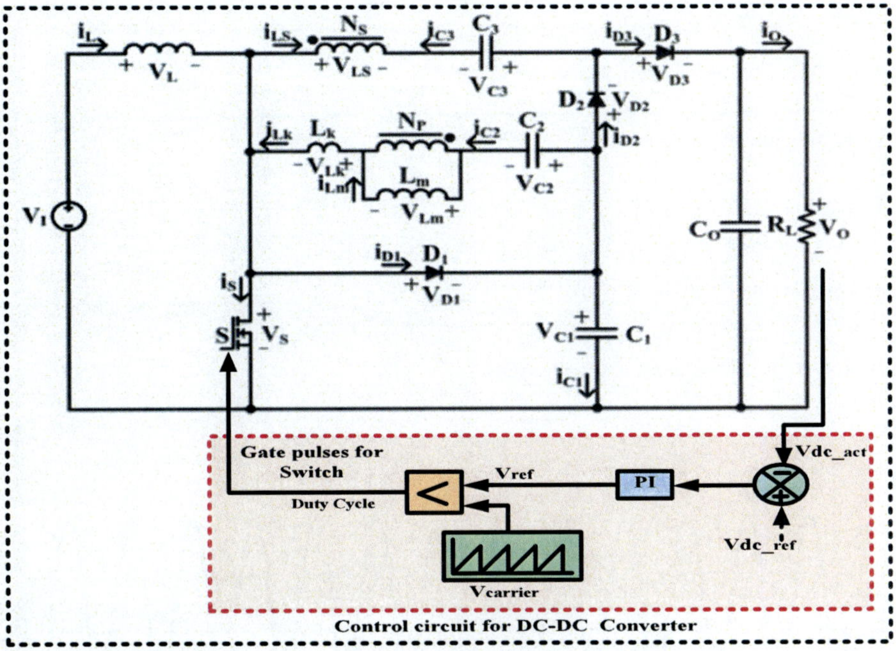

Fig. 7 High-gain step-up converter (Closed loop) configuration-III

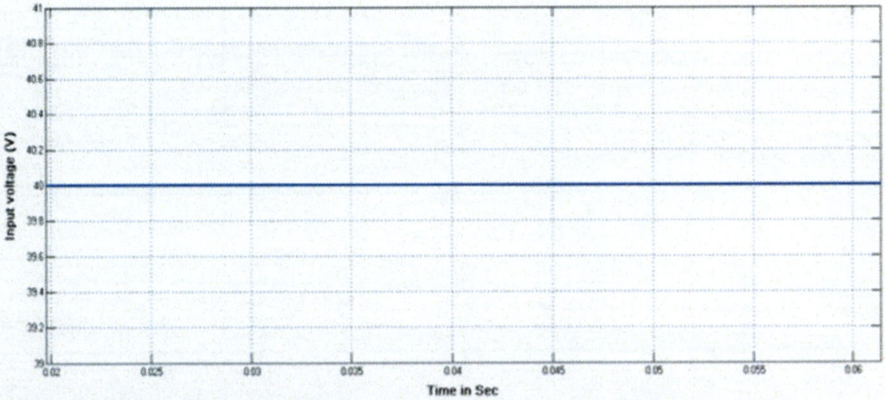

Fig. 8 Input voltage to DC-DC converter

3.2 *Topology-II*

Figure 14 shows the voltage to the converter. Constant 40 V input is given to the converter. The output of the converter in open-loop mode of operation is shown in Fig. 15. Open-loop mode of operating converter gives output slightly less than 400 V.

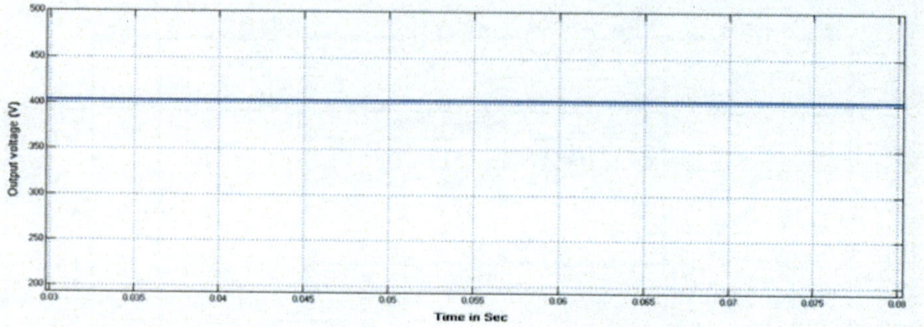

Fig. 9 Output voltage of DC-DC converter

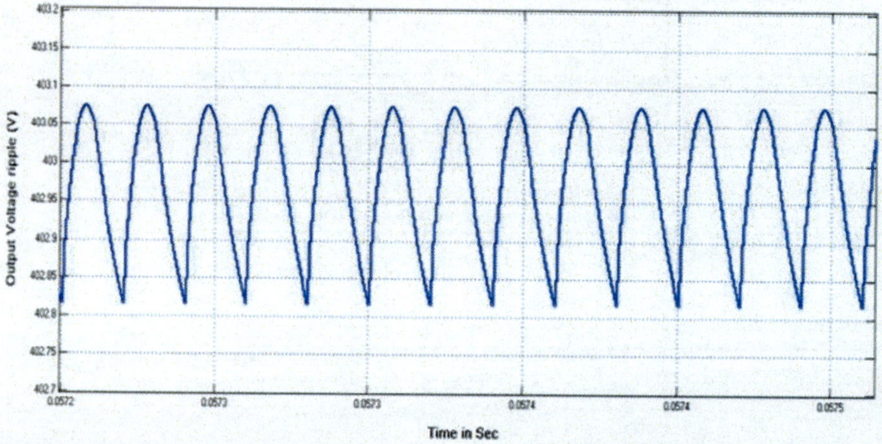

Fig. 10 Output voltage ripple

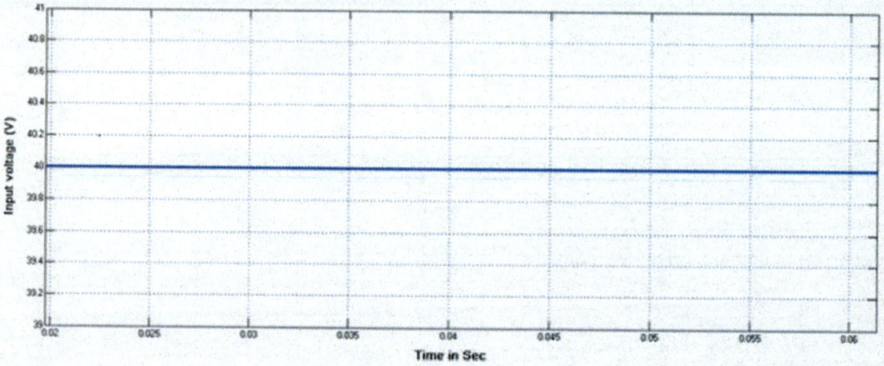

Fig. 11 Input voltage to DC-DC converter

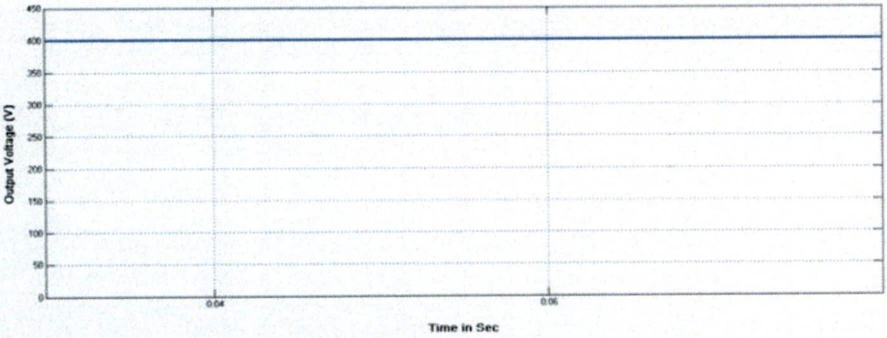

Fig. 12 Output voltage to DC-DC converter

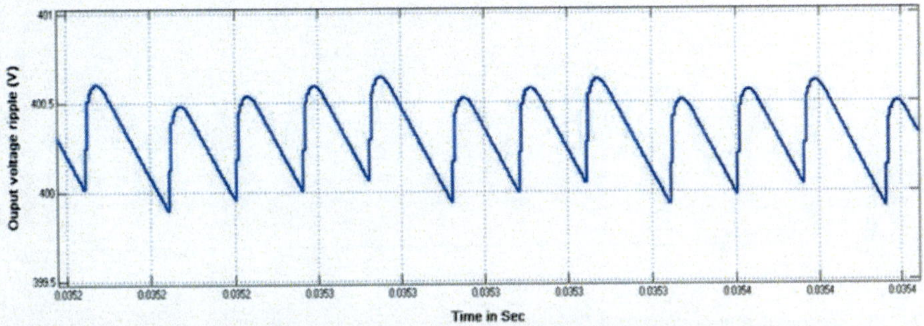

Fig. 13 Voltage ripple of converter

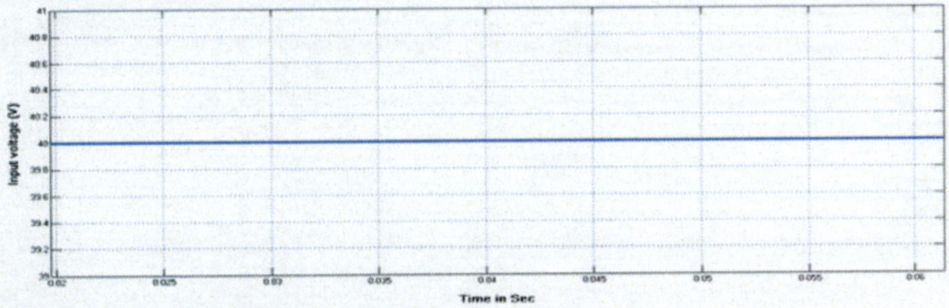

Fig. 14 Input voltage to DC-DC converter

The voltage ripple at output is depicted in Fig. 16, and the waveform shows voltage ripple below 400 V.

Figure 17 shows the voltage to the converter. Constant 40 V input is given to the converter. The output of the converter in open-loop mode of operation is shown in Fig. 18. Open-loop mode of operating converter gives output slightly less than 400 V.

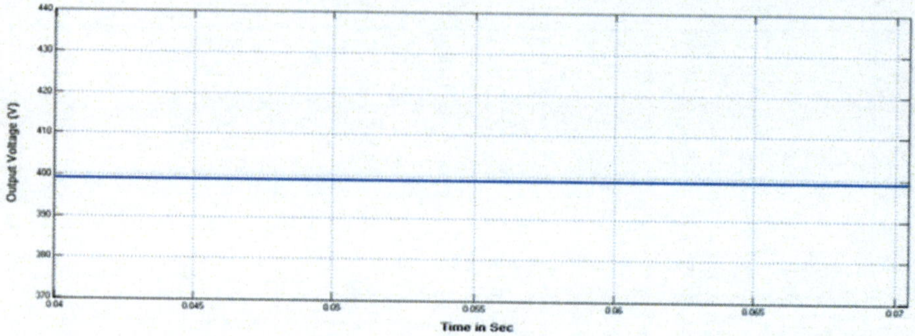

Fig. 15 Output voltage to DC-DC converter

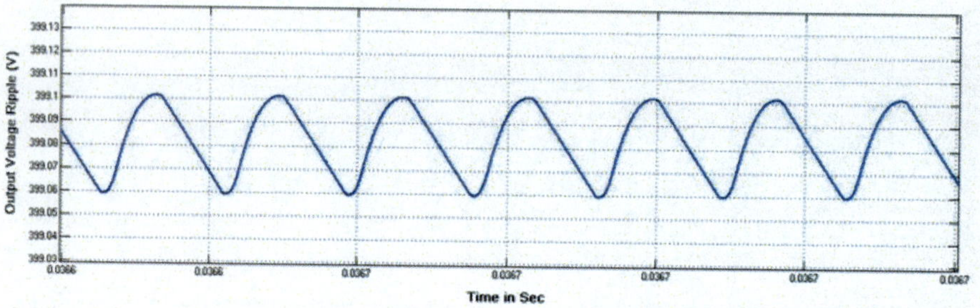

Fig. 16 Voltage ripple of converter

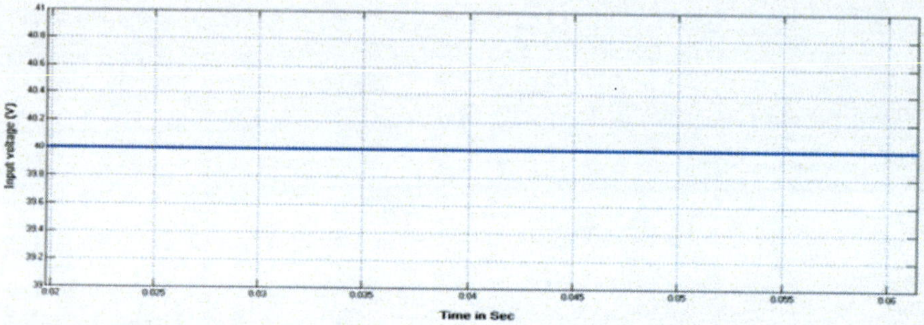

Fig. 17 Input voltage to DC-DC converter

The voltage ripple at output is depicted in Fig. 19, and the waveform shows voltage ripple below 400 V.

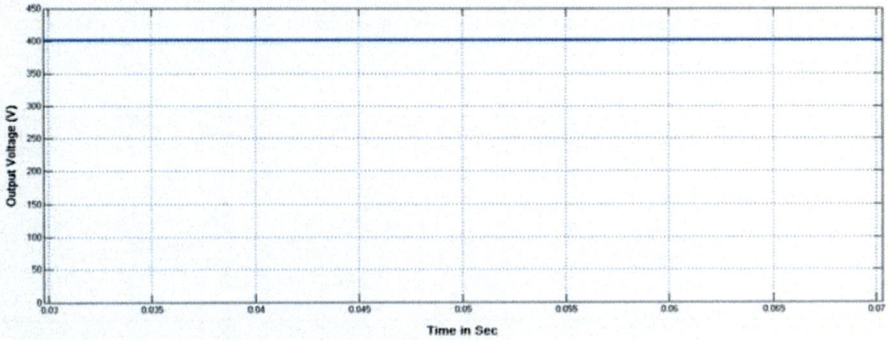

Fig. 18 Output voltage to DC-DC converter

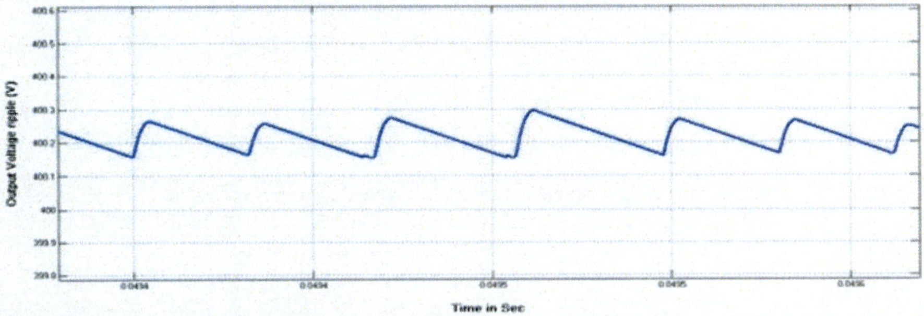

Fig. 19 Voltage ripple of converter

3.3 Topology-III

Figure 20 shows the voltage to the converter. Constant 40 V input is given to the converter. The output of the converter in open-loop mode of operation is shown in Fig. 21. Open-loop mode of operating converter gives output slightly less than 400 V.

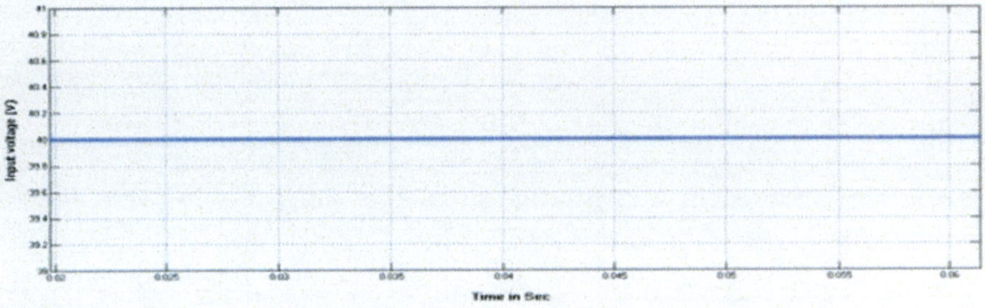

Fig. 20 Input voltage to DC-DC converter

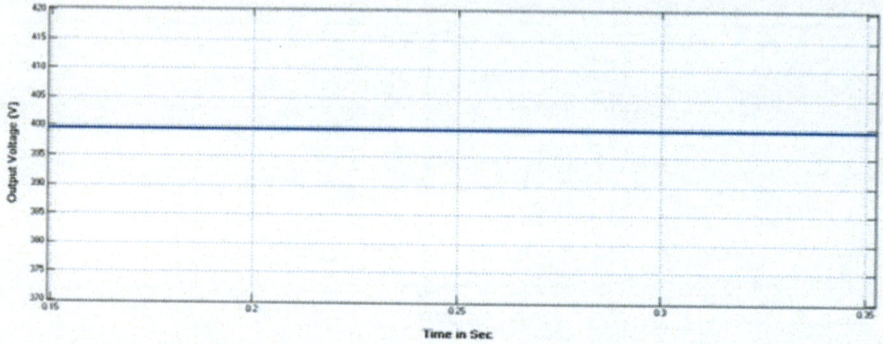

Fig. 21 Output voltage to DC-DC converter

The voltage ripple at output is depicted in Fig. 22, and the waveform shows voltage ripple below 400 V.

Figure 23 shows the voltage to the converter. Constant 40 V input is given to the converter. The output of the converter in open-loop mode of operation is shown in

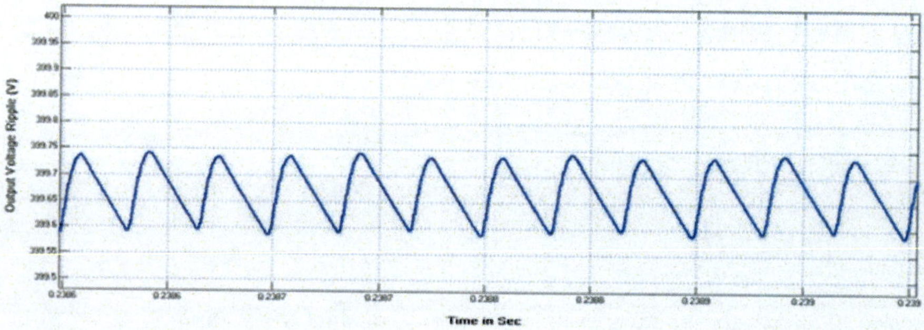

Fig. 22 Voltage ripple of converter

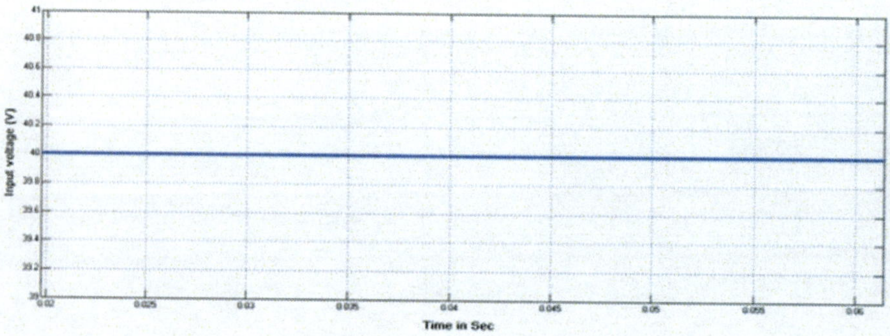

Fig. 23 Input voltage to DC-DC converter

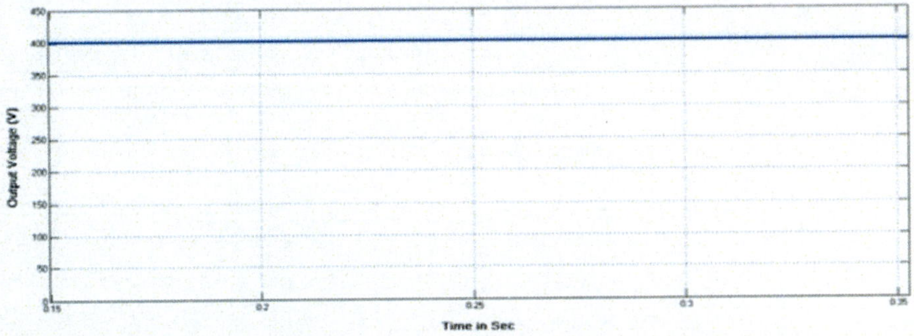

Fig. 24 Output voltage to DC-DC converter

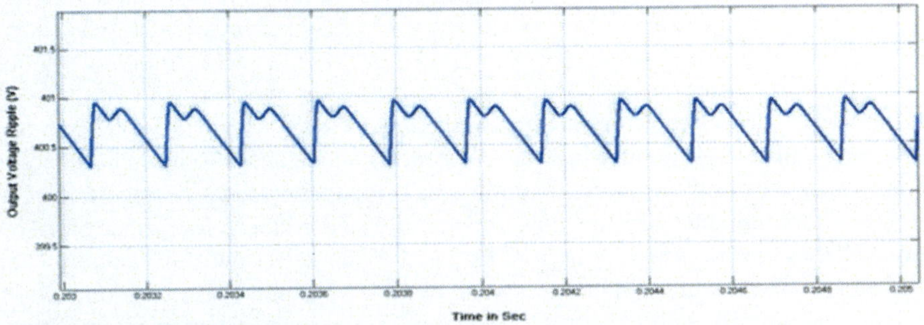

Fig. 25 Voltage ripple of converter

Fig. 24. Open-loop mode of operating converter gives output slightly less than 400 V. The voltage ripple at output is depicted in Fig. 25, and the waveform shows voltage ripple below 400 V.

4 Comparison of Topologies

For simplicity of operation, some assumptions are made for three converter topologies. Comparison is given in Table 1 for three type of DC-DC converters explained before. The three converters are selected and simulated such that they take input of 40 V to give 400 V output voltage. Efficiency is increased by transferring the stored energy in inductors to clamping circuit.

Table 1 Topologies comparison

	Topology-I	Topology-II	Topology-III
Input voltage (V)	40	40	40
Output voltage (V)	400	400	400
Gain formula	$n_2 + \frac{2-D+n_3}{1-D}$	$\frac{2+n+nD}{1-D} \cdot V_{in}$	$\frac{n+2}{1-D}$
Stress on switch (V)	120	80	120
Switch count	1	1	1
Diodes	4	4	3
Capacitors	4	3	3

5 Conclusion

Pollution is a global worrying factor, and the power generation from conventional fuels emits flue gases destroying the environmental balance. Renewable energy sources generate electricity with zero pollution. This paper presents the review of DC-DC converter topologies, and the converters are operated in closed-loop mode of operation. All the converters take input from RES as 40 V and gives out 400 V output. The open-loop analysis yields slightly less or above 400 V output, but closed-loop analysis of converters gives out exactly 400 V output. Comparison of converter topologies is tabulated. DC-DC converters in closed-loop operation will aid in stepping up the voltage in RES applications.

References

1. Dhingra S, Madda RB, Gandomi AH, Patan R, Daneshmand M (2019) Internet of Things mobile–air pollution monitoring system (IoT-Mobair). IEEE Internet Things J 6(3):5577–5584
2. Al-Ali AR, Zualkernan I, Aloul F (2010) A mobile GPRS-sensors array for air pollution monitoring. IEEE Sens J 10(10):1666–1671
3. Cao X, Zhang J, Poor HV (2018) Data center demand response with on-site renewable generation: a bargaining approach. IEEE/ACM Trans Netw 26(6):2707–2720
4. Zhang M et al (2020) Optimal design and operation of regional multi-energy systems with high renewable penetration considering reliability constraints. IEEE Access 8:205307–205315
5. Elgenedy MA, Massoud AM, Ahmed S, Williams BW, McDonald JR (2019) A modular multilevel voltage-boosting marx pulse-waveform generator for electroporation applications. IEEE Trans Power Electron 34(11):10575–10589
6. Maroti PK, Padmanaban S, Holm-Nielsen JB, Sagar Bhaskar M, Meraj M, Iqbal A (2019) A new structure of high voltage gain SEPIC converter for renewable energy applications. IEEE Access 7:89857–89868
7. Minami M, Tomoeda K (2019) An analysis of operation in single-switch high step-up DC-DC converter with three-winding coupled inductor. In: 2019 IEEE applied power electronics conference and exposition (APEC), Anaheim, CA, USA, pp 2135–2137

8. Geetha V, Sivachidambaranathan V (2018) A single switch parallel Quasi resonant converter topology for induction heating application. Int J Power Electron Drive Syst (IJPEDS) 9(4):1718–1724. ISSN 2088–8694
9. Indira D, Sivachidambaranathan V, Dash SS (2011) Closed loop control of hybrid switching scheme for LLC series-resonant half-bridge DC-DC converter. In: Proceedings of the second international conference on sustainable energy and intelligent system (SEISCON 2011). IET Chennai and Dr.MGR University, 20–22 July, pp 295–298
10. Kavitha M, Sivachidambaranathan V (2015) Performance analysis of transformer-less two phase interleaved high gain DC converter using MPPT algorithm. Indian J Sci Technol 8(15):61428:1–8
11. Kavitha M, Sivachidambaranathan V (2017) Power factor correction in fuzzy based brushless DC motor fed by bridgeless buck boost converter. In: IEEE internaltional conference on computation of power energy information and communication (ICCPEIC), 22–23 March 2017, pp 549–553. ISSN 978–1–5090–4324–8/17/$31.00 ©2017 IEEE
12. Sivachidambaranathan V, Dash SS (2010) Simulation of half bridge series resonant PFC DC to DC converter. In: IEEE international conference on recent advances in space technology services and climate change—2010 (RSTS&CC-2010). Sathyabama University in association with Indian Space Research Organisation (ISRO), 13–15 November, Bangalore and IEEE, ISBN 978–1–4244–9184–1, IEEE Explore

Smart Secured Seamless Online Payment Against Counterfeit Transaction Through Aadhaar

P. Ajitha, S. Gowri, J. Jabez, and A. Sivasangari

Abstract To develop a system for achieving digitalization by enabling Aadhaar-linked payment gateways ensuring feasible and authorized access via fingerprint and One-Time Password. The authentication provides greater access control and user identity. Thus, the privacy is maintained throughout the transaction.

Keywords Aadhaar number · Android virtual device · Cash on delivery · Quick response code

1 Introduction

The face of economy has been changing since the inching of twenty-first century. The existence of an average human being has moved from life to lifestyle. With the adept of Internet, this change has been exponential and the value system, buying pattern has really accelerated this shift. Digitization is taking the world by storm, and unique identity-based transaction system is getting more attention and acceptance. Aadhar card-based transaction is one such opportunity to take the e-commerce world on a ride.

2 Existing Work

The existing system of payment has been enhanced in the proposed approach where the Aadhaar card enables secure and cashless payment for all users by verifying the user fingerprint providing authenticated online transaction. The various issues in the existing systems are energy consumption while using the app is higher. Mobile congestion of data is prevalent. [1] discussed about Prone to security risks or attacks. Every approach seems to be costlier compared to Quick Response code. Near-field

P. Ajitha (✉) · S. Gowri · J. Jabez · A. Sivasangari
Sathyabama Institute of Science and Technology, Chennai, India

S. S. Dash et al. (eds.), *Sixth International Conference on Intelligent Computing and Applications*, Advances in Intelligent Systems and Computing 1369,
https://doi.org/10.1007/978-981-16-1335-7_38

communication works only in shorter range [2]. Recognition errors can occur which makes the system proposed not very reliable. Compromised biometric data cannot assure privacy protection [3]. Ethical issues may arise while following the traditional system of purchase [4]. In this paper, we demonstrate that Lee's plan is powerless to different assaults and neglects to give basic security properties. We at that point present our very own plan and play out a casual examination to substantiate the case that our plan can oppose the past plan's shortcomings. In [5], the author demonstrated that Lee's plan is powerless to different assaults and neglects to give basic security properties. In [6], the author shows a procedure to topple the agreement assaults, and the information inspecting component is accomplished by methods for vector responsibility and in reverse unlinkable verifier nearby repudiation amass signature. In [7], the author manufactured a unique finger impression recovery by a measurable Bayesian classifier to produce extra or all the more preparing tests utilizing spatial demonstrating method.

The author in [8] suggested a flexible parameter development, pre- and post-handling stages using nearby-based analysis and worldwide learning-based investigation to improve the accuracy of fingerprint construction strategies. Carsten3 portrays a strategy for estimating edge recurrence and upgrading images using separate curved regions and bended Gabor filters (GFs). Bended GFs increase smoothing control by enabling channel parameter decision. Two introduction field estimation techniques are connected for boisterous images to be specific, line sensor and gradient-based techniques. In [9], they presented a convincing two-organized plot of improvement for low-quality fingerprint images in both spatial and recurrence spaces. The main stage development plot is planned to use neighborhood edge setting data using spatial sifting. The broken edges are connected, and the mixed edges are viable isolated.

The band transfer channels are learned from the enhanced image from both the first and the main stage in the second organize. Using nonlinear successive mean quantization change (SMQT), the authors in [10] proposed a versatile unique image upgrade system. Increases in dynamic range and accurate platforms are used to update the exclusive finger impression picture worldwide environment. The time-related quality of this method can be further enhanced by improving computational undertakings in the pre-preparation phase.

3 Proposed Work

To develop a system to achieve a digitalization of transactions by using Aadhaar payment gateways which will ensure authenticated access and eliminates credit/debit cards. This system would prevent impersonation and provide high-level authentication for online payments through fingerprint scanning or One-Time Password.

3.1 System Architecture

The system architecture shown in Fig. 1 defines the overall structure of the system and functionalities of the entities involved in it. The user needs to sign up or sign in the application and following that the user's record gets automatically stored in the database. This takes the user to the home page of the application where they can make a purchase of the desired product using the product catalog. After selecting the product, the user needs to scan the quick response code present in the Aadhaar card which automatically displays all the user's details. Following that the user needs to select the mode of payment.

1. Cash on delivery (COD)
2. Aadhaar

If the option is COD, then the order will be placed successfully and the product will be delivered to the address of the user scanned from the Quick Response code of Aadhaar. If the mode of payment is Aadhaar, then authentication is made through fingerprint or one-time password and the payment is completed by selecting the desired customer's bank account linked with the 12-digit unique Aadhaar number.

The entire sequence flow is depicted as a functional architecture diagram as shown in Fig. 2.

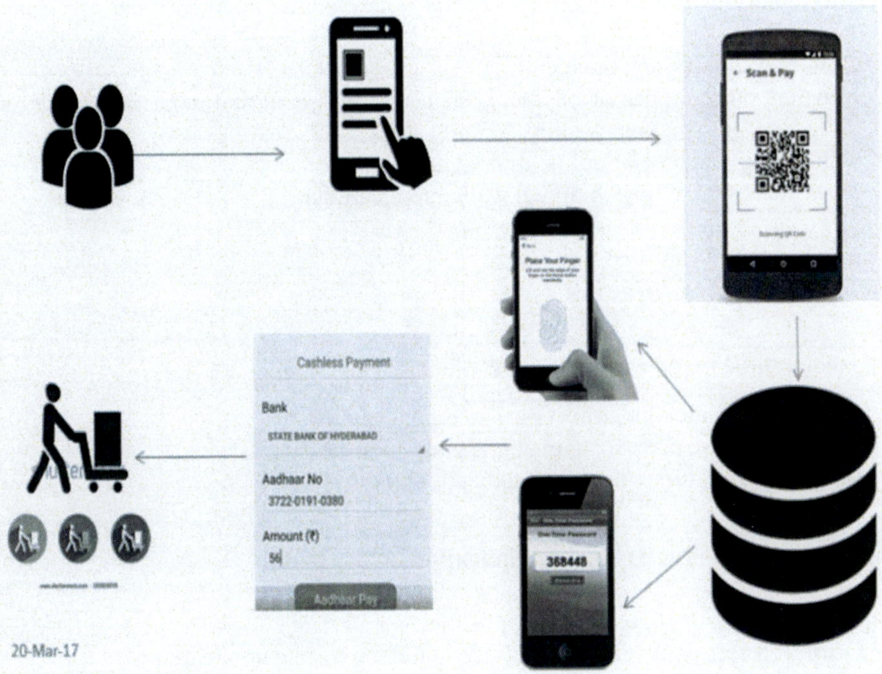

Fig. 1 System architecture

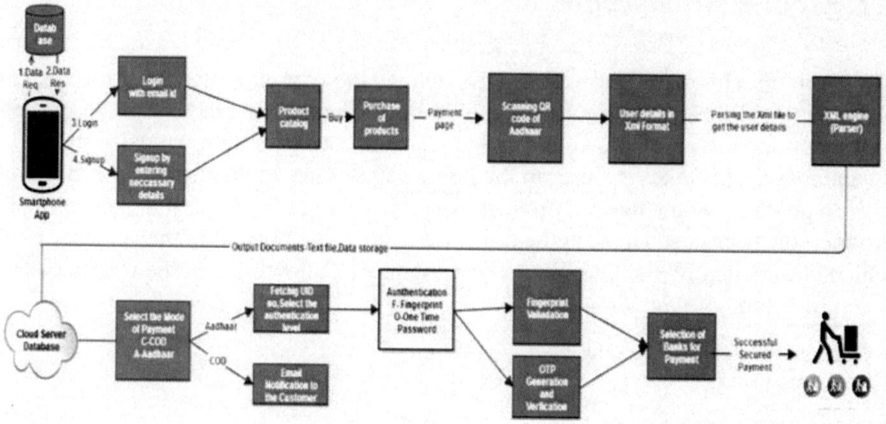

Fig. 2 Functional architecture of the system

After validating these authentication factors, the user will automatically move to their respective payment page, in which the users opt their corresponding bank accounts that has a linkage to Aadhaar and do a successful secured payment. Some of the benefits of using this application are

- To protect our personal data and create a new online payment platform using Aadhaar.
- Cost- and time-saving payment.
- It avoids other modes of payments such as credit card and debit card, thereby protecting our personal wealth.
- Complete avoidance of online fraudulent payment.
- Does away with umpteen digital wallet applications.

3.2 System Modules

The major three modules which constitute the whole system are

- Creation of unique identification-based application.
- Quick Response code scanning and parsing for user's data.
- Fingerprint/one-time password authentication.

3.2.1 Creation of UID-Based Application

Input: Database for E-commerce application.
Output: Successful generation of user interface for the application.
Algorithm:

1. Download and open the Android Studio.

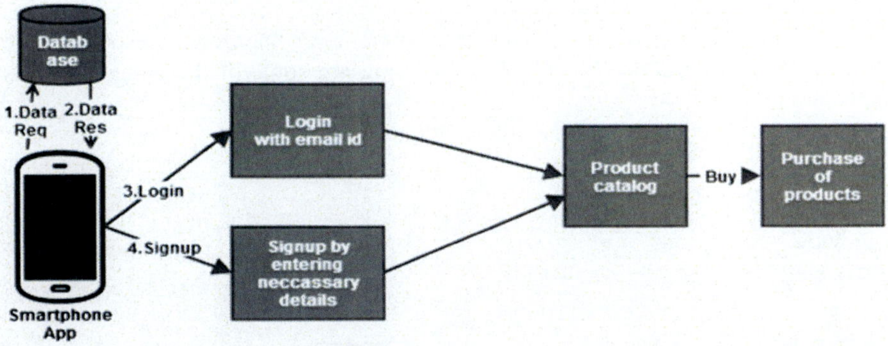

Fig. 3 Creation of UID-based application

2. In Android Studio, create a new project by clicking on "Start a new project."
3. Enter the task name.
4. Enter the layout backend code in MainActivity.java file which contains the class description for the activity created.
5. Enter the source code in activity main.xml which describes the application layout.
6. Describe the basic features of the application.
7. By opening the Android Virtual Device Manager and selecting the appropriate Android Virtual Device, build Android Virtual Device.
8. Choose your project and press Run to check the program from the toolbar as shown in Fig. 3.

3.2.2 Quick Response Code Scanning and Parsing for User's Data

Input: Scanning of Quick Response–code of Aadhaar
Output: User details display
Algorithm:

1. After signing into the app, select the product to purchase listed in product catalog.
2. Scan the QR code present in the Aadhaar card by using the phone's camera which displays user's unique identification number, address, name, date of birth, and PIN code.
3. Retrieval of the contact number is done by accessing the database.
4. Database present in the server enables the user to obtain the contact number in a secured gateway.
5. Select the mode of payment, i.e., M
   a. Cash on delivery (COD) C
   b. Aadhaar A
6. End if

In this module, as in Figs. 4 and 5, user will scan the Quick Response code of Aadhaar to get the user details; the following are some of the features of this application,

User can do the below with the information of Aadhaar card

1. Scan.
2. Save.
3. View.
4. Delete.

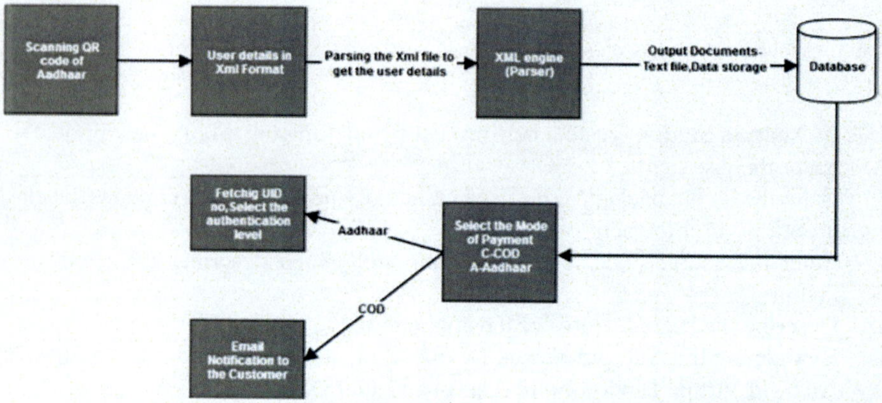

Fig. 4 QR code scanning and parsing for user's data

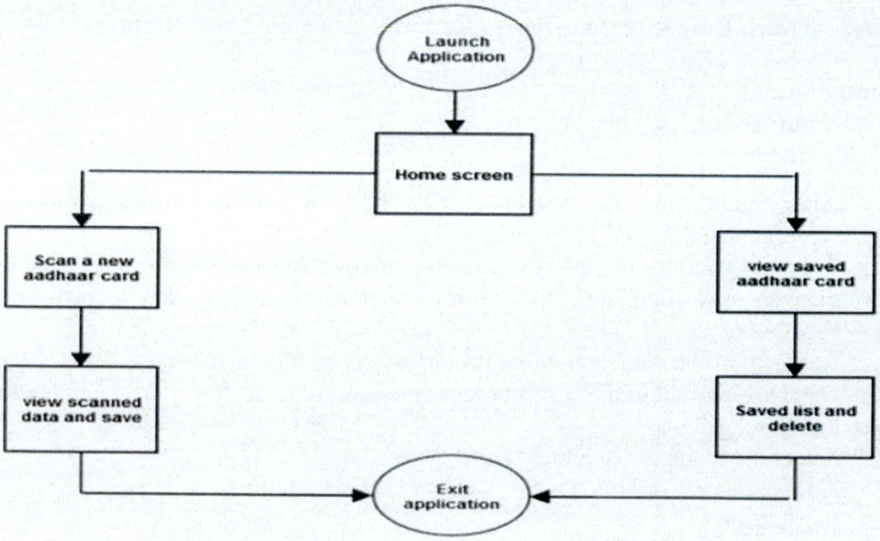

Fig. 5 QR code scanning and parsing for user's data

Home screen: There will be two "actions call" (buttons) on this screen. One to start a new scan (Scan New Aadhaar Card) and the other to display Scan Card Screen saved.

Aadhaar cards: This screen will appear when the user clicks on the "Scan New Aadhaar Card" button. This will allow the user to scan Aadhaar card's QR code.

Scanned Aadhaar card data: The data scanned from QR code will be shown on this device. The scanned data can be saved or canceled, and a new scan can be started.

Saved Aadhaar card: All saved Aadhaar cards will be displayed on this page. The saved Aadhaar cards can be deleted from here by the customer. After receiving user details, the next step is payment; there are two payment methods in this proposed system.

1. Cash on delivery (COD)
2. Aadhaar

If user selects the option COD, then the order will be placed successfully and the product will be delivered to the address of the user scanned from the QR code of the Aadhaar card.

3.2.3 Fingerprint/One-Time Password Authentication

Input: User details displayed with Aadhaar number
Output: Secured online payment
Algorithm:

1. Select fingerprint F OTP O (for one-time password generation)
2. Scan the fingerprint F for user authentication
3. If F or O is true, then
   Select from the list of banks displayed
   Pay the respective amount
4. Else
5. End if

Prompt the error message

In this module as shown in Fig. 6, the more secured way payment is done by using Aadhaar and with the help of authentication factors such as fingerprint and one-time password generation and validation.

Final steps of payment are type the customer Aadhaar card number in the Aadhaar field.

If the customer wants to do Payment, then

1. Use fingerprint scan and OTP as password for completing the transaction.
2. Select the bank to make payment.

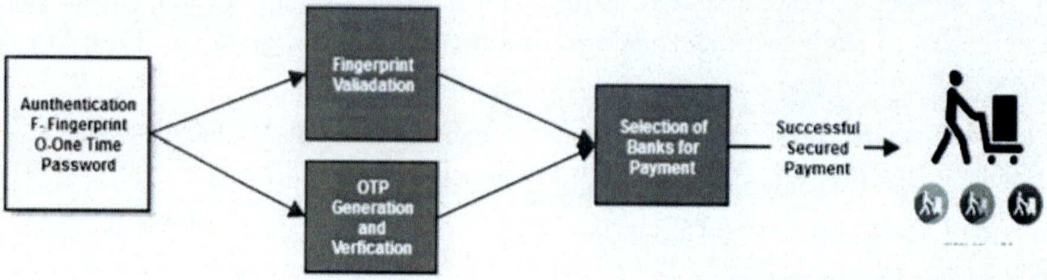

Fig. 6 Fingerprint/OTP authentication

3. Upon entering the Aadhaar number, the app would automatically fetch the bank accounts linked with your Aadhaar number, and then, the customer can make the selection of bank for payment as per his/her choice.

4 Experimental Results

Once the system's design component is finished, the system enters the test phase. Monitoring is an important part of the whole process of development and maintenance. The primary purpose of testing is to identify errors. It is used to ensure quality. The purpose of the test phase is to check that the specification has been correctly completed and integrated into the design and to ensure that the design itself is correct as shown in Table 1.

Research is an inquiry to provide information to customers on the quality of the product or service being evaluated. Testing can also provide an objective, independent view of the software to enable the company to appreciate and understand the risks of implementing software. This is about matching our program with existing online shopping applications using measures such as fingerprint, OTP, Aadhaar card system.

Figure 7 shows the comparison between existing shopping online applications and our smart pay shopping application. *X*-axis consists of the various threats that may result in online shopping, while the *Y*-axis explains the security strength against the online shopping threats on the scale of 10. Of the many online shopping threats, the most important ones, namely credit/debit fraud, identity theft, data breach, adware, are placed in the *X*-axis. Our application overcomes all these problems in a more efficient way than the existing applications by using authentication techniques like fingerprint, OTP, and Aadhaar.

5 Conclusion

We have developed an application for online shopping that facilitates Aadhaar-enabled financial transactions through mobile phones as a part of its drive to convert

Table 1 Test cases

Test case id	Test description	Test input	Expected result	Actual result
TC 1	To login/register	User details	Should store the user details in database	Registration/login successfully completed
TC 2	To create database for e-commerce application	Product details/description	Should store the product details in the database	Updating of product details into the database
TC 3	To create e-commerce application	Necessary data in the database, XML, and Java files	E-commerce application should be created	E-commerce application created
TC 4	To scan the QR code present in the Aadhaar	QR code present in the Aadhaar card	Should provide user details in XML format	User details in XML format is obtained
TC 5	To parse the XML file	User details in XML file	Should parse and return the user details	Parsed the user details from the XML file
TC 6	To authenticate user through fingerprint	User's fingerprint	Should return 1 if authentication is true else 0	Successful user authentication

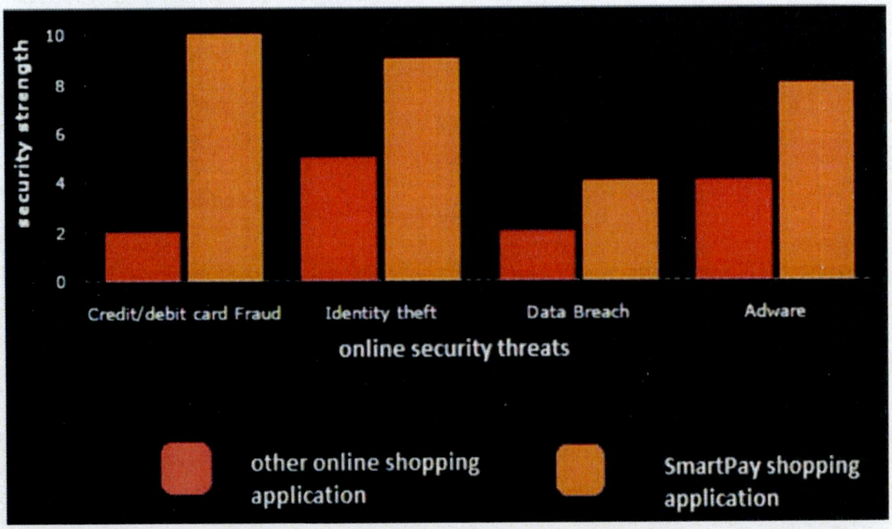

Fig. 7 Comparison between online shopping system and smartpay shopping application system

country into a cashless country. Payment is done here in a highly secured manner by incorporating the techniques of fingerprint and OTP. The use of credit/debit cards is replaced by the system of Aadhaar, preventing impersonation and unauthorized access. Existing work does not won credit/debit cards to make cashless payments. Thus, this system is helpful to all classes of people be it the poor or the rich since almost 99% of the citizens in India own an Aadhaar card.

The future enhancement includes the performance of the system such as efficiency and security. Our system provides a high level of authentication through fingerprint for fingerprint-enabled android phones and through for android phones without fingerprint feature. Thus, the future work can be based on providing authentication through fingerprint alone for all kinds of android phones. Our system provides authentication by comparing the user's fingerprint with the fingerprint stored in the mobile database instead It is more efficient if the fingerprint recorded in the Aadhaar has been compared with the user's fingerprint while making payment to provide a high level of authentication.

References

1. Arunalathaa JS, Tejaswib V, Shailaa K, Anvekarc D, Venugopala KR, Iyengard SS, Patnaik LM (2015) FIVDL: fingerprint image verification using dictionary learning. In: Eleventh international multi-conference on information processing
2. Khan J, Abbas H, Al-Muhtadi J (2015) Survey on mobile user's data privacy threats and defense mechanisms. In: International workshop on cyber security and digital investigation CSDI
3. Badra M, Badra RB (2016) A lightweight security protocol for NFC-based mobile payments. In: The 7th international conference on ambient systems, networks and technologies (ANT 2016)
4. Li Y, Dai W, Ming Z, Qiu M (2016) Privacy protection for preventing data over-collection in smart city. IEEE Trans Comput 65:5.5
5. Pradhan A, Karuppiah M, Niranchana R, Jerlin A, Rajkumar S (2018) Design and analysis of smart card-based authentication scheme for secure transactions. Int J Internet Technol Secured Trans 84:494–515
6. Santhosh kumar P, Parthiban L, Jegatheeswari V (2018) Secured data storage and auditing of data integrity over dynamic data in cloud. Int J Internet Technol Secured Trans 8(4)
7. Huang Y, Huang Z, Zhao H, Lai X (2013) A new one-time password method. In: 2013 International conference on electronic engineering and computer science
8. Leung KC, Leung CH (2011) Improvement of fingerprint retrieval by a statistical classifier. IEEE Trans Inf Forensics Secur 6(1):59–69
9. Turroni F, Maltoni D, Cappelli R, Maio D (2011) Improving fingerprint orientation extraction. IEEE Trans Inf Forensics Secur 6(3):1002–1013
10. Gottschlich C (2012) Curved-region-based ridge frequency estimation and curved gabor filters for fingerprint image enhancement. IEEE Trans Image Process 21(4):2220–2227

Dual-Axis Solar Tracking System Using ANFIS Technique

S. D. Sundarsingh Jebaseelan, V. Sivachidhambaranathan, D. Godwin Immanuel, D. N. S. Ravikumar, and M. Pushpavalli

Abstract The proposed system consists of solar panel which is capable of moving in dual axis that is both in *x*-axis and *y*-axis. A dual-axis tracker allows your panels to move on two axes, aligned both north–south and east–west. This setup is designed to maximize the solar energy. Two stepper motor is used for moving the solar panel. One stepper motor is for *X*-axis, and other stepper motor is for *Y*-axis. Motor driver circuit is used for the function of stepper motor. The system is also capable of tracking power point on the solar panel using ANFIS technique. The output of solar panel is fed into a DC-to-DC converter. This converter is used for boosting up. The solar panels output is boosted up using this converter and given to the load. There are three modes of operation according to the output of converter. The hardware also consists of a battery setup which can be used in the absence of solar panel so constant voltage can be achieved.

Keywords Solar panel · Stepper motor · ANFIS technique · LDR

1 Introduction

Solar is one of the most important renewable resources that are useful for the environment. This energy does not produce greenhouse gas, and it does not cause any kind of pollution to water or air. It is also useful for home applications. Rising cost of energy is one of the most troubling issues of today. In this paper, we are going to propose that the solar panel is capable of moving in a dual axis using. The main purpose is to track the maximum power of solar panel using ANFIS technique. To boost up the output of solar panel using DC-to-DC converter, the voltage from the output can be used to a load and excess voltage is stored using a battery so in the case of low voltage supply by solar panel battery can be used [1–3].

S. D. Sundarsingh Jebaseelan · V. Sivachidhambaranathan · D. Godwin Immanuel · D. N. S. Ravikumar (✉) · M. Pushpavalli
Department of Electrical and Electronics Engineering, Sathyabama Institute of Science and Technology, Chennai, India

S. S. Dash et al. (eds.), *Sixth International Conference on Intelligent Computing and Applications*, Advances in Intelligent Systems and Computing 1369,
https://doi.org/10.1007/978-981-16-1335-7_39

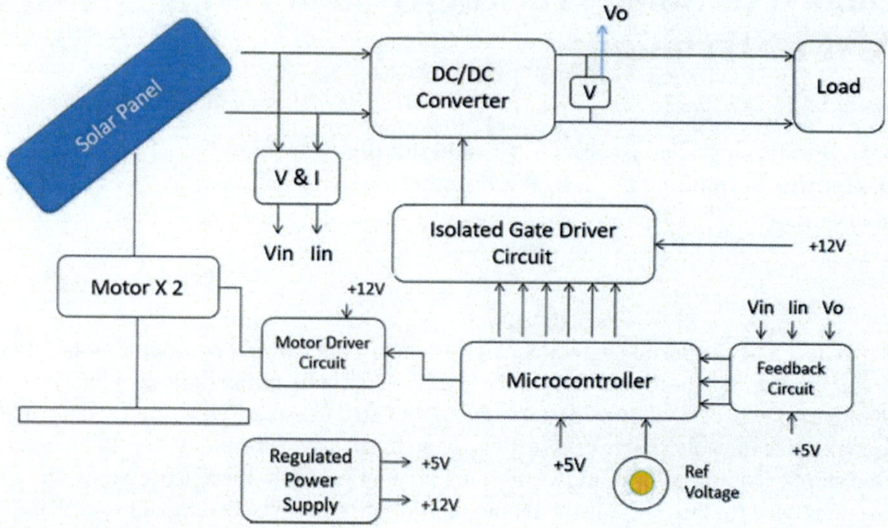

Fig. 1 Block diagram

2 Proposed Method

The above diagram represents the solar conversion. The solar panel is tilted using two-stepper motor: one in *X*-axis and the other in *Y*-axis. The stepper motor is operated using motor driver circuit which is controlled by microcontroller. The input for motor driver circuit is given using a six-tapped 12 V transformer. The output of the dual-axis solar panel setup is now fed into a DC-to-DC converter, where the voltage from the output of solar panel 12 volts is boosted upto 24 volts. Now the boosted up voltage is further used by load or stored in a battery. In the absence of solar energy, this battery can be used up by the load [4, 5] (Fig. 1).

3 Simulation Results

The above fig shows the simulation circuit (Fig. 2).

Figure 3 represents the input voltage graph. It is the output of solar panel.

Figure 4 represents the input current graph of the solar panel.

Figure 5 represents the input power graph of solar panel.

Figure 6 represents the voltage graph of converter which has been boosted up

Figure 7 represents the output current graph of DC-to-DC converter.

Figure 8 represents the output power graph of DC-to-DC converter.

Table 1 represents the comparison of input from the solar panel and the output from DC-to-DC converter which varies for certain time period.

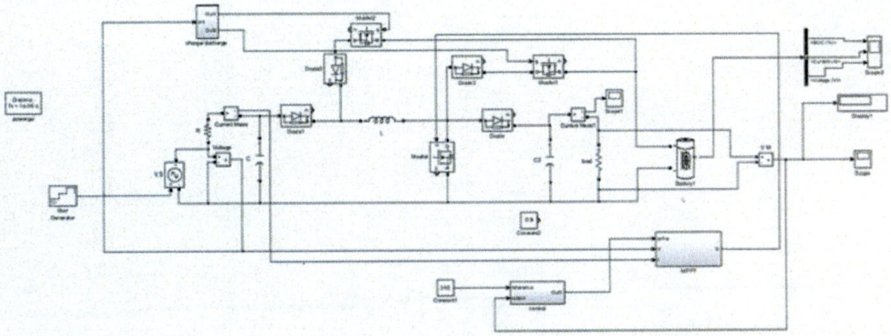

Fig. 2 Simulation circuit

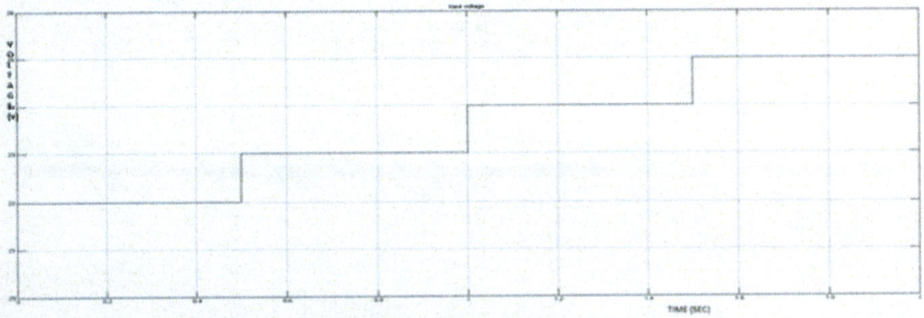

Fig. 3 Input voltage graph

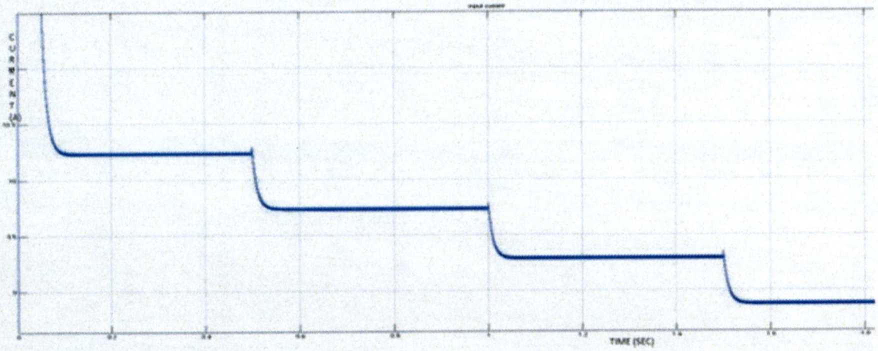

Fig. 4 Input current graph

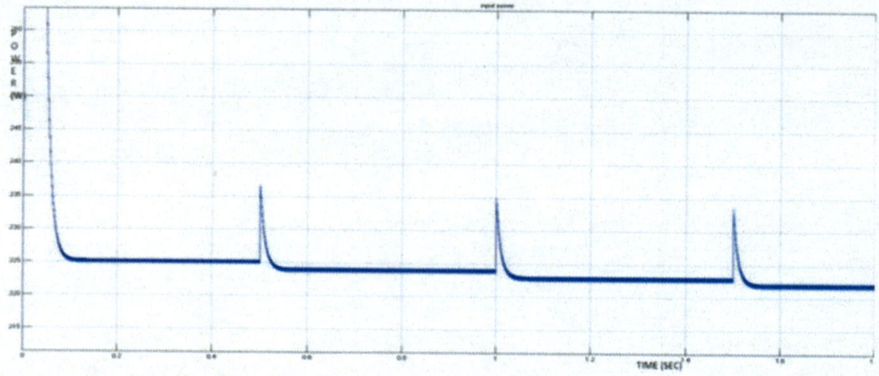

Fig. 5 Input power graph

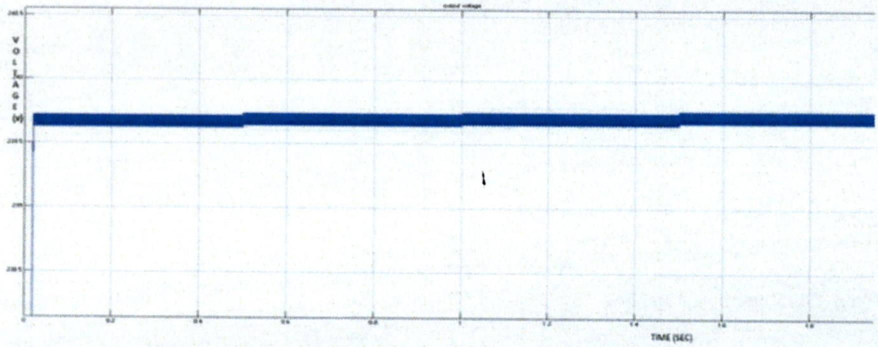

Fig. 6 Output voltage graph

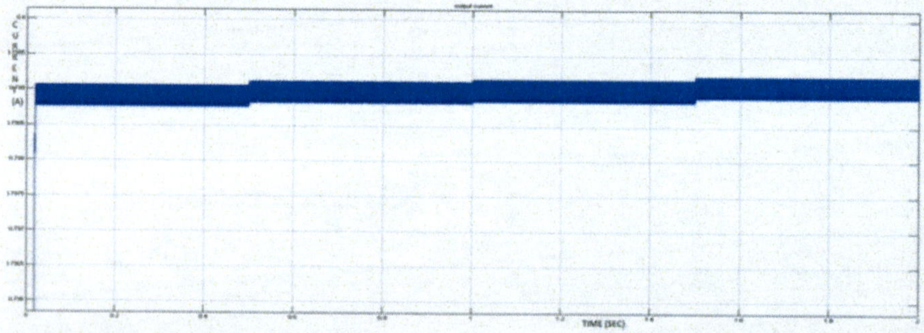

Fig. 7 Output current graph

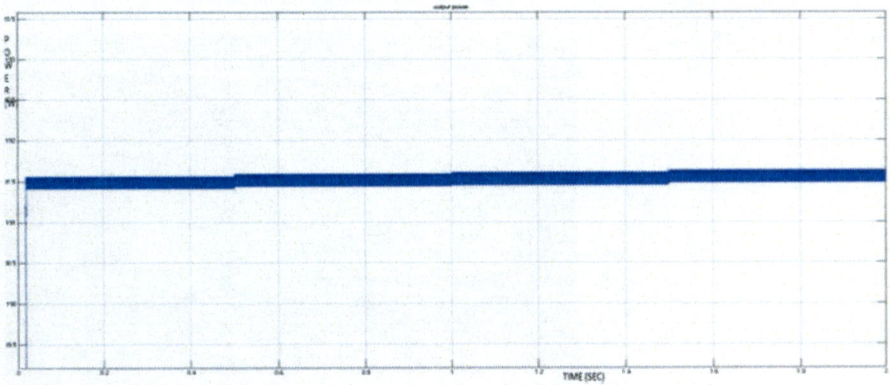

Fig. 8 Output power graph

Table 1 Comparison table

Parameter	0–0.5 s	0–0.5 s	1–0.5 s	1.5–2 s
Input voltage				

4 Hardware

4.1 *Solar Panel*

The solar panel consists of solar cells which is an electrical device that converts light energy into electricity using photovoltaic effort. 5 watts 12 volts solar panel can be used for house appliances. It is compact in size.

4.2 *Motor Driver Circuit*

Stepper motor performance is directly dependent upon the driver circuit. This initiates the action of stepper motor. It gives the required voltage to the stepper motor to perform its action.

4.3 *Working*

The dual-axis solar panel is tilted using two-stepper motor. One stepper motor is for *X*-axis which is bipolar for more torque. The other stepper motor is for *Y*-axis which is unipolar [7, 8]. The stepper motor's is controlled using motor driver circuit who

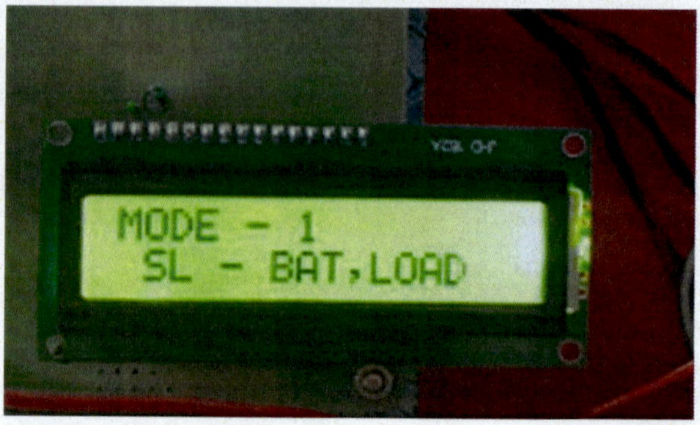

Fig. 9 Output of mode 1

is controlled using PIC microcontroller. The LDR's input is given to the Arduino which decides the action of microcontroller. The input of the solar panel is given to the DC-to-DC converter which boosts up the output of solar panel. The boosted voltage can be used up by the load. There are three modes of operation:

1. Mode 1: [when Vs < 8 volts]:

In mode 1 operation, if the output voltage of the solar panel Vs is less than the voltage needed by the load, then the voltage from the battery is supplied to the load till the solar panel produces the required voltage Vo

2. Mode 2: [when Vs = 10 volts., 11 volts, 12 volts]

In mode 2 operation, the output voltage of the panel and the load is equal so there will not be any voltage supply to the battery to charge up. The supply will be only from the panel to the load.

3. Mode 3: [when Vs >12 volts]

In mode 3 operation, solar panel produces higher voltage than the voltage needed by the load, so the voltage produced by the solar is used by the load and remaining voltage is charged in the battery.

Figure 9 represents the output of mode 1, when Vs is less than 8 volts.

Figure 10 represents the output of mode 3, when the voltage is greater than 12 volts.

Hardware Setup:

Figure 11 shows the hardware setup of the proposed system.

5 Conclusion

The final hardware setup shows that the dual-axis panel is more efficient. The ANFIS technique which is used to find maximum power point tracking has brought the panel to fix in a correct position where the solar light is more. This setup comes to a correct

Fig. 10 Output of mode 3

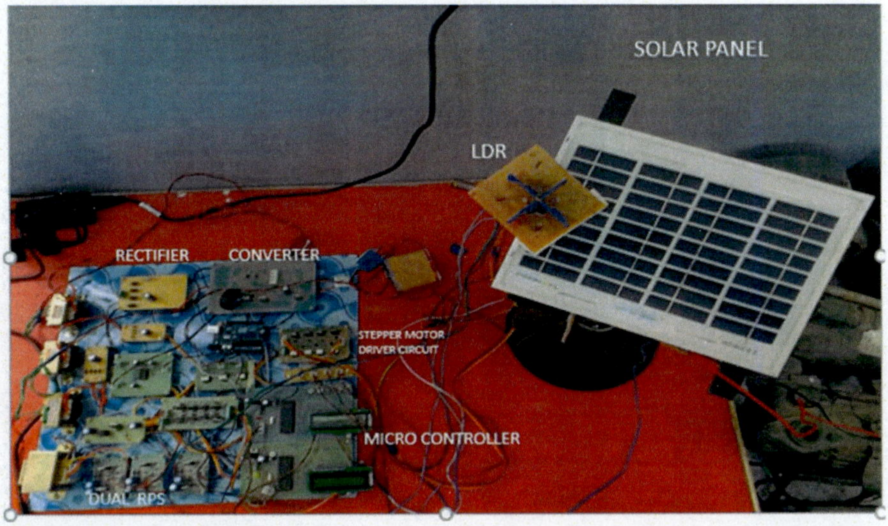

Fig. 11 Hardware setup

position when its required voltage is achieved so that solar panel will find the correct angle in short period of time. The DC-to-DC converter is used to boost up the input voltage so that it can be used for other purpose also.

References

1. Udaokah Y-O, Egwuchukwu (2018) Design and Implementation of a dual axis solar tracker using arduino microcontroller. ELEKTRIKA J Electr Eng 17:41–48

2. Megha JK, Pallavi KS (2018) Arduino based dual axis solar tracking system. In: IEEE international conference on electrical and electronic engineering
3. Gunputh S, Murdan AP (2017) Design and implementation of a low cost arduino based smart home system. In: 9th IEEE international conference on communication software and networks (ICCSN 2017), vol 3, pp 1491–1495
4. Gabe IJ (2017) Design and Implementation of low cost dual axis autonomous solar tracker. In: 2017, IEEE International conference on electrical and electronic engineering
5. Balaji PP, Maharaja K, Xavier RJ (2015) Intensity based dual axis solar tracking system. Int J Appl Eng Res 10:19457–19465
6. Kamble S, Chavan V (2015) Dual axis solar tracker system. IJIERT Int J Eng Technol
7. Geetha V, Sivachidambaranathan V (2018) A single switch parallel Quasi resonant converter topology for induction heating application. Int J Power Electron Drive Syst (IJPEDS) 9(4):1718–1724. ISSN 2088-8694
8. Bharathi ML, Basha RFK, Ramanathan SK, Harikrishnan R (2020) Fuzzy logic controlled maximum power point tracking for SEPIC converter fed DC drive—a hybrid power generation system. Microproce Microsyst

Delegated Authorization Framework for EHR Services Using QR Code-Based Encryption

G. Narendra, G. Sujeeth, and J. Albert Mayan

Abstract The point of this examination is to investigate the intentions belated the appropriation or dismissal of electronic health records (EHR) frameworks in the USA by narcotic workplaces. The goal is to lessen excess and increment quality by sharing medical data among various wellbeing associations like emergency clinics, doctor workplaces, laboratories, and medical organizations. The achievement of such change requires the cooperation and coordinated effort of every one of these substances and their resigned. Earlier testing conveys that the selection of electronic medical records frameworks by clinics and doctor workplaces has been clear, however, at a degree that is slower than in different nations. Besides money-related hindrances, specialized, mental, social/legitimate and hierarchical boundaries exist. So as to understand these effects, we make our task. The primary objective of our undertaking is to plan a versatile application for patients through online application the board that gains secure access to understand records through distributed storage administration. Administrator can remotely store persistent data into the cloud with sheltered and make sure about distributed storage through a patient can without much of a stretch access their information which are put away in cloud dependent on EHR service the board. Remote information respectability of wellbeing record is proposed to ensure the uprightness of the made sure about information put away in the cloud. In some regular distributed storage frameworks, cloud document may contain some delicate data. Encoding the entire shared record can understand the sensitive data concealing, however, will make this mutual record unfit to be used by others. Right now, propose a remote report reference id consequently convert into the QR code then simply filter client module at that point download the specific record respectability that acknowledges information offering to delicate data covering up. Marks are utilized to check the document in the period of honesty wellbeing records.

G. Narendra (✉) · G. Sujeeth
Department of CSE, Sathyabama Institute of Science and Technology, Chennai, India

J. Albert Mayan
School of Computing, Sathyabama Institute of Science and Technology, Chennai, India

S. S. Dash et al. (eds.), *Sixth International Conference on Intelligent Computing and Applications*, Advances in Intelligent Systems and Computing 1369,
https://doi.org/10.1007/978-981-16-1335-7_40

Keywords Electronic medical records · Wellness record · Electronic health record · Attribute-based access

1 Introduction

In the current scenario, health information technology performs in various methods and specifications. So, this paper talks about various obstacles for implementation of electronic record system in doctor practices. The meaning of electronic frame work differs from one specification to other supported quite work and desire from the framework. So, the definition and usage of the framework vary from one applicant merchant to other. Electronic medical records (EMR) and electronic health records are most essential frameworks for storing patient's information. The EHR is about quality, safety, and efficiency. It is great tool for physicians, but cannot ensure these virtues in isolation. Achieving the true benefits of EHR systems requires the transformation of practices, based on quality improvement methodologies, system and team-based care, and evidence-based medicine. The details include meanings of the voltaic frameworks vary starting with one foundation then onto the next dependent on every one's kind of effort includes desires from the framework. The determinations and settling likewise vary starting with one application merchant then onto the next. The orders of these frameworks are discretionary as yet. The most well-known names to human services records the executives' frameworks are electronic medical records (EMR) and electronic health records (EHR). Despite the fact that the pair is regularly utilized conversely, there live a separation in the middle of duo. Comment on [1] has recognized more than decade distinctive advanced wellbeing and preventive put down the executives' frameworks that own exists the issue of scholarly investigations considering dyad decagon. Electronic patient record (EPR) framework and computerized patient record (CPR) framework allude to a computerized or advanced log about a solitary victim midsole specific emergency clinic either more disinfectant establishment [1]. Likewise, it has a particular spotlight on the sort of medical information they stock. The US administration alludes to the thermionic plea for overseeing and following wellbeing and medicinal list as electronic health record (EHR) framework. The administration's desire for an EHR is to bring down the expense of preventive consumption. EHR perhaps bring down consumption in the event that it makes giving and handling medical consideration increasingly effective by disposing of superfluous doctor visits, enabling specialists to analyze a patient remotely, overseeing medicines and medication satisfaction quickly, distinguishing early indications of sicknesses, and giving consideration conveyance associations a complete account about some sufferer to have the option to settle on blow disinfectant choices in the absence of delivering redundant medical methods and managerial capacities. That kind of wellbeing data innovation requires the participation and coordinated effort of fully CDOs from the general population along with personal segments.

Dissimilar to EHR, an EMR is a complete mechanized tonic information archive framework possessed by alone [2]. For dealing with their EMR, a CDO will rely upon a medical information store gave by a solitary association; in the time, an EMR be allowed to get a natural spotlight on cathartic consideration, and it imparts significant normal fact and order to EPR. For improvement and combination plan, the title EMR while utilized right now alludes into some modernized for bearing tonic witness the executive's framework, that is, occupy, refreshed, altered, followed, and oversaw, at a solitary for that association's inward cause. EMR is an entreaty domain made out of the medical information archive, medical choice help, controlled medical jargon, request passage, modernized supplier request section, drug store, and medical proof request [2]. An EHR is an extensive mechanized evidence considering single valetudinarian and permits tolerant info and surge [2]. Interest is made out of numerous alleviating jackets. An EHR's prosperity is firmly subject to information trade between EMR adopters that is continuous and finish. As opposed to the vends comprehension of EMR and EHR framework, differentiation midway the banal is obligatory for appropriate comprehension of the degrees of reception, execution and dispersion of every, which has its effect on assessing an exact advancement of the social insurance change arrangement

A cloud alludes to a domain that is intended with the end goal of remotely provisioning versatile and estimated IT assets, where the system giving remote access to a lot of decentralized IT assets. The thought behind the venture is to plan a versatile application for data sharing through circulated capacity administration customer can remotely store there data to the cloud and comprehend the data offering to different elements which additionally contains some touchy information. The delicate data ought to guarantee it is recovered by the validated client when the cloud record is shared. Right now, sanitizer is utilized to disinfect the data comparing to record and changes these data's stamps in authentic ones for the disinfected document. These marks are used to check the integrity of the cleaned data in the time of uprightness surveying.

Information is sharing between two substances in cloud: one section medical clinic the board to treat as the administrator part and afterward patient to treated as the client part. Administrator to transfer the medical record in our cloud (data and patient ID) at that point search the specific patient id and medical archive reference id naturally convert into the QR code then simply filter client module at that point download the specific report. The principle advantage is to have more made sure about change between different sides, increasingly adaptable, simple access; at long last, we can download the record as a free report. In existing framework, they have utilized RFID innovation which isn't reasonable for progressively reasonable information putting away and not minimized for all divisions. There are numerous associations and people who may get a kick out of the chance to store their information in the cloud. Nonetheless, the information put away in the cloud may be debased or lost because of the unavoidable programming bugs, equipment flaws, and human blunders in the cloud providing security for the medical archive utilizing QR code-based Android application, where user can store their reports in cloud which will be changed over into QR code which can be downloaded by the specific individual who have picked up the entrance to the archive and they can impart it to the relating substances. This brings the trust commendable among the clients to transfer their index in cloud where administrator transfers the archive along the reference id which at that point consequently creates the QR code.

2 Related Work

Dhamodaran et al. [5] This can be credited to the flexibility, significant level of accessibility, and diminished expense of cloud administrations. Right now, there are various cloud-based EHR administrations, together with CureMD 4, operation heating 5 and unhealthy 6. Associations like GE preventive medicine 7 and history energy solutions 8 are additionally putting resources into cloud-based EHR administrations. Different testing endeavors have been recommended with significant spotlight on safe, cloud-based EHR frameworks. Different specialists have additionally proposed confided in registering utilizing SGX processors for cloud security [9, 10]. In any case, most of the bid addresses are insufficient in ensuring a complete out flow charge and inscribe component. Alongside this, most accessible appeal is authorized and therefore costly to receive. Right now, open-source, minimal effort EHR overseeing application should be built up that can ensure refined levels of information security and assurance. Through the EHR manager approach, this exploration exertion attempts to fabricate such an answer by utilizing all open code improvement apparatuses separated from the outsider cloud administrations.

The EHR chief is an open-source device which gives a simple attachment to medical hands just as patients to see and additionally alter the EHR. Inherit, this appeal ensures solid gale domination and information insurance system. The health insurance transportability and reporting Act of 1996 gives information protection, reliability, and shielding represents ensuring electronic medical data of people [5, 6]. HIPAA gives rules to electronic medical record the board for offsetting singular security as for medical records alongside the need to ensure strength of the majority. The health data innovation for monetary and clinical health acts toward keeping up electronic clinical records by ensuring quality, wellbeing, productivity, protection, and security [4].

Conforming to the permissible of overseeing medical proceedings and simultaneously building up a simple to utilize electronic wellbeing record framework turns into a significant innovative work challenge. There exist numerous EHR the board instruments like GE Healthcare, Epic, CureMD, and so on which offer HER types of assistance utilizing distributed computing. Already, we built up linguistics-rich ingress manage model dependent on attribute-based access control (ABAC) [16]. This model assessed an entrance choice dependent on the qualities of the client mentioning an archive and those of the mentioned report. We planned and executed a philosophy to exhibit the utilization of ABAC in an authoritative setting. Access control choices were assessed against a hierarchical classification strategy.

This work showed the utilization of arrangement-based, semantic web approach of actualizing ABAC at a report quantity. The framework has been boost to assess an entrance choice on the fields of an archive instead of the whole record. The refine framework would now be able to classify the allowed access rather than only a twofold choice. Aside from this, the recently evolved framework showed the idea of perimeter registering [17] where the authoritative limit was viewed as the edge of the framework. The cloud specialist organization was examined as an doubting substance and in this way lied underneath the hierarchical edge. All data exchanges in the middle of the hierarchical limit and the cloud were encoded utilizing an unaware storage and the unaffected RAM [18, 19], darkening the entrance designs between the association and the cloud specialist organization.

The EHR manager utilizes the CPABE toolbox to model the examination exertion. ABE has been one of picked advancements for electronic wellbeing record the executives frameworks too Akinyele et al. [20] have introduced a structure and execution of electronic medical records (EMRs) utilizing property-based hide on cell phones. In their framework, they give disconnected help to refresh the medical records with help for inevitable consistency. In any case, their model does not bolster a field-level conceal of the EHR. Analysts at Microsoft built up a patient controlled electronic medical record framework with trait-based cipher. As the name recommends, this framework put all the entrance power lies inpatients hands. The command and dispersion of key sequence were the patients obligation. Be that as it may, this methodology requires an elevated level of authority adopt on the patients end. The EHR manager, be that as it may, does not force any overhead on the patient. The focal framework handles all the secure sequence and circulation of the EHR.

3 Existing System

The existing idea is an appropriate approval assignment system to utilize cloud-based EHR service the executives utilizing attribute-based encryption (ABE) in web advancements that includes the mix of utilizing semantic web method with properties-based plans. Quality-based encryption (QBE) utilized in existing framework. Standard build conceal is a style of open key cipher where the puzzle key of a market and the particular fulfilled are reliant upon characteristics. Address where the EHR information approval must be affirmed by the patient. This isn't pragmatic in each situation, and additionally, the patient may not be in a state to give this approval when need. Subsequently, there is a demand to build up an approval appointment system where by the patient approves the supplier entry to his/her EHR and the supplier thus assigns this approval to suitable representatives or associates to get to the information. Customary job-based access models won't fill in as the cloud-based EHR frameworks can be gotten to from any area and from any gadget. So notwithstanding the guardian's jobs, their different qualities, similar to area, time, obligation period, and so on can likewise impact the appointed approval. We have built up a novel, brought together, quality-based approval component for EHR offices that utilize attribute-based code and take into consideration assigned solid path of serene records. This system moves the administration the board insurance from the tranquil to the healing association and permits simple assignment of cloud established EHRs get to power to sedative suppliers.

4 Proposed System

Right now proposed a remote archive reference id naturally convert into the QR code then simply examine client module at that point download the specific record uprightness that acknowledges information imparting to touchy data stowing away. Marks are utilized to confirm the document in the period of trustworthiness wellbeing records. Information sharing between two substances in cloud: one section medical clinic the executives to treat as the administrator part and afterward patient to treated as the client part. Administrator to transfer the medical record in cloud (data and patient ID) at that point search the specific patient id and medical archive reference id consequently convert into the QR code then simply examine client module at that point download the specific report (Fig. 1).

The principle advantage is to have more made sure about progress between different sides, increasingly adaptable, simple access; at long last, we can download the document as a free report. A sanitizer is utilized to clean the information squares containing any touchy data of the record and changes these information squares' marks into acceptable ones for the sterilized document. These marks are exploit to check the respectability of the strain record in the phase of uprightness of QR code

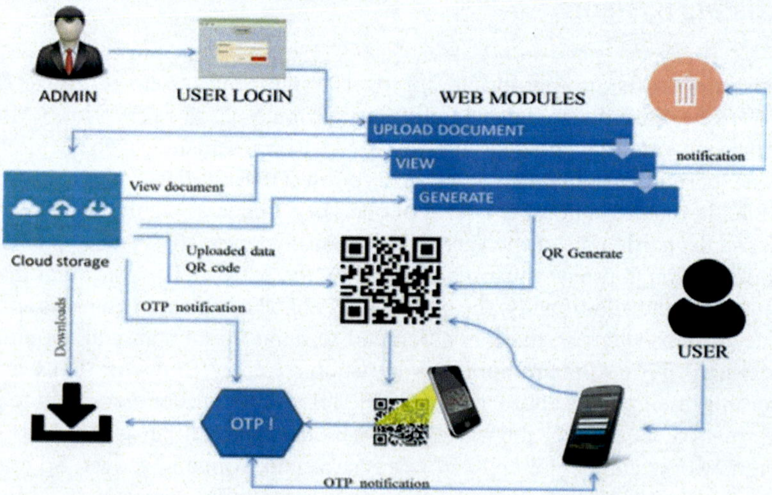

Fig. 1 Overview of the proposed system

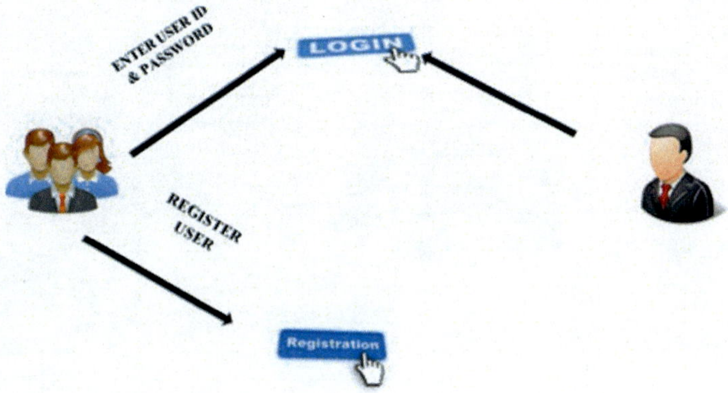

Fig. 2 Login and registration

create Algorithm. QR code ends up being unreasonably costly for some applications when contrasted with other following and recognizable proof techniques, for example, the basic standardized tag.

5 Modules Description

5.1 *Login and Registration*

In these module, there is a login and enrollment structure. In these enrollment structure, we need to enlist by utilizing username or email id. Furthermore, we need to login with the equivalent username or email id, then just it will login else it won't (Fig. 2).

5.2 *Database Creation*

Client email id or name and secret key are put away in these SQLite database for putting away and bringing client application structure (Fig. 3).

5.3 *Medical Information Transfer*

Now, the client data has been transferred to secure information as shown in Fig. 4.

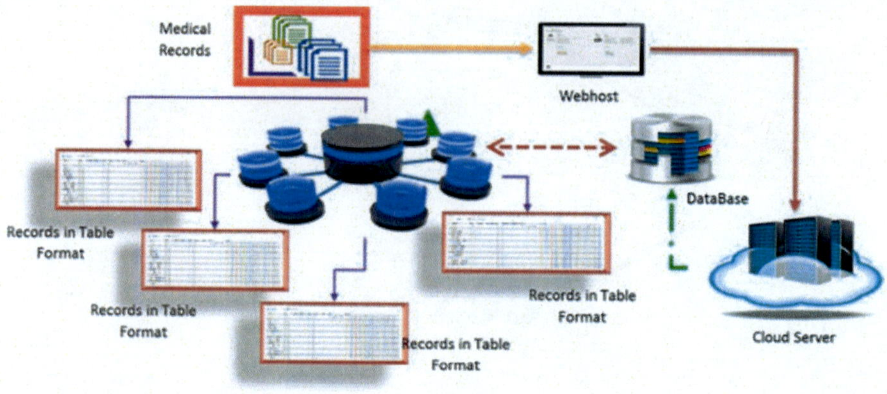

Fig. 3 Database creation

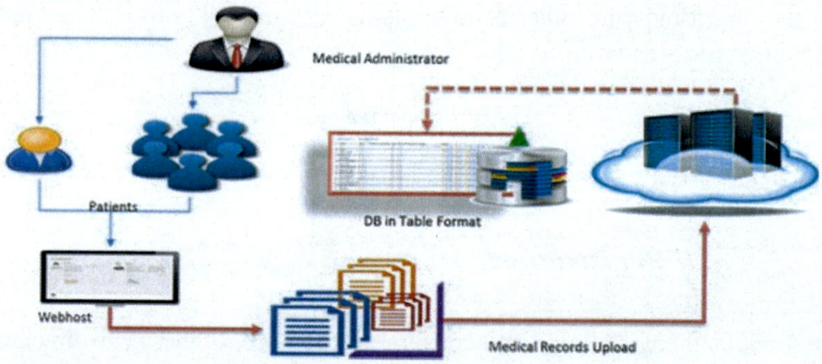

Fig. 4 Medical information transfer

5.4 *QR Code Encryption*

In Fig. 5 a QR code has produced an utilizing encryption which worth like medical records information and patients subtlrties.

5.5 *Medical Data See/Remove*

Right now, information see/expel that procedure can be utilized on logged client see a medical information and incase that information can evacuate. Be that as it may, unique qualities not erase in perpetual (Fig. 6).

Fig. 5 QR code encryption

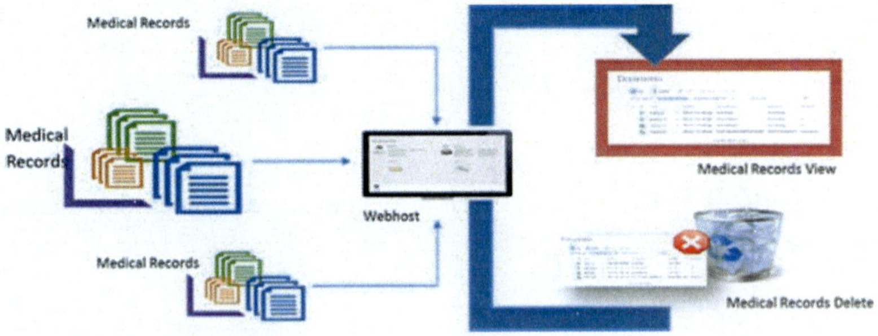

Fig. 6 Medical data see/remove

5.6 Security Scanner

In Fig. 7, security scanner utilized the tasks are security fix, QR code Reader and Make QR code.

Fig. 7 Security scanner

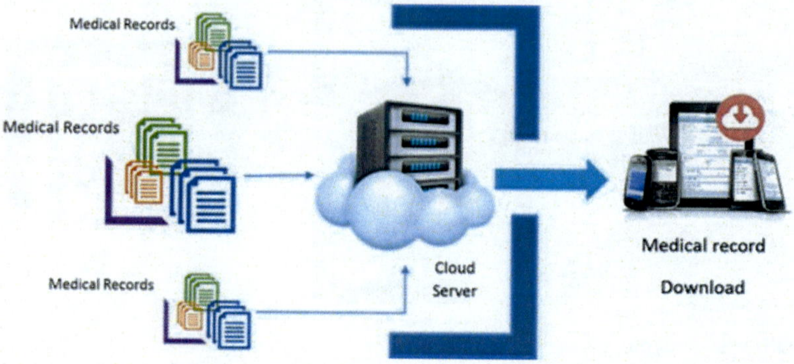

Fig. 8 Medical file downloads

5.7 Medical File Download

We need to make a medical document download are in general information put away in cloud the particular client medical record finds to see and download the specific client records (Fig. 8).

6 Result and Discussion

6.1 Registration Page

In these registration page, the hospital admin is used to enter the details of the hospital name, id, phone number, address and password of the admin, and it is used to store in the database then only the admin can login with the id and password (Fig. 9).

6.2 Creating Patient ID

See Fig. 10.

6.3 Adding a File of a Patient

In this web page, the admin is used to store the file of a patient in a cloud and this file is accessed from anywhere in the presence of patient (Fig. 11).

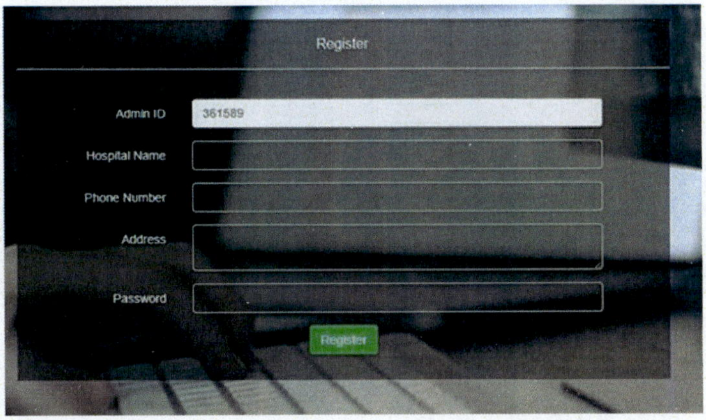

Fig. 9 Registration page

Fig. 10 Creating patient ID

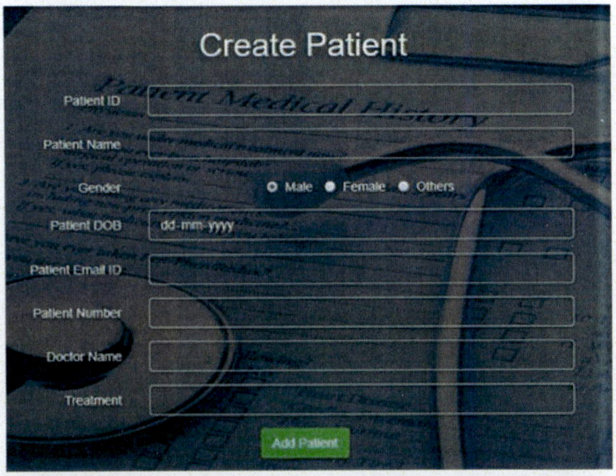

Fig. 11 Adding a file of a patient

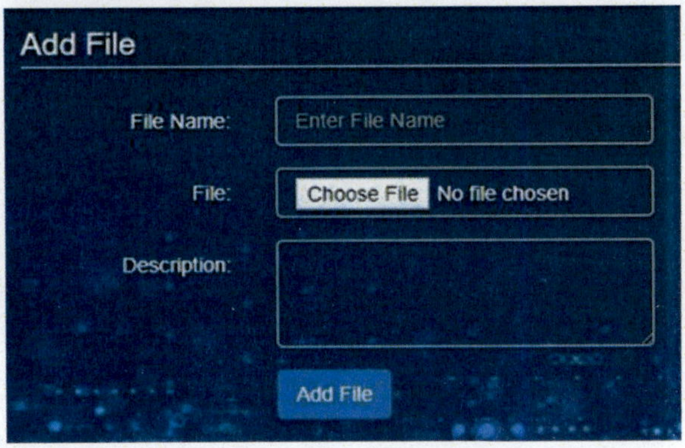

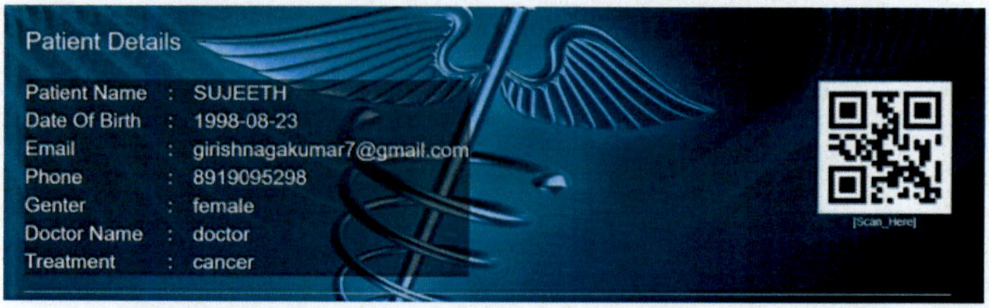

Fig. 12 QR code for every patient

6.4 QR Code for Every Patient

This is used to give all the details of the patient and the QR code for every patient. Whenever the patient scans the QR code, it can access the patient data (Fig. 12).

6.5 QR Code Scanning

In this android application, the patient is used to scan the QR code, then it sends an OTP to that particular patient and by entering it only it can access the data (Fig. 13).

Fig. 13 QR code scanning

Fig. 14 OTP verification

Fig. 15 Patient files

6.6 OTP Verification

In this, the patient is used to enter the correct OTP that which came to his mobile (Fig. 14).

6.7 Patient Files

See Fig. 15.

7 Conclusion

The medical data framework is ceaselessly developing. One significant future advancement is probably going to be the sparing of medical records to electronic media, as opposed to the paper records presently utilized in many medical clinics. In the remedy move framework condition, the patient profile, solution subtleties, and assessment results are spared as content and used during treatment. Plus, it develops the further developed EMR by entering the specialist's solution record just into the PC. Distinguishing proof in the EMR framework, which uses significant individual data, is the most significant security factor that chooses the privilege of access to the EMR framework, and the entrance right of the medical faculty concerned. The open level QR code can be perused by any QR code understanding application. In any case, the private level QR code needs a particular application with explicit info data. Also, this private level QR code is made by supplanting dark modules alongside some particular finished examples. Any standard QR code per user considers the finished example; as a dark modules in this manner, the private level is imperceptible to standard QR code per users, and this private level QR code does not influence any of the open levels.

References

1. Julian Menezes R, Jesu Jeyarin P, Albert Mayan J (2019) A scholarly audit on the traits of enciphering, deciphering algorithms bifurcated under symmetric, assymetric for wired cum wireless environment. J Adv Res Dyn Control Syst 11:1443–1454
2. .
3. Albert Mayan J, Lakshmi Priya K, Yovan Felix A (2015) Empirical study of GUI to reuse unusable test cases. JATIT J Theor Appl Inf Technol 75(2):186–192
4. Bahga A, Madisetti VK (2013) A cloud-based approach forinteroperable electronic health records (ehrs). IEEE J Biomed Health Inform 17(5):894–906
5. Dhamodaran S, Albert Mayan J, Saibharath N, Nagendra N, Sundarrajan M (2020) Spatial interpolation of meteorological data and forecasting rainfall using ensemble techniques. In: AIP conference proceedings, vol 2207, pp 050005
6. Chandra S, Karande V, Lin Z, Khan L, Kantarcioglu M, Thuraisingham B (2017) Securing data analytics on sgx with randomization. In: European symposium on research in computer security (ESORICS), pp 352–369
7. Mayan JA, Priya KL (2015) Novel approach to reuse unused test cases in a GUI based application. In: IEEE international conference on circuit, power and computing technologies, pp 1–7
8. S. Narayan, M. Gagn´e, and R. Safavi-Naini, "Privacy preservingehr system using attribute-based infrastructure",CCSW, 2010
9. Eswar V.O.S,Vinil.B,Ankayarkanni B and Albert, "Integrated Collective Node Behavior Analysis with Onion Protocol for Best and Secured Data Transmission," 2018 International Conference on Communication and Signal Processing (ICCSP), Chennai, 2018, pp. 0918–0921
10. Pandian Asha (2019) Bharathi B, Albert Mayan J, Prem Jacob, Pravin,"A Comprehensive View of Scheduling Algorithms for MapReduce Framework in Hadoop". J Comput Theor Nanosci 16(8):3582–3586
11. Velmurugan A, Albert Mayan J, Niranjana P (2020) Richard Francis,"Expense Manager Application". Journal of Physics: Conference Series (ICCPET) 1712:1–8
12. Mayan J.A, Arifa S, Pavithra R,"Semantic based multi lexical ranking technique for an effective search in protected cloud, International Conference on Control, Instrumentation, Communication and Computational Technologies (ICCICCT), pp. 570–576,2016
13. Mary Posonia A, S. Vigneshwari, Albert Mayan J, D. Jamunarani,"Service Direct : Platform that Incorporates Service Providers and Consumers Directly",International Journal of Engineering and Advanced Technology, Vol.8,No.6, 2019

14. Sandhu RS, Coyne EJ, Feinstein HL, Youman CE (1996) Rolebasedaccess control models. Computer 29(2):38–47
15. Ankayarkanni B, Albert Mayan J, Aruna J,"Support vector machine for effective robust visual tracking",Journal of Computational and Theoretical Nanoscience, Vol.18, No.8, pp. 3571–3575
16. A. Anderson, A. Nadalin, B. Parducci, D. Engovatov, H. Lockhart,M. Kudo, P. Humenn, S. Godik, S. Anderson, S. Crocker et al.,"extensible access control markup language (xacml) version 1.0," OASIS, 2003
17. Sivachidambaranathan.V & S.S.Dash (2010), "Simulation of Half Bridge Series Resonant PFC DC to DC Converter", IEEE International Conference on "Recent Advances in Space Technology Services & Climate Change – 2010" (RSTS&CC-2010), Sathyabama University in association with Indian Space Research Organisation (ISRO), Bangalore and IEEE, ISBN 978-1-4244-9184-1, November 13-15, IEEE Explore pp 146-148
18. M. D. Kamalesh, Albert Mayan. J, Yovan Felix, Dhamodaran S and Mohana Prasad, "Automation of Blood Donation by Data Integration Using Data Mining," *2020 4th International Conference on Trends in Electronics and Informatics (ICOEI)(48184)*, Tirunelveli, India, 2020, pp. 944–948

House Price and Renovation Prediction Analysis Using Different Machine Learning Algorithms

M. Naga Srinivasa Karthik, M. Rahul Sai Krishna, and A. Mary Posonia

Abstract The prediction of accurate housing price has become an essential task for the real-estate parties on demand of houses for splendid population. Selling of old properties is also a challenging task as it is costly to renovate all the property. Renovation of on-demand attributes of the property saves money and time. Machine learning predictive methods play a key role in these scenarios. Different regression methods produce different error values. In this paper, the focus is on finding the relative best predictive method on Boston housing dataset with relative less error, thereby increasing accuracy. The comparison of mean and standard deviation of price among ten techniques has been done and visualized using various plots. Gradient boost algorithm was found to be the best among all the methods analyzed with higher accuracy and very low MSE value at 10.292, RL being a prominent variable. So, the rooms with damaged property have to be renovated in first priority. External and internal factors contributing to price variation have also been studied.

Keywords Decision tree · Gradient boost regression · Housing price prediction

1 Introduction

Be it the prediction of house prices, automobile safety, personal assistants, customer support or commuting predictions, machine learning being a powerful domain is applicable in a wide range of problems, and different models in machine learning can be applied for different scenarios. Using supervised learning, training of labeled data can be done and output is predicted. Through regression model, using available features, a new feature can be predicted. In this particular problem, prediction of the price of a Boston area house has been done using multiple machine learning and the results have been compared. Algorithms such as classification, regression and neural networks have been used. Based on the mean and the standard deviation, the methods

M. Naga Srinivasa Karthik · M. Rahul Sai Krishna · A. Mary Posonia (✉)
Computer Science and Engineering, Sathyabama Institute of Science and Technology, Chennai, India
e-mail: maryposonia.cse@sathyabama.ac.in

S. S. Dash et al. (eds.), *Sixth International Conference on Intelligent Computing and Applications*, Advances in Intelligent Systems and Computing 1369,
https://doi.org/10.1007/978-981-16-1335-7_41

with less mean have been found more accurate and suggested. Linear regression, Lasso, elastic, decision tree, random forest, K-neighbors, AdaBoost, gradient boost, extra tree regression and support vector machine are the methods analyzed to predict house prices. After finding the best method, there is a need to find the most impactful attribute [1–7]. By finding it, prediction of the renovation priority is done. In this paper, not only internal but also external factors that affect the price have been compared. Based on the result obtained on Boston data with more external features, the best technique obtained is applied. The data which is taken in this paper that is Boston data [8] is a standard dataset of fourteen features proposed by Harrison Jr, D. and Rubinfeld, D. L., in 1978. Gradient boosting has been found to be the most accurate method with a lot of internal features which improved accuracy shown in Fig. 1.

The data is analyzed, visualized and implemented using algorithms by train and testing. The Python libraries ***pandas, matplotlib, seaborn and scikit-learn*** are used.

Fig. 1 Phases of prediction cycle

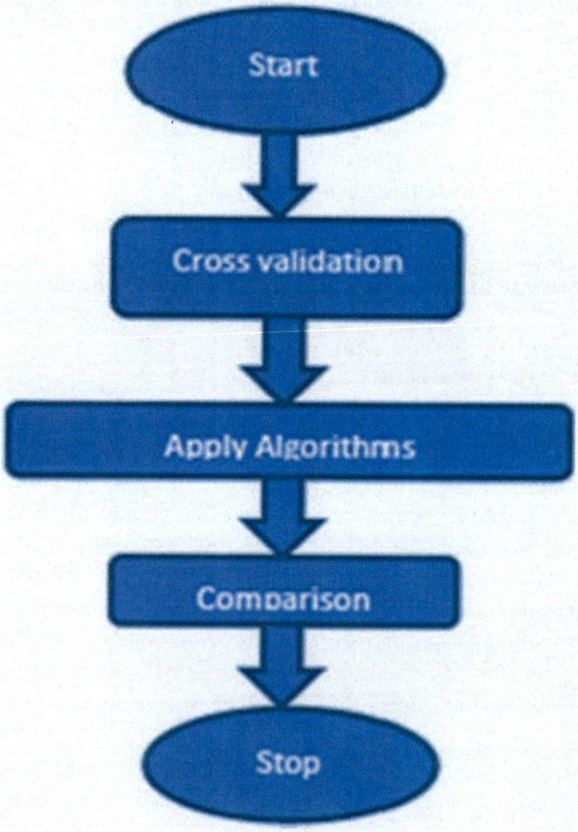

1.1 Correlation and Data Distribution

From Fig. 2, the correlation between variables was visualized and used to understand the relationship using a Pearson map. The outliner analysis can be done.

From Fig. 3, the correlation is determined using Spearman heat map and it is observed that least correlation is in between price and LSTAT.

Similarly, remaining heat maps from Fig. 4 also show a similar correlation using Kendall. It is observed all the three methods are showing similar maps. So, the data is clean without much noise. The univariate data plots are generated using box plot and density plot. The multivariate plot used is scatter plot. All the fourteen variables are visualized, and outliner variables are also determined. The attributes PRICE, B, LSTAT, PTRATIO, DIS, CHAS, CRIM and ZN have outliners in the data.

As from Fig. 5, there are outliners in the data, which are to be removed. Similarly for the price as it from Fig. 6 is visualized. From Fig. 7, it can be seen the data distribution is skewed Gaussian distribution.

Similarly, we can see the bimodal distribution in Fig. 8. Scatter plot is used to determine the relationship between target variable and attributes.

From Fig. 9, you can see the number of rooms and price have a strong correlation. This also determines it as an important attribute. After data exploration, the algorithms are implemented. The language used to execute algorithms is Python.

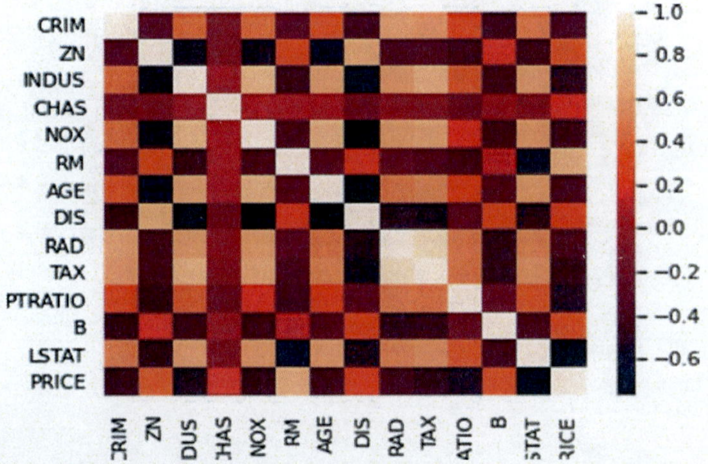

Fig. 2 Pearson heat map

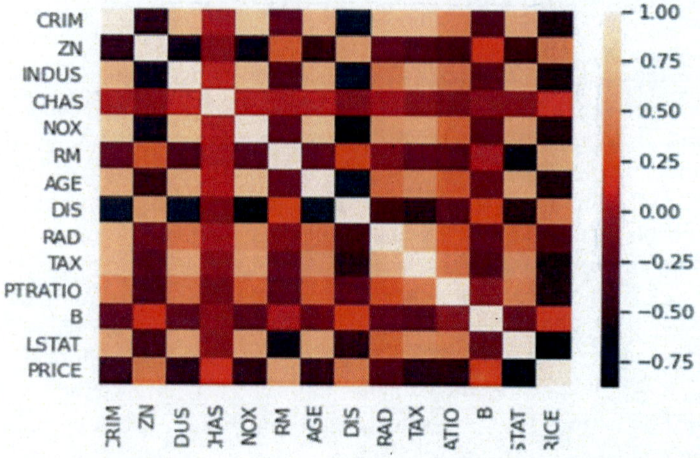

Fig. 3 Spearman heat map

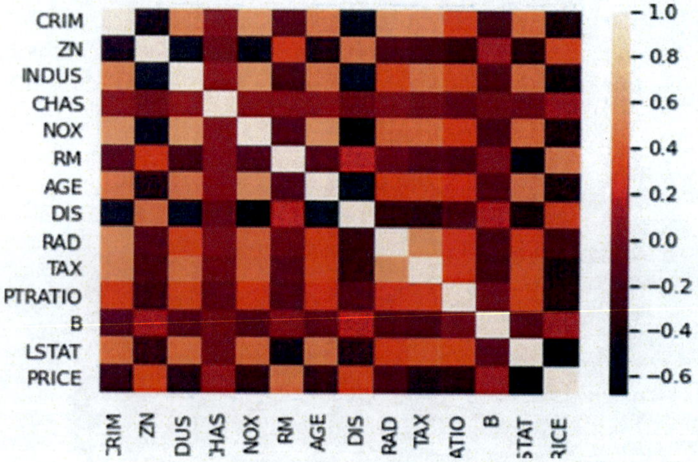

Fig. 4 Kendall heat map

Fig. 5 RM box plot

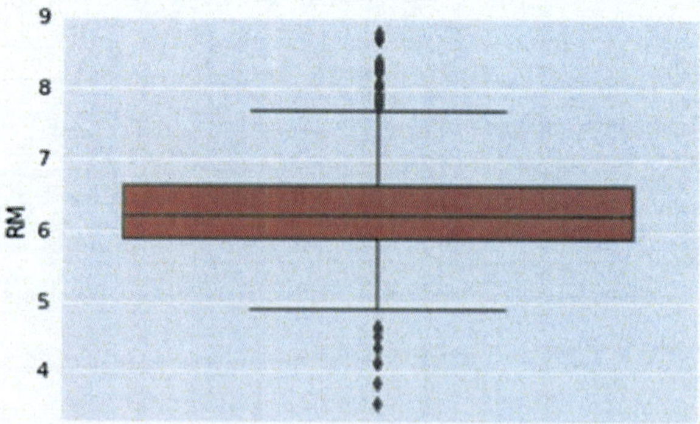

Fig. 6 PRICE box plot

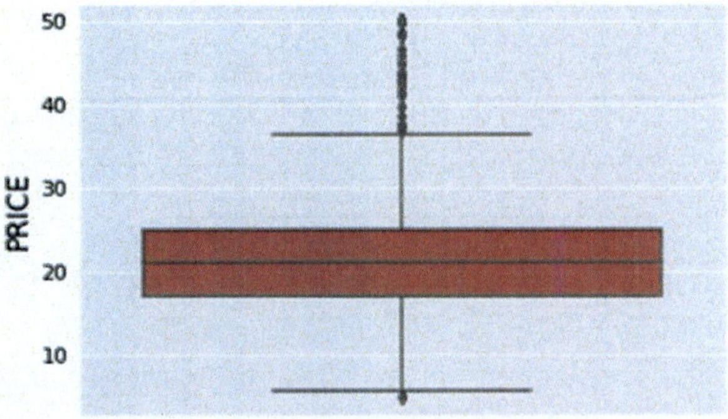

Fig. 7 RM density plot

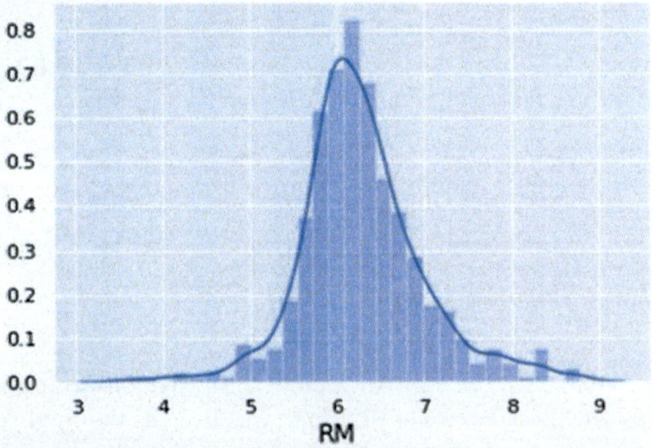

Fig. 8 Price density plot

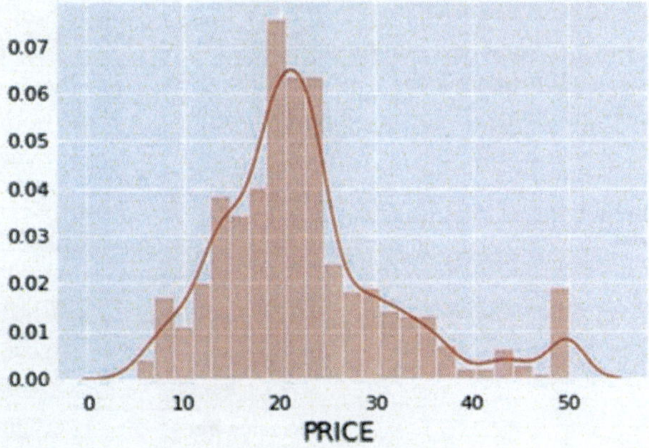

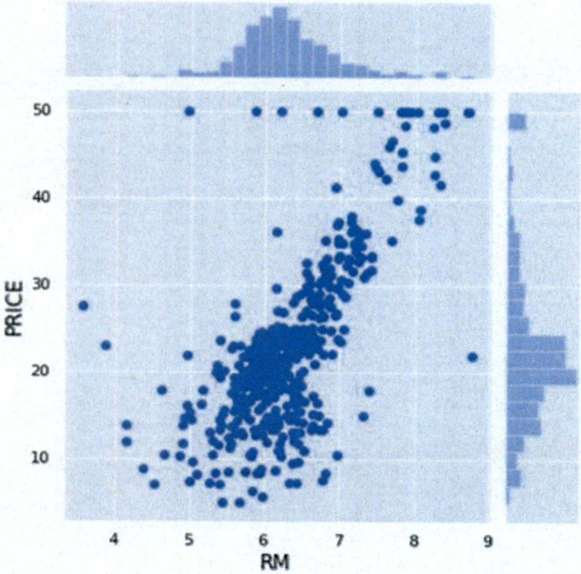

Fig. 9 RM scatter plot

1.2 *Training and Testing*

For the analysis, 404 rows in a tenfold way to train the algorithm and 102 rows to test the model were used. Mean was used as the mode of comparison to compare all the 10 methods and predict the best method. The mean and standard deviation from all the methods have been listed in Table 1.

As per the above results, we see that boosting algorithms by boosting of decision trees got very less standard deviation and larger negative mean. In those, gradient boost algorithm is the one with larger negative mean and it has very less standard deviation. So, it is inferred in most of the cases the boosting algorithms predict well.

Table 1 Mean and standard deviation from all methods

Algorithm	Mean	Standard deviation
Linear regressor	−23.795	12.359
Lasso regressor	−63.83	20.648
Elastic net	−69.363	21.372
KNN method	−26.365	16.168
Decision tree	−25.26	13.482
SVR	−34.845	17.408
AdaBoost	−14.574	6.573
Gradient boost	−10.319	5.284
Random forest	−12.002	6.226
Extra trees	−11.597	7.835

Now, the mean square error for the boosting algorithm is determined. This is an efficient technique to train a decision tree of weak components. The tree is trained until optimal condition is satisfied and MSE is 10.292.

2 Results and Discussion

2.1 Mean Square Error Comparison

As observed from Fig. 10, the box plot obtained from the gradient boost has very less standard deviation and minimum error. The final MSE obtained for gradient boost was 10.292 which is very less compared to linear regression which is around 27.8. The actual y-test values are green dots in Fig. 11. The predictions done by our model are red dots. By gradient boost algorithm, we have the maximum accuracy among all. This shows boosting algorithms are better than the naïve methods. The variable

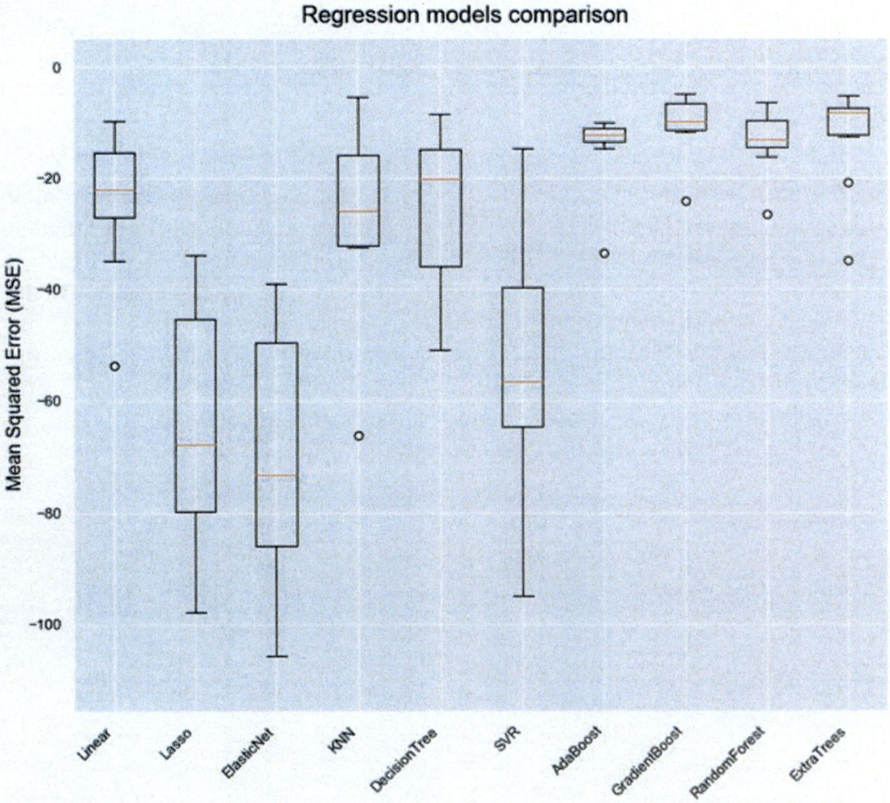

Fig. 10 Regression model MSE comparison

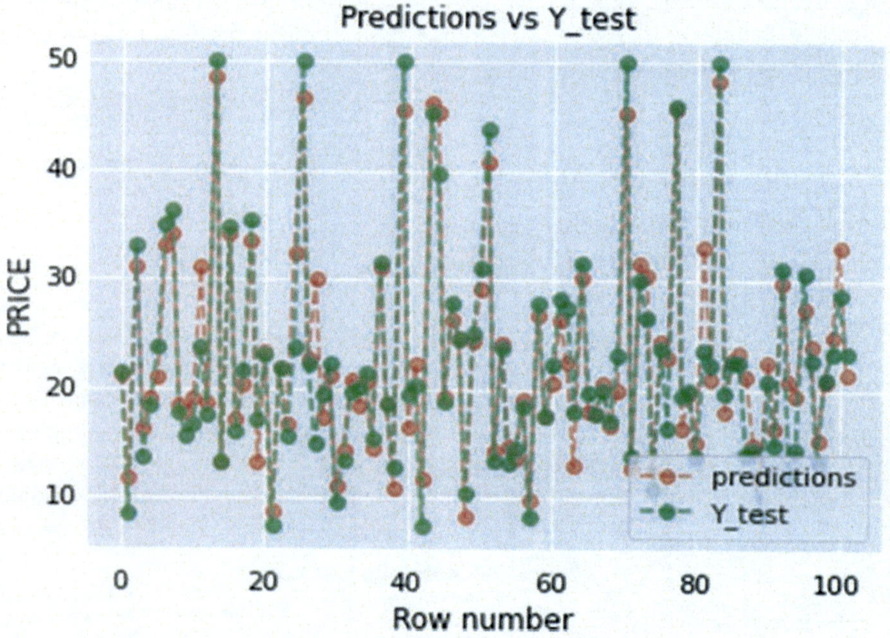

Fig. 11 Predictions versus Y-test

which is most dominant in deciding the price has been analyzed. By comparison of variable importance, RM average number of rooms per dwelling has been found to be the major deciding attribute as shown in Fig. 12, followed by LSTAT, which is low status population area, and DIS, which is distance from the city centers. As the major attributes taken in this Boston dataset are external factors, the deciding factors have been observed to be the internal factors. Gradient boost implemented on the other similar housing dataset also gave less error. Being weaker models, decision tree models have been boosted for accuracy through gradient boost.

From Fig. 10, you can see the gradient boost algorithm at the top of the plot.

2.2 *Predictions Versus Y-Test*

See Fig. 11.

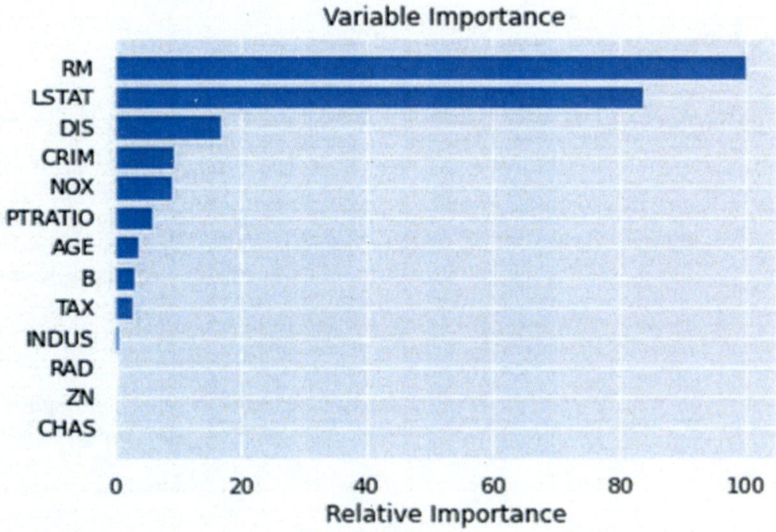

Fig. 12 Variable importance

2.3 *Variable Importance and Renovation Priority*

So from the above figure, it is inferred that the number of rooms (RM) is a internal factor, and it is given first priority for renovation while selling the property followed by other internal factors in the order.

3 Conclusion

We conclude that every problem type has some best available algorithm which reduces the error by exponential factor. So, implementing that sort of algorithm in related applications increases the accuracy in the prediction. These gradient boosting techniques can be applied to wine quality assessment problem which is a weak model by decision tree which can be boosted by gradient boost algorithm. The choice of more investment on the ingredients in wine is also determined in a similar way. Relative predictive analysis is to be implemented on further famous set of problems to find the relative best techniques.

References

1. Harrison Jr D, Rubinfeld DL (1978) Hedonic housing prices and the demand for clean air
2. Lim WT, Wang L, Wang Y, Chang Q (2016) Housing price prediction using neural networks. In: 2016 12th international conference on natural computation fuzzy systems and knowledge discovery (ICNC-FSKD), pp 518–522
3. Shinde N, Gawande K, (2018) Survey on predicting property price. In: 2018 International conference on automation and computational engineering (ICACE), pp 1–7. IEEE
4. Nghiep N, Al C (2001) Predicting housing value: A comparison of multiple regression analysis and artificial neural networks. J Real Estate Res 22(3):313–336
5. Lu S, Li Z, Qin Z, Yang X, Goh RSM (2017) A hybrid regression technique for house prices prediction. In: 2017 IEEE international conference on industrial engineering and engineering management (IEEM), pp 319–323. IEEE
6. Banerjee D, Dutta S (2017) Predicting the housing price direction using machine learning techniques. In: 2017 IEEE international conference on power, control, signals and instrumentation engineering (ICPCSI), pp. 2998–3000. IEEE
7. Varma A, Sarma A, Doshi S, Nair R (2018) House price prediction using machine learning and neural networks. In: 2018 Second international conference on inventive communication and computational technologies (ICICCT), pp 1936–1939. IEEE
8. https://www.cs.toronto.edu/~delve/data/boston/bostonDetail.html
9. Ghosalkar NN, Dhage SN (2018) Real estate value prediction using linear regression. In 2018 fourth international conference on computing communication control and automation (ICCUBEA), pp 1–5. IEEE

Discovery and Categorization of Voice Pathology Using Feature Selection Techniques

Talasu Likhitha, Tanya Chris Elizabeth, and A. Mary Posonia

Abstract Clinical analysis of voice issue and assessment of treatment result vigorously depend on exact measurement of voice quality, which is intently attached to the physiology and capacity of the laryngeal instrument. Considering the assessment strategy of the voice, two principle classes of sound-related perceptual appraisal and acoustic examination can be recognized. This work exhibits another methodology for acoustic examination of voice quality, which carries a few focal points to the field. The proposed approach is nonparametric as it does not require the estimation of the principal recurrence or unearthly reaction of the vocal tract. This lessens the computational multifaceted nature of the estimation and decreases the potential blunders because of wrong estimation of those parameters. Also, the technique does not make any presumption about the phonetic setting and subsequently can possibly be applied to associated discourse. This work focuses on building up an exact and hearty component extraction for recognizing and grouping voice pathologies by researching diverse recurrence groups utilizing auto-correlation and entropy. We extricated greatest pinnacle esteems and their relating slack qualities from each casing of a voiced sign by utilizing auto-correlation as highlights to identify and group neurotic examples. We likewise separated the entropy for each casing of the voice signal after we standardized its qualities to be utilized as the highlights.

Keywords Voice pathology detection · Classification · Frequency investigation · Voice pathology detection · Deep learning · Vocal fold physiology · EGG parameters · Glottal closure instant

T. Likhitha (✉) · T. C. Elizabeth
Department of Computer Science, Sathyabama Institute of Science and Technology, Chennai, India

A. Mary Posonia
Department of Computer Science, Institute of Science and Technology, Chennai, India
e-mail: maryposonia.cse@sathyabama.ac.in

S. S. Dash et al. (eds.), *Sixth International Conference on Intelligent Computing and Applications*, Advances in Intelligent Systems and Computing 1369,
https://doi.org/10.1007/978-981-16-1335-7_42

1 Introduction

The medicinal services industry is not just for bringing in cash yet additionally for giving essential to top of the line well-being administrations to individuals. Remote human services or tele social insurance is required these days. The need emerges on account of a few reasons: (1) Pro-specialists are rare, (11) driving to remote zones is some of the time troublesome, (iii) top hour congested road in the urban zone may forbid heading off to the emergency clinic, (iv) people are reluctant to visit the specialists for the development, and (v) citizens have occupied timetable. Consequently examine on remote, social insurance or portable human services has expanded as of late.

A versatile social insurance system needs a few parts, for example, devices that can gather information from patients, compact preparing units, small and lengthy expand remote correspondence, and edge or potentially cloud servers. Notwithstanding these parts, enlisted specialists and guardians alongside devoted vehicles are important to accelerate the administration.

The enormous improvement remote in correspondence advances and registering preparing power has empowered portable human services to give quick, ease, agreeable, and bother-free administrations. Be that as it may, portable medicinal services are yet to get wide patient acknowledgment because of trust protection and security issues. The modern and remarkable advancements in innovations is identified with precision protection and security increment as patient.

The enormous improvement in remote correspondence advances and registering preparing power has empowered portable human services to give quick, ease, agreeable, and bother-free administrations. The portable human services industries are anticipated to receive income of more than billions of US currency in the following couple of years [1]. Be that as it may, portable medicinal services are yet to get wide patient acknowledgment because of trust, protection, and security issues. The modern and remarkable advancements in innovations is identified with precision, protection, and security increment its patient acknowledgment.

Voice pathology can be serious if there should be an occurrence of disappointment in early location and legitimate administration. Be that as it may, portable medicinal services are yet to get wide patient acknowledgment because of trust, protection, and security issues.

Voice pathology can be serious if there should be an occurrence of disappointment in early location and legitimate administration. This pathology is especially predominant among experts, for example, instructors, artists, and attorneys, who exorbitantly utilize their voice.

The issue with patients encountering voice pathology is that the greater part of them turn away visiting medical clinics after their first treatment; that is, the patients neglect to catch up on medical procedures. Once in a while, this sort of nonappearance exasperates the circumstance. Hence, a shrewd human services structure that requires no physical medical clinic visit for quiet follow-up is required. Right now, propose another portable medicinal services structure that contains a programmed voice pathology recognition framework.

The system comprises keen sensors, distributed computing, and correspondence among patients and partners, for example, clinics, specialists, and medical attendants.

The proposed voice pathology identification framework utilizes convolutional neural system (CNN). The fundamental acoustic parameters have been considered, for example, the Fo, jitter gleam, and HNR. A lucky relapse calculation has been utilized to consequently discover a connection between these parameters, while the Youden investigation has been utilized to discover the edge of an incentive to gauge the nearness or not of a voice issue. The exhibition of DDI has been tried on wide data set made of voices chosen from three distinct databases.

The fundamental acoustic parameters have been considered, for example, the F0, jitter, gleam, and HNR. A lucky relapse calculation has been utilized to consequently discover a connection between these parameters, while the Youden investigation has been utilized to discover the edge of an incentive to gauge the nearness or not of a voice issue. The exhibition of the DDI has been tried on a wide data set made out of voices chose from three distinct databases.

2 Related Work

Another significant class of issues is classification issues, where we wish to appoint an info design x to one of A classes Al, … , AA. Handy instances of characterization issues are manually written digit acknowledgment (where we wish to order a digitized picture of a transcribed digit into one of ten classes (0–9)) and the arrangement of items recognized in galactic sky reviews into stars or cosmic systems (data on the appropriation of the data obtained from the spectrum to determine the hypothesis) [1].

These models present show that grouping issues can either be double (or two class, parallel, multi-class A 2) or multi-class (A > 2). We will concentrate consideration on probabilistic grouping, where test forecasts probabilistically appear as class probabilities [2], this stands out from strategies which give characterization just a theory at the class name, and this qualification closely resembles the distinction between precise circulations and point expectations in the relapse setting. Since speculation to test cases innately includes some degree of vulnerability, it appears to be normal to endeavor to make expectations in a manner that mirrors these vulnerabilities in a handy application one may well look for a class surmise, which can be acquired as the answer for a choice issue, including the prescient probabilities just as a detail of the results of mulling explicit forecasts the misfortune work.

Both order and relapse can be seen as capacity estimate issues. Lamentably, the arrangement of grouping issues utilizing Gaussian procedures is preferably all the more requesting over for the relapse issues. This is on the grounds that we accepted in the past part that the probability work was Gaussian; a Gaussian procedure earlier joined with a Gaussian probability offers ascend to a back Gaussian procedure over capacities, and everything remains diagnostically tractable for arrangement models, where the objectives are discrete class names.

It is conceivable to distinguish a few qualities and shortcomings of the two methodologies. The discriminative methodology is engaging in that it is legitimately displaying what we need, p(ytk) [3, 4]. Likewise, thickness estimation for the class-restrictive circulations is a difficult issue, especially when x is high-dimensional, that we are simply keen on characterization. At that point, generative methodology may imply that we are attempting to take care of a more difficult issue than we have to.

In this case, to manage the missing information like unlabeled information, a principled manner has been used.. it is extremely useful to approach p(x), and this can be obtained from the underestimating the class mark y from the joint as done in the generative methodology. A further factor in the decision of a generative or discriminative methodology could likewise be which one is generally helpful for the fuse of any earlier data which is accessible. See Ripley for additional conversion of the voice spectrum.

The unwavering quality of the DSI has been assessed by assessing the understanding between the watched and anticipated apparent voice qualities communicated by G. Tried on a private data set made out of 387 subjects (319 obsessive and 68 solid), an ideal understanding was acquired in half of the cases. Moreover, an examination between the DSI and Voice Handicap Index (VHI) was assessed, getting a high connection between the two measures. A correlation between CAPE-V and a phantom/cepstral-based acoustic list proposed by Awan et al. is introduced in [3].

These highlights were researched in particular recurrence groups to survey the commitment of each band to the identification and arrangement forms. Different examples of the continued vowel/for/a both ordinary and neurotic voices were separated from three unique databases in English, German, and Arabic. A help vector machine (SVM) [5] was utilized as a classifier. We likewise performed u-tests to research if there is a huge contrast between the methods for the typical and obsessive examples.

This file was built by utilizing a various relapse investigation of the cepstral peak prominence (CPP), the unearthly proportion, and their standard deviations. Tests, performed on a private data set made out of just 32 subjects (24 obsessive and 8 sounds), demonstrated a high connection with CAPE-V, getting a receiver operating characteristic (ROC) region of 0.79, while the affect ability and particularity were, separately, 72 and 80%.

3 Proposed Work

The expectation propagation (EP) calculation (Minka 2001) is a general estimate device with a wide scope of utilizations. Here it is applied to the particular instance of a GP model for paired grouping. This displayed a comparable strategy for double

GPC dependent on the fixed point conditions of the Thouless–Anderson–Palmer kind of mean-field estimate from factual material science. The fixed focuses for the two strategies are the equivalent, in spite of the fact that the exact subtleties of the two calculations are extraordinary; the EP calculation normally fits inadequate approximations.

$$x_2^T F x_1 = 0,$$
$$F = K_2^{-T} S_t R K_1^{-1}$$

The $[X_2]$ in the above formula stands for the sampling frequencies to the power of time $[T]$ multiplied with the modulation frequency $[F]$, and frequency of the given sample $[X_1]$ is always zero. The modulation frequency $[F]$ is given using the modified distances $[K_1]$ and $[K_2]$, sensitivity $[S_t]$, region of CNN $[R]$, and time $[T]$.

We are approximating the probability, for example, a likelihood dissemination which standardizes over the objectives *xi*, by an un-standardized Gaussian conveyance over the inactive factors *di*. This is sensible, on the grounds that we are keen on.

This is how the probability carries on as an element of the idle *fi*. In the relapse setting, we used the Gaussian state of the probability; however more to the point, the Gaussian dissemination for the yields *yi* likewise inferred a Gaussian shape as a component of the inert variable *fi*. So as to process the back, we are obviously principally intrigued by how the probability carries on as a component. The property that the probability ought to standardize over *yi* (for any estimation of *fi*) is not at the same time attainable with the aim of Gaussian reliance on *fi*, in the *EP* guessing the voice pathology tractability.

$$\lambda x = P X^{-1} H X = \hat{P} \hat{X}.$$

The accuracy of the result is given using the above in that $[\lambda]$ is eigenvalue and $[x]$ is the mean frequency which product is found using $[P]$ precision, $[X]$ is the value of sampling frequencies and $[H]$ is the harmonics intensity of the given and user-defined samples.

3.1 Architecture

The architecture of the proposed system is as shown above. The process proceeds as follows. First, the voice sample is taken from the user from the device front end which is developed using HTML and JavaScript, then the voice signals are sent to the back end as waves which are converted to spectrum signal using the API and these are compared using the given algorithms with the stored database samples. The stored database samples include several human voices, bird voices, and also animal

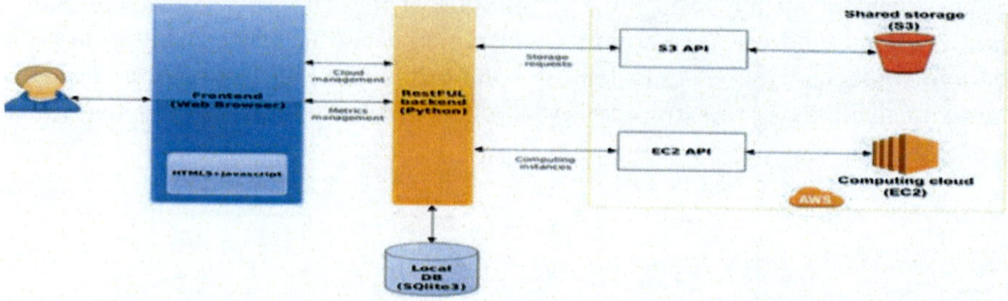

Fig. 1 Architecture of the proposed system

voices such as dogs and many other. Using these sample voices, the system is trained to determine the result of the pathology of the given sample (Fig. 1).

The phenomena of the proposed system mentioned in the following way, first the obtained waves are converted into spectrum then all the useless noises are eliminated from the obtained sample then it is converted to a spectrum from which the Jitter is determined and then further analyzed with the database values.

The phenomenon used in the proposed system is used to obtain high accuracy and better results in aspect of time and efficiency.

3.2 *Advantages*

i. Obtaining a real-time accurate feedback
ii. High accuracy
iii. Easily affordable
iv. Proves sufficient to detect pathology
v. Classification accuracy is very high
vi. Portable
vii. Less time consuming.

3.3 *Results*

The graph of mean frequency of the voice spectrum on y-axis to that of length of the voice jitter recorded is given as shown in Fig. 2. This shows the variation of voice jitter with normal pre-recorded voice and that of the client to determine the cause. The proportionality of the variation is used to determine the diagnosis.

The graph of skew to length of spectrum is derived from the mean frequency graph when converted using spectrographic analysis as described in Fig. 3. This is used to determine the jitter more precisely to increase the accuracy of the jitter analysis and to determine the diagnosis of the client.

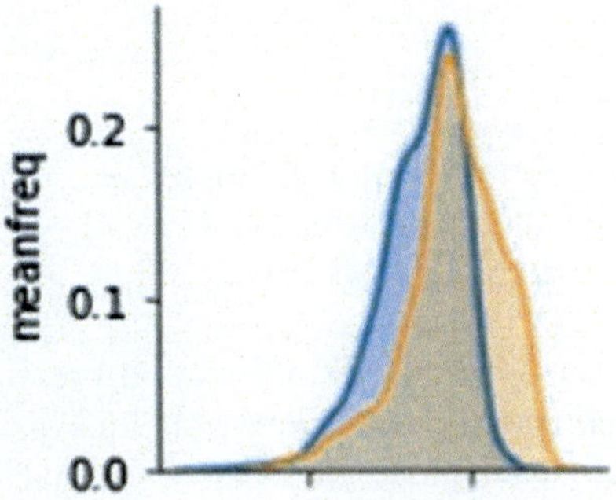

Fig. 2 Mean frequency graph

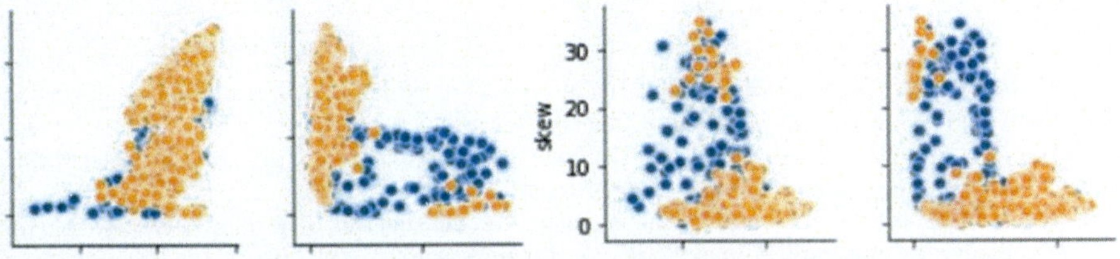

Fig. 3 Skew graph

The accuracy of each result to that of number of training samples is plotted in graph in Fig. 4. The plotted graph given precision of the test data and the pathology of the given client is given up to the precision of 93% using the presented system.

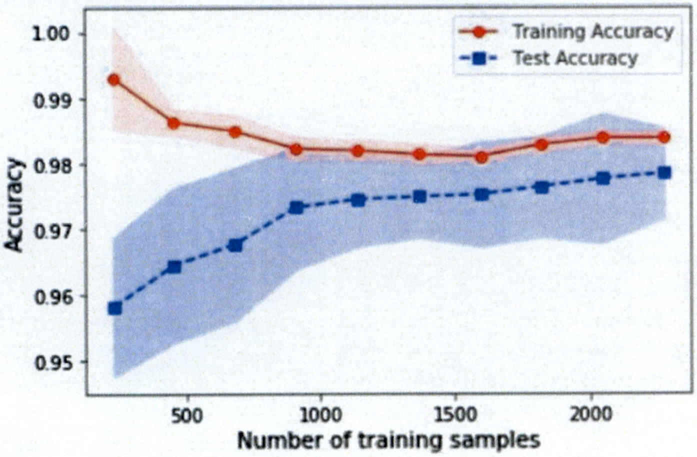

Fig. 4 Accuracy graph

4 Conclusion

The proposed system is to put in use for the topology of glottal source HMM with voice pathology discovery. This frame work exhibited a low-band discourse examination framework for voice pathologies distinguishing proof with persistent discourse signals.

The system utilized for non-straight information pressure with a use with the structure of multilayer perceptron is productive technique that permits a huge learning vector measurement decrease without loss of any sign data.

References

1. Mary Posonia A, Kanmani Rajathi J (2015) Infrequent weighted item set mining using decision making approach algorithm. Int J Appl Eng Res 10(3)
2. Selvan MP, Gupta A, Mukherjee A (2019) Give attention to overlapping network detection in networks for multimedia. J Comput Theor Nanosci 16(8)
3. Foteinos V et al (2013) Cognitive management for the internet of things: a framework for enabling autonomous applications. IEEE Veh Technol Mag 8(4):90–99
4. Viji Amutha Mary A (2019) A random projection approach to strengthen the privacy level of medical images. J Comput Theor Nanosci 16(8)
5. Ali Z et al (2018) New zero-watermarking algorithm using hurst exponent for protection of privacy in telemedicine. IEEE Access 6:7930–7940

The Classification and Segmentation of Plant Disease Using Convolutional Neural Network (CNN)

A. Yovan Felix, S. Shriram, and D. Mohan Kumar

Abstract Indian monetary atmosphere is very dependent on rural proficiency. Thus around the area of farming, a crucial function is played by the detection of disease in plant leaf. In order to identify a plant disease within an extremely first point, the utilization of an automated disease detection strategy is helpful. The feature extraction is grounded on color as well as feel through the laboratory-style room as well as the LBP technique. After that within the subsequent action, segmentation is done by installing a k-means algorithm. The teaching on the SVM classifier is achieved utilizing instruction information. Lastly, with the aid of examining information, the classification of the suggested strategy is going to be validated. Within the suggested method, an algorithm for picture segmentation method can be used for automated detection as well as category of plant leaf disease. The color threshold strategy is utilized after which the convolutional neural network (CNN) is utilized to determine the kinds of disease.

Keywords Segmentation · Classification · Leaf diseases · CNN

1 Introduction

In agricultural land, there are more types of plants and trees, in that, only few could be noticeable and few are not. Plant life is definitely the way of life assistance process for income organisms as oxygen is provided by them. Growing components as well as the products of theirs such as fresh fruits, stem, cereals, flowers, and actually leaves are taken in ways that are different by animals and humans. Although chiefly they are employed for preparing of meals products, medication, and production of mustard oil, which makes of biofuel and so on, you will find a selection of methods within what plant life engages in a significant function. Plants and flowers may also be recognized as well as realized by the features of theirs such as level, form, color, or size of vegetable or maybe the subsequent parts of it. The classification or

A. Yovan Felix (✉) · S. Shriram · D. Mohan Kumar
Department of Computer Science and Engineering, Sathyabama Institute of Science and Technology, Chennai, India

S. S. Dash et al. (eds.), *Sixth International Conference on Intelligent Computing and Applications*, Advances in Intelligent Systems and Computing 1369,
https://doi.org/10.1007/978-981-16-1335-7_43

identification of the vegetable tends to be more frequently accomplished with the aid of foliage visually according to the characteristics such as size, shape, texture, and color of them. Our proposed system will detect the plant disease before it gets to manufacture or it gets affected heavily. This system was tested in the sample plant for seeing the accuracy level. As a result, we've developed a comparing mechanism.

2 Related Works

Machine learning techniques has provided birth to a lot of brand new applications and technologies which are hugely utilized in different areas [1], automatic robot course preparing [2], agriculture plant image detection within a specific area [3], and also average coverage category [4]. It is a procedure for studying a certain undertaking with no man treatment as well as enhancing the overall performance just by a consistent mastering procedure. Machine learning has two types: supervised studying [5], the location where the labeling is provided for the functions on the instruction dataset, and unsupervised studying [6]. SVM classifiers, which can deal with any picture for classification and featuring, are among the most commonly used machine learning methods [7], Naive Bayes [9], Decision Trees [8], as well as Logistic Regression [10]. All techniques are not that much satisfied to this model. When bigger datasets are being utilized within an application program, subsequently the usage of artificial neural networks (ANNs) is chosen for removal as it makes a lot more exact outcomes. Much more usually, the unwanted weeds are experiencing features that are similar because of the harvest and also consequently cannot be differentiated through the naked eye. The nutrient information in dirt is reduced by them, therefore upsetting harvest quality as well as efficiency [11].

3 Proposed Approach

Within the suggested method, an algorithm for picture segmentation method can be used for automated detection as well as category of grow leaf diseases. Additionally, it was proposed with various category methods which may be utilized for grow leaf disorder detection. The style threshold strategy is utilized after which the convolutional neural network is utilized to determine the kinds of disease. Farming efficiency is one thing where economic system extremely is dependent. This is the primary reason that disease detection in vegetation plays a crucial part of farming area as getting health problems in vegetation is very common. Next phase, the characteristics are extracted based upon color as well as feel through the laboratory color room as well as LBP technique. After that within the subsequent action, segmentation is done by installing a k-means algorithm. The teaching on the SVM classifier is achieved utilizing instruction information. Lastly, with the aid of examining information, the classification of suggested strategy is validated.

3.1 Preprocessing

If the input images are colour images, we'll get a grey scope instead of a colour image. After that, it could be changed into binary transformation (zero signifies black color, and one implies white). Image preprocessing refers to the method of enhancing the quality of a picture, these kinds of that the consequential perception is a lot improved when as opposed together with the original form.

3.2 Segmentation

The feature extraction is extracted to determine the leaf area. The characteristic removal incorporates SURF. HARRIS, and FAST. The HSV segmentation is utilized to be able to draw out the disease within the leaf. Segmentation divides an image within the constituent areas of it or perhaps things. The aim of segmentation is to make the representation of a picture simpler as well as to change it into something that is less complex and means a lot more to analyze.

3.3 CNN Algorithm

The action places are extracted plus it is transferred upon the recognition process by which the attributes are extracted, and furthermore, the labeling of the actions is achieved. The outcome of existing covering operates since the responses of approaching covering. The quantity of ranges and, furthermore, the activation might be converted and seen to recognize the top highly effective carrier. If the instructions are finished, a perception is offered toward the method which reads the photograph. When viewing the photo by, the outline on the picture is first observed (Fig. 1).

As a result, the spot segmentation device develops within that the tasks inside the photograph are segmented separately and then the abounding program is bounded near to the joy patterns that are discovered together with the product.

4 Experimental Results

The examinations are performed on MATLAB R2017b adaptation. The calculations are performed utilizing the picture classification learner toolbox that is promptly accessible in MATLAB.

For starters, the type in picture via the scanner repository is provided to the device. Figure 2 shows a test impression from a scanner repository. If this impression is

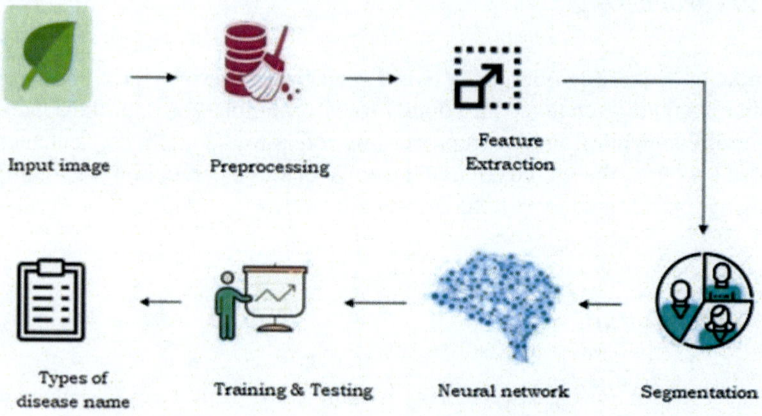

Fig. 1 Architecture diagram

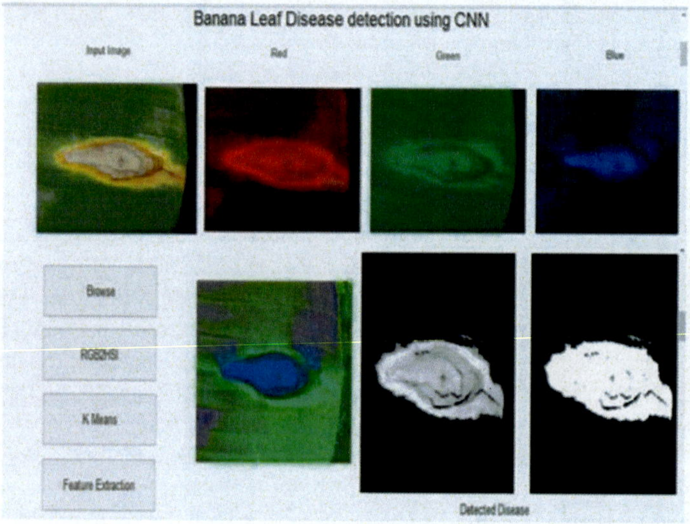

Fig. 2 Disease detection

transferred, the preprocessing comes about in which all of the characteristics are extracted for additional segmentation.

In the following step, the RGB elements are obtained from the picture and then changed to grayscale picture (Fig. 2). Figure 2 displays the picture once the grayscale air filter is used toward the picture. Each image is in contrast to one another to locate the big difference between the characteristics; then, machine learning techniques are utilized toward the pictures to determine the plant leaf disease. The characteristics are then simply taught with CNN that is popular for almost all category as well as regression strategies. Subsequently, the scrutinizing is completed on the quantity of clusters gotten (Fig. 3).

The above output window is the neural network training process which occurs after the successful configuration of the neural box. The input cardiac image acts as input to the neural network. The target is nothing but the segmented portion of the image will be matched with database images. The plots shown in the above window are being explained in the next steps.

In Table 1, we have accomplished the accuracy amount on evaluating and also the identification of disease leaf in the picture is accurate. The precision of the CNN method is found with 99.69% accuracy system.

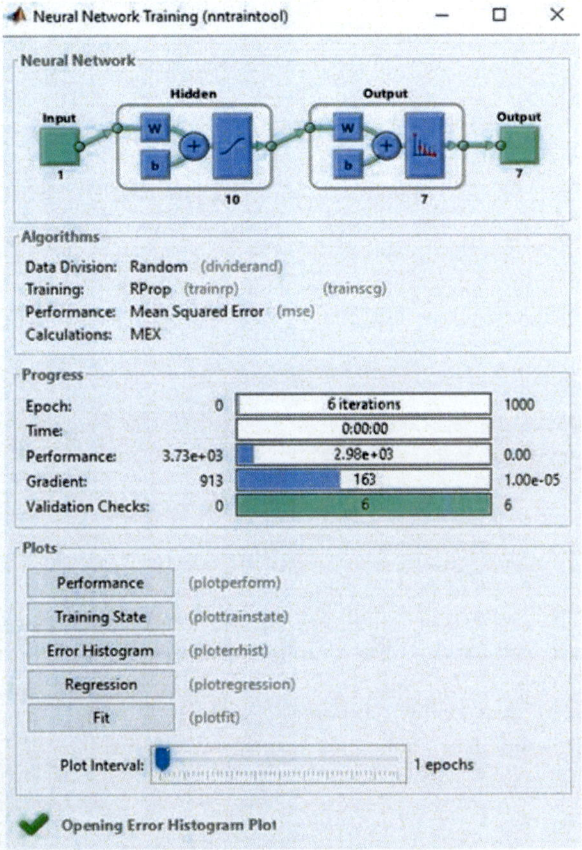

Fig. 3 Output process

Table 1 Accuracy and predicted class

	Predicted class		
Actual class		Class = Yes	Class = No
	Class = Yes	True positive	False negative
	Class = No	False positive	True negative

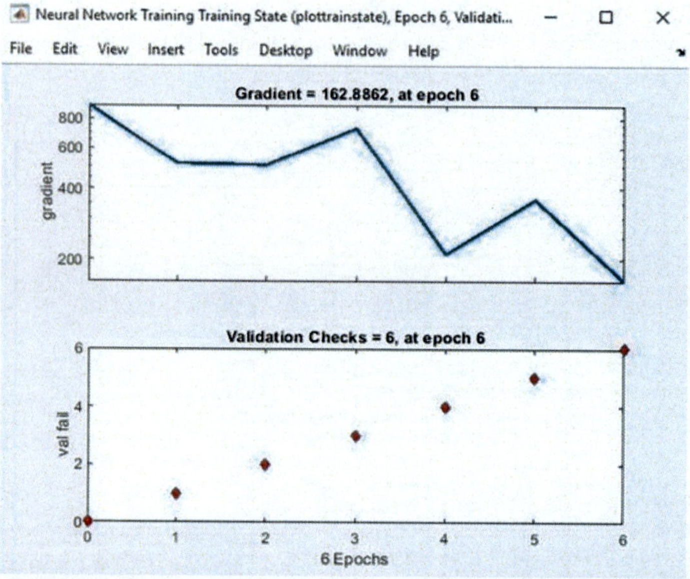

Fig. 4 Training and testing

4.1 Training and Testing

The above plot is nothing but the training states of the given input in the neural network toolbox. The numbers of epochs and the best validated checks are shown clearly in the plot. The epochs are nothing but the number of times the neural network does the training process, and the total number of epochs is 6 the maximum. Depending upon the data size, the number of epochs can be varied (Fig. 4).

4.2 Performance

The above plot is the performance plot in which the training process and testing process are being plotted (Fig. 5).

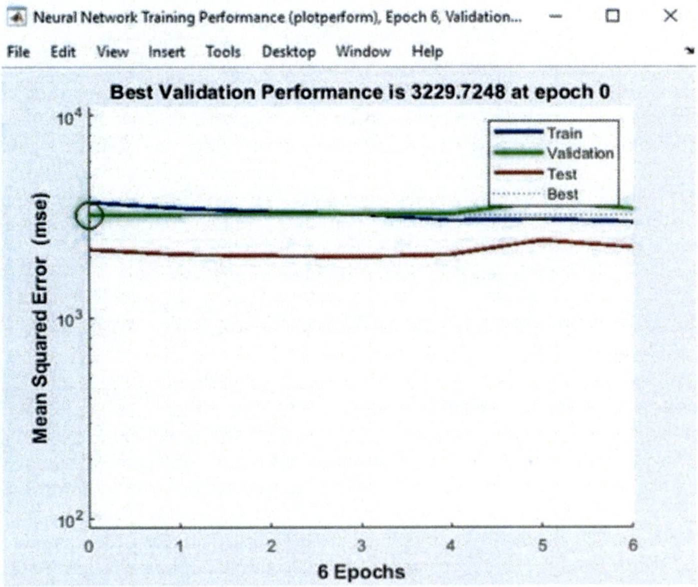

Fig. 5 Performance

5 Conclusion

Several diseases are prevailing within many plants these days. Plant disease is one of the most dangerous illnesses that the majority of agriculture environments around the world are struggling with. Original detection of these disease cells might lessen the danger of loss in plants living to a much better level. With this paper, we have suggested an effective approach to diagnose the cells at a previous phase by utilizing CNN techniques. The pictures from scanners are utilized for fusion wearing in such a manner which yields excessive picture quality. These pictures are after that employed for the additional categories.

References

1. Shruthi U, Nagaveni V, Raghavendra BK (2019) A review on machine learning classification techniques for plant disease detection. In: 2019 5th International conference on advanced computing & communication systems (ICACCS), pp 281–284. IEEE
2. Hungilo GG, Emmanuel G, Emanuel AW (2019) Image processing techniques for detecting and classification of plant disease: a review. In: Proceedings of the 2019 international conference on intelligent medicine and image processing, pp 48–52
3. Poornima S, Kavitha S, Mohanavalli S, Sripriya N (2019) Detection and classification of diseases in plants using image processing and machine learning techniques. In AIP conference proceedings, vol 2095, no. 1, p 030018). AIP Publishing LLC

4. Iniyan S, Jebakumar R, Mangalraj P, Mohit M, Nanda A (2020) Plant disease identification and detection using support vector machines and artificial neural networks. In: Artificial intelligence and evolutionary computations in engineering systems, pp 15–27. Springer, Singapore
5. Wu H, Wiesner-Hanks T, Stewart EL, DeChant C, Kaczmar N, Gore MA, Lipson H (2019) Autonomous detection of plant disease symptoms directly from aerial imagery. Plant Phenom J 2(1)
6. Wiesner-Hanks T, Wu H, Stewart E, DeChant C, Kaczmar N, Lipson H, Gore MA, Nelson RJ (2019) Millimeter-level plant disease detection from aerial photographs via deep learning and crowdsourced data. Front Plant Sci 10
7. Ferentinos KP (2018) Deep learning models for plant disease detection and diagnosis. Comput Electron Agric 145:311–318
8. Shamkuwar DO, Thakre G, More AR, Gajakosh KS, Yewale MO (2018) An expert system for plant disease diagnosis by using neural network
9. Krishnakumar A, Narayanan A (2018) A system for plant disease classification and severity estimation using machine learning techniques. In: International conference on ISMAC in computational vision and bio-engineering, pp 447–457. Springer, Cham
10. Durmuş H, Güneş EO, Kırcı M (2017) Disease detection on the leaves of the tomato plants by using deep learning. In 2017 6th International conference on agro-geoinformatics, pp 1–5. IEEE
11. Aravind KRNVVD, Prayla Shyry S, Felix Y (2019) Classification of Healthy and Rot Leaves of Apple using Gradient Boosting and Support Vector Classifier. International Journal of Innovative Technology and Exploring Engineering 8(12):2868–2872

An Evolutionary Programming-Based Tabu Search Algorithm for Solving Unit Commitment Problem in Smart Grid Environment

C. Christober Asir Rajan

Abstract This paper presents a new approach to solving the unit commitment problem using an evolutionary programming (EP)-based Tabu search (TS) method in smart grid environment. The objective of this paper is to find the generation scheduling such that the total operating cost can be minimized, when subjected to a variety of constraints. This also means that it is desirable to find the optimal generating unit commitment in the power system for the next H hours. This paper proposes distributed sources which include electric vehicles and distributed generation. EP, which happens to be a global optimization technique for solving unit commitment problem, operates on a system, which is designed to encode each unit's operating schedule with regard to its minimum up/down time. The Neyveli Thermal Power Station (NTPS) Unit - II in India demonstrates the effectiveness of the proposed approach; extensive studies have also been performed for different power systems consists of IEEE 10, 26, 34 generating units. Numerical results are shown comparing the cost solutions and computation time obtained by using the EPTS and other conventional methods like dynamic programming, Lagrangian relaxation and simulated annealing and Tabu search in reaching proper unit commitment.

Keywords Unit commitment · Evolutionary programming · Tabu search · Lagrangian relaxation · Dynamic programming

1 Introduction

In power stations, the investment is quite expensive and the resources in operating them are considerably becoming sparse of which focus turns on to optimizing the operating cost of the power station. In today's world, it becomes an utmost necessity to meet the demand as well as optimize the generation. Unit commitment in power systems refers to the optimization problem for determining the on/off states of generating units that minimize the operating cost for a given time horizon.

C. Christober Asir Rajan (✉)
Professor, Department of EEE, Puducherry Technological University, Puducherry, India
e-mail: asir_70@pec.edu

S. S. Dash et al. (eds.), *Sixth International Conference on Intelligent Computing and Applications*, Advances in Intelligent Systems and Computing 1369,
https://doi.org/10.1007/978-981-16-1335-7_44

TS is a powerful optimization procedure that has been successfully applied to a number of combinatorial optimization problems. It has the ability to avoid entrapment in local minima. TS employs a flexible memory system (in contrast to “memory less” systems, such as simulated annealing and genetic algorithm, and rigid memory system such as in branch-and-bound). Specific attention is given to the short-term memory component of TS, which has provided solutions superior to the best obtained with other methods for a variety of problems.

Research endeavors, therefore, have been focused on efficient, near-optimal UC algorithms, which can be applied to large-scale, power systems and have reasonable storage and computation time requirements. A survey of existing literature [1–33] on the problem reveals that various numerical optimization techniques have been employed to approach the complicated unit commitment problem. More specifically, these are the dynamic programming method (DP), the mixed integer programming method (MIP), the Lagrangian relaxation method (LR), the branch-and-bound method (BB), the expert system (ES), the fuzzy theorem method (FT), the hop field method (H), the simulated annealing method (SA), the Tabu search (TS), the genetic algorithm (GA), the artificial neural network (ANN), the cuckoo optimization algorithm (COA) and so on. The major limitations of the numerical techniques are the problem dimensions, large computational time and complexity in programming.

The DP method [1, 2, 13] is flexible, but the disadvantage is the “curse of dimensionality,” which results may lead to more mathematical complexity and increase in computation time if the constraints are taken into consideration. The MIP methods [3, 4] for solving the unit commitment problems fail when the number of units increases because they require a large memory and suffer from great computational delay. The LR approach [5–8] to solve the short-term UC problems was found that it provides faster solution, but it will fail to obtain solution feasibility and solution quality problems and becomes complex if the number of units increased. The BB method [9] employs a linear function to represent fuel cost and start-up cost and obtains a lower and upper bounds. The difficulty of this method is the exponential growth in the execution time for systems of a practical size. An ES algorithm [10, 13] rectifies the complexity in calculations and saving in computation time. But it will face the problem if the new schedule is differing from schedule in database. The FT method [11, 13, 24] using fuzzy set solves the forecasted load schedules error, but it will also suffer from complexity. The H neural network technique [12] considers more constraints, but it may suffer from numerical convergence due to its training process. SA [14–17, 23, 24] is a powerful, general-purpose stochastic optimization technique, which can theoretically converge asymptotically to a global optimum solution with probability one. But it will take much time to reach the near-global minimum. The TS [18–20, 23] is an iterative improvement procedure that starts from some initial feasible solution and attempts to determine a better solution in the manner of a greatest decent algorithm. However, TS is characterized by an ability to escape local optima by using a short-term memory of recent solutions.

GA [13, 21–24] is a general-purpose stochastic and parallel search method based on the mechanics of natural selection and natural genetics. It is a search method to have potential of obtaining near-global minimum. And it has the capability to obtain the accurate results within short time, and the constraints are included easily. The ANN [12] has the advantages of giving good solution quality and rapid convergence. And this method can accommodate more complicated unit-wise constraints and is claimed for numerical convergence and solution quality problems. The solution processing in each method is very unique. The EP [25, 26] has the advantages of good convergent property and a significant speedup over traditional GA's and can obtain high-quality solutions. The "Curse of dimensionality" is surmounted, and the computational burden is almost linear with the problem scale. Electric vehicle (EV) and its impact on the cost and emission of power system are studied on basis of UC model in [28–30]. The significance and feasibility of DR and its role in supply–demand schedule are examined in [31, 32]. Economical operation of distributed generation (DG) and chance-constrained schedule of active network with DG are researched [33].

From the literature review, it has been observed that there exists a need for evolving simple and effective methods, for obtaining an optimal solution for the UCP. Hence, in this paper, an attempt has been made to couple EP with TS for meeting these requirements of the UCP, which eliminates the above-mentioned drawbacks. The application on the NTPS and IEEE systems consists of 10, 26, 34 generating units which shows that we can find the optimal solution effectively, and these results are compared with the conventional methods.

2 Problem Formulation

2.1 Smart Grid Environment

With the progress of smart grid, DR becomes more active. They may play an increasingly essential part in power system operation. In this paper, EV and DG are considered in the UC model.

EV: Smart grid is a perfect platform for the collaborations between the system operators and EVs. With the associated techniques getting settled, it is feasible for EV to sold electricity back to the grid. There is fictional to be an aggregator to link between the system operator and a great number of EV owners [28–30]. If an EV is indolent for a certain period, its owner can sign a contract with the system operator for commitment via the load aggregator. The sum of EV can be preserved as a special unit. Considering there is an increasing peripheral cost to involve more EV proprietors, the cost function of EV is expected to be a quadratic function

$$\mathrm{EVeTC}(\mathrm{EVe}_k) = a_1 + b_1 \mathrm{EVe}_k + c_1 \mathrm{EVe}_k^2 \tag{1}$$

where

EVeTC	EV export total cost
EVe_k	EV export at time k
$\mathrm{a}_1, \mathrm{b}_1\mathrm{EVe}_k, \mathrm{c}_1\mathrm{EVe}_k$	cost coefficients of EV export at time k

Firstly, in case of emergent use of EV's owners, a lower limit of SoC is considered (Eq 2). Secondly, for the sake of safe operation of the gird, an upper limit on total output of EVs at each hour should be stipulated (Eq 4). Thirdly, now that EV may not be connected to the grid all the 24 h, it is sensible to set a time range limit when EV is available for the system operator (Eq 5). Fourthly, the available capacity of EVe at each hour has an upper limit, respectively.

$$\mathrm{SoC}_{k,t} \geq \mathrm{SoC}_{\min} \tag{2}$$

$$\mathrm{SoC}_{k,t} = \frac{C_{o,k} + \sum_{x=1}^{k} \mathrm{EVi}_{x,t} - \sum_{x=1}^{k} \mathrm{EVe}_{x,t}}{C_k} \tag{3}$$

$$\mathrm{EVe}_k = \sum_{t=1}^{m} \mathrm{EVe}_{k,t} \leq \mathrm{EVe}_{\max} \tag{4}$$

$$\mathrm{EVe}_{k,t} = o, k \notin [k_1, k_2] \tag{5}$$

$$\mathrm{EVe}_k \leq \mathrm{EVe}_{k,\max} \tag{6}$$

where

SoC	state of charge of EV
$\mathrm{SoC}_{k,t}$	state of charge of EV t at time k
$\mathrm{SoC}_{\min}$	minimum limit of SoC at each hour
$\mathrm{C}_{o,k}$	initial charging state of EV at time k
$\mathrm{EVe}_{k,t}$	EV export t at time k
$\mathrm{EVe}_{\max}$	maximum limit of EV export capacity
$\mathrm{EVe}_{k,\max}$	maximum available EV export capacity at time k

DG: When more DGs are connected to the grid, then both importing and exporting power from and to the DG's should be taken into consideration in the UC model. Here the power which can be sold to grid is DGe, and the power purchased from the grid is DG_i. The cost function of the DG is expected to be a quadratic equation

$$\mathrm{DG}(\mathrm{DGe}_k) = a_1 + b_1\mathrm{DGe}_k + c_1\mathrm{DGe}_k^2 \tag{7}$$

where

DGe_k	DG export at time k
$a_1, b_1\mathrm{DGe}_k, c_1\mathrm{DGe}_k$	cost coefficients of DG

Firstly, DG's output is subject to natural resource and weather condition, so an upper limit on available DG at each hour is considered. Secondly, DG tends to be intermittent and volatile, an upper limit on its penetration rate should be set to ensure a reliable operation of the power system

$$\mathrm{DGe}_k \leq \mathrm{DGe}_{k,\max} \tag{8}$$

$$\eta_k = \frac{\mathrm{DGe}_k}{\sum_{i=1}^{N}\left(P_{i,k} I_{i,k}\right) + \mathrm{EVe}_k + \mathrm{DGe}_k} \leq \eta\max \tag{9}$$

where

$\mathrm{DGe}_{k,\max}$	maximum available DG export capacity at time k
η_k	penetration rate of DG at time t
$\eta_{\max}$	Max penetration rate of DG at time t
$P_{i,k}$	output of unit i at time k
$I_{i,k}$	on/off status of unit I at time k

2.2 UC Model

The objective is to find the generation scheduling such that the total operating cost can be minimized, when subjected to a variety of constraints [27]. In the UCP under consideration, an interesting solution would be minimizing the total operating cost of the generating units with several constraints being satisfied. The major component of the operating cost, for thermal and nuclear units, is the power production cost of the committed units, and this is given in a quadratic form in (Eq 10).

$$F_{it}(P_{it}) = A_i P_{it}^2 + B_i P_{it} + C_i \ \mathrm{Rs/h} \tag{10}$$

where

A_i, B_i, C_i	the cost function parameters of unit i (Rs./MW^{2}h, Rs./MWh, Rs/h)
$F_{it}(P_{it})$	production cost of unit i at a time t (Rs/h)
P_{it}	output power from unit i at time t (MW)

The start-up cost depends upon the down time of the unit, which can vary from a maximum value, when the unit i is started from cold state, to a much smaller value, if the unit i has been turned off recently. The start-up cost calculation depends upon the treatment method for the thermal unit during down time periods. The start-up cost S_{it} is a function of the down time of unit i as given in (Eq 11).

$$S_{it} = So_i[1 - D_i \exp(-\mathrm{Toff}_i / T\mathrm{down}_i)] + E_i \quad \mathrm{Rs} \tag{11}$$

where

So_i unit i cold start-up cost (Rs)
D_i, E_i start-up cost coefficients for unit i

The overall objective function of the UCP is given in (Eq 12).

$$F_T = \sum_{t=1}^{T}\sum_{i=1}^{N}[(F_{it}(P_{it})U_{it} + S_{it}V_{it})] + \sum_{k=1}^{24}[\text{EVe}(\text{EVe}_k) + \text{DG}(\text{DGe}_k)] \quad \text{Rs/h} \tag{12}$$

where

U_{it} unit i status at hour $t = 1$(if unit is ON) $= 0$ (if unit is OFF)
V_{it} unit i start-up/shut down status at hour $t = 1$ if the unit is started at hour t and 0 otherwise.
F_T total operating cost over the schedule horizon (Rs/h)
S_{it} start-up cost of unit i at hour t (Rs).

2.3 *Constraints*

Depending on the nature of the power system under study, the UCP is subject to many constraints, the main being the load balance constraints and the spinning reserve constraints. The other constraints include the thermal constraints, fuel constraints, security constraints, etc. [27]

Load Balance Constraints

The real power generated must be sufficient enough to meet the load demand and must satisfy the following factors given in (Eq 13).

$$\sum_{i=1}^{N} P_{it}U_{it} = \text{PD}_t - \text{EVe}_k - \text{DGi}_k - \text{DGe}_k + \text{PL}_k \tag{13}$$

where

PD_t system peak demand at hour t (MW)
N number of available generating units
U (0, 1) the uniform distribution with parameters 0 and 1
UD (a, b) the discrete uniform distribution with parameters a and b

Spinning Reserve Constraints

The spinning reserve is the total amount of real power generation available from all synchronized units minus the present load plus the losses. The reserve is considered to be a pre-specified amount or a given percentage of the forecasted peak demand. It must be sufficient enough to meet the loss of the most heavily loaded unit in the system. This has to satisfy the equation given in (Eq 14).

$$\sum_{i=1}^{N} P\max_i U_{it} >= (\mathrm{PD}_t + R_t);\ \ 1 \le t \le T \tag{14}$$

where

$P\max_i$	Maximum generation limit of unit i
R_t	spinning reserve at time t (MW)
T	scheduled time horizon (24 hrs.)

Thermal Constraints

The temperature and pressure of the thermal units vary very gradually, and the units must be synchronized before they are brought online. A time period of even 1 h is considered as the minimum down time of the units. There are certain factors, which govern the thermal constraints, like minimum up time, minimum down time and crew constraints.

Minimum up time:

If the units have already been shut down, there will be a minimum time before they can be restarted and the constraint is given in (Eq 15).

$$\mathrm{Ton}_i \ge \mathrm{Tup}_i \tag{15}$$

where

Ton_i	duration for which unit *i* is continuously ON (h)
Tup_i	unit i minimum up time (h)

Minimum down time:

If all the units are running already, they cannot be shut down simultaneously and the constraint is given in (Eq 16).

$$T\mathrm{off}_i \ge T\mathrm{down}_i \tag{16}$$

where

$T\,\mathrm{down}_i$	unit *i* minimum down time (h)
$T\,\mathrm{off}_i$	duration for which unit *i* is continuously OFF (h)

Must Run Units

Generally in a power system, some of the units are given a must run status in order to provide voltage support for the network.

3 Evolutionary Programming

3.1 Introduction

EP is a mutation-based evolutionary algorithm applied to discrete search spaces. David Fogel (Fogel 1988) extended the initial work of his father Larry Fogel (Fogel 1962) for applications involving real-parameter optimization problems. Real-parameter EP is similar in principle to evolution strategy (ES), in that normally distributed mutations are performed in both algorithms. Both algorithms encode mutation strength (or variance of the normal distribution) for each decision variable, and a self-adapting rule is used to update the mutation strengths. Several variants of EP have been suggested (Fogel 1992).

3.2 Evolutionary Strategies

For the case of evolutionary strategies, D. B. Fogel remarks "evolution can be categorized by several levels of hierarchy: the gene, the chromosome, the individual, the species, and the ecosystem." Thus, while genetic algorithms stress models of genetic operators, evolutionary strategies emphasize mutational transformation that maintains behavioral linkage between each parent and its offspring at the level of the individual. Evolutionary strategies are a joint development of Bienert, Rechenberg, and Schwefel. The first applications were experimental and addressed some optimization problems in hydrodynamics.

3.3 EP General Algorithm

Evolutionary programming [25, 26] is conducted as a sequence of operations and is given below. The schematic diagram of the EP algorithm is shown in Fig. 1.

Algorithm 1 EP General Algorithm

1. The initial population is determined by setting $s_i = S_i \sim U(a_k, b_k)^k$ $i = 1, \ldots, m$, where S_i is a random vector, s_i is the outcome of the random vector, $U(a_k, b_k)^k$ denotes a uniform distribution ranging over $[a_k, b_k]$ in each of k dimensions, and m is the number of parents.
2. Each s_i, $i = 1, \ldots, m$, is assigned a fitness score $\vartheta(s_i) = G(F(s_i), v_i)$, where F maps $s_i \rightarrow R$ and denotes the true fitness of s_i, v_i, represents random alteration in the instantiation of s_i, random variation imposed on the evaluation of $F(s_i)$, or satisfies another relation s_i, and $G(F(s_i), v_i)$ describes the fitness score to be assigned. In general, the functions F and G can be as complex as required.

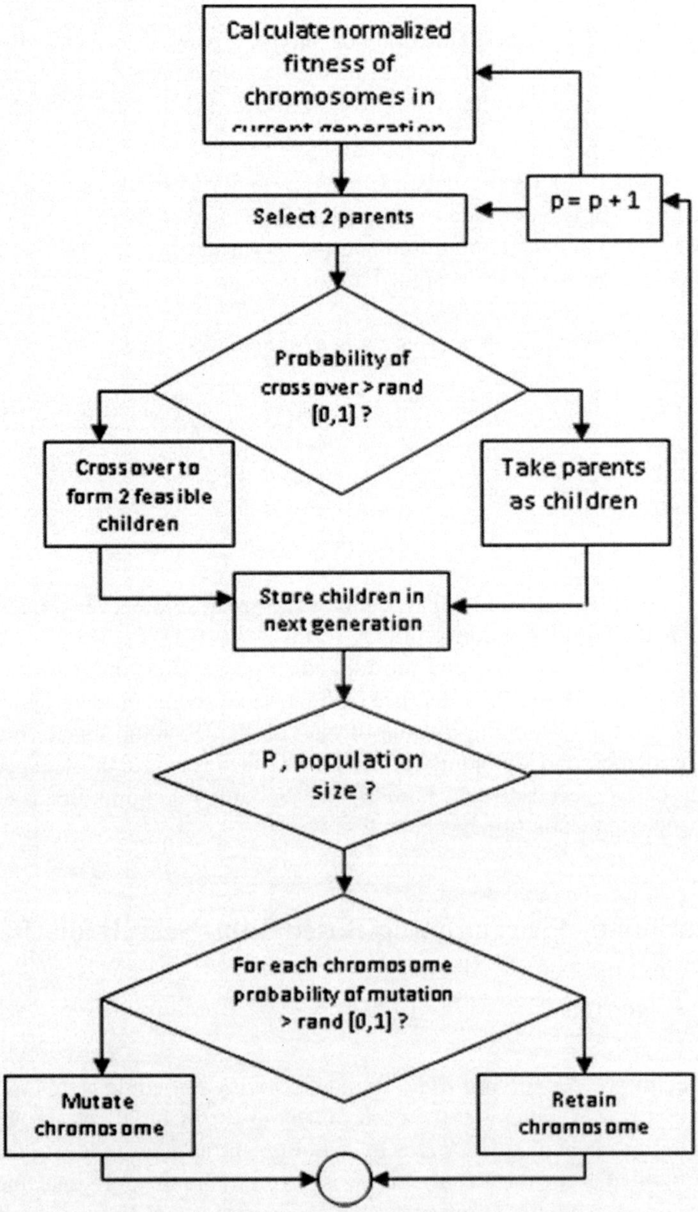

Fig. 1 Flowchart of evolutionary programming algorithm

For example, F may be a function not only of a particular s_i, but also of other members of the population, conditioned on a particular s_i.

3. Each s_i, i = 1,…,m, is altered and assigned to s_{i+m} such that

   $s_{i+m} = s_{i,j} + N(0, \beta j\, \vartheta(s_i) + z_j)$, j=1, … ,k
   $N(0, \beta j\, \vartheta(s_i) + z_j)$ represents a Gaussian random variable with mean μ and variance σ^2, β_j is a constant of proportionality to scale $\vartheta(s_i)$, and zj represents an offset to guarantee a minimum amount of variance,
4. Each s_{i+m}, i = 1,…,m, is assigned a fitness score
   $\vartheta(s_{i+m}) = G(F(s_{i+m}), v_{i+m})$
5. For each s_i, i = 1,…,2 m, a value wi is assigned according to

$$w_i = \sum_{t=1}^{c} w_t^*$$

$$w_t^* = \begin{cases} 1 \text{ if } \vartheta(s_i) \le \vartheta(s_i); \\ 0 \text{ otherwise}; \end{cases}$$

   where $\rho = [2\, mu_1 + 1]$, $\rho \neq$ i, $[x]$ denotes the greatest integer less than or equal to x, c is the number of competitions, and u1 ~ U (0,1).
6. The solutions s_i, I = 1…2m, are ranked in descending order of their corresponding value W_i [with preference to their actual scores $\vartheta(s_i)$ if there are more than m solutions attaining a value of c]. The first m solutions are transcribed along with their corresponding values $\vartheta(s_i)$ to be the basis of the next generation.
7. The process proceeds to step 3, unless the available execution time is exhausted or an acceptable solution has been discovered.

4 Evolutionary Programming-Based Tabu Search for UCP

4.1 EP Based TS

In the TS technique for solving UCP, Initial Operating Schedule status in terms of maximum real power generation of each unit is given as input. As we know that TS is used to improve any given status by avoiding entrapment in local minima, the offspring obtained from the EP algorithm is given as input to TS and the refined status is obtained. And evolutionary strategy selects the final status. The flowchart for EPTS for UCP is shown in Fig. 2.

Algorithm 2 EPTS for UCP

1. Get the demand for 24 h and no. of iterations to be carried out.
2. Generate population of parents (N) by adjusting the existing solution to the given demand to the form of state variables.
3. Unit down time makes a random recommitment.

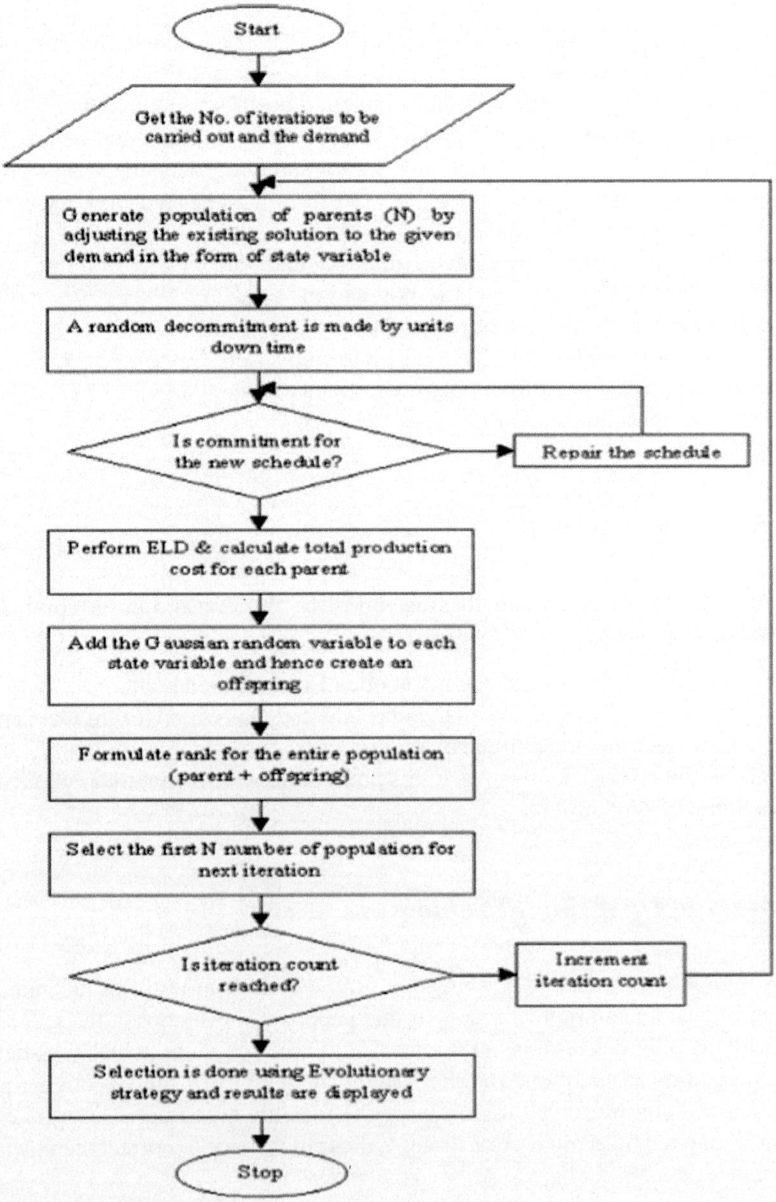

Fig. 2 Flowchart for EPTS for UCP

4. Check for constraint in the new schedule by TS. If the constraints are not met, then repair the schedule as given in Sect. 4.3.
5. Perform ELD and calculate total production cost for each parent.
6. Add the Gaussian random variable to each state variable, and hence, create an offspring. This will further undergo for some repair operations as given in Sect. 4.4. Following these, the new schedules are checked in order to verify that all constraints are met.
7. Improve the status of the evolved offspring and verify the constraints by TS.
8. Formulate the rank for the entire population.
9. Select the best N number of population for next iteration.
10. Has iteration count reached? If yes go to step 11 else go to step 2.
11. Select the best population (s) by evolutionary strategy [25, 26].
12. Print the optimum schedule.

4.2 Repair Mechanism

A repair mechanism to restore the feasibility of the constraints is applied and described as follows:

- Pick at random one of the OFF units at one of the violated hours.
- Apply the rules in Sect. 4.3 to switch the selected unit from OFF to ON keeping the feasibility of the down time constraints.
- Check for the reserve constraints at this hour. Otherwise repeat the process at the same hour for another unit.

4.3 Making Offspring Feasible

While solving the constrained optimization problem, there are various techniques to repair an infeasible solution [21, 26]. In this paper, we have chosen the technique, which evolves only the feasible solutions. That is the schedule, which satisfies the set of constraints as mentioned earlier. Here, in this paper, the selection routine is involved as "culling force" to eliminate the feasible schedules. Before the best solution is selected by evolutionary strategy, the trail is made to correct the unwanted mutations.

4.4 Implementation

Software programs were developed using MATLAB software package, and the test problem was simulated for ten independent trials using evolutionary programming algorithm. The training and identification part as implemented in the Tabu search technique is employed here and considered as a process involving random recommitment, constraint verification and offspring creation.

5 Numerical Results

A NTPS in India with seven generating units, each with a capacity of 210 MW, has been considered as a case study. A time period of 24 hours is considered; the unit commitment problem is solved for these seven units and also compared with IEEE 10, 26, and 34 generating unit power systems. The required inputs for solving the UCP are briefed here. The total number of generating units, the maximum real power generation of each unit and the generation system operation data of each unit is tabulated for a day, respectively, as shown in Tables 1 and 2 for NTPS. The cost coefficients of DR are shown in Table 3. EV and DG are set at buses 5 and 7, respectively, since the load demands at these buses are quite high compared with other buses.

Some other parameters in our model are assumed as follows. The lower limit of SoC is 25%. The average of SoC before dispatching is 75%. The average battery capacity of EVs is 34 kWh. The quantity of EVs that can conduct EVe is 8500. In order to facilitate management, this paper assumes that the system operator will only sign contract of EVe if the EVs are available in the evening. Accordingly, the available periods of all the EVs are the same in this paper, assumed 20–24 o'clock. The available capacity of EVe at these hours is 24, 23, 22, 21, and 20 MW, respectively.

The maximum bearing capacity of the grid for EVe is assumed to be 24 MW at each hour. The upper limit of demand curtailment response is 7% of total load demand in this hour, and the upper limit within a day is 320 MWh. Now that solar power may be the main DG source, the available DG is set higher during the day. Specifically, it is 40 MW between 7 o'clock and 17 o'clock and 20 MW during the other time, among which two-thirds are used by consumers themselves and one-third would be sold to the grid. The upper limit of DG penetration rate is 6%. This paper considers four scenarios as shown in Table 4.

The status of unit i at time t and the start-up/shut-down status obtained are the necessary solution for SA, TS, EPTS, DP, LR methods for NTPS. The comparison of the total costs and central processing unit (CPU) time is shown in Table 5 for NTPS, and IEEE 10, 26, and 34 generating unit power systems. Figure 3 represents the total production cost obtained by each parent for 25 iterations in EPA. Similarly, for 50 and 100 iterations are obtained. Figure 4 gives the plot of EPTS average performance from 1000 runs. Figure 5 gives the plot of no. of iteration versus the time taken to complete those iterations and the maximum production cost obtained under each iteration.

The proposed EPTS approach was compared to the related methods in the references indented to serve this purpose, such as the DP with a zoom feature, the SA, and the GA approaches. And with the use of Tabu search method, the status is improved by avoiding the entrapment in local minima. By means of stochastically searching

Table 1 Daily generation of seven units in MW

Hour	P_{max}	I	II	III	IV	V	VI	VII
1	840	60	80	100	101	149	150	200
2	757	60	0	100	100	147	150	200
3	775	60	0	100	115	150	150	200
4	773	60	0	100	113	150	150	200
5	770	60	0	100	110	150	150	200
6	778	60	0	100	118	150	150	200
7	757	60	0	100	100	147	150	200
8	778	60	0	100	118	150	150	200
9	770	60	0	100	110	150	150	200
10	764	60	0	100	104	150	150	200
11	598	60	0	99	97	142	0	200
12	595	60	0	100	96	139	0	200
13	545	0	0	100	99	146	0	200
14	538	0	0	99	97	142	0	200
15	535	0	0	100	96	139	0	200
16	466	0	0	0	116	150	0	200
17	449	0	0	0	101	148	0	200
18	439	0	0	0	97	142	0	200
19	466	0	0	0	116	150	0	200
20	463	0	0	0	113	150	0	200
21	460	0	0	0	110	150	0	200
22	434	0	0	0	95	139	0	200
23	530	60	0	0	120	150	0	200
24	840	60	80	100	101	149	150	200

multiple points at one time and considering trail solutions of successive generations, the EPTS approach avoids entrapping in local optimum solutions. Also, disadvantages of huge memory size required by the SA method are eliminated. Moreover, intellectual schemes of encoding and decoding entailed by the GA approach are not needed in the proposed EPTS approach. The problem of power unbalance previously existing in the solution of GA is circumvented as well in this paper. In comparison with the results produced by the referenced techniques, the EPTS method obviously displays a satisfactory performance with respect to the quality of its evolved solutions and to its computational requirements.

Table 2 Generation system operation data

Unit	$P_{\min}$ (MW)	$P_{\max}$ (MW)	Running cost			Start-up cost		
			C_i (Rs)	B_I (Rs/MWh)	A_i (Rs/MWh^2)	SO_i (Rs)	D_i (Rs)	E_i (Rs)
1	15	60	750	70	0.255	4250	29.5	10
2	20	80	1250	75	0.198	5050	29.5	10
3	30	100	2000	70	0.198	5700	28.5	10
4	25	120	1600	70	0.191	4700	32.5	9
5	50	150	1450	75	0.106	5650	32	9
6	50	150	4950	65	0.0675	14,100	37.5	4.5
7	75	200	4100	60	0.074	11,350	32	5.5

Table 3 Cost coefficients of DR

Running cost			
DR	C_i	B_i	A_i
	(Rs)	(Rs/MWh)	(Rs/MWh^2)
EVe	750	70	0.255
DGb	1250	75	0.198

Table 4 Operation settings

Operation settings	
Operation 1	EV not connected, DG not connected
Operation 2	EV connected, DG Not connected
Operation 3	EV not connected, DG connected
Operation 4	EV connected, DG connected

Table 5 Comparisons of cost and CPU time for NTPS, 10, 26, 34 unit systems

System	Methods	Total cost (p.u.)	CPU time (Sec)
7—Unit utility system	DP	1.00000	605
	LR	0.96481	578
	SA	0.95000	570
	TS	0.95239	575
	EPTS	0.94045	525
	DP [1]	1.00000	325
	LR [6]	0.94123	279
10—Unit IEEE system	SA [17]	0.93210	285
	TS [18]	0.93435	290
	EPTS	0.92261	233
	DP [1]	1.00000	509
	LR [6]	0.95968	495
26—Unit IEEE system	SA [17]	0.94570	489
	TS [18]	0.94750	494
	EPTS	0.93605	467
	DP [1]	1.00000	1452
	LR [6]	0.99910	1368
34—Unit IEEE system	SA [17]	0.98015	1370
	TS [18]	0.98291	1376
	EPTS	0.97135	1341

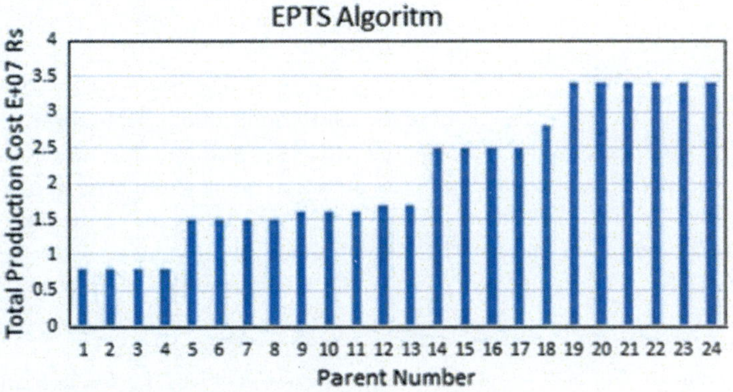

Fig. 3 Total production cost for 25 iterations

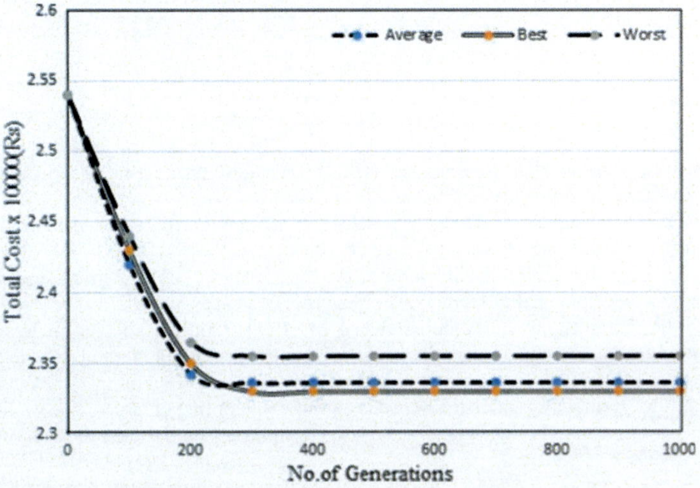

Fig. 4 EPTS average performance from 1000 runs

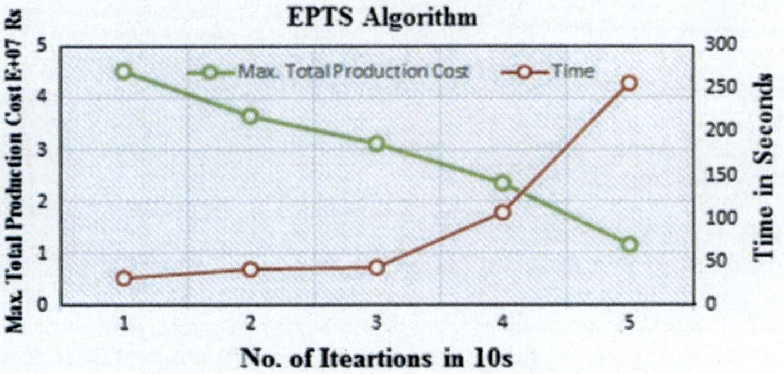

Fig. 5 No. of iterations vs time taken and max. production cost

6 Conclusion

This paper presents an EPTS to solve the unit commitment problem in smart grid environment. In this method, the essential processes simulated in the procedure are mutation, competition, and selection. The mutation rate is computed as a function of the ratio of the total cost by the schedule of interest to the cost of the best schedule in the current population. In comparison with the results produced by the referenced techniques (EP, DP, LR and SA & TS), the EPTS method obviously displays a satisfactory performance.

Acknowledgements The author is to acknowledge the support of AICTE, New Delhi, for the RPS project titled as "Development of Hybrid Algorithms to Solve the Unit Commitment Problem in Smart Grid Network using Soft Computing Techniques" sanctioned during 2017–2020.

References

1. Snyder WL, Powell HD, Rayburn C (1987) Dynamic programming approach to unit commitment. IEEE Trans Power Syst 3(2):339–350
2. Hsu SuYY, Liang CC, Lin CJ, Huang CT (1991) Dynamic security constrained multi-area unit commitment. IEEE Trans Power Syst 6:1049–1055
3. Takriti S, Birge JR (2000) Using integer programming to refine lagrangian—based unit commitment solutions. IEEE Trans Power Syst 15(1):151–156
4. Mohan MR, Kuppusamy K, Abdullah Khan M (1992) Optimal short-term hydro-thermal scheduling using decomposition approach and linear programming method. Int J Electr Power Energy Syst 14(1):39–44
5. Ngundam JM, Kenfack F, Tatietse TT (2002) Optimal scheduling of large—scale hydrothermal power systems using the lagrangian relaxation technique. Int J Electr Power Energy Syst 22(1):237–245
6. Bakirtzis AG, Zoumas CE (2000) Lambda of lagrangian relaxation solution to unit commitment problem. IEE Proc Gener Transm Distribu 147(2):131–136
7. Chang C-P, Liu C-W, Liu C-C (2000) Unit commitment by legrangian relaxation and genetic algorithms. IEEE Trans Power Syst 15(2):707–714
8. Nieva R, Inda R, Guillen I (1987) Lagrangian reduction of search range for large scale unit commitment. IEEE Trans Power Syst 2:465–473
9. Cohen AI, Yoshimura M (1983) A branch and bound algorithm for unit commitment. IEEE Trans Power Apparatus 102(2):444–451
10. Ouyang Z, Shahiderpour SM (1990) Short term unit commitment expert system. Int J Electr Power Syste Res 20:1–13
11. Su CC, Hsu YY (1991) Fuzzy dynamic programming: an application to unit commitment. IEEE Trans Power Syst 6(3):1231–1237
12. Sasaki H, Watanabe M, Kubokawa J, Yorino N (1992) A solution method of unit commitment by artificial neural networks. IEEE Trans Power Syst 7(3):974–981
13. Padhy NP (2000) Unit commitment using hybrid models: a comparative study for dynamic programming, expert systems, fuzzy system and genetic algorithms. Int J Electr Power Energy Syst 23(1):827–836
14. Zhuang F, Galiana FD (1990) Unit commitment by simulated annealing. IEEE Trans Power Syst 5(1):311–318
15. Kirkpatrick S, Gelatt CD, Vecehi MP (1983) Optimisation by SA. Science 220:4598

16. Selim SZ, Alsultan K (1991) A simulated annealing algorithm for the clustering problem. Pattern Recogn 24(10):1003–1008
17. Mantawy AH, Abdel-Magid YL, Selim SZ (1998) A simulated annealing algorithm for unit commitment. IEEE Trans Power Syst 13(1):197–204
18. Mantawy AH, Youssef L, Abdel-Magid, Selim SZ (1998) A unit commitment by tabu search. IEE Proc Gener Transm Distrib 145(1):56–64
19. Lin Whei-Min, Cheng Fu-Sheng, Tsay Ming-Tong (2002) An improved tabu search for economic dispatch with multiple minima. IEEE Trans Power Syst 17(1):108–112
20. Bai X, Shahidehpour M (1996) Hydro-thermal scheduling by tabu search and decomposition method. IEEE Trans Power Syst 11(2):968–975
21. Yong-Gang Wu, Ho Chun-Ying, Wang Ding-Yi (2000) A diploid genetic approach to short-term scheduling of hydro-thermal system. IEEE Trans Power Syst 15(4):1268–1274
22. Hong Y-Y, Li C-Y (2002) Genetic algorithm based economic dispatch for cogeneration units considering multiplant multibuyer wheeling. IEEE Trans Power Syst 17(1):134–140
23. Mantawy AH, Youssef L, Abdel-Magid, Selim SZ (1999) Integrating genetic algorithm, tabu search and simulated annealing for the unit commitment problem. IEEE Tans Power Syst 14(3):829–836
24. Wong KP, Suzannah, Wong YW (1996) Combined genetic algorithm/simulated annealing/fuzzy set approach to short—term generation scheduling with take-or-pay fuel contract. IEEE Trans Power Syst 11(1):128–135
25. Juste KA, Kita H, Tanaka E, Hasegawa J (1999) An evolutionary programming solution to the unit commitment problem. IEEE Trans Power Syst 14(4):1452–1459
26. Yang HT, Yang PC, Huang CL (1996) Evolutionary programming based economic dispatch for units with Non-smooth fuel cost functions. IEEE Trans Power Syst 11(1):112–117
27. Wood AJ, Woolenberg BF (1996) Power Generation and control. 2nd edn, Wiley, New York (1996)
28. Lu L, Wen F, Xue Y, Xin J (2011) Unit commitment in power systems with plug-in electric vehicles. Autom Elect Power Syst 35:16–20
29. Khodayar ME, Wu L, Shahidehpour M (2012) Hourly coordination of electric vehicle operation and volatile wind power generation in SCUC. IEEE Trans Smart Grid 3(3):1271–1279
30. Ikeda Y, Ikegami T, Kataoka K, Ogimoto K (2012) A unit commitment model with demand response for the integration of renewable energies. In: IEEE power energy socciety general meeting, San Diego, CA, USA, pp 1–7
31. Du P, Lu N (2011) Appliance commitment for household load scheduling. IEEE Trans Smart Grid 2(2):411–419
32. Moghimi H, Ahmadi A, Aghaei J, Rabiee A (2013) Stochastic techno economic operation of power systems in the presence of distributed energy resources. Elect Power Energy Syst 45:477–488
33. Goleijani S, Ghanbarzadeh T, Nikoo FS, Moghaddam MP (2013) Reliability constrained unit commitment in smart grid environment. Elect Power Syst Res 97:100–108

Coaxial-Fed Rectangular Microstrip Patch Antenna for WLAN

Kolli Venkatrao and G. Merlin Sheeba

Abstract In recent communication systems, microstrip patch antennas play a vital role because of their compact size and cost and easily fabricated. A new proposed antenna has been intended for wireless LAN applications. This proposed antenna is planned for operation within the bandwidth of 5.2 GHz from 3 to 7 GHz. The antenna with a dimension of 17.56*12.56*1.6 mm^2 and dielectric constant of 4.4 is designed with FR4 substrate. The simulated results show a perfect impedance match with the bandwidth. Thus, it covers wireless local area network (WLAN) of 5.2/5.8 GHz band. The various parameters, i.e. return loss, voltage standing wave ratio and current distribution for E and H fields, are analysed to characterize the proposed antenna.

Keywords Rectangular microstrip patch antenna · Coaxial probe feeding · HFSS · FR4 substrate · VSWR · Return loss

1 Introduction

Microstrip patch antennas are widely under research for the present wireless technology, due to their low profile, cost effective, easy to fabricate and conformability [1]. The upper conducting layer of the antenna acts as a radiating patch, and the lower one acts as a perfect reflecting ground plane. Today, communication system has exclusively focussed its research on miniaturization of antennas for transmitting and receiving the signals. Microstrip antennas have numerous advantages like low profile planar configuration, can be easily integrated with circuit elements, low fabrication cost and capable of multiple frequency of operation [2]. The wireless mesh WLAN networks have enormous applications on E-health monitoring and services in rural regions, and thereby, multiple frequency antenna plays a major role in these real-time scenarios [7, 11]. The coaxial feeding in microstrip antennas can be performed typically in conducting and non-conducting ways, as shown in Fig. 1. In the conducting method, the patch is coupled directly with RF power using a microstrip line, whereas

K. Venkatrao · G. Merlin Sheeba (✉)
School of Electrical and Electronics Engineering, Sathyabama Institute of Science and Technology, Chennai, India

S. S. Dash et al. (eds.), *Sixth International Conference on Intelligent Computing and Applications*, Advances in Intelligent Systems and Computing 1369,
https://doi.org/10.1007/978-981-16-1335-7_45

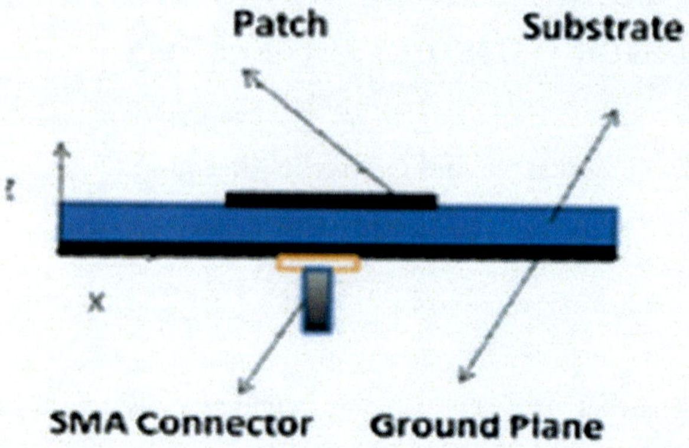

Fig. 1 Coaxial feed antenna inside view

in non-conducting scheme, an electromagnetic field coupling is performed to the patch [3].

There are many different methods of feeding like microstrip line, coaxial probe f, aperture coupling and proximity coupling feed [4, 5]. In coaxial feeding method, the exterior of the probe is connected to the ground pane, and inner coaxial probe is connected to the radiation patch. To match input impedance, the probe is fixed at a desired location on the patch. This technique is simple, and adjustment of input impedance makes this method popular [6, 7]. As the substrate thickness increases, the coupling between the nearby feeds may decrease the performance.

2 Antenna Geometry

The rectangular patch of prospective antenna is etched on one face of FR4 with dimensions 17.56 mm × 12.56 mm × 1.6 mm and permittivity of 4.4. The ground plane has dimensions of 35.9 mm × 58.94 mm. The patch is fed with 50Ω load impedance at position 4.185 mm × 2.823 mm and is shown in Table 1. The frequency range used is 3–7 GHz [8], and proposed antenna has a resonant frequency of 5.2 GHz [9]. The design of proposed antenna is suitable for WLAN applications and is shown in Figs. 2, 3 and 4.

Table 1 Simulation antenna parameters

Freq	ε_r	H	L	W	L_g
5.2 GHz	4.4	1.6 mm	12.56 mm	17.56 mm	35.9 mm
W_g	X	Y	L_r	W_r	h_r
58.94 mm	4.52 mm	2.15 mm	35.9 mm	58.94 mm	6.18 mm

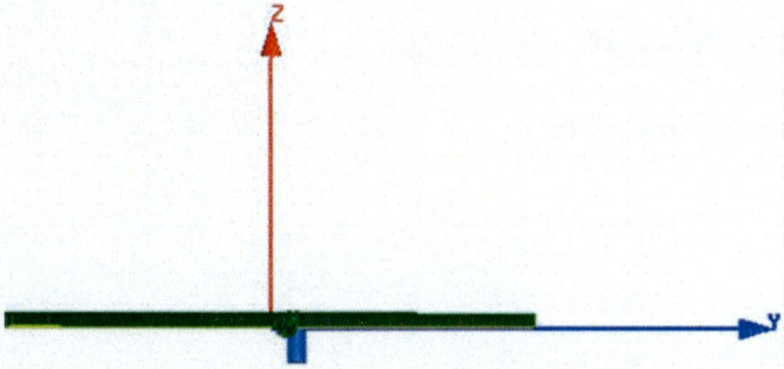

Fig. 2 Rectangular patch coaxial feed antenna in front view

Fig. 3 Rectangular patch coaxial feed antenna in top view

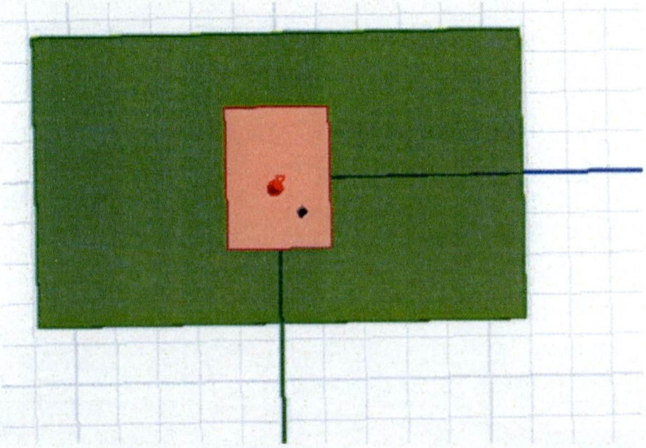

Fig. 4 Rectangular patch coaxial feed antenna with radiation box

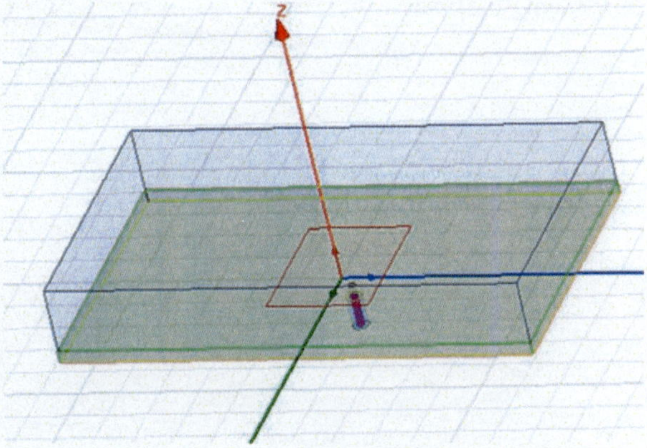

3 Design Equations

3.1 Patch Dimensions

The width of patch can be measured by [10]

$$W = \frac{c}{2f}\sqrt{\frac{2}{(\epsilon_r + 1)}} \tag{1}$$

The dielectric constant measured by

$$\epsilon_{\text{reff}} = \frac{(\epsilon_r + 1)}{2} + \frac{(\epsilon_r - 1)}{2}\sqrt{\frac{1}{1 + 12\left(\frac{h}{w}\right)}} \tag{2}$$

The effective length can be measured by

$$L_{\text{eff}} = \frac{c}{2f\sqrt{\epsilon_{\text{reff}}}} \tag{3}$$

The change in length is

$$\Delta L = 0.412h\left(\frac{(\epsilon_{\text{reff}} + 0.3)\left(\frac{W}{h} + 0.264\right)}{(\epsilon_{\text{reff}} - 0.258)\left(\frac{W}{h} + 0.8\right)}\right) \tag{4}$$

The length of patch is

$$L = L_{\text{eff}} - 2\Delta L \tag{5}$$

3.2 Ground Dimensions

The parameters of the antenna can be calculated by

$$\begin{aligned} L_g &= L + 6H \\ W_g &= W + 6H \end{aligned} \tag{6}$$

where $H = \frac{0.0606\lambda}{\sqrt{\epsilon_r}}$.

3.3 Feed Positions

The feed positions of the coaxial cable antenna are

$$X = \frac{L}{2\sqrt{\epsilon_{\text{reff}}}} \tag{7}$$

$$Y = \frac{W}{3\sqrt{\epsilon_{\text{reff}}}} \tag{8}$$

3.4 Radiation Box Dimensions

The dimensions of radiation box can be calculated by

$$L_r = L_g + \left(\frac{\lambda_g}{3}\right) \tag{9}$$

$$W_r = W_g + \left(\frac{\lambda_g}{3}\right) \tag{10}$$

$$h_r = h + \left(\frac{\lambda_g}{3}\right) \tag{11}$$

where $\lambda_g = \frac{\lambda}{\sqrt{\epsilon_{reff}}}$.
Here, f = resonant frequency of antenna.
ϵ_r = dielectric constant.
h = height of substrate.
λ_g = guided wavelength.
X = feed point distance from origin in x-axis.
Y = feed point distance from origin in y-axis.
L_r = radiation box length.
W_r = radiation box width.
h_r = radiation box height.

4 Simulated Results

The simulated results are obtained by using HFSS software. The S11 parameter of the proposed antenna is −20.5869 dB at 5.2 GHz frequency as shown in Fig. 5. For the incident power of 1 W, the accept power is 0.98461 W, and radiated power is 0.64469 W.

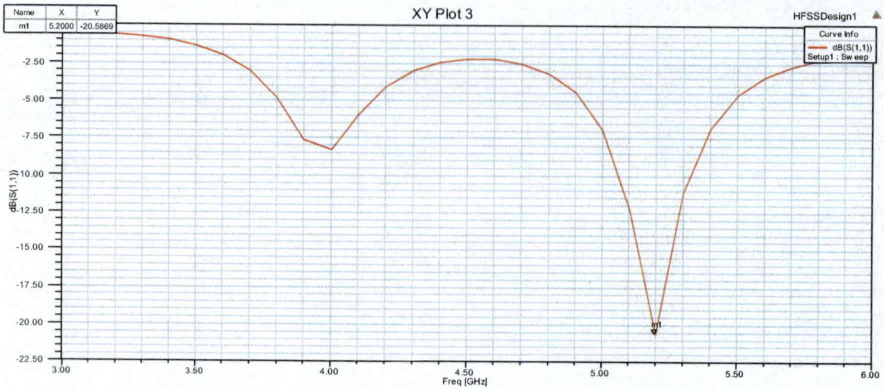

Fig. 5 S11 distribution plot

The designed antenna shows a VSWR of 1.2. However, the VSWR value for any perfectly resonating antenna is 1.1. This infers that a very small fraction of the power reflects back through feed cable. But this small difference is negligible and the VSWR value can be approximated to 1.1 and hence the designed antenna is perfectly matched to the port. The VSWR plot is shown in Fig. 6.

The actual antenna impedance and S parameters are plotted using the smith chart as shown in Fig. 7.

The gain of 3D polar plot is 4.13 dB, and the directivity of 3D polar plot is 5.64 as shown in Fig. 8a, b.

The flow of current distribution and surface current distribution is presented in Fig. 9a, b. Current distribution is normal and equally distributed along surface of the antenna.

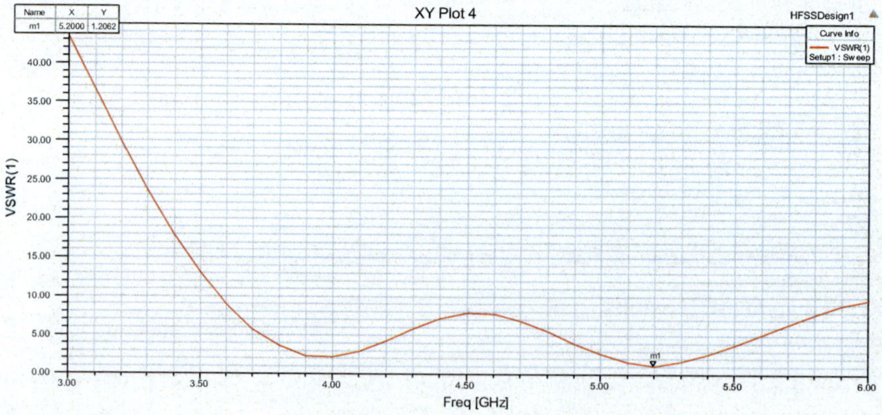

Fig. 6 VSWR

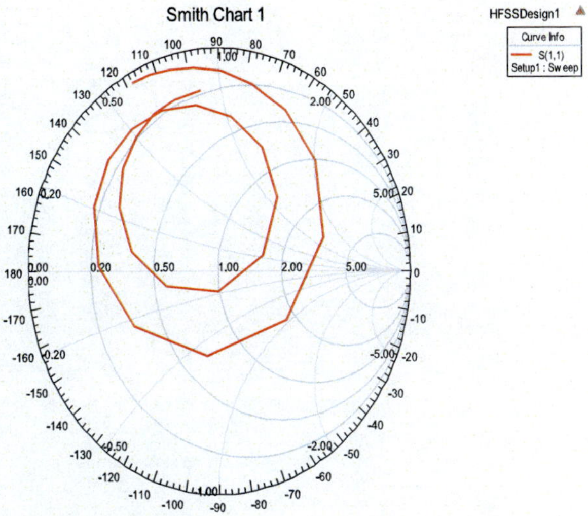

Fig. 7 Smith chart

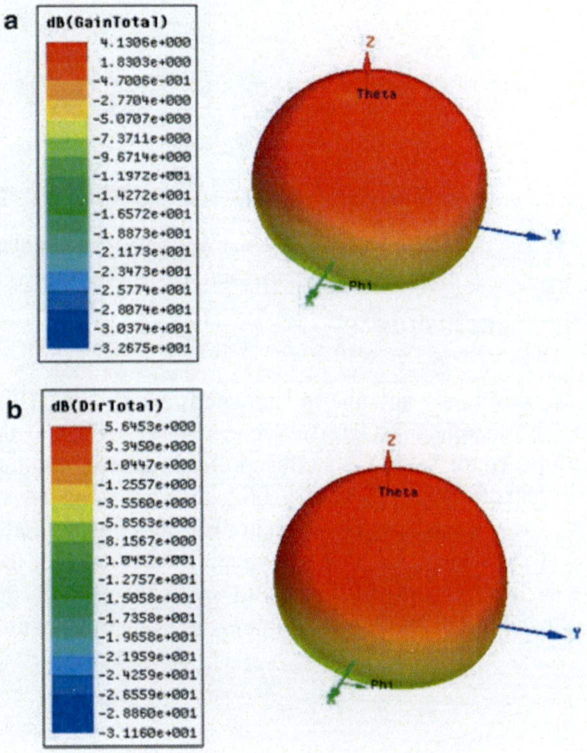

Fig. 8 **a** Gain in 3D polar plot. **b** Directivity in 3D plot

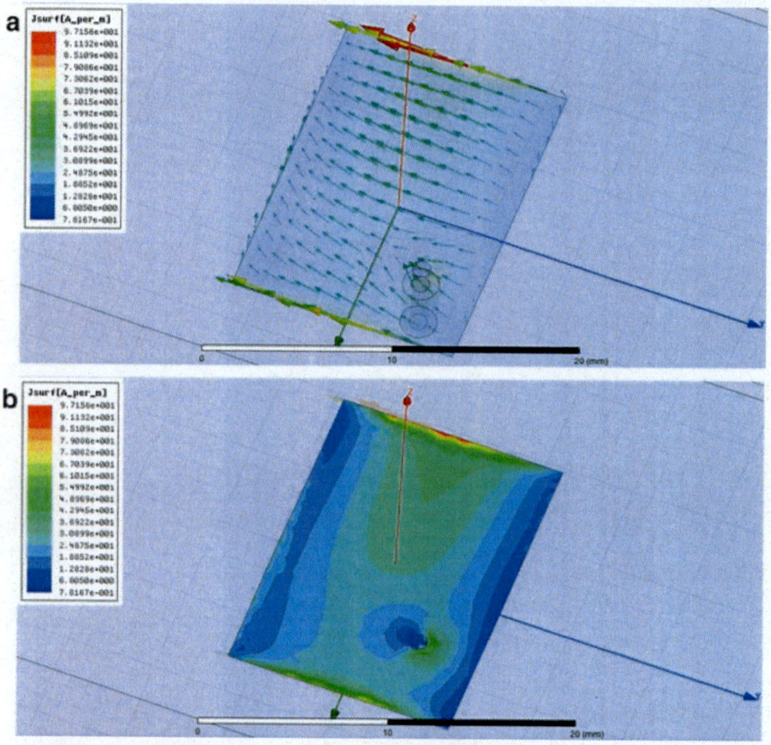

Fig. 9 **a** Flow of current distribution. **b** Surface current distribution

5 Conclusion

The developed antenna is simulated using Ansys HFSS, and the antenna is found to resonate at 5.2 GHz frequency. Based on the simulated results, the antenna is proposed for WLAN-specific applications. The simulated results of return loss are − 20.586 dB, and VSWR is 1.2062 at centre frequency. The radiation pattern is stable at centre frequency. The current distribution is normal and equally distributed along surface of the patch antenna. Here, the efficiency of the antenna is 0.7055. The gain and directivity of 3D polar plot are 2.5886 and 3. 668. However, to enhance the efficiency, the feeding techniques can be changed with minimizing the antenna size.

References

1. Tiwari RN, Singh P, Kanaujia BK (2019) A modified microstrip line fed compact UWB antenna for WiMAX/ISM/WLAN and wireless communications. AEU Int J Electron Commun 104:58–65

2. Afrough M, Fakharian MM, Tavakol-Hamedani F (2015) Compact dual-band suspended microstrip slot antenna with an antipodal parasitic element for WLAN applications. Wireless Pers Commun 83(1):571–579
3. Mishra B, Singh V, Singh R (2017) Dual and wide-band slot loaded stacked microstrip patch antenna for WLAN/WiMAX applications. Microsyst Technol 23(8):3467–3475
4. Gan Y, Wang S, Yu J (2012) Design of a dual-band microstrip antenna for WLAN. In: Wang X, Wang F, Zhong S (eds) Electrical, information engineering and mechatronics 2011. Lecture notes in electrical engineering, vol 138. Springer, London. https://doi.org/10.1007/978-1-4471-2467-2_52
5. Sinha S, Rana B, Ghosh CK, Parui SK (2012) A CPW-fed microstrip antenna for WLAN application. Procedia Technol 4:417–442
6. Deshmukh AA, Shah M, Kudoo C, Mhatre A (2018) Variations of wide band E-shape and U-slot cut rectangular microstrip antennas. Procedia Comput Sci 143:94–100
7. Sheeba GM, Nachiappan A (2017) Performance evaluation of fuzzy DE based node placement in WMN. J Eng Res 5(4):106–120
8. Sung Y (2011) A printed wide-slot antenna with a modified L-shaped microstrip line for wideband applications. IEEE Trans Antennas Propag 59(10):3918–3922
9. Saadh AM, Poonkuzhali R (2019) A compact CPW fed multiband antenna for WLAN/INSAT/WPAN applications. AEU Int J Electron Commun 109:128–135
10. Balanis CA (2016) Antenna theory: analysis and design. Wiley
11. Sheeba GM, Nachiappan A (2019) Computation of mesh node placements using DE approach to minimize deployment cost with maximum connectivity. Wirel Pers Commun 107(1):291–302

Design of Single Switch Voltage Equalizer for Partially Shaded Photovoltaic Modules

J. Ajay Daniel, V. N. Ganesh, M. Maria Joseph, S. Balaji, D. Siva Kumar, and B. Pavithra

Abstract Photovoltaic modules are the main elements of solar panels. During the operation of solar panels, they are placed in a certain area which may have a chance of partial shade on the panel which creates shaded and un-shaded regions in it. This decreases the effective output of the solar panel due to the uneven production in the string. The shaded regions produce maximum MPPs as well as local and global MPPs. To minimize the issues of partial shaded regions in the panel, we use single switch voltage equalizer comprising of multi-stacked buck-boost converters, and single switch also reduces the complexity in the circuit. In this paper, we use SEPIC-based buck-boost converter, study is done, and results are shown which increases the efficiency of PV modules by using voltage equalizer.

Keywords Partial shading · Photovoltaic system · Buck-boost converter · Voltage equalizer · SEPIC · MPP-Maximum power point

1 Introduction

With the increased usage of fossil fuels in the world, there is a threat of extinction of some of the conventional fuels in the coming future. To save them and as a way of conserving them, we need to use renewable energy resources. Among renewable energy, solar energy is the most abundant and never-ending source, one of the disadvantage is that solar energy extraction needs more area to extract more power; in this due to some geographical or constructional issues, there is a possibility of forming shaded regions on the panels. On a photo voltaic module or a string, the shaded regions, therefore, produce less current which mismatches the voltage produced from the other materials strings. This reduces the total efficiency of the plant. This is a major problem that needs attention and to be eliminated, to extract more power from solar energy.

J. A. Daniel (✉) · V. N. Ganesh · M. M. Joseph · S. Balaji · D. S. Kumar · B. Pavithra
Department of Electrical and Electronics Engineering, Faculty of Engineering and Technology, SRM Institute of Science and Technology, Ramapuram 600089, India

S. S. Dash et al. (eds.), *Sixth International Conference on Intelligent Computing and Applications*, Advances in Intelligent Systems and Computing 1369,
https://doi.org/10.1007/978-981-16-1335-7_46

When partial shading occurs the voltage, mismatches happen in between the substrings creating multiple MPPs. Each substring working at an individual maximum power point track (MPPT) can eliminate this problem. This can be obtained by using converters for each module, but this needs number of switches increasing cost. A voltage equalizer can minimize the number of switches used.

In this paper, we discuss on the design of a single-switch voltage equalizer built on a multi-stacked CLD filters on SEPIC-based buck-boost converter with two inductors. In Sect. 2, we discuss about the existing system, Sect. 3 on proposed system and voltage equalization strategy, and Sect. 4 on operational analysis and V conclusion.

2 Existing Systems

The solar plates collect the energy from the sun, then it is here sent to the Boost converter where the voltage of the generated voltage is increased up to the needed level, the boost converter output is sent to the Inverter where the DC voltage is converted into AC voltage and sent to the load.

The duty cycle needed for the operation of the boost converter and inverter is generated using a gate driver, whose output is controlled by the programming done in the microcontroller as needed (Fig. 1).

When there is partial shading issues in the above system, the voltage generated is not constant through each substring, therefore decreasing the efficiency of the total power produced. The whole system is maintained with the same duty cycle neglecting the changes happening in the substrings due to shading issues.

There are many ways to eliminate this problem, one of which is when there is a shaded region on a module that module is neglected, and the current is passed through the bypass diode (as in Fig. 2) which is connected in parallel to it, and this method

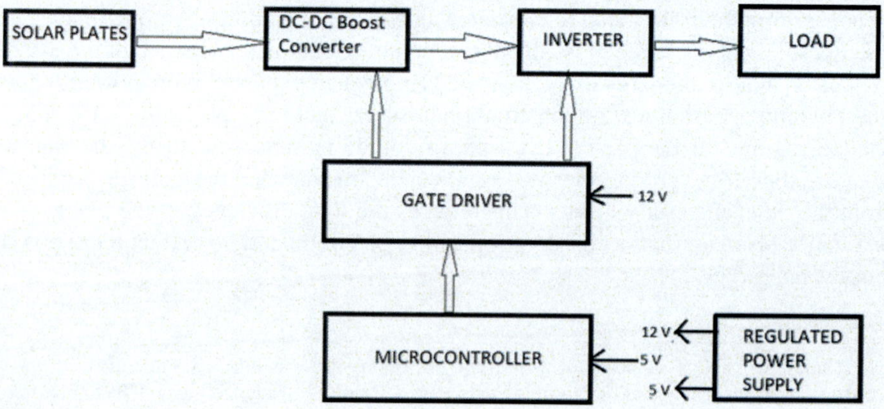

Fig. 1 Existing block diagram

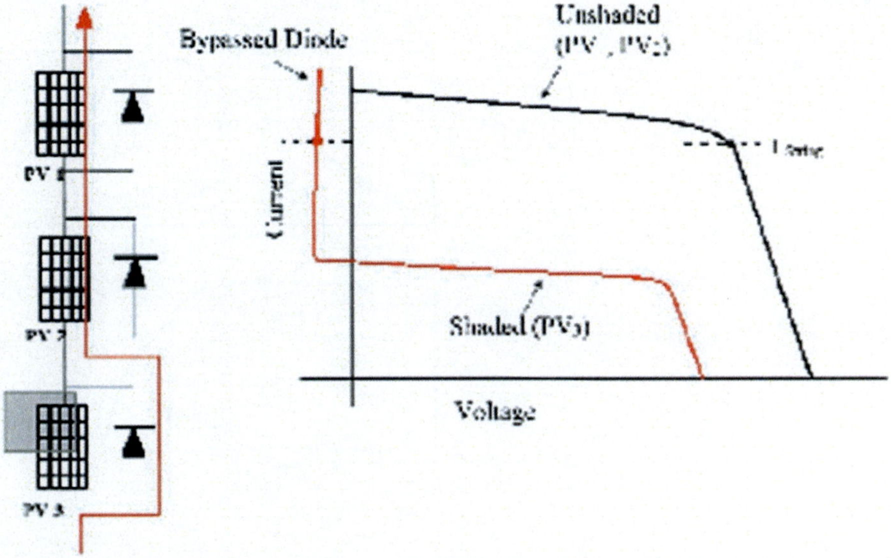

Fig. 2 Bypassed PV module and its characteristics

reduces the total power extracted from the string. The bypass diode comes into play when there is a voltage mismatch in that substring. This voltage mismatches create multiple MPP's local MPPs and maximum MPPTs, which needs a MPP tracking (MPPT) to track a global MPPT. Even though a global MPPT is obtained, all PV modules cannot reach to this maximum point due to the voltage mismatches which eventually reduces the extracted power. For eliminating this problem, we need to work on individual MPPTs for each PV module which increases the number of converters and switches increasing the total cost with increase in number of PV modules that are connected (Fig. 3).

A single switch voltage equalizer is an alternate way than using a converter for each PV module reducing the number of switches therefore reducing circuit complexity and many duty cycle variations.

3 Proposed System

In this system, there comes a voltage equalizer and a MPPT tracker which decreases the partial shading issues of PV modules.

Apart of conventional buck-boost converter, here SEPIC-based buck-boost converter is proposed.

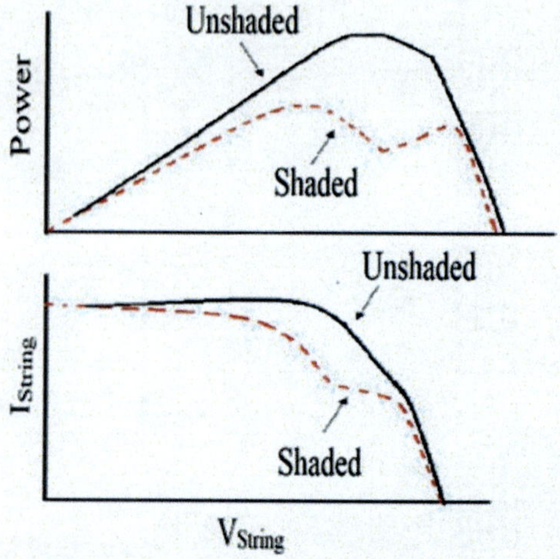

Fig. 3 Total string characteristics

3.1 *Voltage Equalization Strategy*

When an issue related to shading occurs over a photovoltaic module output produces different voltages creating multiple MPPs (Fig. 4).

The proposed concept eliminates the multiple MPPs, and this voltage equalization takes an equivalent current produced from all the substrings and divides them equally and sent.

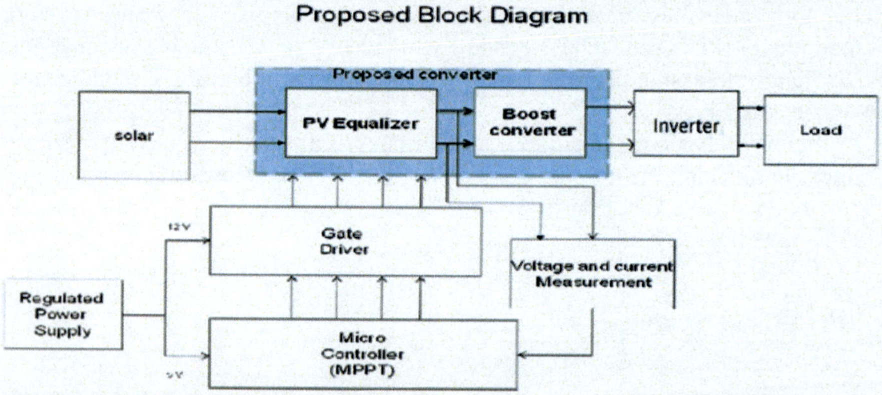

Fig. 4 Proposed block diagram

3.2 Circuit Diagram

Three topologies can be used for proposed single-switch voltage equalizer using multi-stacked buck-boost converters for partially shaded series—connected PV modules. (a) SEPIC based (b) ZETA based (c) CUK based (Fig. 5).

Lack of floating gate driver and transformer made the SEPIC-based topology most promising, inspite floating gate driver is needed for ZETA based, and without transformer operation of CUK based, topology is not possible.

In circuit diagram, there are three substrings for which three CLD filters are used; with increase in the number of substrings, CLD filters can be extended to each substring.

Proposed voltage equalizer can be operated in two modes: continuous conduction mode (CCM) and discontinuous conduction mode (DCM). In the existing system, voltage equalizers were designed to work in DCM in which currents are limited to the certain level without any feedback control. With this work, there is a possibility of designing a voltage equalizer which can operate in CCM without limiting the current to a certain level without feedback control.

The proposed voltage equalizer is like existing one, and each CLD filter has different DC current level which depends on degree of shading on PV modules.

$$D_{\text{Critical}} = \frac{V_{PV_i} + V_D}{V_{\text{String}} + V_{PV_i} + V_D}$$

For getting the duty cycle that is given to the switches in the gate driver is known as D_{Critical} where D denotes duty cycle.

where

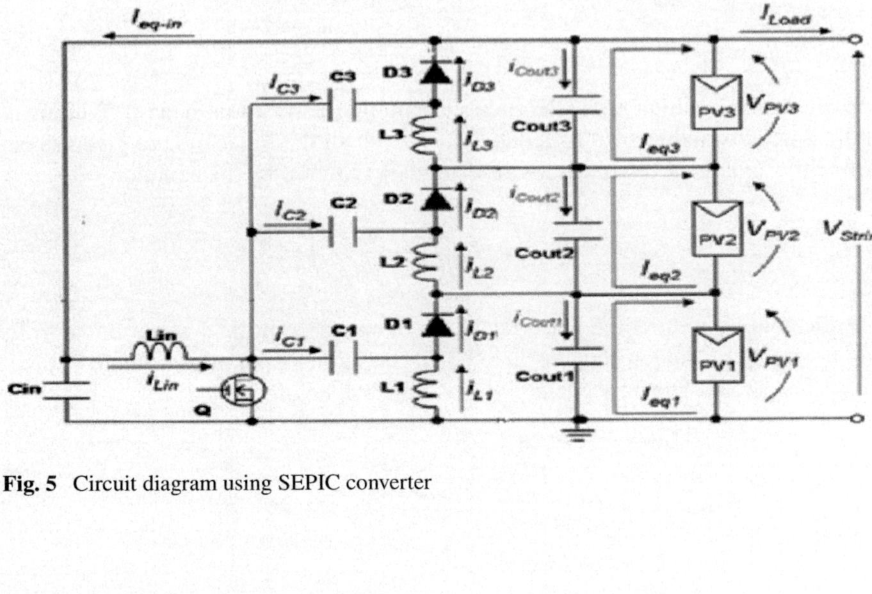

Fig. 5 Circuit diagram using SEPIC converter

V_{PV_i} is input voltage of photo voltaic module,
V_D is bypass diode voltage, and V_{String} is total string voltage.
Output voltage and input voltage are obtained from

$$V_{O/P} = \frac{1}{1 - D} \times V_{I/P}$$

where $V_{O/P}$ is Output voltage controlled by MPPT and $V_{I/P}$ is total string voltage

$$I_{O/P} = \frac{V_{O/P}}{R_L}$$

where $I_{O/P}$ is the output current controlled by MPPT and R_L is load resistance (100 Ω).

3.3 Mathematical Formulae

The total current that is passing through the inductors that are connected in series is given by

$$i_{L_{tot}} = i_{L_1} + i_{L_2} + i_{L_3}$$

In the same way, the equivalent currents are divided by the voltage equalizer with respect to each PV panel, and the total equivalent current passed through them is equal to the total inductor current.

$$I_{eq\text{ - tot}} = I_{eq_1} + I_{eq_2} + I_{eq_3} = I_{L_{tot}}$$

In all the CLD filters used, the inductor among them, when made it to equivalent that inductor inductance is denoted by L_{Tot} as all the inductors are given same asymmetric square wave voltage, whose value is acquired by the equation

$$L_{tot} = v_L\left(\frac{dt}{di_{L_1}} + \frac{dt}{di_{L_2}} + \frac{dt}{di_{L_3}}\right) = \frac{L_i}{3}$$

While coming to the time period analysis of working of the system, in the Toff period, ceasing of diodes happens here as all the inductors applied voltage will be zero in this period. The duty cycle in this period will be given by

$$D_a = \frac{DV_{String}}{V_{pv_i} + V_D}$$

Here, D is the duty cycle and V_D is the forward voltage drop of the diodes. The average current passing through L_{in} and L_{tot} is given as

$$I_{L_{\text{tot}}} = \frac{V_{\text{String}} D D_a T_S}{2}\left(\frac{L_{\text{in}} + L_{\text{tot}}}{L_{\text{in}} L_{\text{tot}}}\right)$$

Here, T_s is switching period.

As in the SEPIC converter, the relation between voltage conversion ratio and current conversion ratio is given by

$$V_{PV_i} = \frac{D}{1 - D} V_{\text{string}} - V_D$$

$$\frac{I_{L_{\text{tot}}}}{I_{L_{\text{in}}}} = \frac{1 - D}{D}$$

4 Operational Analysis

The circuit consists of three PV modules of given voltage as 12 V, keeping partial shading issues in consideration an MPPT tracker and voltage equalizer as CLD filters for each substring is used. With the presence of capacitors in parallel with the solar cells, the output is not at rated level but will be of 11.5 V. The PV cells that are connected are taken as a string whose individual voltages will be as in Fig. 7 (Fig. 6).

Figure 7 gives the input of each PV substring whose value taken in work is 11.5 V which is V_{PV_i}.

From Fig. 8, VD is 0.263 V which is bypass diode voltage, and this diode voltage denotes the critical voltage that have it during the operation, and pulse width modulation (PWM) output shows the switch ON mode showing the ON operation of the voltage equalization.

In Fig. 9, using MPPT as a closed-loop average, duty cycle value can be obtained. MPPT contains PID controller, and A&O algorithm of MPPT is used for the closed loop, thus giving required duty cycle.

Using Eq. (1) with the values $V_{\text{PV}_i} = 11.5\,\text{V}$.

In Fig. 10, obtaining the fixed value of net duty cycle is not possible due to the PID controller usage in the circuit of MPPT.

$V_D = 0.8$ V, $V_{\text{String}} = 35$ V giving D_{Critical} as 0.26.

To justify the output obtained with this work which is not possible by using MPPT due to its pulsed duty cycle, by using a pulse generator (as in Fig. 11), duty cycle is fixed, and the analytical calculation can be obtained for the output.

Using open loop is a disadvantage due to the change in duty cycle is needed to be changed manually which lead to decrease in efficiency. Here, usage of open loop is limited only to justify the work done.

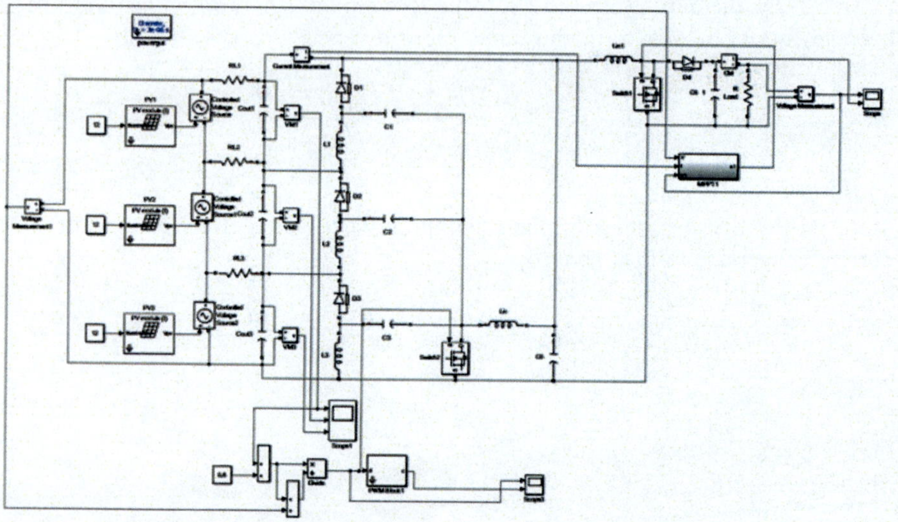

Fig. 6 Simulation diagram of closed loop circuit

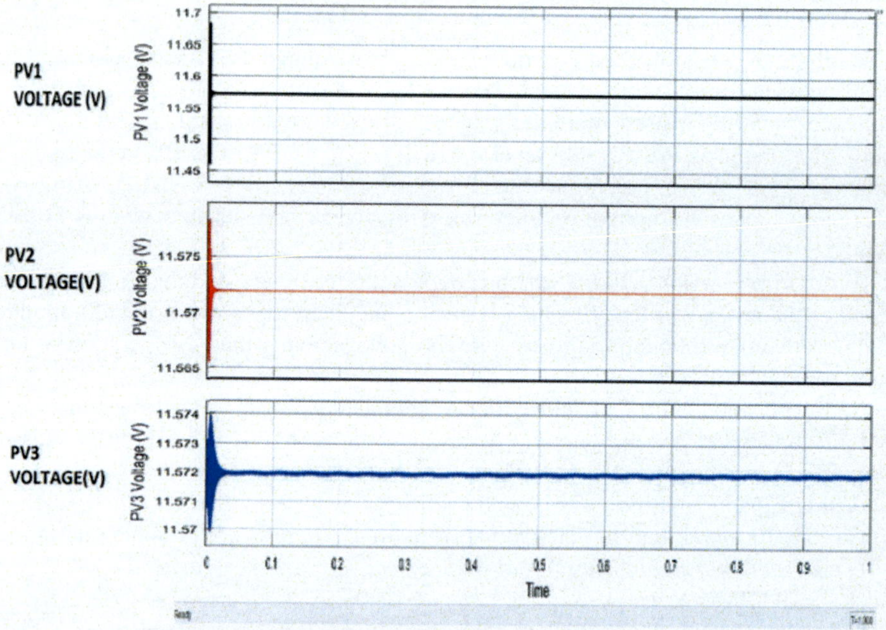

Fig. 7 Input voltage of photo voltaic substrings

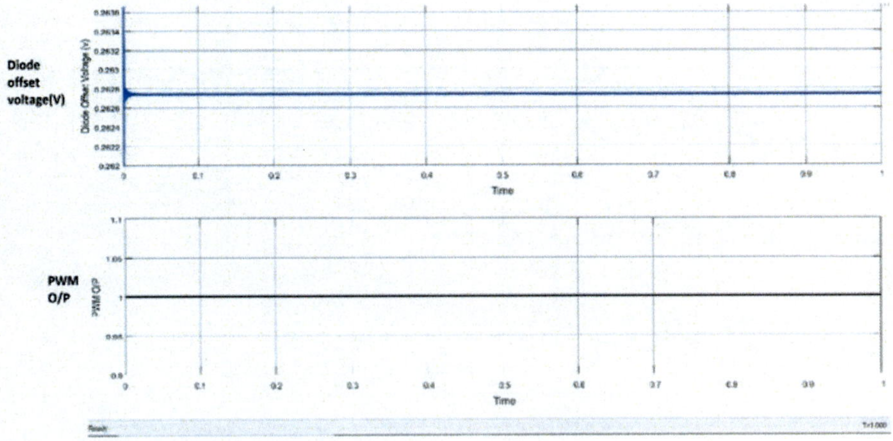

Fig. 8 Diode offset voltage and PWM o/p

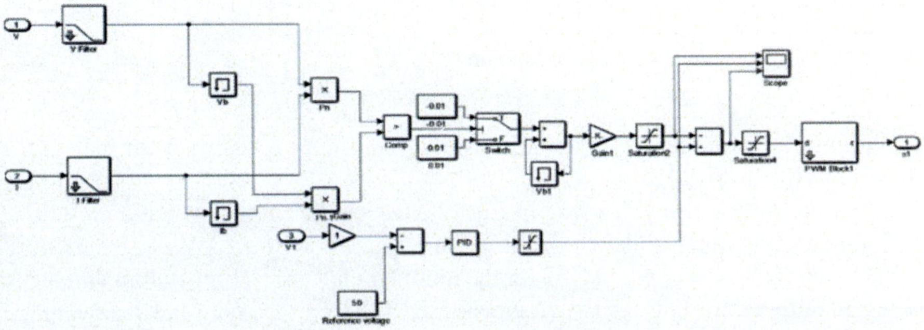

Fig. 9 Simulation circuit of MPPT

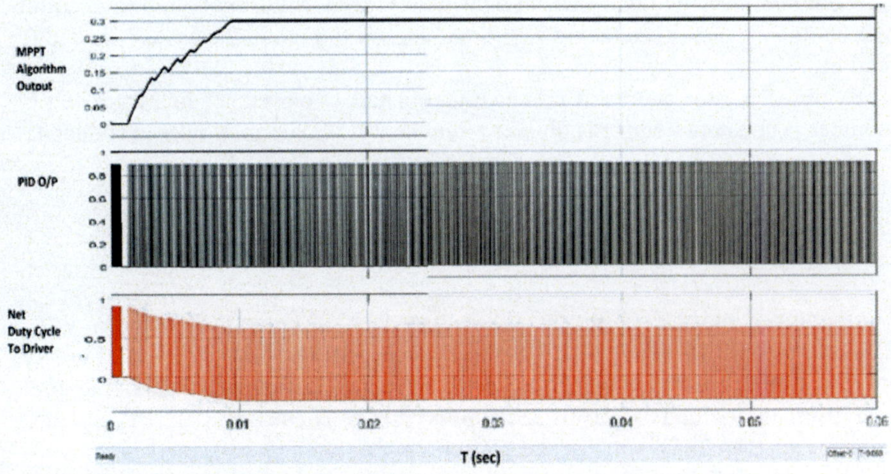

Fig. 10 Simulation output of MPPT for closed loop

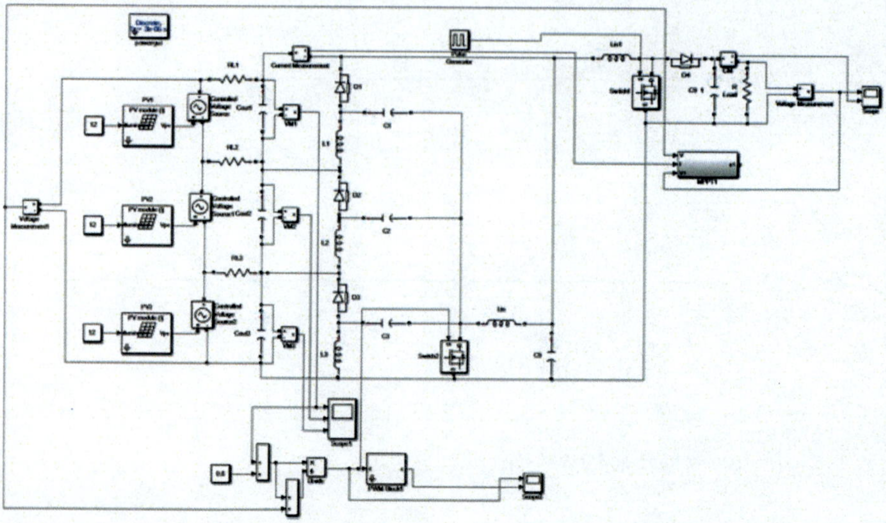

Fig. 11 Simulation circuit in open loop condition

With giving variable duty cycle (D) as 0.6, the output is justified for the values used in the work (Fig. 12).

Using the duty cycle (D) value in Eq. (2), we get the output voltage as $V_{\mathrm{o/p}} =$ 87.5 V, and substituting this value in Eq. (3), we get $I_{\mathrm{o/p}} = 0.875$ A.

While with the proposed system, for the input voltage of 35 V, the output voltage is 51 V, and output current is 0.51 A. The proposed system overcomes the disadvantage of open-loop system by avoiding the need in the change of duty cycle amplitude manually for every operation cycle. Figure 13 shows the output employing MPPT algorithm for getting duty cycles with the help of feedback from voltage and current measurements, hence producing a constant voltage though in the case of shaded PV modules.

By observing the results, it is shown that behalf of partial shading issues on PV modules, a constant voltage is obtained, and hence, maximum power is obtained.

5 Hardware Analysis

The hardware block consists the following specified components. SMPS I/P = 230 V/10 mA, SMPS 0/P = 12 V/200 mA. Microcontroller unit I/P = 5 V. Microcontroller used PIC 16F883 (8 bit). Voltage and current measurements LM324 Op-amps, gate driver unit with TLP250 IC's, 7812 and 7912 switches, and 12 V A.C supply. Voltage equalizer 225 μF major and 220 μF/25 V minor capacitors and 470 μF/50 V as, 20 A/200 V diodes, IRF540 switches. Boost converter 200 μH, IRF540 switch, and 1000 μF/20 V capacitor. 12 V/10 W solar panels and load of 22 kΩ/5 W (Fig. 14).

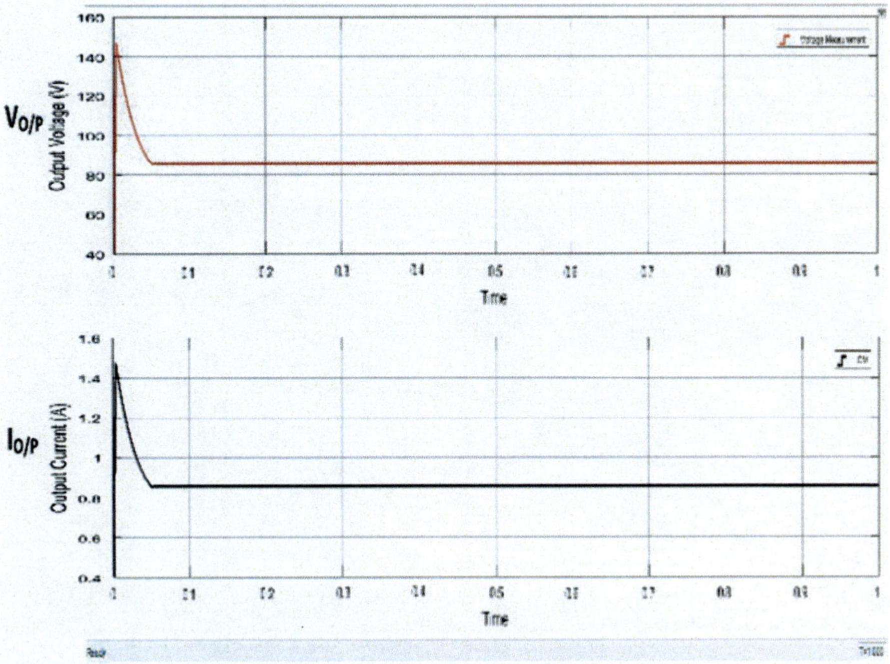

Fig. 12 Output voltage and current characteristics in open loop condition

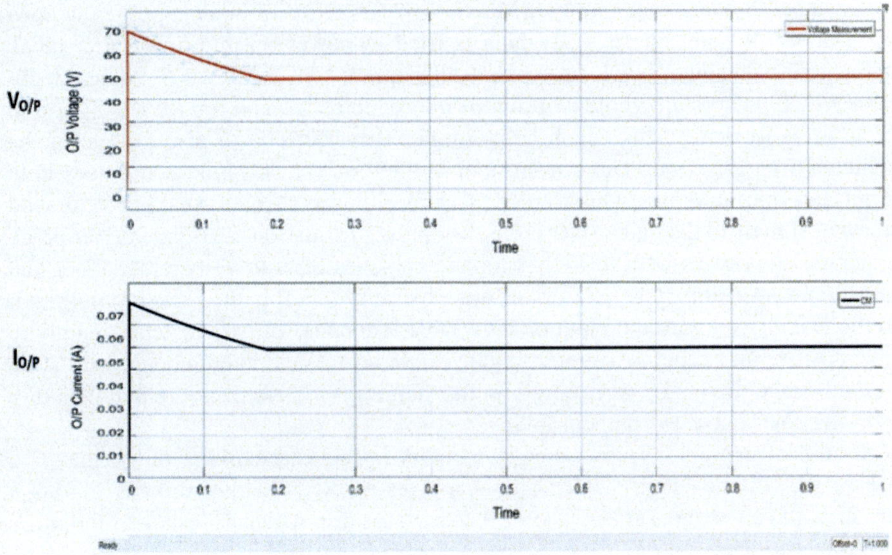

Fig. 13 Output characteristics in closed loop condition

Fig. 14 Overall hardware picture

Switched mode power supply (SMPS), the 210–230 V of input supply is stepped down to 12 V constant supply, which is used to power up microcontroller block and gate driver power block. Microcontroller block consists of a 5 V voltage regulator which powers up whole working of microcontroller, and the MPPT algorithm program is loaded in this which helps in gate driver circuit to give pulses for the switches for operation. The current and voltage measuring blocks measure both input and output values which are sent to the microcontroller, making it to send relevant signals to the gate driver circuit for pulses formation. This gate driver block generates the pulses with these ICs for the switches of both voltage equalizer and boost converter unit. The gate driver power supply block is the power supply unit to the gate driver, with the input of 12 V D.C. It consists of two switches mainly to give positive and negative sources to gate driver unit. When there comes a situation of the voltage dip in any working PV cells, then this block makes sure that the other working solar panel to compensate it.

Behalf of using 12 V solar panels, to demonstrate the working of the idea, we used a 12, 12, and 9 V transformers. This gives us the benefit of directly getting a dip in voltage production because of 9 V transformer, which is less than the other two transformers.

Figures 15, 16, and 17 show the digital CRO output of first 12 V, second 12 V, and third 9 V transformer.

Figure 18 shows the pulse wave form produced by gate driver circuit when o/p 50 V is needed (Fig. 19).

Fig. 15 CRO o/p of input 12 V transformer

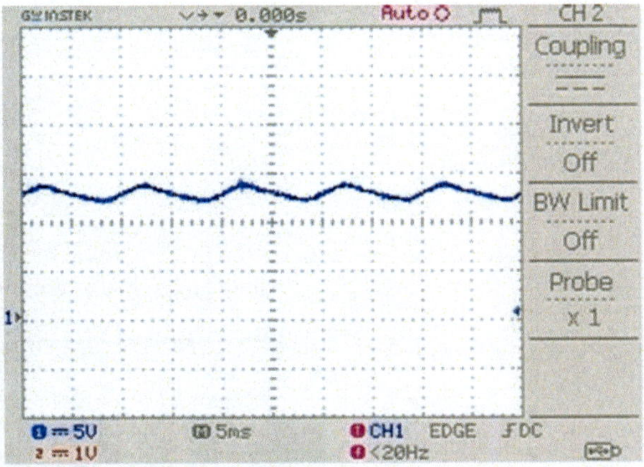

Fig. 16 CRO o/p of input 12 V transformer

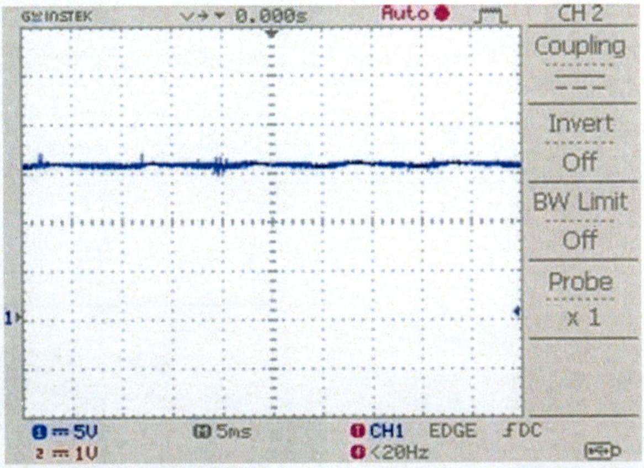

Fig. 17 CRO o/p of input 9 V transformer

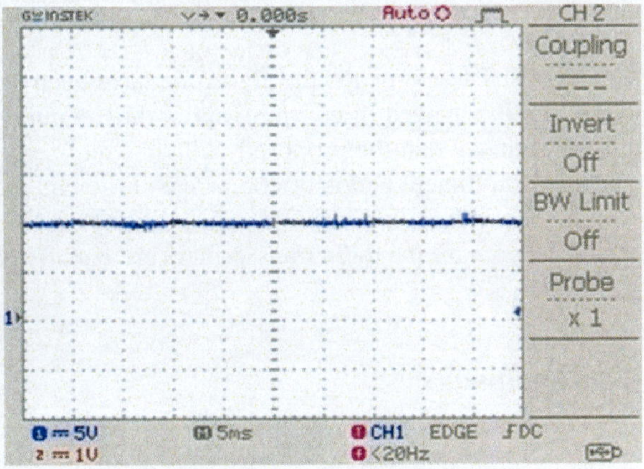

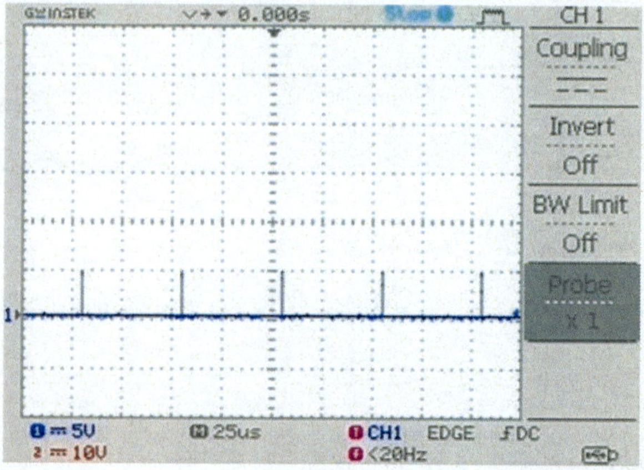

Fig. 18 PWM o/p pulses when VRf is 50 V in CRO

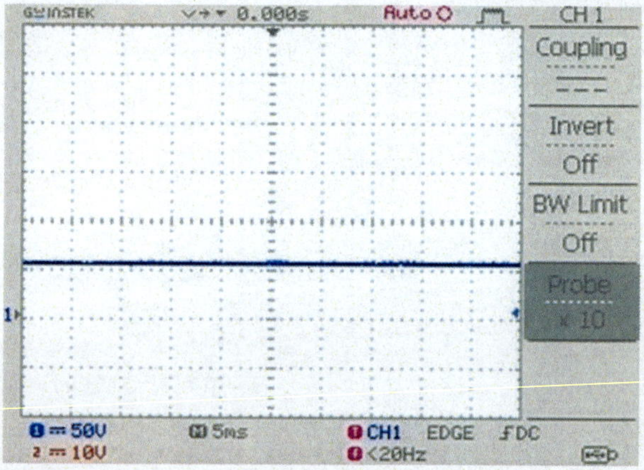

Fig. 19 O/p from load when VRf is 50 V

The MPPT tracking and the algorithm have been worked in efficient manner, and the output is received in constant value, thus maintaining the desired output. With this, we can say that the proposed voltage equalizer with single switch mechanism has been so efficient and is of course cost efficient because of the usage of a single switch. The pulses produced switches the thyristors accordingly that the currents produced from all the three transformers are equally divided and distributed.

6 Conclusion

In this paper, discussion on the effects of partial shading issues on the PV modules is done. Using converter for each substring to eliminate voltage mismatch, increases cost. A method to eliminate voltage mismatches among the substrings and to reduce the number of switches is proposed with the concept of single switch voltage equalizer, using multi-stacked CLD filters on traditional SEPIC-based buck-boost converter.

Three photo module substrings are taken, and the proposed system is employed whose operational analysis and results are shown proving the increase in the efficacy of whole system using voltage equalizer.

References

1. Rai GD (2010) Non-conventional resources of energy, 4th ed. Khanna Publishers
2. Rao S, Parulekar (2005) Energy technology, 4th ed. Khanna Publishers
3. Khan BH (2009) Non-conventional energy resources, 2nd ed. The MC Graw Hills
4. Chandra S, Chauhan DS, Agrawal S, Rajoria CS (2016) An enhancement in electrical efficiency of photovoltaic module. In: 2016 IEEE 7th power india international conference (PIICON)
5. Uno M, Yamamoto M (2017) Switched capacitor-based PWM converter integrating string converter and voltage equalizer for photovoltaic strings under partial shading. In: 2017 IEEE region 10 conference, TENCON 2017
6. Shinohara T, Uno M (2017) Micro-inverter based on quasi quasi-Z source inverter integrating switchless voltage equalizer for photovoltaic panels under partial shading. In: 2017 IEEE 3rd international future energy electronics conference and ECCE Asia (IFEEC 2017—ECCE Asia)
7. Uno M, Kukita A (2015) Single-switch voltage equalizer using multistacked Buck-Boost converters for partially shaded photovoltaic modules. IEEE Trans Power Electron 30(6)
8. Nakane T, Shinohara T, Uno M (2017) Single-switch voltage equalizer based on forward-fly back resonant voltage multiplier for partially-shaded series connected photovoltaic modules. In: TENCON 2017—2017 IEEE region 10 conference
9. Zhang W, Li J, Mao P (2017) A novel isolated-port voltage equalizer for photovoltaic systems under mismatch conditions. In: IECON 2017—43rd annual conference of the IEEE industrial electronics society
10. Chikh A, Chandra A (2015) An optimal maximum power point tracking algorithm for PV systems with climatic parameters estimation. IEEE Trans Sustain Energ 6(2)
11. Chan Z, Yan Z (2017) Improved maximum power point tracking algorithm with CUK converter for PV systems. J Eng

Design and Fabrication of Object Sorting Conveyor System Based on Shape and Colour

V. Krithika, M. Nandhini, C. Subramani, A. Hima Koteswara Rao, and R. Rahul

Abstract Sorting of objects is extensively used in many industries like food processing industries, toy industries, etc. To ensure that the quality of the product is up to the mark, this process is simplified by the use of automation. Automation is the use of control systems like computers or robots for handling different process and machineries to replace a human being and provides mechanical assistance. Automated systems generally use more complex algorithms which increase the cost of the design and the power consumed. This not only reduces manual efforts, time consumed, gives more time for marketing, but also prevents danger which might occur when human beings work in hazardous environments. Automation greatly improves the productivity and is highly scalable. Here, OpenCV is used which is a computer vision library used extensively in the industry. The Python programming language is used as it is easy to use and has sufficient speed for our task. A simple USB camera is used to capture the real time video. The camera is engaged in sensing the object for shape and colour by means of image processing and after this the conveyor moves and the flaps are actuated by means of servomotor in order to sort the objects based on their respective shape and colour.

Keywords Conveyor belt · SMPS · DC−DC converter · Camera · Python · OpenCV

V. Krithika (✉) · M. Nandhini · A. Hima Koteswara Rao · R. Rahul
Department of Mechatronics Engineering, Faculty of Engineering and Technology, SRM Institute of Science and Technology, Kattankulathur 603203, India

M. Nandhini
e-mail: nandhini.m@ktr.srmuniv.ac.in

C. Subramani
Department of Electrical and Electronics Engineering, Faculty of Engineering and Technology, SRM Institute of Science and Technology, Kattankulathur 603203, India

S. S. Dash et al. (eds.), *Sixth International Conference on Intelligent Computing and Applications*, Advances in Intelligent Systems and Computing 1369,
https://doi.org/10.1007/978-981-16-1335-7_47

1 Introduction

This project is the research work done with an aim to automate the Object Sorting Conveyor System which is both economical and much more simplified to use. The object sorting conveyor brings in more efficient way of sorting the objects based on the predefined parameters as they result in proper way of identification of the shape and colour of the objects and separate them accordingly, the image processing system is capable of recreating the objects captured by the camera and process it by means of OpenCV to identify its shape and colour. This way of sorting mechanism reduces time, human labor and the running cost is also comparatively low and its gives more accurate results as compared to other conventional systems which uses colour sensor as in the place of camera module to detect the only the colour of the object alone but in this system the camera is capable of detecting both the shape and colour of the object [1].

The object sorting conveyor brings in more efficient way of sorting the objects based on the predefined parameters as they result in proper way of identification of the shape and colour of the objects and separate them accordingly, the image processing system is capable of recreating the objects captured by the camera and process it by means of OpenCV to identify its shape and colour. This way of sorting mechanism reduces time, human labour and the running cost is also comparatively low and its gives more accurate results as compared to other conventional systems which uses colour sensor as in the place of camera module to detect the only the colour of the object alone but in this system the camera is capable of detecting both the shape and colour of the object. The object sorting conveyor has been widely used in automobile, Pharmacy and confectionary industries in order to sort the objects based on different shapes and colours, as the objects of all the shape and colours are transported in the conveyor the sorting mechanism helps in separating them according to their respective shape and colour Fig. 1.

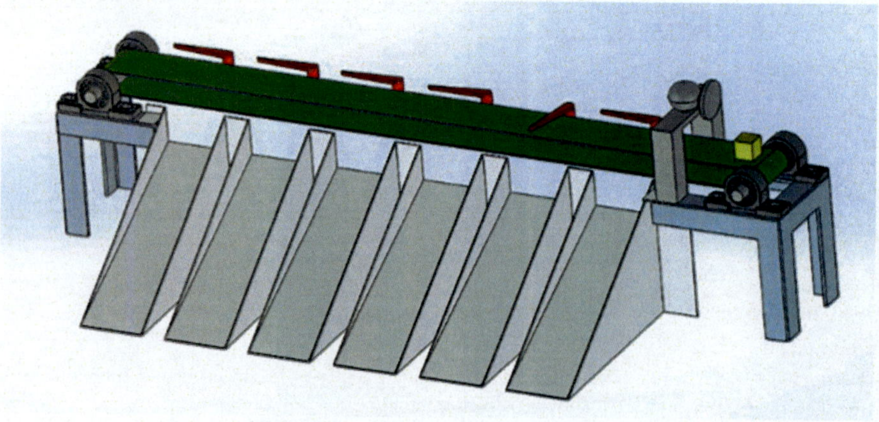

Fig. 1 Design of proposed conveyor system

It helps in reducing the time, results in accurate way of object sorting method and to sort object easily based on different shapes and colours, to minimize human labour. This is ideal for sorting identical objects with different colours. It is cost efficient and has given are life to the labour efforts. In order to increase the level of productivity rapidly and it also helps to stack the transported materials at the end of the production line. It saves the product from getting damaged and also prevents injuries. It is an ingenious way to increase the process of production effectively and it consumes less power and is durable in nature.

1.1 Design and Components Used

The design of the conveyor system and the object sorting setup is ideally obtained using the designing software SolidWorks, in which the 3-D model was sketched and visualized, further the entire system built and developed as per the computational design. It includes the frame design, belt design, position of camera, motor, trays, etc. The Schematic of a pulley is shown below (Fig. 2).

Design calculation of the sorting conveyor system:

$$V = \frac{(\pi \times d \times N)}{60}$$

where

- V is the linear velocity of the belt,
- N is the speed of the driver pulley, $N = 60$ rpm.
- ω is the angular velocity of the driver pulley,
- d is the diameter of pulley, $d = 32$ mm.

$$V = 83.73\ \text{mm/s}$$

Hence,
Angular Velocity of the driver pulley is,

$$\omega = \frac{V}{r}$$

$$\omega = 5.23\ \text{rad/s}$$

Linear velocity of belt,

$$V = \frac{(\pi \times d \times N)}{60} \times \left[1 - \frac{S_1}{100}\right]$$

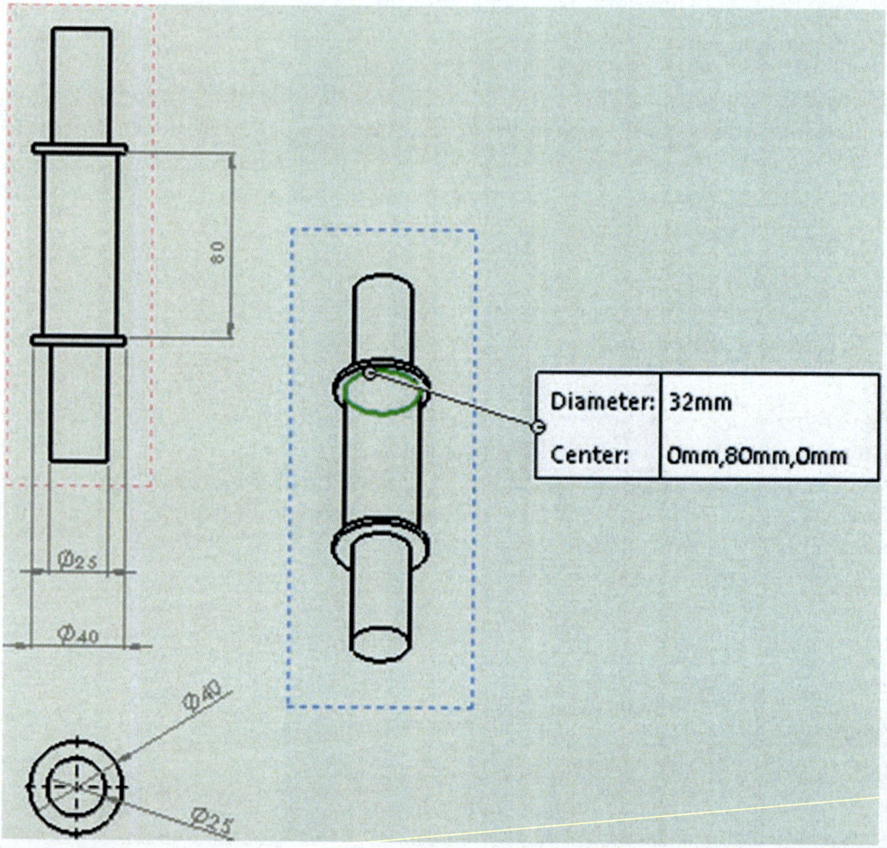

Fig. 2 Schematic of a pulley

where, S_1 is the slip between driver pulley and belt. Since, there is no slip acting and it is negligible. The equation of the linear velocity of the driver pulley becomes same as belt.

Linear velocity of the driver pulley,

$$V = 83.73\ \text{mm/s}$$

In view of the fact that the driver pulley and driven pulley have the same diameter, the velocities are same.

Length of the belt,

$$\text{LO} = \pi/2(d_1 + d_2) + 2C + 1/4C(d_1 - d_2)^2$$

where

- LO is the length of the belt.
- $d_1 = d_2 = d = 32$ mm, diameter of the driver or driven pulley.
- C is the distance between the centre of driver and driven pulley [$C = 1080$ mm].

Thus taking the above mentioned facts into consideration, the equation for the Length of the belt is further reduced to,

$$\text{LO} = \pi(d) + 2C$$

Therefore, substituting the values of d (diameter of pulley) and C (distance between the centre of driver and driven pulley),

$$\text{LO} = \pi(32) + 2 \times 1080$$

the value of length of the belt is,

$$\text{LO} = 2260.48\ \text{mm}.$$

1.2 Methodology

The object sorting conveyor has been powered using a 12 V, 5 A SMPS (Switch mode power supply) capable of providing 2 outputs. One output of SMPS is connected to Geared DC motor and the other with the DC–DC Buck converter. When the power supply is switched ON, the SMPS powers the DC motor and the Buck converter. The DC motor gets actuated resulting in driving the pulley attached to the shaft of the DC motor. Due to this the belt attached to the pulley moves by driving the driven pulley. The conveyor movement is achieved by fixing the pulley with pillow block bearings. The other output connected to the buck converter gets input of 12 V, 5 A supply. The pillow block, pulley and DC motor used are shown in Fig. 3.

An acrylic rectangular flap is fixed to the servo motor in order to push the objects into the respective slots, these flaps move as the servo motor is actuated to an angle of 90°. There are six such servo motors with flaps as shown in Fig. 4, are fitted in the frame in order to segregate three different colours and two different shapes.

The arrangement of conveyor belt (PVC) with pillow block with pulleys, dc motor with frames is shown in below Fig. 5.

The DC–DC power converter steps down voltage (while stepping up current) from its input (supply) to its output (load) which is adjusted to 5 V. It powers the Arduino Sensor shield which is connected with 6 servomotors fixed with Arduino Uno board. Servomotors are used for sorting the objects which consists of a flap made of acrylic. Camera which is fixed to a stand is connected using USB cable to the Laptop along with Arduino Uno in the other port of laptop Fig. 6.

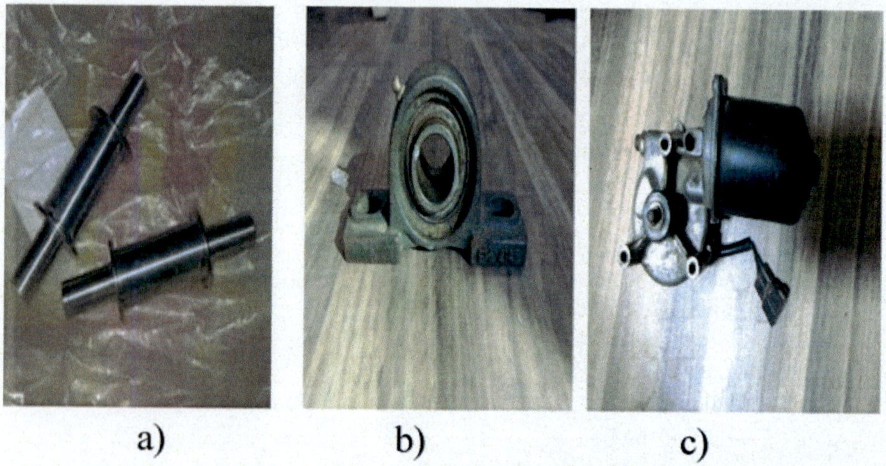

Fig. 3 Images of **a** pulley, **b** pillow block, **c** DC motor

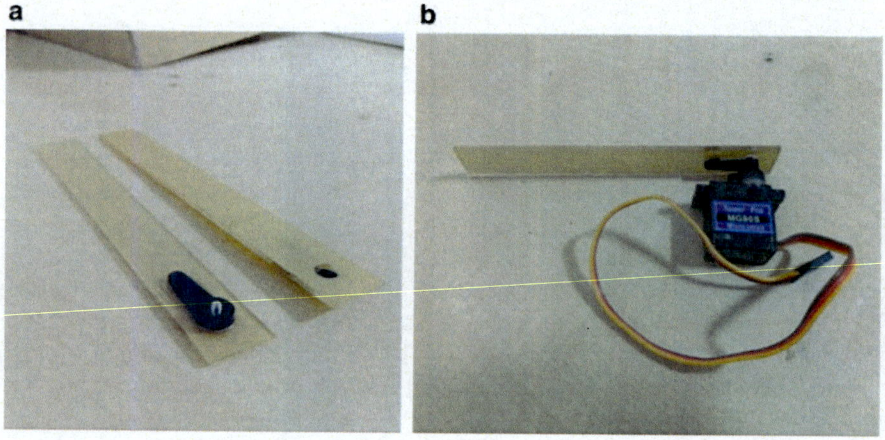

Fig. 4 **a** Acrylic flaps **b** servo motor fixed in acrylic flaps

After the power is switched ON, Python code using OpenCV for Image processing is made to run. Arduino code for servomotor actuation and serial communication with Python is uploaded.

The conveyor is switched ON and an object is placed for example: Red Square is placed manually on the belt. The camera senses the object Shape and colour using the contour edge detection for detecting shapes and masking technique for Colour detection using OpenCV platform in Python software [2–10]. Both the shape and colour values are combined giving a unique value. For every Object (shape colour) a value is assigned. Let us say the value for Red Square is 11. These are taken as serial values and are sent to Arduino using serial communication code. Using these serial

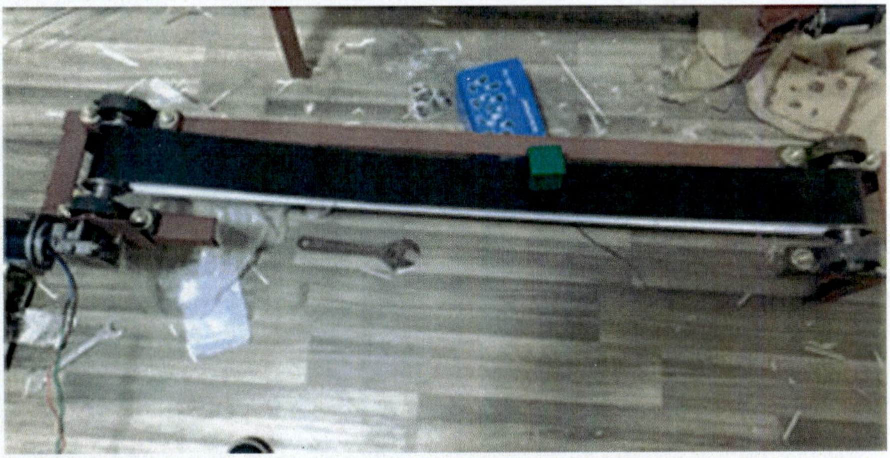

Fig. 5 Conveyor belt

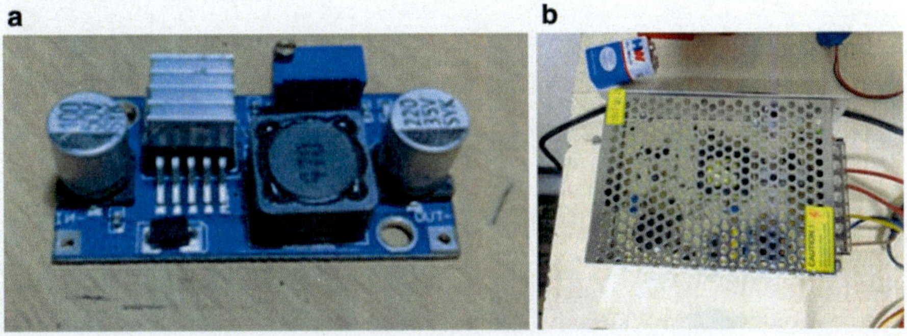

Fig. 6 **a** DC–DC Step down converter. **b** Switch mode power supply

codes obtained by the Arduino software from Python, based on the values obtained one of the six servo motors are actuated. For Red Square, Servo S1 is actuated and so the object moving on the belt is the flap of the actuated, at an angle and falls on one of the six slides and so the object gets sorted. This is the same for other objects but they get sorted by different servos and in different slides. Thus, these objects based on shape and colour are sorted using Object sorting conveyor Fig. 7.

2 Results and Discussion

Python is used for colour and shape sensing of the object, initially OpenCV 2, NumPy and Math from library is imported into python, then the camera is initialized after that while loop gets started with the frames initializing is done simultaneously using

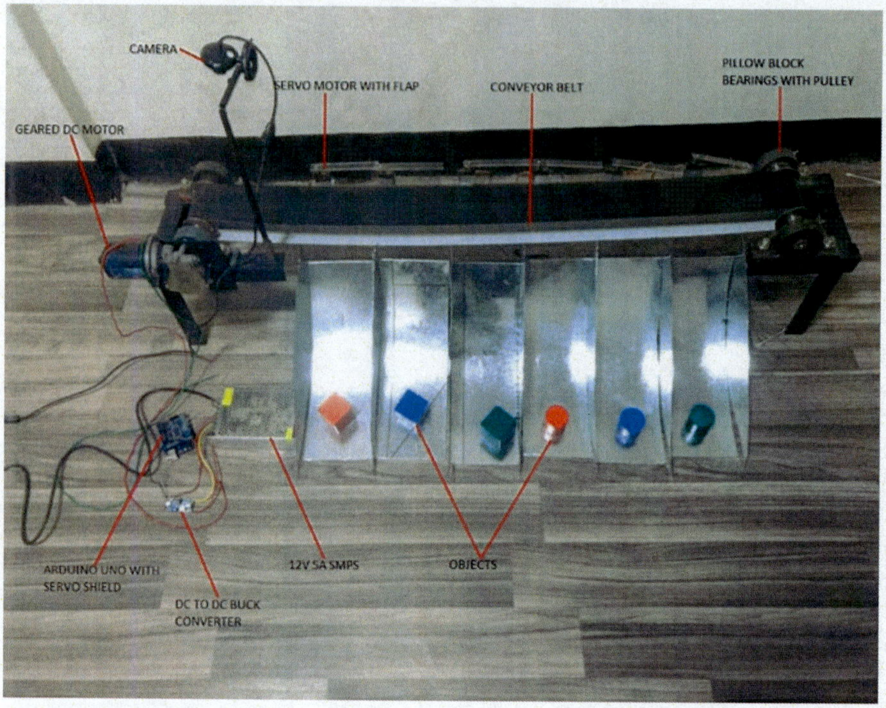

Fig. 7 Proposed object sorting conveyor system

the return value of camera. After this the HSV values which are obtained from the camera are converted to RGB colours.

The ranges of each colour with their higher and lower RGB values are mentioned. If the values is between (0, 50, 80) and (20, 255, 255) then it is RED colour. The values between (65, 100, 39) and (90, 255, 100) are GREEN colour and the values between (110, 100, 100) and (130, 255, 255) are BLUE colour. For shape sensing, edge detection and point mapping has been used to identify shape of the object. This code searches for edges of the objects and measures the length on each sides using contour mapping, finding the area [5, 6].

For circle by using the arc length it recognizes it as a circular object. Similarly, by area calculation and mapping, square shape is detected. This method can also be used to find different shapes such as rectangle, triangle, rhombus etc.

The real-time recognized images of objects are shown below. Figure 8 represents the coding by which RED colour with circle shape is sorted. The corresponding Flap gets activated and slides the selected object through the tray.

Similarly, in following figures, the detection of a blue colour square (in Fig. 9), green coloured square (in Fig. 10) and green coloured circle (in Fig. 11) are shown.

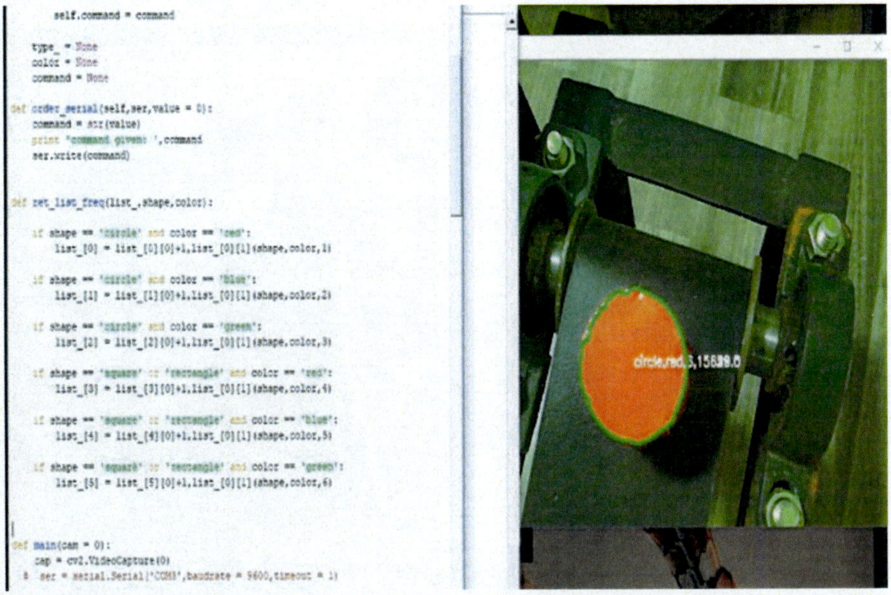

Fig. 8 Object detection: circle red

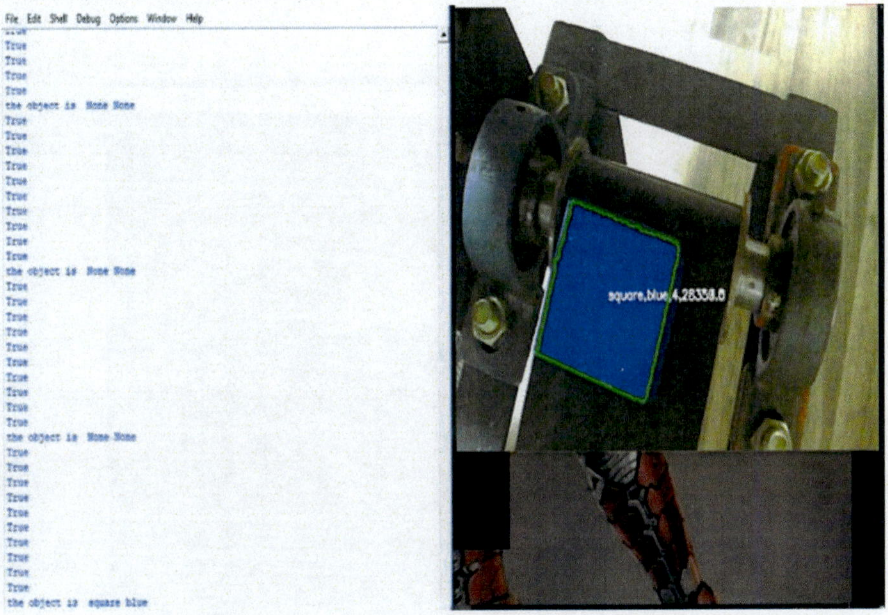

Fig. 9 Object detection: square blue

Fig. 10 Object detection: square green

Fig. 11 Object detection: circle green

3 Conclusion

From the experimental results conducted, thereby it is concluded that the object sorting conveyor which is used to sort the objects based on shape and colour is successfully accomplished and the objects are sorted and pushed into the respective slots. Since the objects are sorted on the basis of image processing, the accuracy level of the sorting process is comparatively high and chance of error is also less, the images of the objects are being captured by the camera and then it is being processed in Python and OpenCV and the colour is being identified within the given ranges. Accordingly, the servo motor flaps get actuated in order to push the objects into their respective slots.

References

1. Sheela S, Meghashree S, Monica L, Prathima A, Kumar SM (2016) Automation for sorting of objects using Raspberry Pi 3. Int J Adv Electron Comput Sci (Sep-2016):71–74. ISSN: 2393-2835
2. Kruglov AV, Chiryshev YV (2016) Detection and tracking of the objects in real time applied to the problem of the log volume estimation. AIST
3. Tho TP, Thinh NT (2015) Design and development of the sorting system based on robot. In: 2015 15th International conference on control, automation and systems (ICCAS), Busan, pp 1639–1644. https://doi.org/10.1109/ICCAS.2015.7364620
4. Pereira V, Fernandes VA, Sequeira J (2014) Low-cost object sorting robotic arm using Raspberry Pi. In: 2014 IEEE global humanitarian technology conference-South Asia Satellite (GHTC-SAS), pp 1–6
5. Kulkarni VA, Kale VR (2014) Automation of object sorting system using image processing
6. Parlak MU (2012) Camera based product counting of belt conveyors. Int J Electron Mech Mechatronics Eng 4(2):771–785
7. Krithika V, Subramani C (2020) Fuel enhancement of parallel hybrid electric two-wheeler motorcycle. Int J Electr Comput Eng (IJECE). 10:6180. https://doi.org/10.11591/ijece.v10i6.pp6180-6188
8. Krithika V, Chittal K, Nandhini M, Kumar MM, Balaji S, Kumar TR (2017) Design and fabrication of tidal energy harvester. In: 2017 IEEE international conference on power, control, signals and instrumentation engineering (ICPCSI), Chennai, pp 2542–2547. https://doi.org/10.1109/ICPCSI.2017.8392176
9. Krithika V, Subramani C, Sridharan R, Geetha A (2018) Multi bus DC–DC converter in electric hybrid vehicles. J Phys Conf Ser 1000:012105
10. Nandhini M, Krithika V, Chittal K (2017) Design of four pedal quadruped robot. pp 2548–2552. https://doi.org/10.1109/ICPCSI.2017.8392177

Design and Development of Waste Management System

M. Nandhini, V. Krithika, M. Mohamed Rabik, and V. Karthik

Abstract The goal of this proposal is to assess and develop effective schemes for integrated management of solid waste that is environmentally effective and economically affordable in order to eliminate adverse impacts on the environment and support the ecological and socio-economic development of the cross-border rural areas. This work will support the local authorities and stakeholders in establishing a dialogue platform for assessing and developing effective management schemes and providing efficient route mapping. It helps us to avoid dumpster overflow and thereby reduces pollution and accumulation of hazardous wastes. It provides easy management of solid waste and dumpster overflow tracking. The information of location where the dumpster fills is given to the municipal corporation where they analyze and give instruction to nearest dump truck drivers to clean the dumpster. So shortest route planning is done to reach the nearest dumpsters that are filled. This ensures that the dumpsters are not overflown and saves unnecessary time and money wastage that happen when full inspection are to be done. Design of a dumpster can be controlled by an Arduino Uno. The dumpster automatically opens the lid when the ultrasonic sensor that is placed on the top of the lid detects anything that is in its field of view and within its working range. This provides hand-free interaction with the bin while disposing the garbage. Another ultrasonic sensor that is placed inside the dumpster detects the level of trash in the bin. The distance value measured by the ultrasonic sensor is inversely proportional to the amount of trash in the dumpster, i.e., the ultrasonic sensor reads maximum distance value if the dumpster is empty as the trash increases the value measured decreases. If this value gets to a very low value, it implies that the dumpster is full and this triggers the GSM module to send a message to the municipal authorities regarding to clean the dumpster. This message also includes information about the location of the bin, time when it is filled, and the specific **number that is encoded to it.**

M. Nandhini (✉) · V. Krithika · M. Mohamed Rabik · V. Karthik
Department of Mechatronics Engineering, SRM Institute of Science and Technology, Kattankulathur, India
e-mail: nandhinm@srmist.edu.in

S. S. Dash et al. (eds.), *Sixth International Conference on Intelligent Computing and Applications*, Advances in Intelligent Systems and Computing 1369,
https://doi.org/10.1007/978-981-16-1335-7_48

Keywords Smart bin · Smart trash · Dumpster · Wastage collector · Garbage collector

1 Introduction

Over the decades, population has been increasing at a very faster rate. Many people are migrating from villages to the cities in pursuit of better living opportunities. But in urban cities due to over population and ineffective waste management, the garbage overflows from dumpster. Different types of diseases are spread via various breeds of germs, insects, bacteria, and viruses [1, 2]. The above-mentioned spread of disease is caused either because the dustbin is already overloaded and the waste is spilled out or it may be due to improper disposal of waste in the bin. All these may lead to the outspread of various virus-borne diseases. Cleanliness is the need of the hour for a smart urban lifestyle. In order to have a smarter urban lifestyle, different waste management strategies are required worldwide in accomplishment with the smarter technologies to enhance the environment to be much cleaner, greener, and healthier. This work illustrates a substitute strategy for waste management in order to solve the issue. This issue has raised owing to drastic increase in waste due to urban habitants. This alternate waste management strategy is achieved by proposing a bin designed with different sensors to monitor the waste level time to time in the bin. In other words, real-time waste management strategy is proposed for smarter city [3]. To achieve the time–time cleaning process, a threshold level is fixed and once this level is reached, an intimation is passed as a message to clean the bin. Hence, this strategy solves two issues simultaneously, such as cleaning the bin once it is overfilled or making the user to not less utilize the bin (which may also create bacteria borne diseases). In smart urban cities, the proposed strategy can serve as a benchmark for waste management [4].

The main issue that we face in urban cites is garbage overflow in dumpsters. The present waste management schemes are less efficient to manage the consistent increase in the waste level that are being generated [5]. The excess amounts of trash overflow from the dumpster and pollute the environment. This causes many health-related diseases and inconvenience for the surrounding people. Also, the garbage truck is used to go around the city and collect the garbage from every dumpster. This method is very inefficient since some dumpster may not be even fully filled or some already overfilled before the truck arrives [6]. This leads to waste of time and fuel in traveling. This work helps to eradicate or minimize the garbage disposal problem. The dumpster indicates the level of garbage and the level is accurately measured with the help of the sensing system. Once the dumpster is filled, it restricts the opening of the dumpster, also it sends message to the officials of municipality to collect the trash by using the microcontroller and it helps the truck driver to move in the direction of filled dustbins without wasting time in checking all the dustbins and by this the areas are segregated where more trash are collected so that the plastic and non-biodegradable wastes reduced [7]. Measures such as providing paper bags in areas where the trash accumulates more are a very good strategy to help in promoting the safety of the environment. These technologies can provide visibility on solid waste management, route planning for garbage collection, resource optimization, efficient asset management, efficient maintenance, visibility of waste bins, etc.

Few researchers have contributed research in wireless sensors networks (WSNs) to handle various waste management strategies. To achieve this, wireless monitoring unit (WMU) is initially installed in all the bins that are to be monitored. This data collected from the bins (either unfilled level or overfilled level of bin) is transmitted to WAPU. The data from WAPU in term is passed to central monitoring station via an application. A comparative analysis was made between various parameters between WMU and WAPU. A prominent waste management is identified from the results obtained for better implementation [5, 8–12].

2 Methodology and Work Flow

2.1 *Flow Diagram*

Dumpster is been equipped with two ultrasonic sensors, one Arduino Uno board, one GSM module, and an H-bridge. The flow diagram is shown in Fig. 1.

2.2 *Working of the Module*

When a person hand is detected by first ultrasonic sensor which is placed on the top of the bin, it makes the dustbin lid open through an actuator and H-bridge programmed via Arduino Uno. Approximately, the lid opens up for at least 20–30 s to dump the waste and once the waste is dumped the lid automatically closes. To avoid littering and to maintain a cleaner environment, the bin is designed with a closed lid. This helps to avoid arise of unnecessary problems. The sensor has a limited working range, within this range the sensors detects and gives output to Arduino Uno and thereby opening and closing the lid.

The second ultrasonic sensor is placed inside the dustbin, which detects the level of trash. Once the lid of the bin closes, the ultrasound range sensors get activated only once, and it triggers a pulse which then reflects back containing the information regarding the filled level. An ultrasonic sensor is placed on the lid, such that the sensor faces the waste. As trash increases, the distance between the sensor and the former reduces. The sensor is activated once the lid is closed. This in turn generates a pulse which acts as a feedback trigger pulse consisting of the required information (filled level). The sensor is in the sleep mode in the remaining duration. In other words, the sensor does not get activated if the lid is not open. The design calculation used for the conversion of distance to fill percentage is shown below.

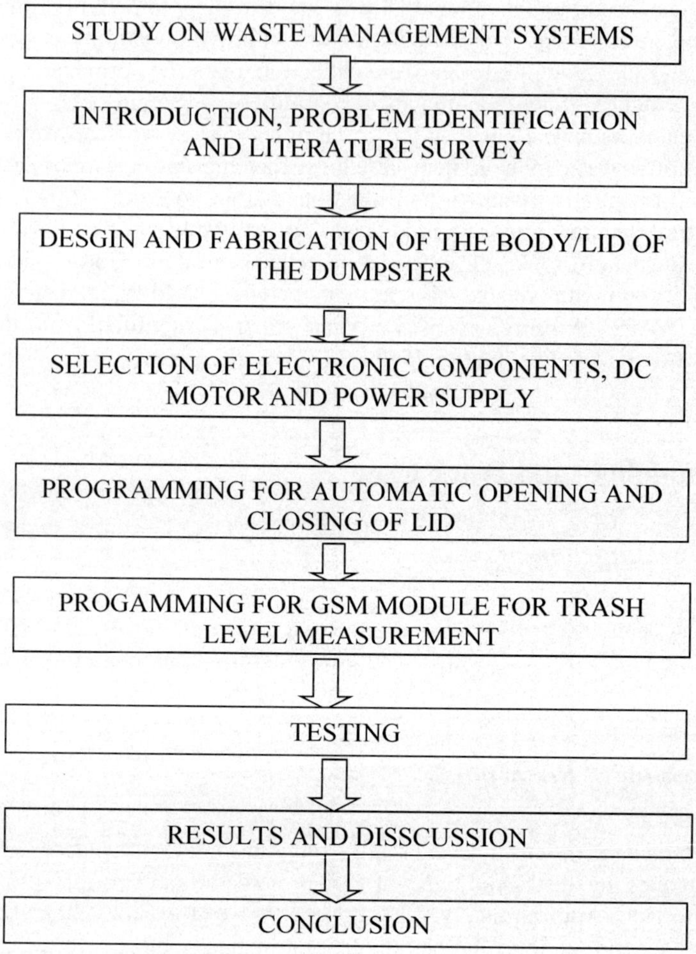

Fig. 1 Work flow of waste management system

$$v_t = v_0 + C \times T \tag{1}$$

where

- v_t velocity of sound w.r.t temperature,
- v_0 velocity of sound at 0 °C,
- T Temperature,
- C rate of change of velocity with per degree rise in Temperature.

$$\text{Distance } (d) = v_t \times \left(\frac{\text{Time}}{2}\right) \tag{2}$$

where

Time ping time from sensor

$$\text{Fill Percentage} = 100-(100/h) \times d \tag{3}$$

where

h height of the bin

The microcontroller receives live data. Once it reaches certain limit as given in the program, a message will be sent to Municipal Corporation (concerned authorities) through GSM module which is connected with Arduino Uno, i.e., in case the fill percentage attains a threshold level of 85–90% which is a variable one decided according to the need. The fill percentage level as mentioned earlier is a variable factor which can be changed accordingly in the program, after which permanent closure of lid occurs. This is to avoid excessive littering and overflow in the bins.

The bin after permanent closure can be opened through a proximity sensor. The proximity sensor is none other than an employee assigned with a unique code. For example, if the unique code assigned to employee A is 1010, then the sensor should be capable enough to sense the digital values 1010 within prescribed time interval. As the input fed to the sensor is a digital input, there is only two probabilities either true or false. In case of true value, the lid opens and the cleaning is done. The final stage is to close the lid which is achieved by making the proximity sensor to read a value of 1 for 5 s. Once it is achieved, the lid closes and simultaneously range sensor is activated which in turn starts measuring the fill percentage. Until the next fill cycle, the bin is now ready for reuse. Thus, trash is disposed time to time without overflowing. This can be used in smart cities for IoT ecosystem.

3 Comparison with Previous Works

The current dumpsters that are used around the city does not have any control or monitoring system. These can only monitor manually via the authorities that are appointed by the municipal board. This process requires more human work and time and money consuming. There is also another method for monitoring implemented by BioEnable [5]. They are using separate control system for Wi-Fi networks. If the network is not proper, they will not give accurate results and some personals are required to monitor over the systems.

This work consists of two ultrasonic sensors; one for opening of the lid and another for monitoring the trash. In this method, no human monitoring is required as the sensors are used for that work. They open and close automatically in the presence of a person, and after the dumpster is full, it does not open for further

Fig. 2 Operation flow

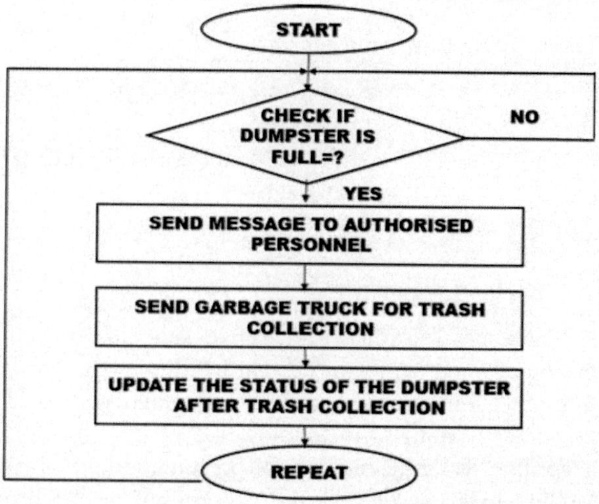

use and sends message to the municipal authorities for the waste disposal process thus ensuring regular maintenance and cleanliness of the environment. To avoid overfilling of garbage in dumpsters, thereby providing for a clean, healthier, and safer environment.

4 Operation of the System

This is operational diagram for sensing the level of the trash in the dumpster. First the sensor inside the bin detects the level of waste. As the waste level in the dumpster increases, the distance between the trash and the ultrasonic sensor inside the sensor decreases, thereby checking if the dumpster is full. If the dumpster is full, it sends a message to the authorized personnel via the GSM module. After cleaning of the dumpster by the personnel, the status of bin is updated and continuous monitoring starts again. Figure 2 represents the operational flow of the setup.

5 Experimental Setup

The work is divided into two parts: the base and the lid. The project was entirely made of plywood because of ease of construction, robust and easy to work on and can be dismantled into several parts for easily assembling the electronic components inside the dumpster. The lid contains of two vertical plates. In one end of the vertical plate, an ultrasonic sensor is mounted. The ultrasonic sensor can measure a maximum distance that is less than the distance of separation of the two plates. This is the working range of the senor. Within this working range, if it detects an obstruction, it opens the slider. The slider is attached on rails below the top surface of the lid. The rails give mechanical support to the slider and also ensure smooth movement of the sider. The bottom surface of the slider consists of a rack. In this rack, a pinion rotates by the help of a motor, whose shaft is attached to the pinion. This is the topmost construction of the project and helps in automatic opening and closing of the lid.

The overall system is shown in Fig. 3. The coding is done by using the Aurdino software. We have used library functions in our coding. The library functions such as "NewPing" and "SerialSoftware" are used. The GSM and the ultrasonic were interfaced such that if the ultrasonic detects the trash in the dumpster it commands the GSM to send a message to the designated number. The other ultrasonic sensor through the commands given senses a person's hands and opens the lid of the dustbin

a) Base/Bin

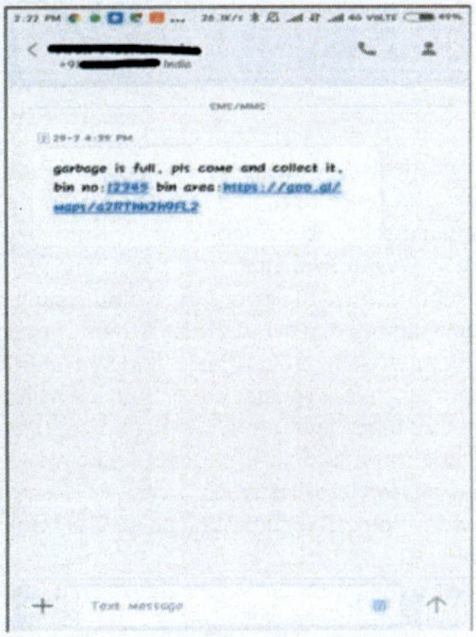

b) Notification to the user

Fig. 3 **a** and **b** waste management system

through a motor actuation and also closes the lid through the reverse actuation. We have used conditional statements to run the codes continually and thereby sensing the trash level at all times. It leaves the loop once the trash gets filled in the dumpster.

6 Conclusion

Waste management can be well-defined as the collection, processing, and disposal of materials considered as garbage. To monitor and solve the issues in large scale, conventional waste management strategies are inefficient. It is high time that recent advancements in technology have to be incorporated in waste management system to achieve a smart city and also to have a control over the consistent increase of waste. The alternative waste management strategy proposed effectively uses technology to develop a strategy much smarter and far more efficient than the existing one. This work aids in waste management by monitoring of the dumpsters. Various debates have been done to overcome the disaster made due to the waste dumped in the environment. Owing to various factors such as economic background of people, climatic conditions, and areas, it would ease the work if the waste management is concentrated properly in cities. This will produce a great result for a better environment, to protect the environment from waste overfilling for reducing the work on local authorities to monitor the waste and effectively resolve the issue of waste collection. By this work, the waste collection and monitoring methods are made simple, and if followed correctly, it ensures the dumpsters are not overflown.

References

1. Nightingale F (1954) Selected writings of Florence Nightingale. In: Seymer LR (ed) The Macmillan Co., New York, pp 38287
2. Herbert L (2007) Centenary history of waste and waste managers in London and South East England. Chartered Inst Wastes Manage
3. Chadwick E (1842) Report...from the Poor Law Commissioners on an inquiry into the sanitary conditions of the labouring population of Great Britain. London, pp 369–372. Via Bearing in mind the complexity of nature and cities, the ecological conditions of urban areas are viewed as the necessary measures for environmental protection and recovery. Land, water, vegetation and other living organisms fill cities and interact with people. In consequence their protection and care reflect upon the quality of human life. Laura Del Col (11 Oct 2002). Chadwick's Report on Sanitary Conditions. The Victorian Web
4. National Waste and Recycling Association History of solid waste management. Washington, D.C. Retrieved 09 Dec 2013
5. United nations statistics division-environment statistics. unstats.un.org. Retrieved 3 Mar 2017
6. BIOENABL. https://www.bioenabletech.com/
7. IFC (2014) Editorial board/aims and scope. Waste Manage 34(3). https://doi.org/10.1016/S0956-053X(14)00026-9
8. Davidson G (2011) Waste management practices: literature review (PDF). Dalhousie University-office of sustainability. Retrieved 3 Mar 2017

9. Handbook of solid waste management and waste minimization technologies, pp 337–465 (2003)
10. Albert R (4 Aug 2011) The proper care and use of a garbage disposal. Disposal Mag. Retrieved 03 Mar 2017
11. United Nations Environmental Programme (2013) Guidelines for national waste management strategies moving from challenges to opportunities (PDF). ISBN 978-92-807-3333-4
12. Barbalace RC (2003) The history of waste. EnvironmentalChemistry.com. Retrieved 09 Dec 2013

Design and Implementation of an Intelligent Health Monitoring System for Variable Frequency Drive in Manufacturing Industry

M. Nirmala, R. Latha, and G. Thenmozhi

Abstract Variable frequency drives (VFD) are acquiring popularity in the effective way of controlling the high-power industrial motor drives. The bewildering failure of these drives may affect the motor performance, and to build unscheduled downtime in the industry, it leads to high loss of production cost. Fault detected at the earlier stage at any part of the three-phase induction motor may protect the machine from catastrophic failure and also result in protection of the system from complete shutdown. In this paper, an intelligent health monitoring system of an induction motor has been developed using measurement of ambient temperature, relative humidity, temperature at heat sink, speed of the cooling fan, the cable length, input voltage and output current to monitor external parameters such as voltage, current and temperature and to reduce the downtime of an industry by monitoring the drive externally. The initial phase fault exposure plays an important role in cataclysmal anticipatory maintenance of three-phase induction motor. To monitor the overall health of the induction motor with the all-parameter, settings of several parts in the machine are composed. The effectiveness of proposed system is verified by both in MATLAB/Simulink software and implementation of real-time hardware setup with a PIC microcontroller based 2.2 kW motor drive in manufacturing industry.

M. Nirmala (✉)
Department of Electrical and Electronics Engineering, Kumaraguru College of Technology, Coimbatore 641049, India
e-mail: nirmala.m.eee@kct.ac.in

R. Latha
Department of Electrical and Electronics Engineering, PSG College of Technology, Coimbatore 641049, India
e-mail: rla.eee@psgtech.ac.in

G. Thenmozhi
Department of Automobile Engineering, Kumaraguru College of Technology, Coimbatore 641049, India
e-mail: thenmozhi.g.auto@kct.ac.in

S. S. Dash et al. (eds.), *Sixth International Conference on Intelligent Computing and Applications*, Advances in Intelligent Systems and Computing 1369,
https://doi.org/10.1007/978-981-16-1335-7_49

Keywords Variable Frequency Drive(VFD) · Frequency Converter(FC) · Pulse Width Modulation(PWM) · Voltage Source Inverter (VSI) · Current Source Inverter (CSI) · Graphical User Interface (GUI)

1 Introduction

Power electronics has gone through a hurried technical expansion in the recent years. This is owing to the technical advancement in drive system mainly due to arrival of high-quality power electronic switches and advanced microcontrollers. The main emphasis of controlling the AC motor by variable frequency drives is possible to always operate in the area to the right of the peak torque point. The drive could be operated at maximum efficiency, maximum torque per ampere and inherent stable operation and independent of speed. Figure 1 shows the key components diagram of variable frequency drive.

During operation, the motor gets trips off from the drive due to fault or malfunction in the system. A monitoring system does not encompass with control or protection system. Signals allied with protection system may be monitored to trip the drives when exceeding it limits. Protection system is intended to protect the device from consequence of its internal faults, and this system has its own limit. To identify faults, bunches of strategies have been concentrated to enhance the unwavering quality and strength of the motor drive. Condition monitoring is one such strategy in which 24 × 7 constantly screen and diagnostics the variable frequency drive and that guarantee quick reaction time with better administration [1, 2]. Condition monitoring can also be done through online and can easily predict the faults originated in motor drive for harsh industrial application [3, 4].

An autonomous state of monitoring the motor drive through online can able to control the drive from remote location, condition monitoring of current in the capacitor during fault condition [5]. Software is linked together with sensors and other

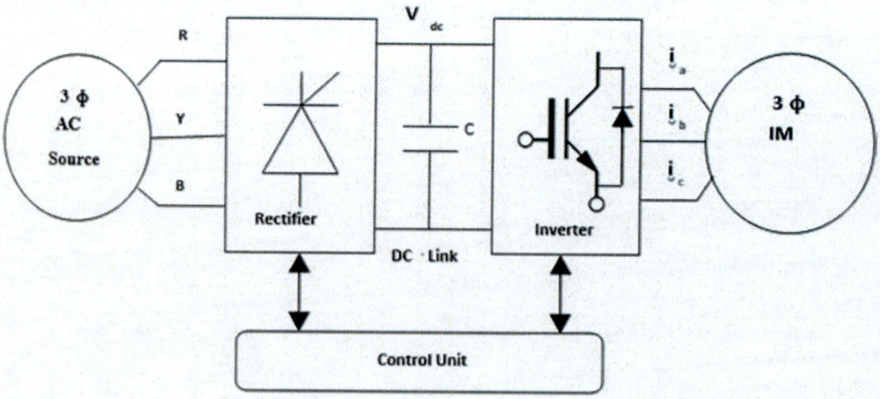

Fig. 1 Key components diagram of variable frequency drive

instruments needed to ensure the monitoring platform for different user applications. Additional sensors are required to monitor the air handling units to crisscross the deposition of dust or unwanted materials in their flow. Particularly, textile industries fault may get arise due to accumulation of fine dust particles in the air handling unit [6]. Fault may arise due to failure in drive operation or lack of sustained load. So that motor gets disturbed and could produce the fault current in immeasurable form. It leads to injuring its coils, insulation and bearing of the motor [7, 8]. Inverter used in the motor control consists of freewheeling diode and get back the current generated from the motor during regenerative operation. Due to fast switching operation of power electronics devices, insulation in the winding gets damage due to the high stress [9]. Mechanical disturbance in the shaft of the motor may also create certain faults and collapse the entire system during close loop operation of the motor drive [10, 11]. Motor health state can be monitored and measured by using different analysis to predict the faults. Impedance analysis is one such analysis which can able to monitor the motor at standstill [12, 13]. The intention of the proposed method is to monitor the external parameters which affect the performance of the drive and create unnecessary actions. By monitoring the parameters, it is easy to protect and prevent the device from impairment or mutilation.

1.1 Variable Frequency Drive

1.1.1 Rectification Module

The proposed method is novel technique of monitoring the external parameters of the variable frequency drive, implemented in MATLAB/Simulink platform and Arduino controller which helps to access the real-time problem into the Simulink environment and can able to trips the drive during transient conditions. Once the proposed system is implemented in the drive equipment, the drive located in the remote area can be easily predicted and controlled. This method provides a cost-effective solution to access the real time data and can prevent the drive during fault condition. The drive operation and performance are get reduced by the external parameters like temperature from atmosphere (ambient temperature), heat dissipated in the system module (IGBT—inverter and diode bridge rectifier), humidity, vibrations, lose connection and so on. By using microcontroller, it is easy to interface the real-time input to the software platform. In this system, the Arduino controller is used, and it takes the real-time temperature input to the MATLAB/Simulink environment. Depending on the value obtained from the sensor, the inverter can be able to send the input current to the motor. By this method, the user can be able to protect motor along with its drive when the external parameters get crosses its threshold limits. Figure 2 shows the overall block diagram of the proposed intelligent health monitoring system.

The first part of the converter is supplied with a rectification unit. The input AC power is converted to DC power output. The obtained DC power is pulsating in nature and is converted to a steady (VDC) operation by means of an intermediate system. For

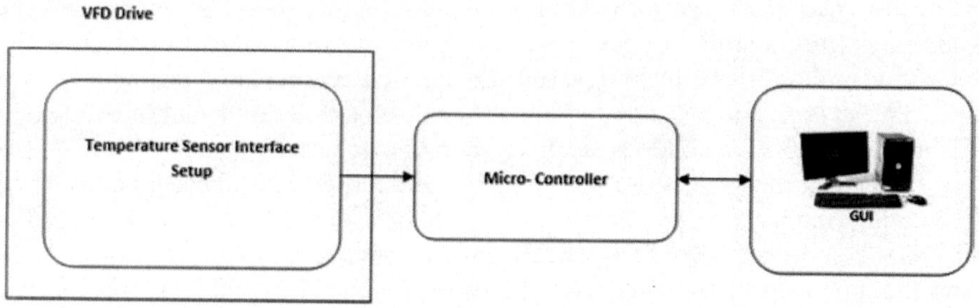

Fig. 2 Block diagram of the proposed health monitoring system

Table 1 Different types of switches and their parameters for rectification module

Parameters	SCR	Diode (12 pulse)	Diode with filter circuit	Active front end converter
THD	<35%	<15%	<10%	<5%
PF	~0.8	~0.8	~0.9	~0.95
Efficiency	>95%	>95%	>95%	>98%
Cost	Low	Medium	High	Very high
Operating efficiency	No impact	Medium	Good	Good

rectification module, industries may use diodes, SCR—silicon-controlled rectifiers or active front-end converter to convert the AC to DC power. Each has its own advantages and disadvantages. Table 1 shows the comparison overview of types of switches used in the rectification unit.

On comparing Si or Ge, semiconductors like silicon carbide—SiC and gallium arsenic (GaAr) will have shorter lifetime. Shorter lifetime implies quick switching capacity. The main objective of converting AC to DC and DC to AC again is to reduce the input disturbance in the power supply such as Harmonics, voltage sag and swell, clamping, transients, etc., and also the output obtained can differ in magnitude and frequency compared to the AC input utility. When input AC is changed over into DC implies, user can change over the DC to AC for their reliance.

1.1.2 Intermediate Module

Generally, the converter without DC link is not ideal, since it contains pulsating DC voltage. Rather than not keep up the constant DC voltage at the inverter input, it will break down its execution. The primary purpose of the intermediate circuit is to detach the rectifier from the inverter and also to reduce the interference of the input. Variable frequency drive is made up of two intermediate circuit types. The intermediate circuit is constant and the intermediate circuit variable. The variable intermediate circuits contain only inductor or play a role for some time with DC-DC converter. On the other hand, constant intermediate circuits are linked in parallel to the inductor and capacitor. The variable frequency drive comes with or without a DC link circuit in an industry. In contrast to these two converters, the DC link converter is generally the best. Variable type intermediate circuit has one inductor or also named as DC Choke. Another type of circuit is the removal of the DC choke/inductor by the replacement of a DC-DC. The converter input is pulsating DC, and the output is obtained as a steady/constant DC voltage. Buck, Boost, Buck-Boost are the type of converters used to keep constant voltage, and it depends purely on our use. The output can be controlled by controlling the switch's gate pulse. Then, the second most common type in the frequency converter called the intermediate circuit of constant. This type has two inductors with a single parallel capacitor or a single inductor parallel with capacitor. Although the capacitor has a minimum lifecycle, it has the pulsating DC to constant DC smoothing.

1.1.3 Inverter Module

Inverter is a last stage converter, changing over the constant DC to AC voltage, and it can be ready to buck or boost the voltage as per necessity. The function of the inverter is to change the DC voltage into AC with wanted extent. The output voltage of a perfect DC-AC inverter is ought to be sinusoidal. Be that as it may, the waveforms of the inverter are non-sinusoidal in nature and contain certain harmonics.

Single stage inverter is ordinarily for low-power applications and where three-stage inverters for high-power applications. The switching rate of the inverter can be varied which relies upon the application utilized. On the off chance that the yield of the inverter is associated with three-stage motor stack, the revolution of the motor can be ready to change its headings by changing its input phase sequence. The conduction method of the inverter is typically in 180° or 120° conduction. Figure 3 demonstrates that the circuit diagram of three-phase voltage source inverter with free-wheeling diodes.

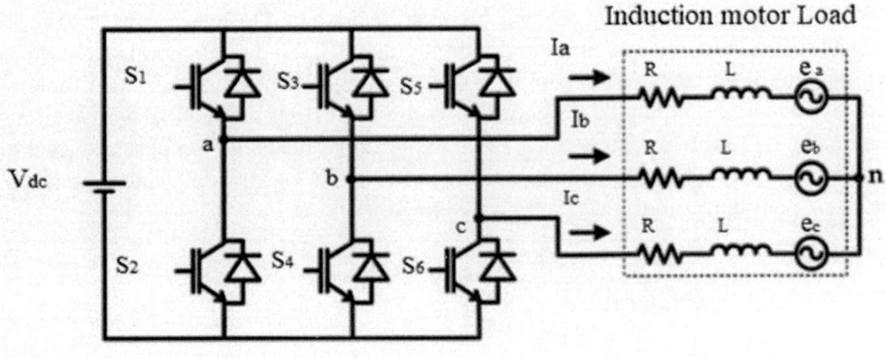

Fig. 3 Three-phase voltage source inverter with free-wheeling diodes

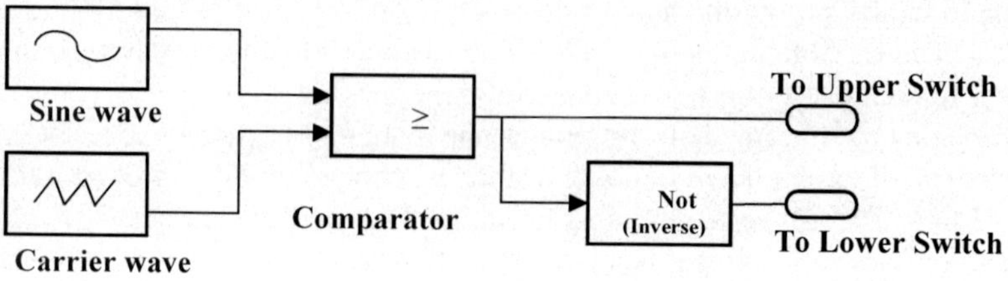

Fig. 4 SPWM switching topology for voltage source inverter

Voltage source inverter is equipped with six switches and three leg topologies. However, three switches stay on at any time. Current source inverter contains six thyristors, diodes and operating capacitor. The diodes need to prevent the load from discharging the capacitor. The use of thyristor in the current inverter type has reverse blocking capacity, and this type of inverter can operate in inherent four-quadrant operation without additional power. The obtained output can vary its pulse width by changing the modulation index. The most helpful topology is PWM—pulse width modulation. By changing the width of the pulse, it can perform variable voltage and variable frequency of the inverter. There are a few PWM strategies, the vital ones are as follows: sine PWM, space vector PWM, hysteresis current control based PWM.

Figure 4 shows that the SPWM-based switching topology. The topology utilized in this undertaking is SPWM—sine pulse width modulation.

1.2 External Monitoring System Parameter

A variable frequency drive includes a rectifier, a DC link (intermediate circuit) and an inverter together with its control unit. During operation, energy is released like heat. In addition, external parameters form this drive equipment and create an unplanned downtime in the industry. The list of external VFD parameters for ambient temperature and relative humidity. Some problems may arise due to internal disturbances such as excess heat release, input and output voltage, cable length, cooling unit and external motor vibration. The main goal of this paper is to monitor these parameters and to reduce the unplanned downtime thereby protecting the device against transient or fault. Ambient temperature, relative humidity, heat sink temperature, brake resistance, input and output voltage, cable and its length, cooling units and vibrations are some parameters in the variable frequency drive.

1.2.1 Ambient Temperature

Every single VFD has its very own base and most extreme temperature limits. To boost the framework security, it is smarter to stay away from extraordinary surrounding temperature, and subsequently, in this way, it delays the life of the equipment's. As a standard guideline, the lifetime of electronic parts diminishes by half for each 10 °C when worked over its plan temperature. Indeed, even gadgets with high assurance evaluations must cling to the predetermined surrounding temperature ranges. Extra cooling of the bureau or establishment site might be required. So, it is mandatory to monitor the ambient temperature to safeguard the devices inside the system.

1.2.2 Relative Humidity

VFD is structured so that can be ready to work legitimately at high moistness (up to 95%). There is an explicit danger of buildup when the recurrence converter is colder than wet encompassing air. Dampness noticeable all around can likewise consolidate on the electronic parts and cause short circuits. So, it is worthwhile to install a heater attached to VFD operated in highly moisture or cooled areas. It is better to avoid installation of VFD in areas subject to frost. Ambient temperature and relative humidity are external monitoring parameters, and remaining things explained below are internal disturbance created due to overrating of its limits.

1.2.3 Heat Sink

Each device in the VFD requires heat sink for its operation. Occasionally, heat sink get overheated due to over rise in ambient temperature or excess liberation of heat form device in the VFD's like rectifier or inverter. Sometimes, it is obligatory to monitor the heat sink. Obstructions in the air flow or cooling fan failure, dust or dirty particle deposited on the heat sink may fail to liberate the heat from rectifier or inverter. So, it is predominant to monitor the temperature in the heat sink. Certain sensors are needed to monitor the heat sink and give the control signal for operation.

1.2.4 Brake Resistor

Brake resistor is introduced in the middle of the inverter unit, and reason for existing is to oppose the turnaround power spill out of inverter to rectifier. It contains a resistor, and it assumes an essential job in opposing the reverse power. So as to oppose the turnaround power, the rating of the resistor ought to have high. Before associating a resister, the accompanying variables should have been considered: 1. The most extreme constant power rating for the resistor(s) 2. The rate at which the resistor changes temperature (resistor warm time steady) 3. Most extreme braking time ON—If the recovery (braking) control is larger than the resistor appraised control, and there is a limit to the ON time, or the resistor overheats before the OFF time frame starts. 4. Least braking time OFF—If the recovery (braking) control is bigger than the resistor evaluated control, the OFF time must be sufficiently expansive for the resistor to cool between ON periods.

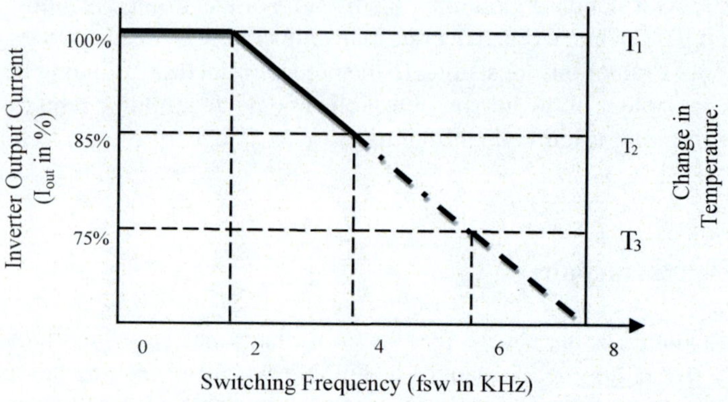

Fig. 5 Change ininverter output current with change in temperature and switching frequency

1.2.5 Input Voltage and Output Current

The VFD can able to withstand ± (10%) of the input voltage fluctuation and ± (3%) of the phase to phase imbalance. It is seen that variation in the input voltage for three-phase AC drives has variation over (380–550) V and may have ± (2%) input frequency variation. Figure 5 shows the graph for variation in the inverter output voltage due to change in temperature along with switching frequency. It is clearly seen that temperature inside the drive gets changes which may get result in reducing the output current entering into the motor. When the motor is not supplied with rated current for prolonged time, it may get damages . So, it is important to monitor the input voltage and the output current.

2 Results and Discussion

A series of MATLAB simulations were applied to verify the outputs made for the quantitative theoretical analysis. The different types of temperature sensor used in the proposed system. In this proposed system, the temperature sensors are interface with MATLAB/Simulink platform for experimental analysis using sensor interface module. The sensor interface module, it provides a connection between the temperature value taken from the different parts in the VFD's and interface with microcontroller for real-time data acquisition. The experimental setup includes temperature sensor unit along with Arduino microcontroller with a PC with MATLAB software. Figure 6 shows that the complete hardware setup of the proposed monitoring system in laboratory environment.

Figure 8 demonstrates that the SPWM switching waveform for the proposed system. The gate pulse for the six switches in the three-phase voltage source inverter

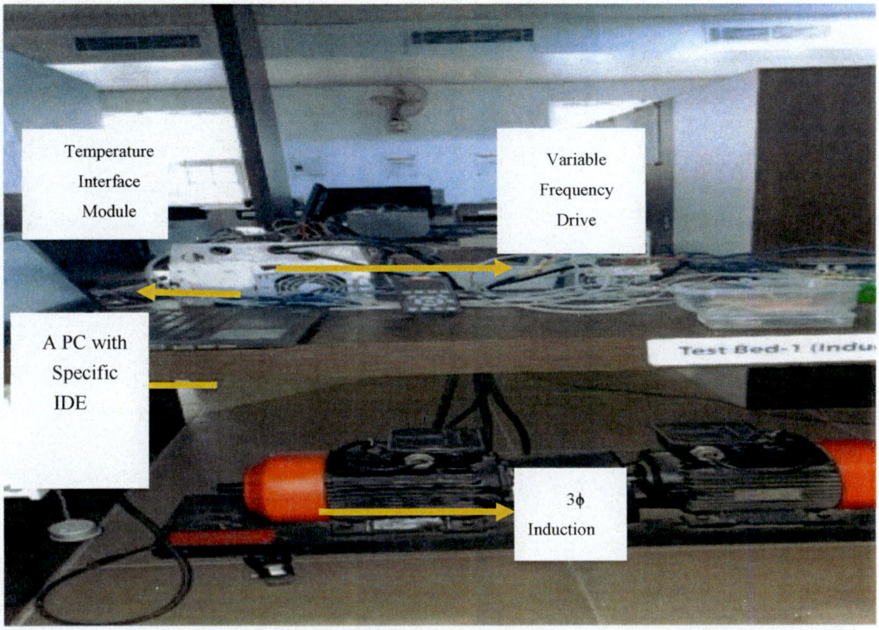

Fig. 6 Hardware setup of the proposed monitoring system in laboratory environment

is depending on the real-time sensor input connected to the interface module. The technical specifications of the Arduino UNO microcontroller are listed in Table 2.

From the specifications, among 14 digital I/O pins, 6 pins are provided with PWM outputs and has 32 kB of flash memory. Figure 9 shows the input current waveform for the motor load. The result of these non-sinusoidal waveforms shows the absence of filter circuit in the inverter. Current waveform gets changes depend on the real time temperature values. Microcontroller is programmed in such a way that during operation in sensing the real-time temperatures, on exceeding it limits, the current entering to the motor should get turned off. On the other hand, the microcontroller needs to trip the inverter from the motor load. Tripping can be done in MATLAB/Simulink model.

Figure 10 shows that the motor speed and torque when the necessary current is supplied to it. In the initial stage, motor possesses high torque and has huge oscillations from (0–0.5) s. The proposed methodology is implemented by interfacing more than one sensor to acquire the real-time temperature and humidity. Humidity and temperature sensor can be easily interface with any microcontrollers like Arduino and Raspberry Pi. These sensors possess low cost and provides high reliability and long-term stability. The sensor senses the temperature and humidity from the variable frequency drive and interfaces with PC. Form this PC, either the user can send the measured temperature to place in remote location or stored for lateral analytical use. Here, the sensor readings are taken for drive environment and protect the inverter form simulink environment.

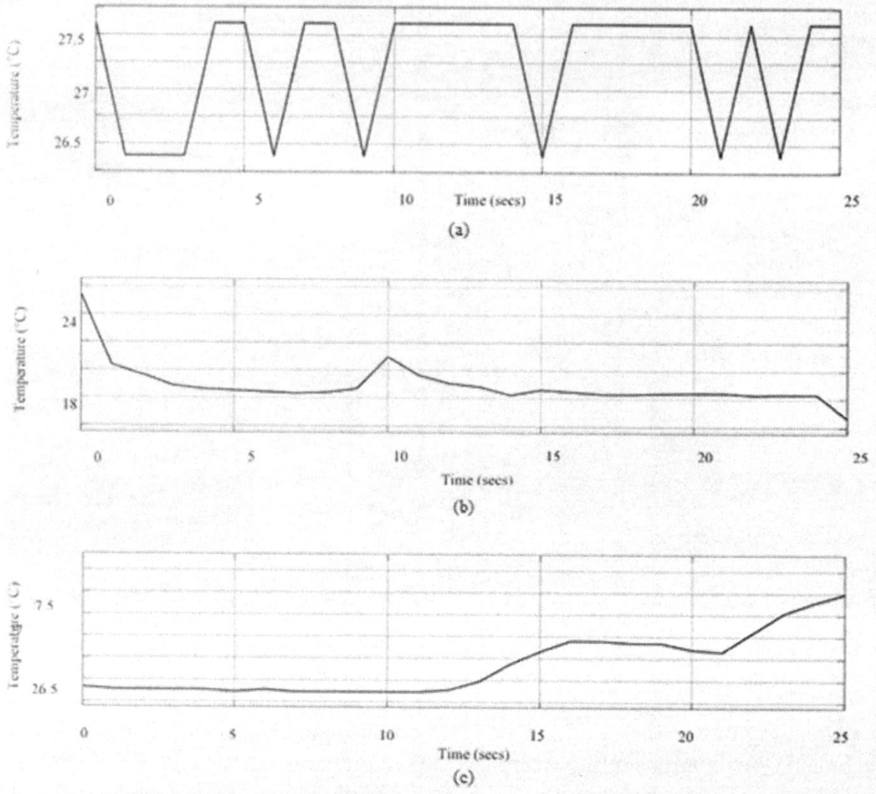

Fig. 7 Real-time sensor output waveforms. **a** From LM-35 temperature sensor. **b** From DHT11 temperature and humidity sensor. **c** From thermistor

3 Conclusion

Sinusoidal pulse width modulation-based inverter is designed by taking real-time temperature input and simulated using MATLAB/Simulink software. Arduino board which helps to take the temperature input to MATLAB environment and depends on the range of the input the Simulink shows the necessary waveforms. The data which is taken from the sensor is for diagnostic and monitoring purpose which is discussed in the report. The information which is at no extra cost and a little hardware investment. This paper has focused on the technique for processing the data and further analysis. Fluctuation of temperature at the fundamental frequency and unbalances among power devices can be especially important in the drive which must be operated the drive under heavy load condition. Difference between the temperatures from average to actual value which could eventually jeopardize the integrity of the devices in the power module. This methodology leads to the storage of the measured value for data analysis and the estimation of the fault that would occur in the future. In this

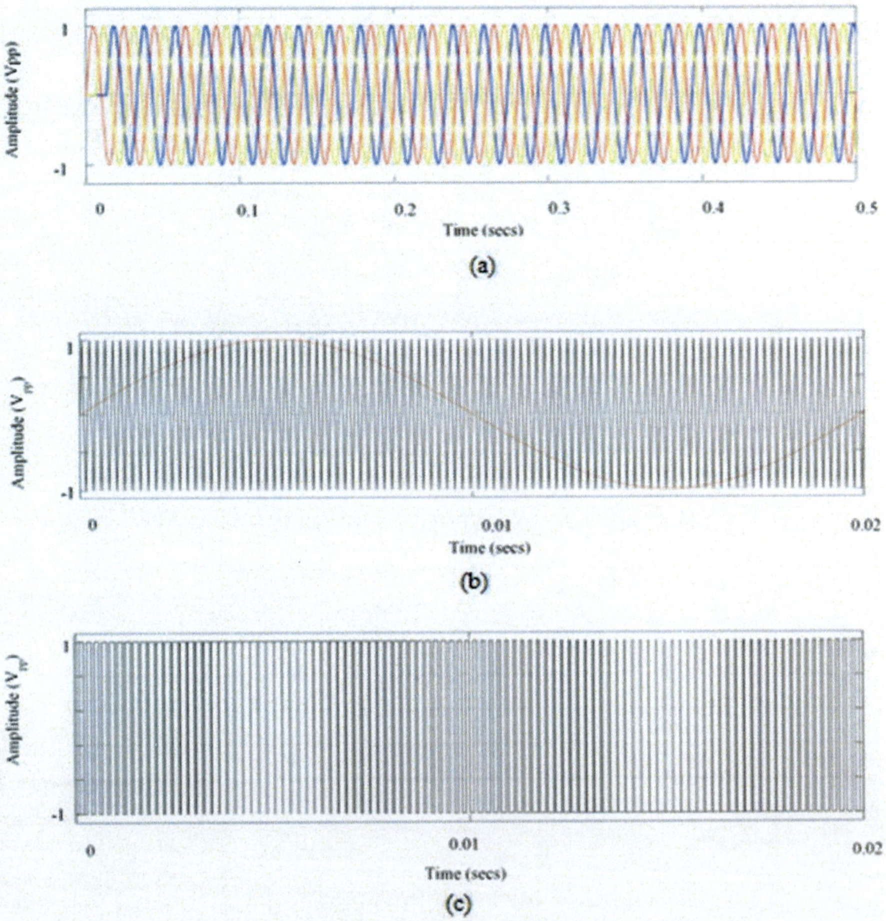

Fig. 8 SPWM technique. **a** Three-phase sine wave. **b** Formation of SPWM pulse with comparison of sine and triangle wave. **c** SPWM pulse for an inverter circuit

Table 2 Technical specifications of controller VFD drive

Microcontroller	UNO REV3
Operating voltage	5 V
Input voltage limit	7–20 V
Digital I/O	14 (6 pins provide PWM output)
Analog I/O	6
DC current per I/O	20 mA
Clock speed	16 MHz
Flash memory	32 kB
SRAM	2 kB
EEPROM	1 kB

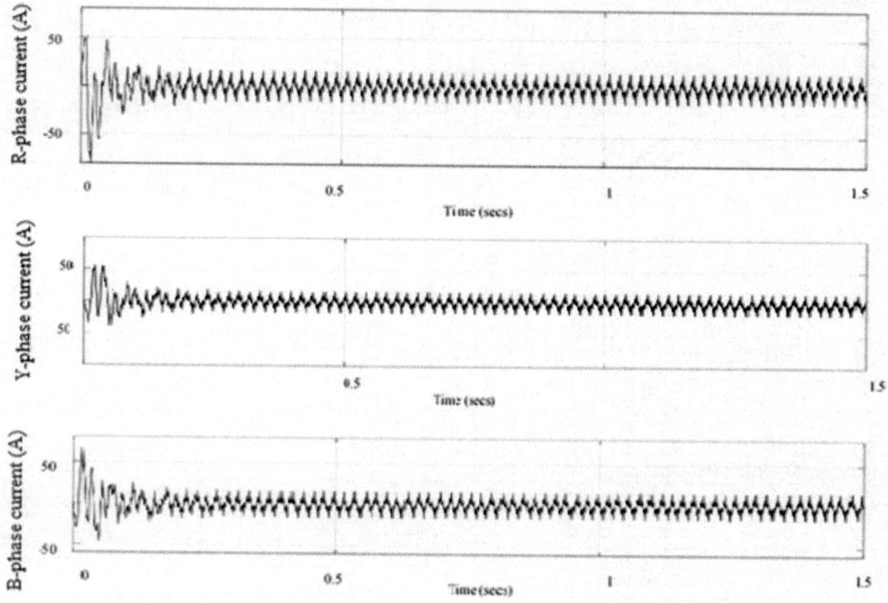

Fig. 9 Motor input current without filter

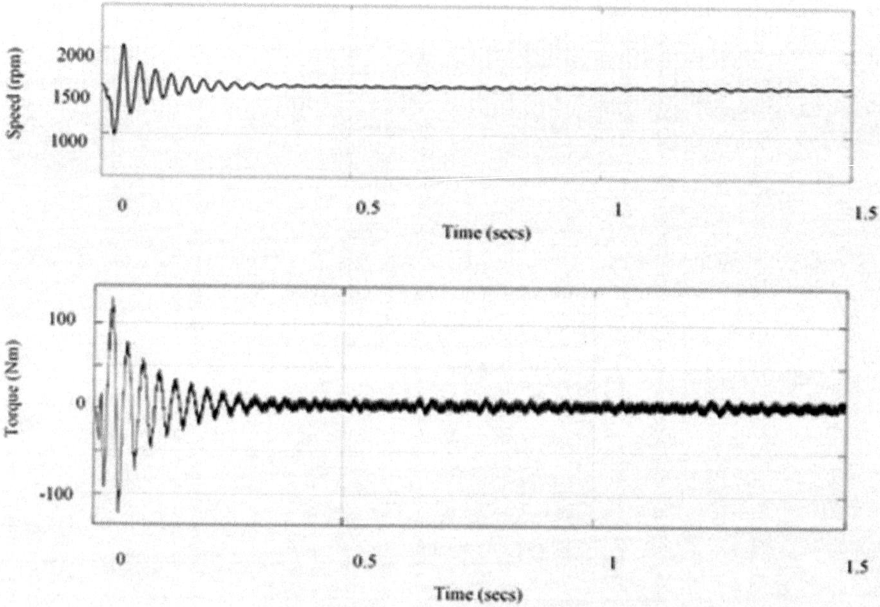

Fig. 10 Control parameters of induction motor such as speed and torque

proposed work, only the temperature sensor values are taken for analysis and monitor in MATLAB/Simulink software.

References

1. Melecio JI, Mohammed A, Djurović S (2018) Towards embedded intelligence enabled autonomous condition monitoring systems for AC motor drives. In: 2018 7th mediterranean conference on embedded computing (MECO), Budva, 2018, pp 1–5
2. Sravani C, Chaitanya VLK, Abitha MW (2017) Condition monitoring and sensor based robust protection for inverter fed induction motor drive system. In: 2017 3rd international conference on science techno-logy engineering and management (ICONSTEM), Chennai, 2017, pp 649–652
3. Hey J, Malloy AC, Martinez-Botas R, Lampérth M (2016) Online monitoring of electromagnetic losses in an electric motor indirectly through temperature measurement. IEEE Trans Energy Convers 31(4):1347–1355
4. Choi S, Pazouki E, Baek J, Bahrami HR (2015) Iterative condition monitoring and fault diagnosis scheme of electric motor for harsh industrial application. IEEE Trans Industr Electron 62(3):1760–1769
5. Nguyen TH, Lee D (2015) Deterioration monitoring of DC-link capacitors in AC machine drives by current injection. IEEE Trans Power Electron 30(3):1126–1130
6. Jiang C, Habetler TG, Cao W (2016) Improved condition monitoring of the faulty blower wheel driven by brushless DC motor in air handler unit (AHU). In: 2016 IEEE energy conversion congress and exposition (ECCE), Milwaukee, WI, 2016, pp 1–5
7. Romanenko A Muetze, Ahola J (2017) Incipient bearing damage monitoring of 940-h variable speed drive system operation. IEEE Trans Energy Convers 32(1):99–110
8. Zoeller C, Vogelsberger MA, Fahrner W, Wolbank TM (2016) Influence of current transducer transfer properties on stator insulation condition monitoring of inverter-fed drives. In: 2016 international symposium on power electronics, electrical drives, automation and motion (SPEEDAM), Anacapri, 2016, pp 55–60
9. Nussbaumer P, Vogelsberger MA, Wolbank TM (2015) Induction machine insulation health state monitoring based on online switching transient exploitation. IEEE Trans Industr Electron 62(3):1835–1845
10. Li J, Sumner M, Padilla JA, Zhang H (2017) Fault signal propagation through the PMSM motor drive systems. IEEE Trans Ind Appl 53(3):2915–2924
11. Akin B, Choi S, Orguner U, Toliyat HA (2011) A simple real-time fault signature monitoring tool for motor-drive-embedded fault diagnosis systems. IEEE Trans Industr Electron 58(5):1990–2001
12. Shin SM, Choi BH, Kang HG (2016) Motor health monitoring at standstill through impedance analysis. IEEE Trans Industr Electron 63(7):4422–4431
13. Popov NZ, Vukosavic SN, Levi E (2014) Motor temperature monitoring based on impedance estimation at PWM frequencies. IEEE Trans Energy Convers 29(1):215–223

Real-Time Monitoring with Labview of the Battery Management System and the Estimated Electric Vehicle Battery SoH

Ensar Kilic, Ramazan Bayindir, and Samet Ayik

Abstract In battery management systems of electric vehicles, the estimation of parameters such as state of charge (SoC) and state of health (SoH) of a battery plays a very important role in preventing battery failure and preserving its life. In this study, a battery state of health estimation and real-time monitoring for electric vehicle batteries were made using the STM32F4 microcontroller, and a novel battery management system was proposed. In the proposed system, the Thevenin equivalent model of the battery is simulated in MATLAB Simulink by applying a Kalman filter, and the results are programmed in LabVIEW real-time software. Also in the study, a graphical driver interface (GDI) was created using the UDP protocol. With the GDI, a real-time monitoring and battery replacement warning can be received.

Keywords Battery management system (BMS) · Battery model · State of health · Graphical user interface · Kalman filter

1 Introduction

Due to global energy scarcity and environmental pollution, countries around the world attach great importance to the development of electric vehicles today [1]. This has led to batteries also gaining great significance. The widespread use of portable electronic devices (mobile phones, computers, audio and video devices, etc.) has greatly increased the need for batteries. In industrial and commercial applications, the use of battery-powered power systems is widespread during mains electricity outages. To examine the behavior of a battery under different usage and ambient conditions, a battery model should be created. By using the battery model, the performance of the battery is analyzed, where the battery charge and battery health statuses

E. Kilic · R. Bayindir (✉) · S. Ayik
Faculty of Technology, Gazi University, Ankara, Turkey
e-mail: bayindir@gazi.edu.tr

E. Kilic
e-mail: ensar.kilic@gazi.edu.tr

S. S. Dash et al. (eds.), *Sixth International Conference on Intelligent Computing and Applications*, Advances in Intelligent Systems and Computing 1369,
https://doi.org/10.1007/978-981-16-1335-7_50

can be estimated. With the development of battery technology, battery management systems have begun to be employed in ensuring the highest performance from the battery. Determination of battery health status (battery state of health (SoH)) is also an important function of battery management systems. Battery health is a concept that expresses the maximum usable capacity of the battery. The factory output capacities of batteries are nominally written on the battery. Depending on usage and ambient conditions, battery capacity decreases over time, and in other words, battery health deteriorates [2]. Information about the health status (SoH) of the battery can be obtained from battery impedance measurements. While the battery is discharged, the high battery impedance reduces the terminal voltage, decreasing the output power. The SoH indicator is required to determine whether a battery replacement is required [3]. In [4], a state of health estimator for commonly used lithium iron phosphate batteries is presented. An online estimator using incremental capacity analysis and a bandpass filter is studied. In [5], the internal losses of high-voltage lithium-ion batteries were investigated, which shows that losses can vary significantly depending on the battery charge states, power curve and temperature. They also made a comparison between historical data and forecast data and mentioned the importance of using battery simulation models to estimate internal battery losses to minimize the uncertainty of the useful remaining life of the battery. In [6], the Kalman filter algorithm has been developed and applied to improve the real-time and accurate estimation of the IMS and to reduce the computational burden. In this study, the SoH estimation algorithm was run by creating a battery model in MATLAB Simulink and using the Kalman Filter. Also, a real-time interface was created by communicating with the Simulink and LabVIEW programs using the UDP protocol.

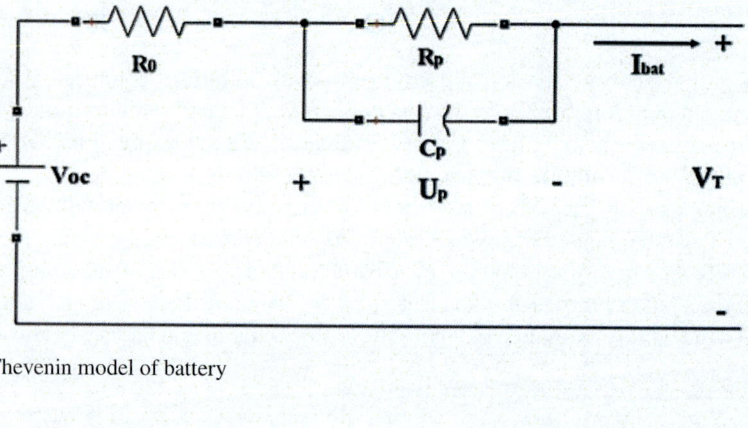

Fig. 1 Thevenin model of battery

2 Battery Model

Figure 1 shows the first-order Thevenin equivalent circuit of a battery. In [7], it was concluded that the first-order RC model of the battery model is the best battery equivalent circuit option considering its low complexity and high accuracy.

The open-circuit voltage V_{oc} in Fig. 1 is the measured battery internal voltage when the battery is not connected to the load. U_p is the voltage of parallel R_p and C_p. R_0 is the internal resistance of the battery, R_p and C_p are the polarization resistance and capacitance, and I_{bat} is the battery current.

Equation (1) shows the terminal voltage when the terminals of the battery are open, in other words when the load is not connected.

$$V_T = V_{OC} \tag{1}$$

When a load is connected to the battery terminal, the terminal voltage is expressed in Eqs. (2) and (3).

$$U_P = -\frac{U_P}{C_P R_P} + \frac{I_{BAT}}{C_P} \tag{2}$$

$$V_T = V_{OC} - U_P - I_{BAT} R_0 \tag{3}$$

3 Battery's State of Health

Changes in the battery model are used to predict battery health. One of these variables is the battery capacity that changes over time and usage. Based on the changes in battery capacity, SoH is defined as in Eq. (4).

$$\text{SoH} = \frac{C_2}{C_1} \times 100\% \tag{4}$$

The C_2 parameter given in Eq. (3) is the current maximum capacity of the battery, while the C_1 parameter is the first maximum capacity of the battery written on the factory datasheet. Battery management systems indicate that the battery should be replaced if the maximum capacity of the battery falls below 80% of its initial capacity [8].

The internal resistance of the battery is another parameter that changes during the deterioration process. The internal resistance value of the battery will increase over time and with the use of the battery [9].

When the SoH value is equal to 100%, the voltage falling on the reference load will be equal to the internal voltage of the battery, meaning that the battery impedance is zero. This is not the same as a new battery because the internal impedance of a new battery is not equal to zero ohms. The voltage falling on the reference load will decrease with aging of the battery because the internal impedance of the battery will increase. Therefore, the battery health status in this study has been estimated as;

$$\mathrm{SoH} = \frac{V_{\text{Reference\_load}}}{V_{\text{internal resistance of battery}}} \times 100 \tag{5}$$

based on the equation given in Eq. (5).

4 Kalman Filter

The Kalman filter is a very powerful method for estimating process states [10–12]. It consists of two parts: the prediction phase and update phase. The process status is estimated in the forecast section. In the update part, the filter receives feedback from the measurements exposed to noise.

$$\widehat{X} = K_{\mathrm{K}} Z_{\mathrm{K}} + (1 - K_{\mathrm{K}})\widehat{X}_{K-1} \tag{6}$$

In Eq. (6),

- $\widehat{X}_{\mathrm{K}} \Rightarrow$ Predictive value
- $K_{\mathrm{K}} \Rightarrow$ Kalman gain
- $Z_{\mathrm{K}} \Rightarrow$ Measured value
- $\widehat{X}_{K-1} \Rightarrow$ Previous forecast value

The basic principle of the Kalman filter is to find the $\widehat{X}_{\mathrm{K}}$ estimated value for each value of . The Z_{K} measurement values here are not always correct, so the K_{K} Kalman coefficient is used for each measurement value.

If the measurement is reliable, $K_{\mathrm{K}} = 1$. In this case, Eq. (5) will result in $\widehat{X}_{\mathrm{K}} = Z_{\mathrm{K}}$. In cases where the measurement value is unreliable, the $\mathrm{K_K}$ coefficient approaches zero. In this way, starting from Eq. (6), $\widehat{X}_{\mathrm{K}}$ approaches $\widehat{X}_{K-1}$, that is, the previous prediction. If $K_{\mathrm{K}} = 0.5$ is selected and the solution of Eq. (5) is examined, it means that Z_{K} has been averaged. ($\widehat{X}_{\mathrm{K}} = 0.5Z_{\mathrm{K}} + 0.5$)

The Kalman model:

$$x_k = Ax_{k-1} + Bu_k + w_{k-1} \tag{7}$$

$$z_k = Hx_k + v_k \tag{8}$$

Here, x_k shows the situation. u_k is the control signal and is not available in all situations. w_{k-1} is the process noise, and v_k is the observational noise. During the application, noise will occur during the measurement and estimation phase. In this study, the Kalman filter is used to filter noisy data.

The Kalman filter generally continues with two different processes as "time renewal" and "measurement renewal." These are estimation and renewal processes, respectively.

Time and measurement formulas can be listed as follows;

Time renewal:

$$\hat{x}_k = A\hat{x}_{k-1} + Bu_k \tag{9}$$

$$P_k^- = AP_{k-1}A^T + Q \tag{10}$$

Measurement renewal:

$$K_k = P_k^- H^T \left(HP_k^- H^T + R\right)^{-1} \tag{11}$$

$$\hat{x}_k = \hat{x}_k^- + K_k\left(z_k - H\hat{x}_k^-\right) \tag{12}$$

$$P_k = (I - K_k H)P_k^- \tag{13}$$

$\hat{x}_k$ is prediction, and P is error covariance. Here, R is the covariance of observational noise, and is the covariance of the process noise.

The Kalman filter is a calculation method in the form of iterations. The flow diagram of the Kalman filter algorithm is given in Fig. 2.

In this study, the Kalman filter is used to filter a system that gives a noisy output instead of a fixed output and to obtain an optimal result.

5 LabVIEW

LabVIEW is a software development environment provided by National Instruments. The LabVIEW environment provides a visual data flow programming language in which functions are organized in a flowchart fashion with "cables" connecting the output of one node to the input of the other. As soon as the data is available, it is run on each node, and the compiler identifies the code segments that can run in parallel and automatically splits the application into multiple threads. LabVIEW offers the ability to work at a higher abstraction layer than text-based programming languages such as *C*. For example, there is no need for manual memory allocation by the user. It allows the user to create an interface. Even users with minimal coding knowledge can code quickly with LabVIEW [13, 14]. The entire algorithm can be developed in this environment, or in embedded applications, and real-time code can be loaded to run on a hardware target. Embedded LabVIEW real-time applications do not have a user interface, so it is necessary to have a host to create the user interface [15]. In

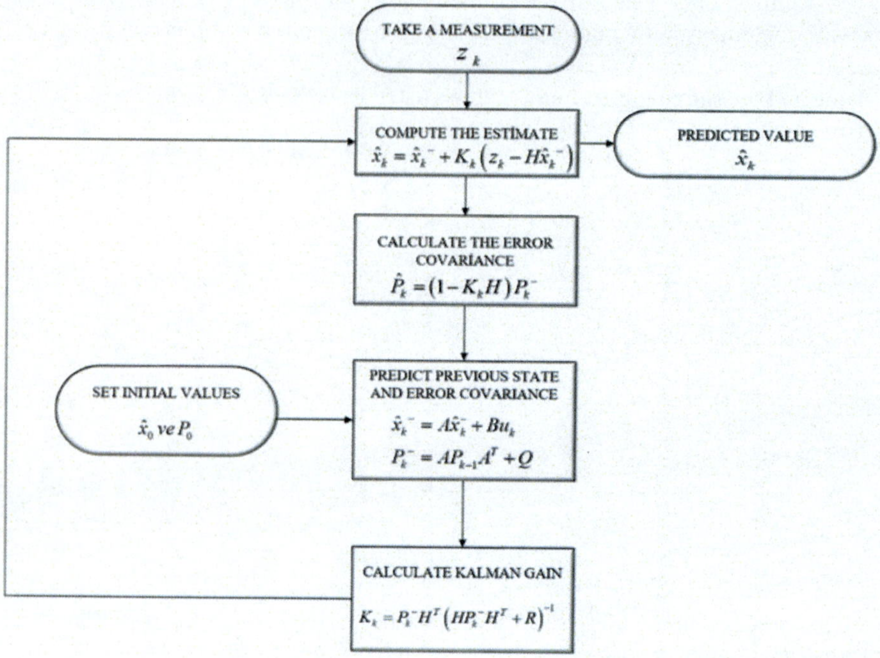

Fig. 2 Flowchart of the Kalman filter algorithm

this study, an interface was created with LabVIEW to monitor the health status of a battery in real time.

6 MATLAB Simulink, Waijung Blockset and UDP Protocol

Simulink is a block diagram environment for multi-area simulation and model-based design. It supports system-level design, simulation, automatic code generation and continuous testing and validation of embedded systems. Simulink provides a graphic editor, customizable block libraries and solvers to model and simulate dynamic systems. It integrates with MATLAB, allowing you to include MATLAB algorithms in models and export simulation results to MATLAB for further analysis [16]. Waijung is a Simulink block set that can be used to easily and automatically generate C code from MATLAB Simulink simulation models for many types of microcontrollers. The Waijung blockset is specially designed to support the STM32F4 microcontroller family, a Hi Performance and DSP MCU of STMicroelectronics [17]. In this study, communication was performed with the Simulink and LabVIEW programs using the UDP protocol. The user datablock protocol (UDP)

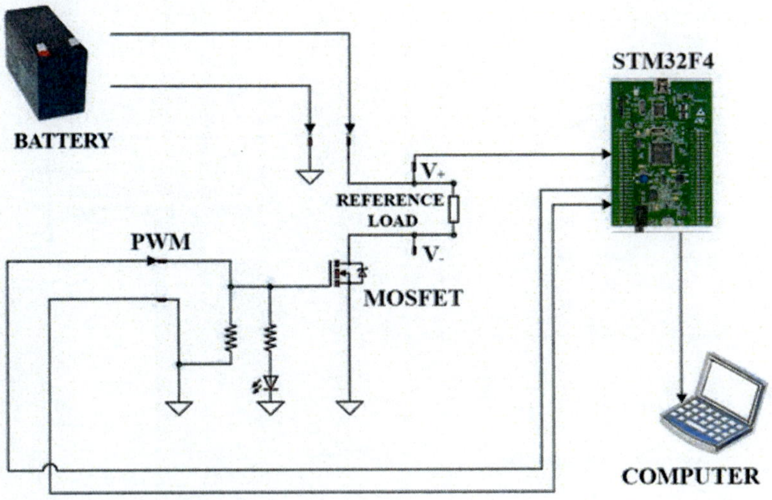

Fig. 3 Wiring diagram of application

is a communication protocol primarily used to establish low latency and loss tolerance connections between applications on the Internet. It speeds up transmissions by enabling data transfer before a confirmation which is provided by the recipient. UDP can be used as an alternative to the TCP protocol as a communication protocol [18].

7 Implementation Phase

The proposed system as shown in Fig. 3 consist of an STM32F4 with the aim of getting data from the battery and the obtained data sent to Kalman filter and computer. Algorithm & Flowchart of process for SoH estimation shown in Fig. 4. The Kalman filter is used for battery SoH estimation and optimization based on simulink block diagram represented in Fig. 5. Figure 6 shows the LabVIEW block diagram to monitor the health status of the battery.

8 Experimental Results

The reason for adding the adjustable resistor in Fig. 7 to the circuit is to observe the situations that may arise from the change of battery internal impedance at the interface. As seen in Fig. 8, the adjustable resistance is short-circuited, and the voltage falling on the internal impedance of the battery used in the application is seen in the LabVIEW interface. Accordingly, using Eq. (5), the SoH value is found.

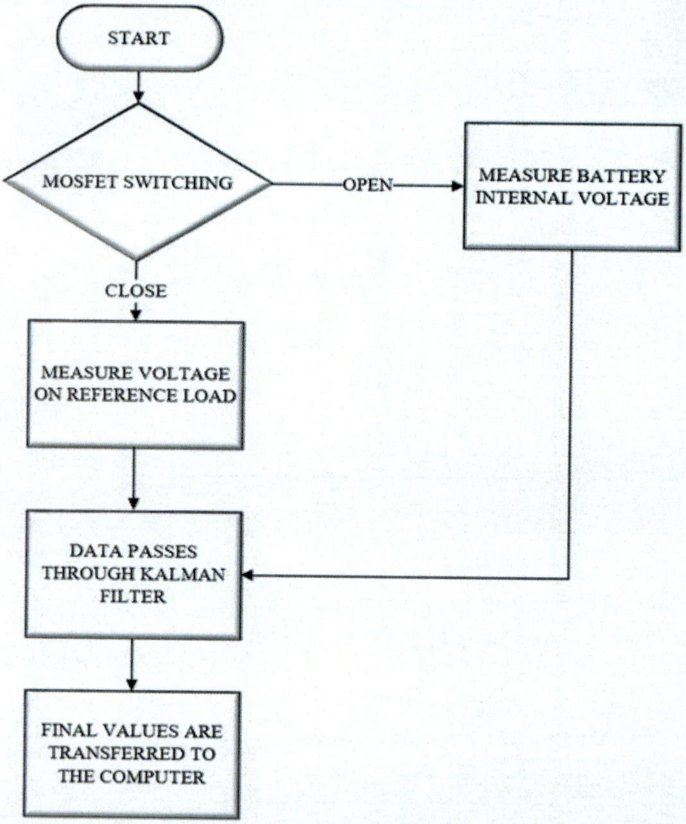

Fig. 4 Flowchart of process

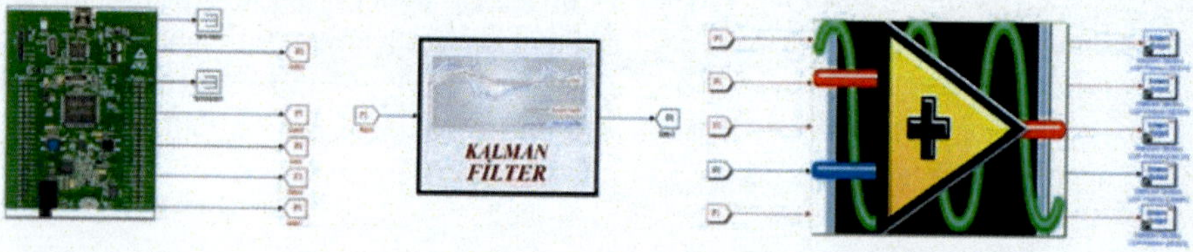

Fig. 5 Simulink block diagram

$$\text{SoH} = \frac{V_{\text{Reference\_load}}}{V_{\text{internal resistance of battery}}} \times 100 \tag{14}$$

As seen in Fig. 8, the health status of the battery is the curve drawn in blue on the graphic screen. However, the fact that the data originates from the switching and is received from the analog channel of the microprocessor makes it noisy, which prevents us from fully obtaining the value obtained in Eq. 5. Here, the Kalman filter

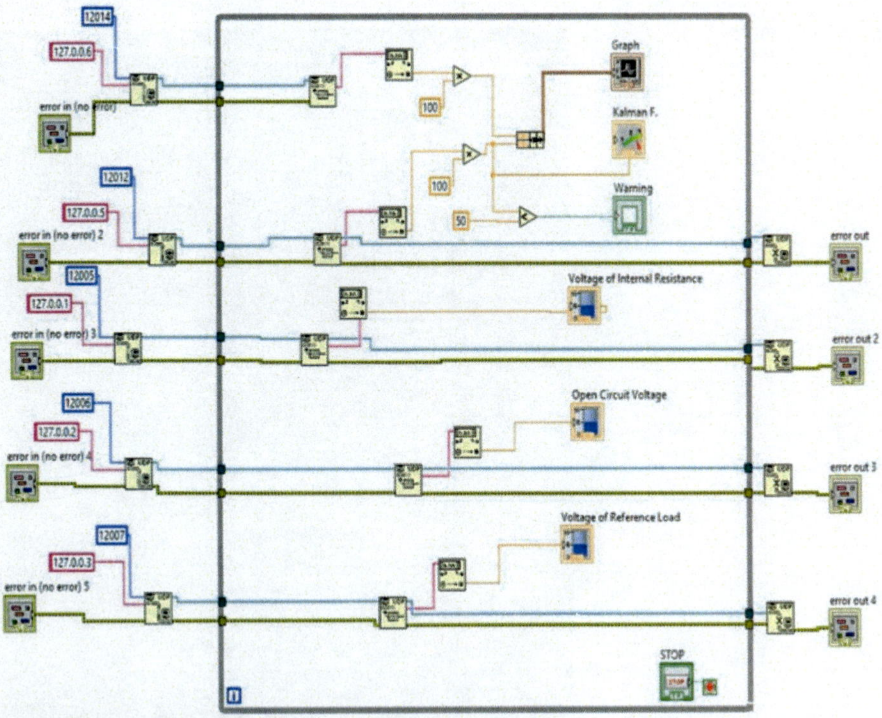

Fig. 6 LabVIEW block diagram

described above comes into play and eliminates this noise, giving the interface an optimal result, and this is the curve drawn in red on the graphic screen.

When the adjustable resistance is brought to half of its maximum value, the health status of the battery decreases, starting from Eq. 5, as seen in Fig. 9. This is because the higher the internal impedance value of the battery, the less voltage it provides to the load on the terminal.

When the adjustable resistance is brought to its maximum value, the voltage that the battery can give to the load at its terminal has further decreased. Therefore, as seen in Fig. 10, the interface warns the user that the battery should be changed. (In this study, a warning is provided when SoH is below 50%).

9 Conclusions

One of the key factors in electric vehicle battery management systems is the battery SoH value. A BMS for the estimation of battery health, which is one of the most important issues of electric vehicle battery management systems (BMS), has been designed, and a method has been developed for real-time monitoring. In order to

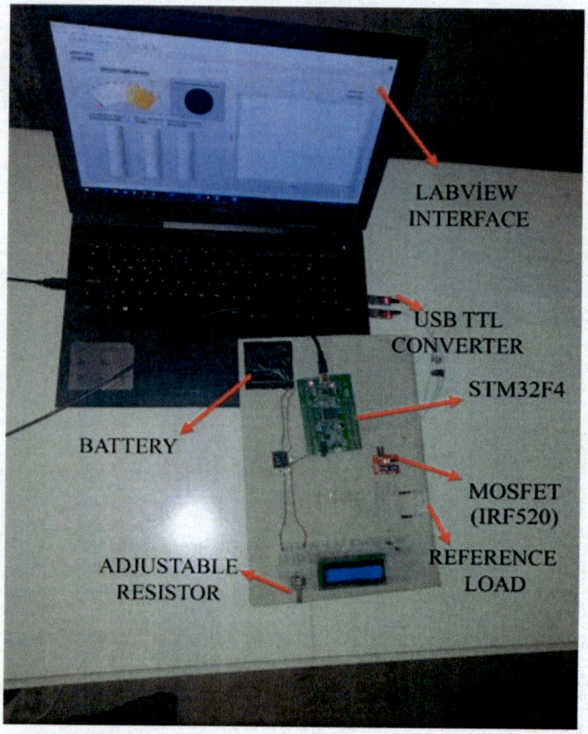

Fig. 7 Layout of implementation

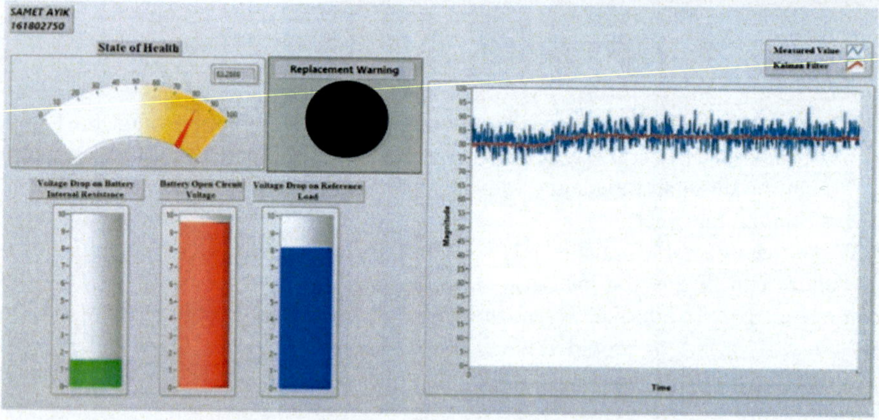

Fig. 8 LabVIEW interface while adjustable resistor is short circuited

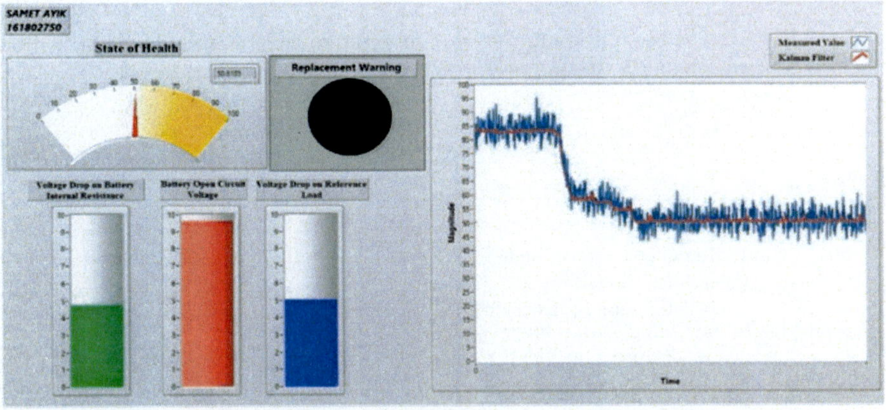

Fig. 9 LabVIEW interface while adjustable resistance is half of its maximum value

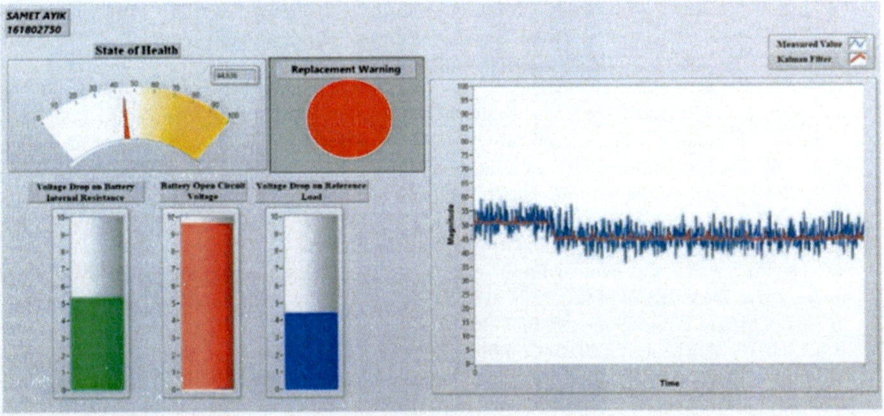

Fig. 10 LabVIEW interface while adjustable resistance is at its maximum value

estimate the state of battery health, firstly, a battery model is considered in this study. The proposed battery model is similar to the electrical circuit models in the literature but is not identical to any model. The most important feature of the proposed circuit model is that it is a general model valid for all battery types. The Kalman filter was initially applied in linear systems, but later, the extended Kalman filter and odorless Kalman filter were developed for nonlinear systems. The expanded and odorless Kalman filters have a more complex structure. In this study, the Kalman filter used in the system considered for the estimation of battery health status is for filtering the noises that will occur in analog reading and the noises that will arise from switching. Therefore, the basic Kalman filter has been used. It is important to examine the health of a battery in real time so that action can be taken beforehand. Since many factors affect battery performance during charging/discharging of a battery, the SoH value of the battery has been optimized with the recommended method. The application platform is designed to verify the performance of the proposed method.

References

1. Ma X, Qiu D, Tao Q et al (2019) State of charge estimation of a lithium ion battery based on adaptive Kalman filter method for an equivalent circuit model. Appl Sci 9:2765
2. Barlak C (2009) Estimation of battery model parameters, state-of-charge, state-of-health and applications to Ni–Mh batteries, Ph.D. thesis, Hacettepe University
3. Juang LW, Kollmeyer PJ, Jahns TM, Lorenz RD (2010) System identification-based lead-acid battery online monitoring system for electric vehicles. In: 2010 IEEE energy conversion congress and exposition, Atlanta, GA, pp 3903–3910. https://doi.org/10.1109/ecce.2010.5617776
4. Venet RP, Sari A, Meniere F, Bultel Y (2015) LiFePO4 battery state of health online estimation using electric vehicle embedded incremental capacity analysis. In: 2015 IEEE vehicle power and propulsion conference (VPPC), Montreal, QC, pp 1–6. https://doi.org/10.1109/vppc.2015.7352972
5. Enthaler A, Gauterin F (2013) Significance of internal battery resistance on the remaining range estimation of electric vehicles. In: 2013 international conference on connected vehicles and expo (ICCVE), Las Vegas, NV, pp 94–99. https://doi.org/10.1109/iccve.2013.6799776
6. Topan PA, Ramadan MN, Fathoni G, Cahyadi AI, Wahyunggoro O (2016) State of charge (SOC) and state of health (SOH) estimation on lithium polymer battery via Kalman filter. In: 2016 2nd international conference on science and technology-computer (ICST), Yogyakarta, 2016, pp 93–96. https://doi.org/10.1109/icstc.2016.7877354
7. Hu X, Li S, Peng H (2012) A comparative study of equivalent circuit models for Li-ion Batteries" Department of Mechanical Engineering, The University of Michigan, Ann Arbor, MI 48109, USA, Department of Mechanical Engineering, Beijing Institute of Technology, Beijing 100081, China. J Power Sour 198:359–367
8. Zou Y, Hu X, Ma H, Li SE (2015) Combined state of charge and state of health estimation over lithium-ion battery cell cycle lifespan for electric vehicles. J Power Sour 273:793–803
9. Shahriari M, Farrokhi M (2010) State of health esimation of VRLA batteries using fuzzy logic. In: 18th Iranian Conference on Electrical Engineering
10. Tasan Y (2019) Model development for state of health estimation in lithium-ion batteries, Master thesis, Gebze Technical University, Institute of Science
11. Smiley J (2015) Estimation of battery aging using an interacting multiple model kalman filter" Master's thesis, University of Colorado
12. Cayiroglu I (2012) Kalman filter and programming. Karabuk University Science and Technology Information Sharing-1
13. Shrenika RM, Chikmath SS, Kumar AVR, Divyashree YV, Swamy RK (2017) Non-contact water level monitoring system implemented using LabVIEW and Arduino. In: 2017 international conference on recent advances in electronics and communication technology (ICRAECT), Bangalore, pp 306–309. https://doi.org/10.1109/icraect.2017.51
14. Mason A, Parkes S (2014) Using spacewire with LabVIEW. In: 2014 international spacewire conference (spacewire), Athens, pp 1–4, https://doi.org/10.1109/spacewire.2014.6936275
15. Liu J, Zhang P, Wang F (2009) Real-time DC servo motor position control by PID controller using labview. In: 2009 international conference on intelligent human-machine systems and cybernetics, Hangzhou, Zhejiang, pp 206–209. https://doi.org/10.1109/ihmsc.2009.59
16. MathWorks. https://www.mathworks.com/help/simulink/index.html?s_tid=srchtitle
17. Aimagin Co.,Ltd. https://www.aimagin.com/waijung-1-stm32-target.html
18. UDP (User Datagram Protocol). https://searchnetworking.techtarget.com/definition/UDP-User-Datagram-Protocol

Design and Implementation of Three-Phase SEPIC-Based Photovoltaic System

P. Srinivasan, D. Sivakumar, P. Rayavel, M. Mohammed Ishaq, A. Arumuga Nainar, and K. Saravanan

Abstract A three-phase SEPIC-based photovoltaic system has been anticipated as an inventive system to condense the rate, involvedness, size and switching fatalities of the three-phase DC to AC alteration structure over customary six switch three-phase inverter-based photovoltaic system. This paper proposes a system to track the maximum power from photovoltaic cell and convert that to three-phase AC in cost-effective way. The output voltage of proposed system is pure sinusoidal with no need for output filter. The proposed system also good to deal with three-phase asymmetrical load. The sliding mode organize is being worn through the anticipated system to optimize its working and to make certain rigidness of the scheme through diverse in service modes. Cradle of the equations recitation the constraints intends, works evaluation, and the maneuver of the planned photovoltaic system is offered in this manuscript. Simulation replica and prototype untried association are used to authenticate the projected perception. Simulations and investigational results show the usefulness of the anticipated photovoltaic system.

Keywords Maximum power point tracking (MPPT) · Single-ended primary inductance converter (SEPIC) converter · Sliding mode control · Six switch three-phase inverter (SSTP)

1 Introduction

Solar energy became popular in recent days as this is renewable energy and widely available energy on earth. There are various research work done in this area to tap the solar energy in effective ways. A SEPIC with the PV element input and the

P. Srinivasan (✉) · D. Sivakumar · M. Mohammed Ishaq · A. Arumuga Nainar · K. Saravanan
Department of Electrical and Electronics Engineering, SRM Institute of Science and Technology, Ramapuram, India

P. Rayavel
Department of Electrical and Electronics Engineering, Sairam Institute of Technology, Chennai, India
e-mail: rayavel.cse@sairamit.edu.in

S. S. Dash et al. (eds.), *Sixth International Conference on Intelligent Computing and Applications*, Advances in Intelligent Systems and Computing 1369,
https://doi.org/10.1007/978-981-16-1335-7_51

summit current approaches manage has been urbanized including small signal model of SEPIC, the PV voltage regulator and the MPPT director. Key requirements are normally necessary in photovoltaic system as reduced losses, high power, less cost of system, low weight and volume. The high-efficiency operation is important to deliver the more power in effective ways [12]. Six pulses three-phase voltage source inverter is widely used for conversion of DC to DC in various industrial application such as motor drives, DC to AC conversion in photovoltaic system and wind energy conversion. Conventional SSTP-based photovoltaic system as shown in Fig. 1 has the drawback as the presence of harmonics in output, and AC filters are needed to suppress these harmonics. Other drawback is that it does not draw the maximum power from solar cell, and power losses are more (Fig. 2). A novel technique for efficiently maximizing the output power of a photovoltaic system supplying to a load under varying operational conditions is proposed in this paper. Recently, SEPIC-based inverter has fascinated the most wellbeing concerning its performance, manage and application [1, 2]. To weigh against to SSTP inverter, the SEPIC-based inverter has a few compensation such as non-inverting output polarity of SEPIC converter, reduced cost and increased reliability due to less number of switch. PERTs and Observe algorithm used to draw maximum power from solar panel [3–5]. The SEPIC

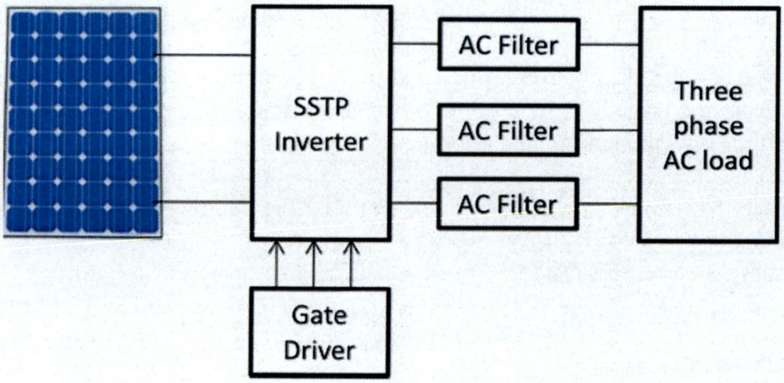

Fig. 1 Conventional SSTP-based photovoltaic system

Fig. 2 Conventional six switch three-phase inverter (SSTP)

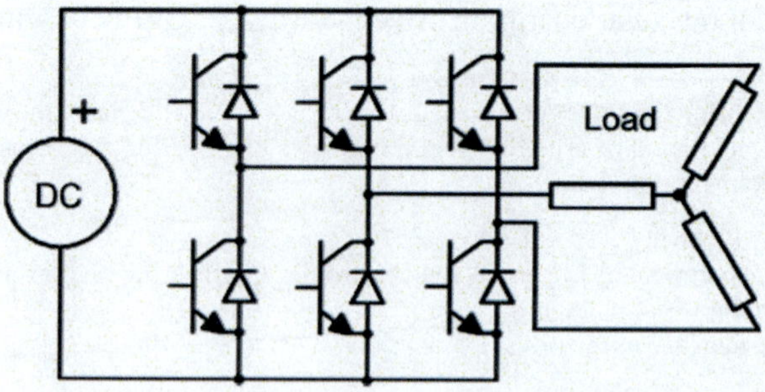

converter is a fourth-order nonlinear system and needs robust control strategy. Sliding control is being used in proposed system to optimize its working and to guarantee rigidness of the structure through unusual in commission modes. An extra stunning attribute of the projected system is the yield voltage which is unadulterated sinusoidal signal [6], consequently no need of any filtering requirement. Also the proposed system is good to deal with unbalanced three-phase load.

2 Proposed Photovoltaic System

The proposed photovoltaic system consists of photovoltaic panel, two SEPIC converter, MPPT, sliding control modules (SM1 and SM2) and regulated power supply (RPS) as shown in Fig. 3.

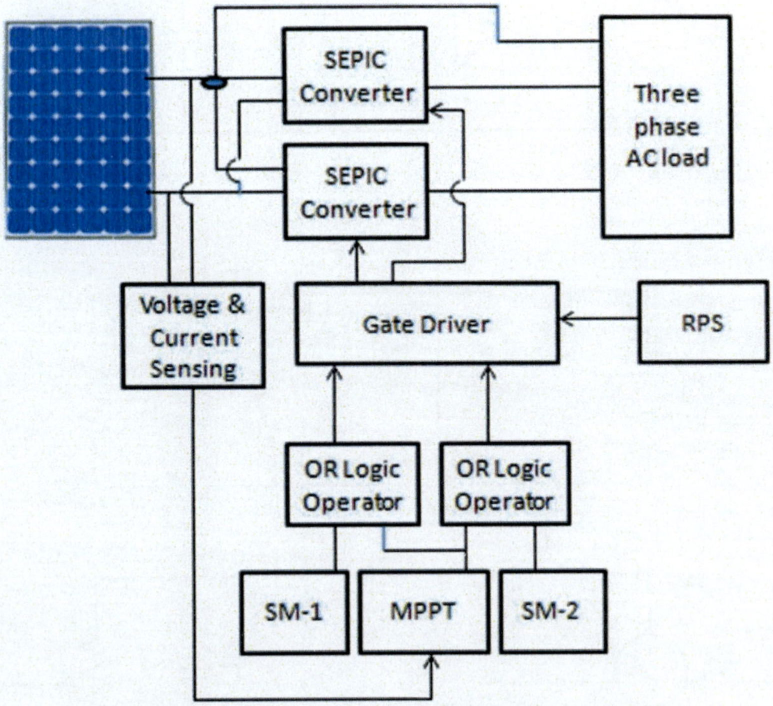

Fig. 3 Proposed photovoltaic system

2.1 SEPIC Convertor

As revealed in Fig. 4, SEPIC converter comprises DC input voltage VDC.

$$V_0 = \frac{D}{1 - D}$$

where D is the duty cycle the counterpart circuits of the SEPIC converter during ON and OFF position are publicized in Fig. 5a, b

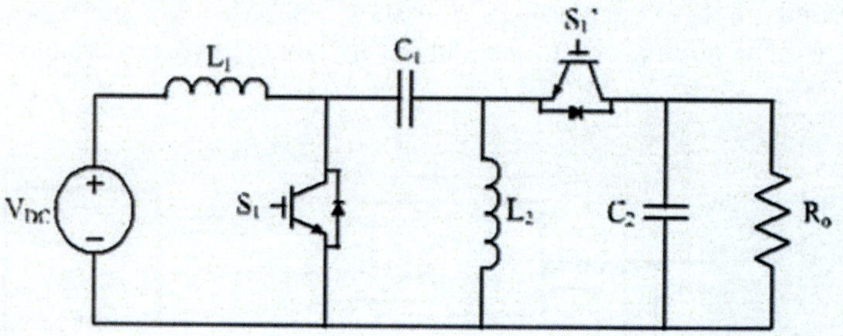

Fig. 4 Bi-directional SEPIC converter

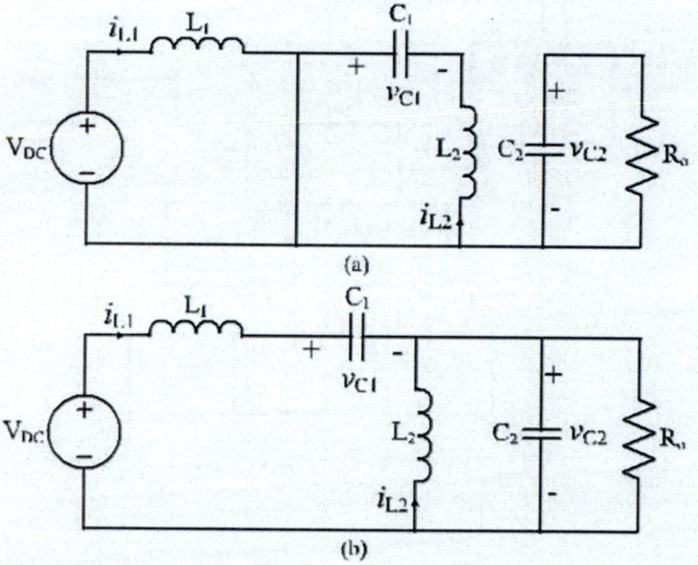

Fig. 5 **a** Switch ON **b** switch OFF

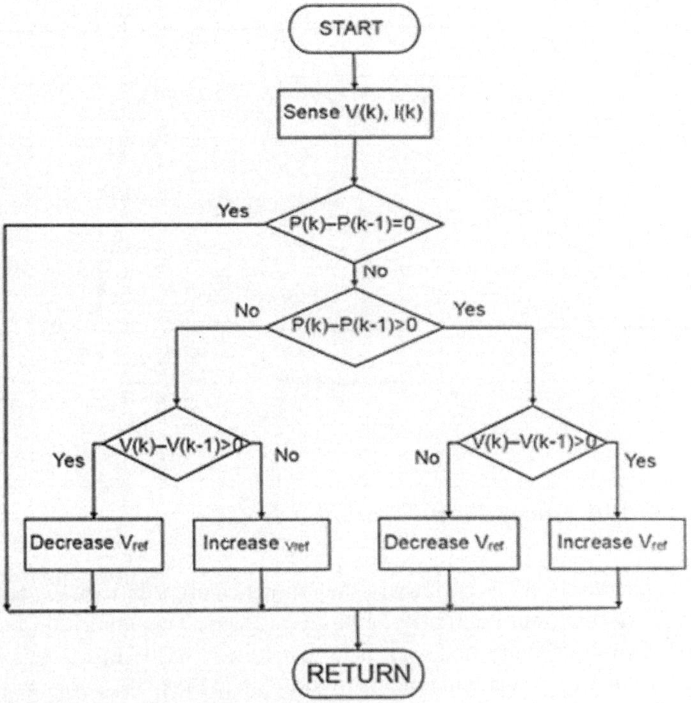

Fig. 6 P&O maximum power tracking algorithm

$$\left.\begin{aligned} L_1\frac{di_{L_1}}{dt} &= V_{DC} - iL_1 rL_1 \\ C_1\frac{dv_{C_1}}{dt} &= -i_{L_2} \\ L_2\frac{di_{L_2}}{dt} &= vc_1 - i_{L_2} r_{L_2} \\ C_2\frac{dv_{C_2}}{dt} &= -\frac{vC_2}{R_o} \end{aligned}\right\} \tag{1}$$

$$\left.\begin{aligned} L_1\frac{di_{L_1}}{dt} &= V_{DC} - i_{L_1} r_{L_1} \\ C_1\frac{dv_{C_1}}{dt} &= -i_{L_2} \\ L_2\frac{di_{L_2}}{dt} &= v_{C_1} - i_{L_2} r_{L_2} \\ C_2\frac{dv_{C_2}}{dt} &= -\frac{v_{C_2}}{R_o} \end{aligned}\right\}\text{Switch ON} \tag{2}$$

$$\left.\begin{aligned} L_1\frac{di_{L_1}}{dt} &= V_{DC} - v_{C_1} - v_{C_2} - i_{L_1} r_{L_1} \\ C_1\frac{dv_{C_1}}{dt} &= -i_{L_1} \\ L_2\frac{di_{L_2}}{dt} &= -v_{C_2} - i_{L_2} r_{L_2} \\ C_2\frac{dv_{C_2}}{dt} &= i_{L_1} + i_{L_2} - \frac{v_{C_2}}{R_o} \end{aligned}\right\}\text{Switch OFF} \tag{3}$$

Here r_{L_1} and r_{L_2} are the corresponding sequence resistance of the inductors L_1 and L_2

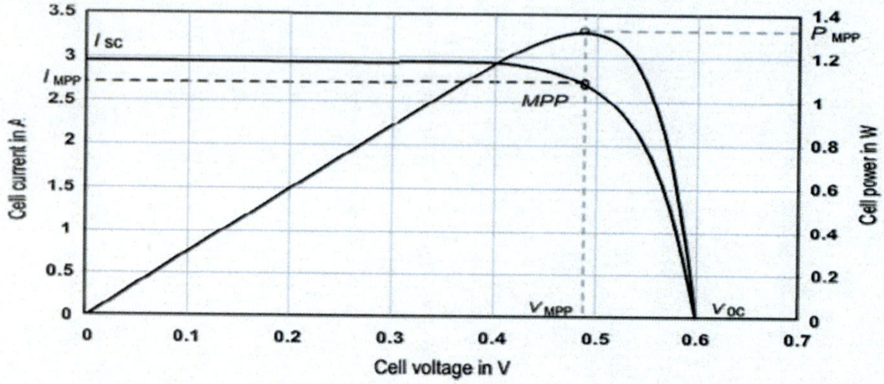

Fig. 7 VI characteristics of a typical solar cell

2.2 *Maximum Power Point Tracking (P&O)*

Perturb and observe (P&O) is solitary of the renowned algorithm owing to its effortlessness used for maximum authority summit pathway. This algorithm pedestal on voltage and current sagacity pedestal used to pathway MPP. Inside this organizer involves reckoning for power and voltage to footpath MPP. In this, voltage is anxious in solitary trend, and if power is unremitting to augment, then algorithm remains on perturb in equivalent course. If innovative power is fewer than previous power, then perturbed in contradictory route. When component power achieves at MPP, there is swinging roughly MPP summit.

The VI characteristics of a typical solar cell are as shown in Fig. 7. The PV characteristics are obtained by the voltage and the current characteristics as shown in below. The point indicated as MPPT is the point at which the panel power output is maximum. P&O algorithm as in Fig. 6 is followed to track the maximum power.

2.3 *Sliding Mode Control*

SMC is nonlinear manage which widen the properties of hysteresis manage to multi-erratic milieu. SMC has the fast dynamic control with the agreement of solidity for hefty discrepancy of organism constraint and aligned with perturbations [7, 8].

Even though the yield voltage V_{c_2} of every SEPIC converter is the concluding manage goal, it will be impracticable for the closed-loop proscribed arrangement to achieve unwavering proposition on the downhill facade if V_{c_2} is simply preferred to be the unswerving organize objective, and as a result, the supplementary variables are supposed to be selected. Afterward, it is projected to raise the quantity of utter variables as squat as probable in the downhill facade. To shun a hefty number of amendments grows, a facade surrounding the say current in totaling to the harvest voltage may perhaps be elected as prearranged by Eq. (4)

$$s\left(i_{L_1}, v_{c_2}\right) = \alpha_1 e_1 + \alpha_2 e_2 \tag{4}$$

where coefficient α_1 and α_2 is gains while e_1 and e_2 is the criticism error of the position variables i_{L_1} and v_{c_2} and given by Eq. (5)

$$\begin{aligned} e_1 &= i_{L_1}\text{ref} - i_{L_1} \\ e_2 &= v_{c_2}\text{ref} - v_{c_2} \end{aligned} \tag{5}$$

Due to the sturdy state errors in inductor and capacitor, an vital phrase of these inaccuracy is introduced by Eq. (6)

$$s = \alpha_1 e_1 + \alpha_2 e_2 + \alpha_3 e_3 \tag{6}$$

where α_1, α_2 and α_3 are the sliding coefficients while e_1, e_2 and e_3 are as

$$\begin{aligned} e_1 &= i_{L_1}\text{ref} - i_{L_1} \\ e_2 &= vc_2\text{ref} - vc_2 \\ e_3 &= \int (e_1 + e_2)\mathrm{d}t \end{aligned} \tag{7}$$

The time derivative of above equations

$$\begin{aligned} \frac{\mathrm{d}e_1}{\mathrm{d}t} &= \frac{d(i_{L_1}\text{ref} - iL_{11}}{\mathrm{d}t} \\ \frac{\mathrm{d}e_2}{\mathrm{d}t} &= \frac{d(vc_2\text{ref} - vc_2}{\mathrm{d}t} \\ \frac{\mathrm{d}e_3}{\mathrm{d}t} &= e_1 + e_2 \end{aligned} \tag{8}$$

The inductor reference is difficult to evaluate as it depends on consignment power stipulate, deliver voltage and consignment voltage. The inductor current allusion be able to be preferred as [9, 10]

$$i_{L_1}\text{ref} = K(vc_2\text{ref} - vc_2) \tag{9}$$

where K is the expand factor of the converter yield voltage slipup. Substituting (9) in (8) furnish

$$\frac{\mathrm{d}e_1}{\mathrm{d}t} = \frac{d(i_{L_1}\text{ref} - i_{L_{11}}}{\mathrm{d}t} = \frac{d\left[K(vc_2\text{ref} - vc_2) - i_{L_1}\right]}{\mathrm{d}t} \tag{10}$$

From Eq. (3) and by introducing the discrete variable ($D = 1$ during ON state and $D = 0$ during OFF state)

$$\frac{\mathrm{d}vc_2}{\mathrm{d}t} = \frac{(1-D)}{C_2} i_{L_1} + \frac{(1-D)}{C_2} i_{L_2} - \frac{1}{R_0 C_2} vc_2 \tag{11}$$

Time derivative of the output capacitor voltage

$$\frac{\mathrm{d}vc_2}{\mathrm{d}t} = \frac{iC_2}{C_2} \tag{12}$$

Assuming vc_2 ref is constant and substituting Eqs. (11) and (12) into (10)

$$\begin{aligned}
\frac{\mathrm{d}e_1}{\mathrm{d}t} &= -\frac{K}{C_2} ic_2 - \left[\frac{(D-1)(vc_1 + vc_2]}{L1} - \frac{r_{L_1}}{L_1} i_{L_1} + \frac{V_{\mathrm{DC}}}{L_1}\right] \\
\frac{\mathrm{d}e_2}{\mathrm{d}t} &= \frac{d(vc_2\mathrm{ref} - vc_2}{\mathrm{d}t} = \frac{-ic_2}{C_2} \\
\frac{\mathrm{d}e_3}{\mathrm{d}t} &= e_1 + e_2 = (K+1)(Vc_2\mathrm{ref} - vc_2) - i_{L_1}
\end{aligned} \tag{13}$$

where D is equivalent to duty cycle of the converter, which can be calculated by setting the derivative of Eq. (6) as zero

$$\frac{\mathrm{d}s}{\mathrm{d}t} = \alpha_1 \frac{\mathrm{d}e_1}{\mathrm{d}t} + \alpha_2 \frac{\mathrm{d}e_2}{\mathrm{d}t} + \alpha_3 \frac{\mathrm{d}e_3}{\mathrm{d}t} = 0 \tag{14}$$

Solving for the corresponding control gesture yields

$$\begin{aligned}
D = \frac{1}{vc_1 + vc_2} & K_1(vc_2\mathrm{ref} - vc_2) - K_2 ic_2 \\
& - K_3 i_{L_1} + r_{L_1} i_{L_1} + (vc_1 + vc_2 - V_{\mathrm{in}})]
\end{aligned} \tag{15}$$

where $K_1 = \frac{\alpha_3}{\alpha_1} L_1(K+1)$, $K_2 = \frac{L_1}{C_2} L_1\left(K + \frac{\alpha_2}{\alpha_1}\right)$ and $K_3 = \frac{\alpha_3}{\alpha_1} L_1$ are the fixed gain parameters of sliding control (Fig. 8).

2.4 *Three-Phase Topology of SEPIC Inverter and Its Working Principle*

The projected topology for inverter comprises of two SEPIC converters and accomplishes DC–AC adaptation as enlightened in Fig. 9

The output line voltage across the load is given by

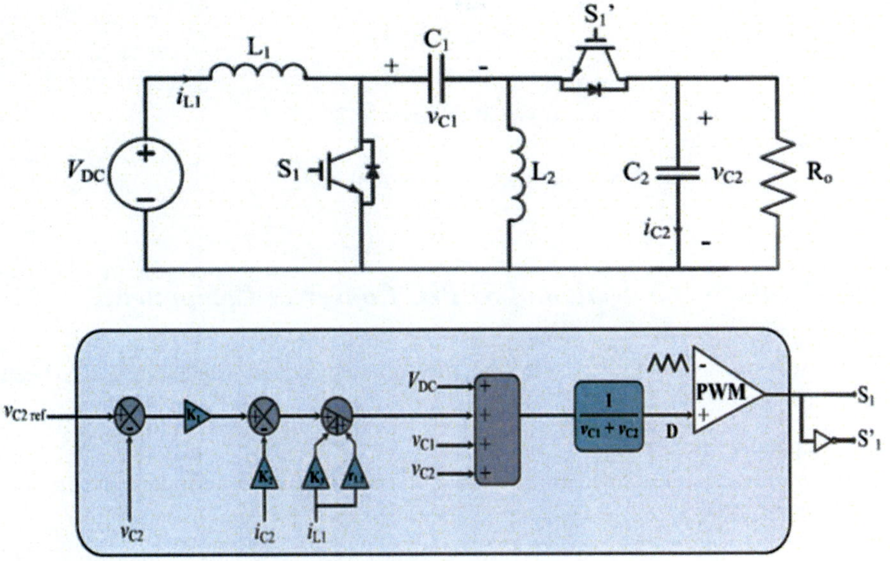

Fig. 8 Sliding mode control for SEPIC converter

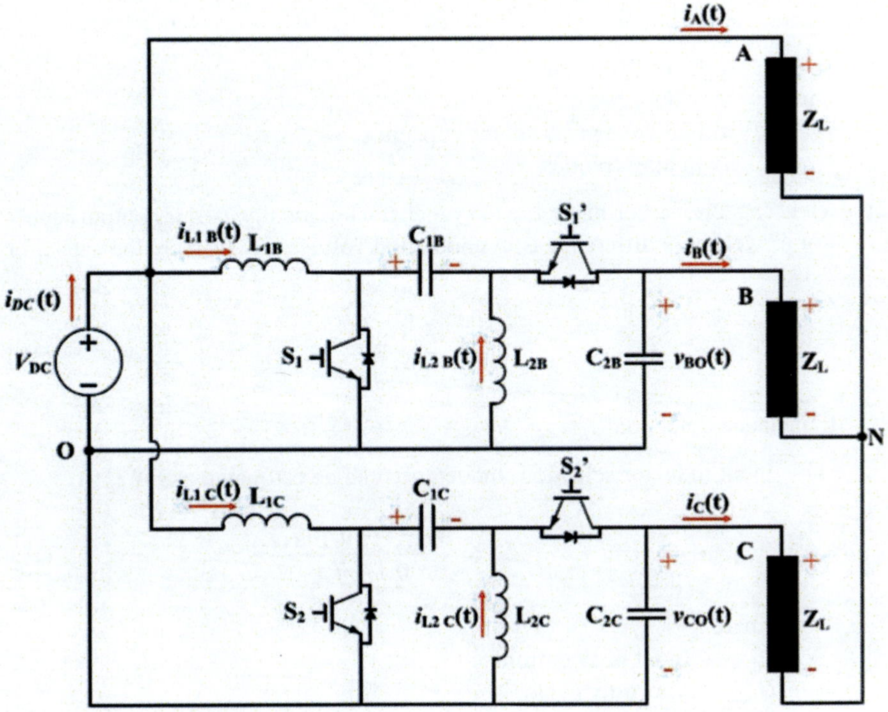

Fig. 9 Three-phase SEPIC inverter topology

$$\begin{aligned} V_{AB}(t) &= V_{ml}\mathrm{Sin}(\omega t) \\ V_{BC}(t) &= V_{ml}\mathrm{Sin}(\omega t - 2\pi/3) \\ V_{BC}(t) &= V_{ml}\mathrm{Sin}(\omega t + 2\pi/3) \end{aligned} \tag{16}$$

where V_{ml} is the peak of desired line to line output voltage.

2.5 Parameters Selection of SEPIC Convertor Components

In this section, a method to selecting the inductance and capacitance has been discussed.

A. Capacitors Selection

(1) Blend capacitor assortment: Pedestal on preferred voltage ripples, the capacitance can be as

$$C_1 = \frac{I_m D\mathrm{max}}{\Delta V f_{\mathrm{sw}}} \tag{17}$$

where

I_m	Input peak current
Dmax	Max duty cycle
ΔV	Ripple voltage across the capacitor
f_{sw}	Switching frequency

(2) Yield capacitor assortment: Eq. (17) uncharacteristically used for output capacitor as well only difference is in undulation voltage transversely the capacitor ΔV_{o}.

$$C_2 = \frac{I_m D\mathrm{max}}{\Delta V_{\mathrm{o}} f_{\mathrm{sw}}} \tag{18}$$

B. Inductance Selection

(1) Input inductor selection: Inductance can be estimated as Eq. (19)

$$L_1 = \frac{V_{\mathrm{DC}}(1 - D\mathrm{max})}{0.1 I_m f_{\mathrm{sw}}} \tag{19}$$

where

I_m	Input peak current
D_{max}	Max duty cycle
V_{DC}	Input DC voltage
f_{sw}	Switching frequency

(2) Output inductor selection: Inductance can be estimated as Eq. (20) by [11].

$$L_2 = \frac{V_{\mathrm{DC}} D\mathrm{max}}{0.3 I_m f_{\mathrm{sw}}} \tag{20}$$

where

I_m	Input peak current
$D_{\max}$	Max duty cycle
V_{DC}	Input DC voltage
f_{sw}	Switching frequency

C. Duty Cycle

$$V_0 = \frac{D}{1-D} V_{\mathrm{in}}$$

$$D\mathrm{max}\ \mathrm{con}\ v_1 = \frac{V_{\mathrm{DC}} + V_{ml}}{2V_{\mathrm{DC}} + V_{ml}}$$

$$D\mathrm{max}\,\mathrm{con}\ v_2 = \frac{V_{\mathrm{DC}} + \left(\sqrt{3}/2\right) V_{ml}}{2V_{\mathrm{DC}} + \left(\sqrt{3}/2\right) V_{ml}}$$

3 Simulation and Experiment Results

The proposed photovoltaic system is simulated by using the MATLAB/Simulink to authenticate the systematic fallout and to verify the sliding mode control strategy. Also the proposed system is proved by practical experiment using hardware.

3.1 Simulation

The recital of anticipated system is validated through the MATLAB/Simulink on 12 V input voltage as shown in below graphs from Fig. 10a–h.

3.2 Simulation Parameters

Input DC Voltage	12 V
Output AC Voltage	3 phase 12 V (Line Voltage)
Inductor	$L_1B = 6$ mH, $L_2B = 6$ mH

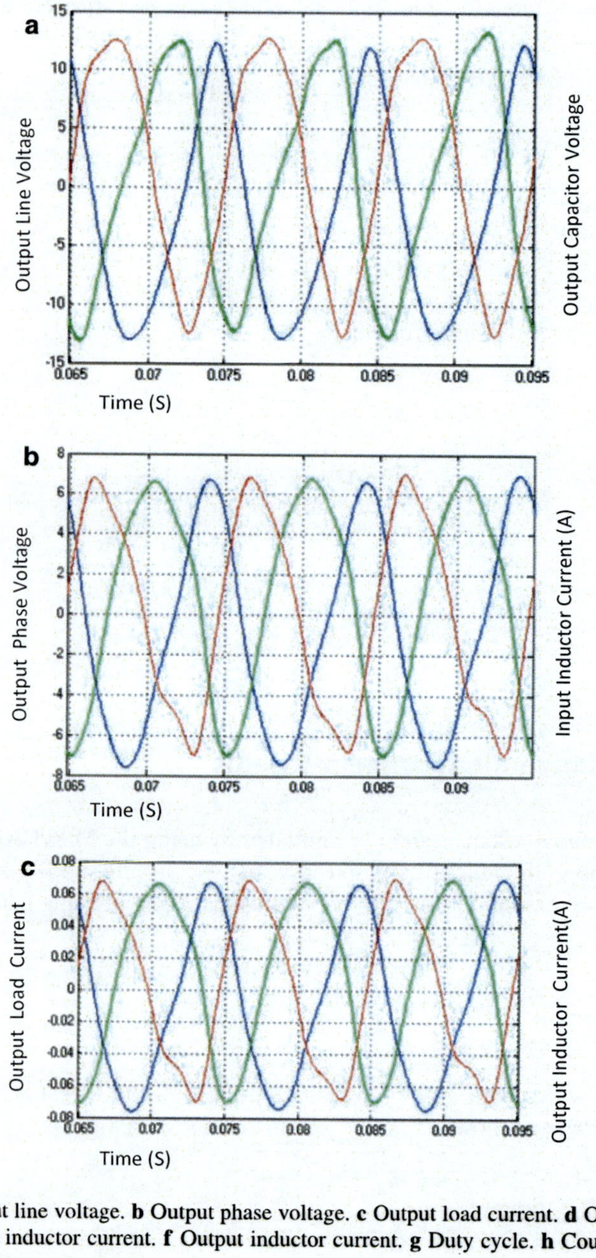

Fig. 10 **a** Output line voltage. **b** Output phase voltage. **c** Output load current. **d** Output capacitor voltages. **e** Input inductor current. **f** Output inductor current. **g** Duty cycle. **h** Coupling capacitor voltage

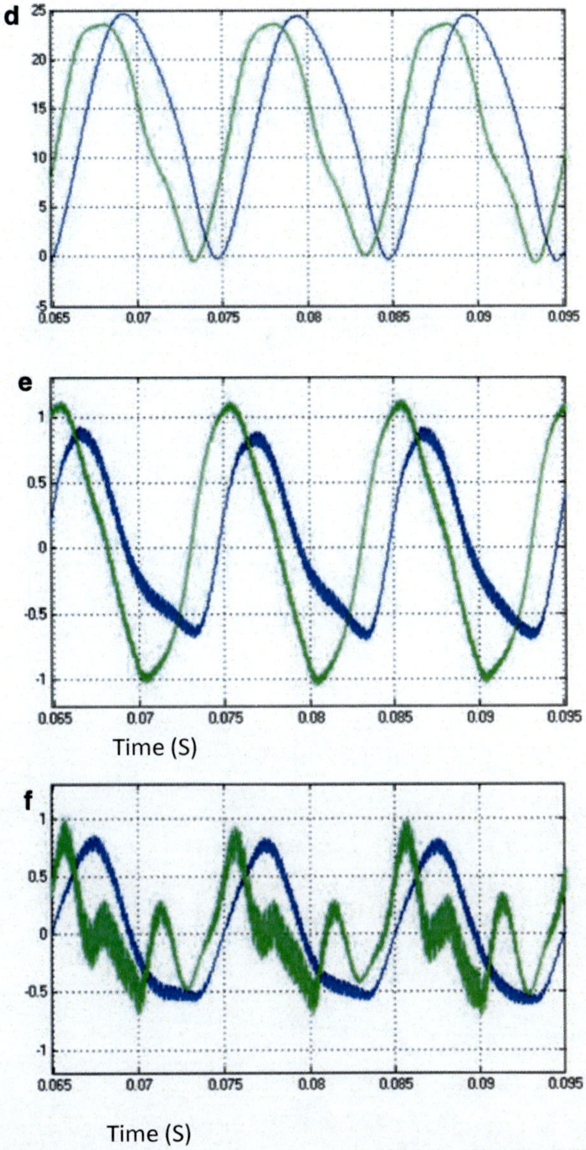

Fig. 10 (continued)

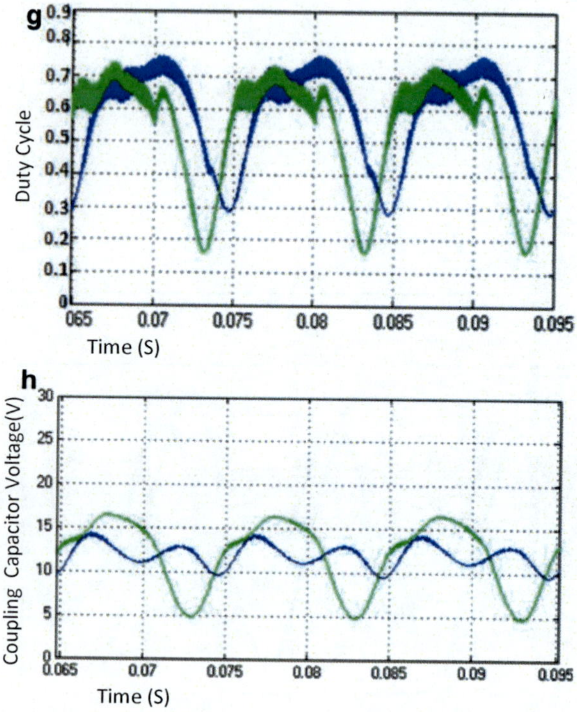

Fig. 10 (continued)

	$L_1C = 9$ mH, $L_2C = 2.36$ mH
Capacitors	$C_1B = 106$ μF, $C_2B = 95$μF
	$C_1C = 105$ μF, $C_2C = 88$ μF
Gain Factor	$K_1 = 2$, $K_2 = 8$
Load	100 Ω star connected

4 Experiment

To review the performance of the proposed system, a prototype system develops in laboratory with the constraints and workings rating as mentioned in this paper. Output has been verified on Oscilloscope as shown in Fig. 11.

Fig. 11 Prototype model prepared in laboratory

5 Comparision of Proposed Photovoltaic System to Conventional Photovoltaic System

There are various advantages of proposed system over the conventional photovoltaic system as

1. Proposed system used four switch for three-phase DC to AC conversion; however, conventional photovoltaic system used six switches.
2. Not required any output filter for proposed system, and however, conventional SSTP-based photovoltaic system required LC filter in output.
3. As the number of switches are less in proposed system, the power losses are less compare to conventional system.
4. Overall system is cheap as one lag of three-phase inverter is completely removed in new system.
5. In proposed system, % utilization of DC bus is 57.74% compare to 50% in conventional SSTP-based photovoltaic system.
6. Stress on switches is more in proposed system.

7. Robust control strategy is needed for SEPIC-based photovoltaic system due to nonlinear characteristics of four passive components in each converter.

6 Conclusion

A SEPIC inverter-based photovoltaic association planned in manuscript. The projected scheme recovers the exploitation of the DC bus. In addition, the construction of three-phase productivity potentials AC waves lacking necessitate acquiesce filter. Distinct predictable FSTP converter pedestal photovoltaic, the projected topology not experienced voltage variation transversely strained from DC supply devoid of passage in a few unreceptive constituent. SMC among unchanging desired occurrence was premeditated and functional to the projected SEPIC converter. Simulation and investigational fallout confirmed the presentation of the planned photovoltaic system with the optional manage tactic.

References

1. de Rossiter Correa MB, Jacobina CB, da Silva ERC, Lima AMN (2006) A general PWM strategy for four-switch three-phase inverters. IEEE Trans Power Electron 21(6):1618–1627
2. El Badsi B, Bouzidi B, Masmoudi A (2013) DTC scheme for a four switch inverter-fed induction motor emulating the six-switch inverter operation. IEEE Trans Power Electron 28(7):3528–3538
3. Veerachary M (2005) Power tracking for nonlinear PV sources with coupled inductor SEPIC converter. IEEE Trans Aerosp Electron Syst 41(3):1019–1029
4. Chiang SJ, Hsin-Jang S, Ming-Chieh C (2009) Modeling and control of PV charger system with SEPIC converter. IEEE Trans Ind Electron 56(11):4344–4353
5. El Khateb A, Abd Rahim N, Selvaraj J, Uddin MN (2014) Fuzzy-logic-controller-based SEPIC converter for maximum power point tracking. IEEE Trans Ind Appl 50(4):2349–2358
6. Caceres RO, Barbi I (1999) A boost DC-AC converter: analysis, de-sign, and experimentation. IEEE Trans Power Electron 14(1):134–141
7. Fadel M, Llor A (2006) Fixed frequency sliding mode control for boost converter. In: Proceedings power electronics and motion control conference, pp 957–960
8. Tan SC, Lai YM, Tse CK (2006) A unified approach to the design of PWM-based sliding-mode voltage controllers for basic DC-DC converters in continuous conduction mode. IEEE Trans Circ Syst 53(8):1816–1827
9. Diab MS, Elserougi A, Abdel-Khalik AS (2014) Non-linear sliding-mode control of three-phase buck-boost inverter. In: Proceedings IEEE international symposium on industrial electronics, pp. 600–605
10. Tan S-C, Lai YM, Tse CK, Martinez-Salamero L, Wu C-K (2007) A fast-response sliding-mode controller for boost-type converters with a wide range of operating conditions. IEEE Trans Ind Electron 54(6):3276–3286
11. Falin J (2008) Designing DC/DC converters based on SEPIC topology. Power Manage Texas Instrum Analog Appl J 4Q:18–43
12. Tomar A, Nguyen PH, Mishra S (2020) SEPIC-MISO converter based PV water pumping system- an improved performance under mismatching conditions. In: 2020 IEEE 9th power India international conference (PIICON), 28 Feb, 1 Mar 2020

Author Index

S. S. Dash et al. (eds.), *Sixth International Conference on Intelligent Computing and Applications*, Advances in Intelligent Systems and Computing 1369,
https://doi.org/10.1007/978-981-16-1335-7